SONET/SDH

Third Edition

WALTER **GORALSKI**

McGraw-Hill/Osborne

New York   Chicago   San Francisco
Lisbon   London   Madrid   Mexico City   Milan
New Delhi   San Juan   Seoul   Singapore   Sydney   Toronto

McGraw-Hill/Osborne
2600 Tenth Street
Berkeley, California 94710
U.S.A.

To arrange bulk purchase discounts for sales promotions, premiums, or fund-raisers, please contact **McGraw-Hill**/Osborne at the above address. For information on translations or book distributors outside the U.S.A., please see the International Contact Information page immediately following the index of this book.

SONET/SDH, Third Edition

1234567890 CUS CUS 0198765432

ISBN 0-07-222524-6

Publisher
   Brandon A. Nordin
Vice President & Associate Publisher
   Scott Rogers
Acquisitions Editor
   Lisa McClain
Senior Project Editor
   Betsy Manini
Executive Project Editors
   Mark Karmendy
   Madhu Prasher

Acquisitions Coordinator
   Martin Przybyla
Technical Editor
   Thomas E. Gordon
Production, Art, and Editorial Services
   Anzai! Inc.
Series Design
   Peter F. Hancik

This book was composed with Corel VENTURA™ Publisher.

Dedicated to my wife Camille, who understands why I write,
and to our children Chris, Alex, Ari, and Clay,
who are still struggling with the concept.

ABOUT THE AUTHOR

Walter J. Goralski has spent more than 30 years in the data communications field, including 14 years with AT&T. He is currently a staff engineer with Juniper Networks, a router vendor in Sunnyvale, California, and an adjunct professor of Computer Science at Pace University Graduate School in New York. He is also the author of several books on DSL, the Internet, TCP/IP, and SONET/SDH, as well as articles on data communications and other technology issues.

ABOUT THE REVIEWER

Thomas E. Gordon has been involved in the aerospace and semiconductor industries for more than 15 years as an optical engineer, program manager, and product manager. In particular, he played a lead role in developing the technology and manufacturing the mirrors for the Chandra X-ray telescope, the third of NASA's "Great Observatories." Mr. Gordon earned BS and MS degrees in Optics from the Institute of Optics at the University of Rochester, and an MBA from the Darden Graduate School of Business at the University of Virginia.

CONTENTS

Part II
How SONET/SDH Works

Part III

SONET/SDH Network Equipment

Part IV

SONET/SDH Advantages

PREFACE

have been working with, teaching, and writing about SONET since 1991. In 1996, because I didn't like any of the available books on the topic, I wrote the first edition of this book. It was just another "royalty opportunity" for me at the time, and it was not until the second edition followed in 2000 (a first for a book of mine) that I realized how popular this particular work was. People I met professionally often knew me from "the SONET book," and when I started a new job at Juniper Networks in 2000, people told me that most technical people there had my book on their shelves. They weren't kidding, I soon found. Early in 2002, returning to SONET/SDH matters after a long absence, I complained that it was hard to do research into alternative ideas and perspectives about SONET/SDH. I was told, "Well, everybody reads your book ..."

So I therefore undertook this third edition with a new sense of responsibility. If my book is to be a sort of standard introduction to SONET/SDH (I hesitate to call it a reference, given that so much more could be said on the topic), then so be it. I have tried to make this the best book I have ever written, and I hope that it is not plagued by the missing figures and meaning-altering typos that somehow made it into print the first two times around.

This edition completely integrates SDH into the text and title. In fact, today it should probably be SDH/SONET, given the increasing importance of SDH in global networking. But most of the chapters give a "pioneer preference" to SONET and treat SDH second. I have made every effort to avoid repetition, and I tried to find every place where SONET and SDH differ in detail, at least above the tributary level.

What else is new? The core chapters on the SONET/SDH frame structures, overhead, payloads, and pointers have been completely rewritten, updated, and expanded. Frame structures now include STS-768/STM-256. The FEC, media-dependent bytes, and tandem connections are explained. Overhead now occupies a chapter on its own, and the network management chapter has more and new information on alarms, plus a split-off chapter on network management trends in SONET/SDH. Payloads now include virtual concatenation and generic framing procedure (GFP). The chapter on SONET/SDH rings has been completely written, updated, and expanded as well, adding considerations for router connectivity and linear protection. Every chapter was considered word-by-word for updating, expansion, or deletion.

At every step I agonized about the need to be accurate and yet concise. Most of the negative comments I received about the earlier editions were along the lines of "that's not the whole story about …" and I admit this was and is true. I am not trying to *replace* the specifications. It will be enough if, after reading this book, people are able to pick up the raw specifications, often written by those who are so familiar with the terminology and operation of a feature that they sometimes forget that others are not, and *understand* what they are reading. To offer a simple explanation of complex features was not an easy thing to do, I found. For example, my section on virtual concatenation is many times the size of that found in the standards themselves. But if readers can gain an insight into how this key aspect of SONET/SDH operates, then the time and effort spent was well worth it.

I still must recommend two books, because they have been reliable guides throughout the making of this edition.

Understanding Optical Communications by Harry J. R. Dutton (Prentice-Hall, 1998) is the best book on pure fiber-optic issues. It is very long but very readable, and is part of the well-known IBM Redbook series. It is even downloadable as a *very* large file from www.redbooks.ibm.com (the link to the book itself changes, so it is best to search for it).

Understanding Fiber Optics, Fourth Edition by Jeff Hetch (Prentice-Hall, 2001) has all the facts on modern optical networks in one package. Written as a self-paced study guide, it has questions at the end of each chapter.

ACKNOWLEDGMENTS

Medieval scribes would write out text in a manuscript, illuminate the book with figures as they pleased, and sometimes use paper or vellum they had made themselves. Publication often consisted of reading the work in public, and distribution was carried out by the author passing the new book around. Advertising was done by the author as well, usually by word of mouth. Sales were limited in most cases to the transfer of ownership of the sole existing copy, but the proceeds went entirely and directly to the author. And some well-educated people could even read to themselves without moving their lips.

Book production has changed somewhat lately. The modern scribe calls upon a host of functionaries to review, edit, produce, distribute and advertise a new book, all with the aid of computers and networks. This book could not have happened without the vision of Tracy Dunkelburger, Lisa McClain, and Martin Przybyla at McGraw-Hill, and the sure hand of the editing and production people, in particular Ann Fothergill-Brown, Tom Anzai, Lee Musick, Allan Shearer, Linda Shearer, Debby Schryer, and Bert Schopf.

Information was provided on many essential matters by many corporations, in particular Nortel and Alcatel. In fact, without the numerous white papers available from Nortel, this book would have been quite different. Alcatel provided information with astonishing speed. The encouragement I receive from my colleagues and superiors at Juniper Networks is invaluable.

I would also like to thank the reviewers. Tom Gordon saw the initial chapters and made several helpful suggestions. Thanks to him, this is a better book. Of course, any errors that remain I must take full responsibility for.

Some things never change with books, though. Some well-educated people can even read to themselves without moving their lips.

INTRODUCTION

For the 1996 Olympics, the Olympic Stadium in Atlanta was wired with twelve miles of SONET/SDH (synchronous optical network/synchronous digital hierarchy) fiber-optic cable. Every Boeing 777 jetliner is wired with several miles of SONET/SDH-capable fiber. Either installation can handle literally gigabits of information per second (a gigabit is one thousand megabits), which is the whole point of using fiber in the first place. Gigabits of generated information must be transported, processed, and presented to concerned users in adequate time to be useful. At the Olympics, the users were officials and judges and members of the press. In an aircraft, the users are pilots and, after the fact, potential investigators. Thankfully, most flight abnormalities do not result in disaster. But all of them must be investigated to avoid future risks. SONET/SDH can possibly be used to record not just cockpit sounds and instrument readings, but ultimately a video record of the cockpit operations and even the whole flight from a tail-mounted digital camera.

Admittedly, most users do not encounter SONET/SDH in Olympic competition nor on an airplane. However, the need for higher bandwidth is an unpleasant fact of life in all types of networks today. Until now, several newer technologies have helped to remedy this situation in a local area network (LAN) environment. Once limited to the choice of cabling (which kind of unshielded twisted pair?) for 10 Mbps Ethernet-type solutions, the LAN implementer today is faced with a wide range of choices for linking desktops. Newer technologies, such as 100 Mbps Ethernet, Fibre Channel, switched Ethernets, Gigabit Ethernet, and the much-anticipated 10 Gigabit Ethernet have started to appear or be planned in many organizations needing faster connectivity between end systems.

These faster networks are needed to link the more powerful—and less expensive and therefore more common—desktop systems, running client/server applications and built-in multimedia capabilities. All of these reasons are explored more fully at the beginning of this book. For now, it is enough to point out that, for a variety of reasons, faster LANs are being built to address bandwidth limitations.

Although faster LANs have addressed the need for desktop bandwidth within a building or campus environment, LANs do nothing to increase the bandwidth available for networking dispersed buildings or campuses over distances of more than a few miles. For this situation, a wide-area network (WAN) is needed. Until relatively recently, there was no easy way to link even 10 Mbps Ethernet LANs with so much as a fraction of the bandwidth that the LAN represented.

Fortunately, SONET/SDH provides welcome relief from this growing bandwidth problem in the WAN environment. SONET/SDH is capable of linking LANs at separate sites not at a mere fraction of 10 Mbps, or even a full 10 Mbps. Rather, SONET/SDH can link several 10 Mbps Ethernets at a single site to other Ethernet LANs across the country. SONET/SDH links usually operate at speeds of 155 Mbps or better, and into the multi-Gbps range.

SONET/SDH ADVANTAGES

SONET/SDH is more than just a lot of bandwidth. SONET/SDH has numerous advantages in a number of areas that will be explored more fully in this book. However, this may be a good place to briefly outline some of the more important and obvious advantages of SONET/SDH over other high-bandwidth networking schemes. These are summarized in Table I-1.

All of the reasons for the current level of interest in and popularity of SONET/SDH are fully explained later. For now, it is enough to note that SONET/SDH enjoys many advantages over other networking methods.

This book explores SONET/SDH from top to bottom. It explains where SONET/SDH came from, what it is used for, and how it is used. The book explores all aspects of SONET/SDH in some technical detail, not to overwhelm the reader, but to point out how the SONET/SDH digital fiber standard addresses some of the problems that have plagued other WAN technologies in the past. All of the advantages of SONET/SDH for users and service providers alike are detailed, and the services deliverable on SONET/SDH are fully

Technology	Unprecedented speeds available on fiber-optic networks
Economics	Best interface for fiber-optic networks Economical adding and dropping of channels
Flexibility	Modular equipment design Adequate overhead for network management
Compatibility	Works well with existing network hierarchies Allows multiple vendors' equipment to interoperate Worldwide standard

Table I-1.   Advantages of SONET/SDH

described. Finally, the book examines general features of SONET/SDH offerings from service providers and products from equipment providers. The goal is to make this work a comprehensive guide to SONET/SDH that is at the same time understandable to non-engineers and interesting to technicians.

Bandwidth

This book is about SONET/SDH. But just as important is the previous (and, of course, still existing) digital transmission system hierarchy in the United States, known as T-carrier. The rest of the world pretty much follows the E-carrier digital hierarchy, which is different from T-carrier. Japan uses a variation of T-carrier usually called J-carrier. At this point, it is necessary only to note the associated bit rates of these schemes and not necessarily to understand the designations and differences. These will be covered later.

This book constantly mentions terms such as DS-1 and DS-3, and speeds of 1.5 Mbps (1.544 Mbps to purists) or 45 Mbps. To minimize the amount of flipping back and forth, and folding down of pages to mark spots and the like, this may be an appropriate place to set out a couple of tables (refer to Tables I-2 and I-3) of T-carrier, E-carrier, and SONET/SDH speeds. Here, at the front of the book they will be somewhat easier to locate.

A Word on Terminology

The nice thing about words is that they can take on almost any meaning a writer wants. But this can be a problem when one writer uses a word in one sense, and other writers do not immediately recognize that this particular sense is far superior to any other. Because it is difficult to write a technical book without using certain common terms over and over, I thought it best to define exactly what I mean by these terms when they are used in this work. I admit each term and its meaning can be debated, but I am not so much concerned about which use is proper. It is more important that readers understand just what is meant by each of these key terms.

Digital Multiplexing Level	Number of Equivalent Voice Channels	Bit Rate (Mbps)		
		N. America	Europe	Japan
DS-0/E0/J0	1	0.064	0.064	0.064
DS-1/J1	24	1.544	–	1.544
E1	30	–	2.048	–
DS-1C/J1C	48*	3.152	–	3.152
DS-2/J2	96	6.312	–	6.312
E2	120	–	8.448	–
E3/J3	480	–	34.368	32.064
DS-3	672	44.736	–	–
DS-3C	1344*	91.053	–	–
J3C	1440*	–	–	97.728
E4	1920	–	139.264	–
DS-4	4032	274.176	–	–
J4	5760	–	–	397.200
5	7680	–	565.148	–

*Intermediate multiplexing rates.

Table I-2.   Digital carrier hierarchy used in North America, Europe, and Japan

Optical Level	Electrical Level	Line Rate (Mbps)	Payload Rate (Mbps)	Overhead Rate (Mbps)	SDH Equivalence
OC-1	STS-1	51.840	50.112	1.728	STM-0
OC-3	STS-3	155.520	150.336	5.184	STM-1
OC-12	STS-12	622.080	601.344	20.736	STM-4
OC-24	STS-24	1244.160	1202.688	41.472	NA
OC-48	STS-48	2488.320	2405.376	82.944	STM-16
OC-192	STS-192	9953.280	9621.504	331.776	STM-64
OC-768	STS-768	39813.120	38486.016	1327.104	STM-256

Table I-3.   SONET/SDH digital hierarchy

First, I use the term "organization" in preference to "company" or "business." By using the term in this generic sense, discussions are not limited to a corporate environment. Corporations are all organizations, but so are federal, state, and local government departments, schools, hospitals, and private law firms, to name a few. An organization can even be a subsidiary or large department of a major corporation. All organizations need more bandwidth to link higher-speed PCs and LANs over a wide area.

Next, I use the term "service provider" in preference to "telco" (telephone company), "carrier," or anything else. By using the term in this generic sense, discussion are not limited to local exchange or inter-exchange telephone companies. All telephone companies are service providers, but so are cable TV companies, Internet service providers, power companies, and almost anyone who wants to enter today's deregulated telecommunications world. All service providers need more bandwidth to service more customers and to provide each customer with the needed bandwidth over a wide area.

Also, I use the term "user" in preference to "customer" or "end-user." By using the term in this generic sense, discussions are not limited to bill-payers or people sitting in front of PCs. All customers are users, but so are the members of a department on a corporate network, LAN administrators (who rarely have time to sit down, let alone use a computer), and many others. All users need more bandwidth to accomplish what they set out to do with a network in the first place.

A minor point about the digital hierarchy itself should be mentioned here as well. Network engineers and educators will often insist on the use of the term "digital signal" (as in DS-3), when most network installers and customers would use the term "T" (as in T-3). As has been pointed out innumerable times, the terms are *not* technically interchangeable, but people often use them as if they were. This book will not attempt to decide whether any harm is done when calling a DS-1 a T-1, but will try to use the terms in the proper fashion consistently.

For some reason, many network folks will allow and feel comfortable with mentioning the speed of a T-3 or DS-3 as "45 Mbps." However, 45 Mbps is not the correct speed. The speed is actually 44.736 Mbps, and yet the rounding to 45 Mbps seems universal and acceptable. But any rounding of the T-1 or DS-1 speed to 1.5 Mbps is likely to be greeted with a comment like "you mean 1.544 Mbps," which is indeed the actual bit rate of the link. Thus, in the interest of conserving 4's, this book will consistently round T-1 speeds to 1.5 Mbps, except where more accuracy is warranted to avoid confusion and misinterpretation, such as in mathematical expressions. I hope this practice offends no one.

Finally, many other terms are defined along these same lines throughout the book. Where a term is confined to one particular aspect of a discussion, the term is defined at that point in the book (such as "client/server"). It is tempting to use such terms as if everyone had a common understanding of what they mean, but this is not the case. An undefined term is just jargon, and an indiscriminate use of jargon has probably caused more network problems than any other factor over the years.

PART I

Introduction to Fiber Optics and SONET/SDH

It sometimes seems that the pace of change in the world is always accelerating. The perception is especially common in the intimately related fields of computers and telecommunications. To a large extent, an accelerating pace of change is somewhat of an illusion. The fact is that sweeping technological change has been a characteristic of world culture, particularly in the United States since the beginnings of the Industrial Revolution. In spite of the vast changes that computerization and advances in telecommunications have wrought in the past 50 years or so, these changes are arguably less than the impact of the automobile or airplane on the world's population. After all, it was only fifty years from the first flight at Kitty Hawk to common commercial aviation.

There are significant differences between normal technological evolution and the evolution of the computer and telecommunications industries, however. Sometimes there is a deep affection for older technology, like classic cars. Every once in a while, a Model T appears on a superhighway. The attachment of a dedicated group to things like steam locomotives even seems somehow noble and uplifting. Never mind that these huffing giants required water every fifty miles and coal every hundred miles. Diesel locomotives do not seize the spirit like the hiss of steam.

With computers and telecommunications, however, the situation is different. No one treasures a classic IBM PC from the early 1980s and runs an old word processor just for fun. Groups do not form to explore the vanished world of 300 bit-per-second modems and marvel at the simple functional elegance of the teletype machine that once pounded away in the background of every news broadcast. Progress in computers and telecommunications has yielded not a treasure of memories but a mountain of worthless junk.

The telecommunications needs of businesses and residential users alike seem to be always one step ahead of the capability of technology to satisfy these needs. This is entirely the point. There would be a huge benefit to coming up with a technology that would be essentially bulletproof with regard to the future, especially the near future. It seems to be impossible with computers themselves. Advances in architectures, hardware, and software always appear to outstrip any one element in the threesome. For instance, a state-of-the-art hardware board will always succumb to a new backplane architecture that is incompatible with it, or a new software advance that makes the hardware function redundant and unnecessary.

The strange thing, and the whole intent of this book, is to describe an area of telecommunications technology where it actually seems that obsolescence may be decades away, instead of eighteen months (which is the average life-span of a major CPU chip architecture like the Pentium).

The application of standards to fiber optic transmission offers a glimmer of hope to leapfrog end-system requirements at least for the immediately foreseeable future. The implications are large and the benefits many.

This section of the work describes the evolution of the SONET standard and how it came to overcome a variety of problems with older transmission methods. No knowledge of fiber optic transmission or digital communication is assumed or needed. The material is introductory and not intended to be challenging in nature; however, more advanced readers may still wish to read these early chapters. There is much to be learned about fiber optics and digital techniques that is not usually covered in standard texts on either topic. They are brought together here and examined with a historical perspective to prepare the reader for the changes that SONET will bring to the world of telecommunications (and perhaps even computing itself) in the next few years.

A special note is needed here to prepare for the introduction later on in this book of the international version of SONET, the Synchronous Digital Hierarchy (SDH). SONET was invented for use in North America and is very well adapted for this environment, but SDH was better suited for the types of networks found around the world. In the early days of SONET/SDH, SDH was a variation of what SONET equipment, chipsets, and protocols did. But as time goes by, SDH becomes more important as the world transitions to fiber optic links, and SONET becomes more a variation of SDH. This point will be raised again in this book, although the emphasis will always be on SONET. Fortunately, the differences between SONET and SDH are more a matter of terminology today and less in substance. So almost anything that applies to SONET also applies to SDH. Whenever there is a substantial difference in operation between SONET and SDH, this aspect will be clearly detailed. Chapter One describes how the concepts of SONET (and SDH), Asynchronous Transfer Mode (ATM), and Broadband Integrated Services Digital Network (B-ISDN) were invented and intended to be used together for not only computer networks, but any network: from voice to cable TV.

The chapter then explores how the related concepts got "de-coupled" and, in a sense, went their separate ways. It concludes with a consideration of whether it is too late to bring them all back together again, and even if it is wise to do so.

Chapter Two forms a tutorial of fiber optic communication. This chapter offers a fiber optic "survivor's guide" for readers who may not be as familiar as they would like to be with basic fiber optic transmission terminology and concepts. The key concepts of fiber cable characteristics, manufacturing, and usage are fully explained. The chapter includes a description of how typical fiber transmitters (i.e., lasers, LEDs) and receivers operate and surveys the current applications for fiber optic cable.

Chapter Three was new to the second edition and introduces the optical concepts that make wavelength division multiplexing (WDM) possible.

The chapter explores all of the ramifications and issues involved with WDM: scattering, other dispersion effects, fiber amplifiers, solitons, and so on. Then dense wavelength division multiplexing (DWDM) is investigated, along with all of the reasons that DWDM is one of the most exciting developments in the optical world today. State-of-the-art terabit systems are introduced here as well. This chapter is not SONET/SDH specific, but applies to optical networking in general.

Chapter Four is an introduction or review of the digital transmission hierarchy, not only in the United States, but around the world. This chapter explores the existing T-carrier and E-carrier digital hierarchies around the world, again for readers who may not be as familiar as they would like with these terms and concepts. Every effort is made not to limit discussion to North American methods (T-carrier), but the emphasis is here. The chapter includes an extensive discussion of the limitations of the current digital hierarchy (such as the lack of a true "mid-span meet") and how SONET addresses them.

CHAPTER 1

B-ISDN, ATM, and SONET/SDH

Computer networks have been around since the mid-1960s. Telephone networks carrying voice and specialized applications, such as fax, are even older: since the 1870s in the case of the public switched telephone network (PSTN). Yet, the need to develop high-speed, low-delay networks has never been as intense as it is today. The key to understanding the relationship among broadband-integrated service digital network (B-ISDN), asynchronous transfer mode (ATM), synchronous optical network (SONET), and synchronous digital hierarchy (SDH) is to understand how the standards-making bodies around the world predicted the direction of computer networking specifically and networking in general at the beginning of the 1980s. It was actually an impressive application of foresight, rare enough in any field.

Those in a position to do something about the future of networking looked mainly at three trends: the growth of powerful desktop systems linked at the endpoints of these networks; the rise of multimedia applications; and the trend toward what is usually called *distributed computing.* It is more common to encounter distributed computing used almost as a synonym for *client/server computing* (defined later). There is nothing wrong with this, in spite of definitions and "network speak." However, purists point out that distributed computing is a more encompassing term than is client/server; that is, client/server is only one way to accomplish distributed computing, although client/server architectures are certainly the most prevalent forms of distributed computing today.

This book begins by placing SONET (and SDH) in its proper context, which is no less than this: SONET/SDH is a high-bit-rate fiber-optic-based transport method that provides the foundation for linking high-speed ATM switches and multiplexers and providing users with B-ISDN-compliant services. There is a lot going on in this statement—thankfully; otherwise, this would be a very short book. In fact, the way SONET/SDH is deployed and marketed by service providers today has enhanced the position intended for SONET/SDH by international standards bodies and embodied in the above statement in many cases.

Before analyzing the statement about SONET/SDH's position, it may be a good idea to discuss details about the rising power of network end systems, multimedia issues, and trends in client/server computing. This detail is important because SONET/SDH includes a provision that tries to make networks based on SONET/SDH much more "future proof" than ever before. SONET/SDH is nearly unique in this ability to "scale" itself to networking needs and end-system technology.

POWER ON THE DESKTOP

It is sometimes said that if you really want to get the best deal on the most powerful computer you can get, you will never buy anything. This argument applies to desktop and laptop personal computers (PCs) for the most part, but it is at least partially true today for UNIX-based workstations, minicomputers, and even mainframes (which are still around). This statement is often made because computer prices have been falling rapidly at the same rate computer capabilities have been growing. Why buy a computer when the price next month will be less—sometimes much less—and the performance will be enhanced as well?

Many good examples have been cited by authors Daniel Burstein and David Kline in their book, *Road Warriors.* For instance, a "singing" birthday greeting card has more processing power than existed in the entire world in 1950. A common video camcorder has more processing power than an IBM System/360 computer had in 1964. (An IBM engineer added this: "Well, the camcorder has an accurate clock. That means it's *already* more powerful than a System/360 in 1964." I will not comment further.) A simple $100 video game player found in many households today is more powerful than a Cray Supercomputer was in 1976; and the Cray cost about $4 million. These numbers were appropriate for the first edition of this book, written in 1997. Today, birthday cards talk, camcorders are all-digital, and video games are in a class by themselves when it comes to graphics.

The accumulation of information that must be managed in organizations is staggering. Networks will form an increasingly crucial element as the reliance on information management technologies continues to increase. Organizations will seek to get more and better use from their capital investments and employee productivity (because, in many cases, fewer and fewer of them exist). Management technology is always finding better ways to integrate such previously diverse information technologies, such as optical storage, hypertext, neural networks, scientific visualization, and virtual reality.

In the area of optical storage, the emergence of inexpensive and abundant CD-ROM writers has addressed issues of inadequate tape and (in many cases) disk storage capabilities. It is almost inconceivable today to consider distributing compound documents (i.e., text and images merged with voice annotations), interactive graphics, and video files on diskettes. In the near future, such content will have to be distributed across a network to satisfy a number of application environments, such as computerized design applications and desktop publishing. These applications will require very high-capacity networks that bring together large amounts of information from diverse sources at a single, networked location.

Hypertext has been mainstreamed recently with the popularity of the World Wide Web on the Internet. The Web enables users to access truly staggering amounts of information in a nonlinear manner. This is the essence of hypertext. Anyone who has used a Windows-based "Help" file has experienced the power of hypertext. Instead of having to sequentially access the information stored on a computer or in a database, a user has the power to move through the information in a more intuitive and useful manner. Of course, the promise of hypertext, initially experimented as long ago as the 1960s, was fulfilled only with the development of powerful networks that allow the hypertext link to lead anywhere, not just to another portion of the same file on the same computer. Many organizations are increasingly relying on hypertext for online training and documentation, text management, and other forms of information distribution.

Neural networks are composed of a large number of microprocessors, all working together to solve a particular problem. Neural networks are particularly well-suited for information-processing-intensive industries, such as banking, securities trading, and insurance. All of these industries are distinguished by the fact that they seek to use information to gain a competitive advantage over other companies in the same field. The needed information must be quickly gathered and analyzed for trends and relationships. In these fast-moving fields, mere seconds can result in the gain (or loss) of several millions of dollars. In the world of companies characterized by downsized employee staffs, lack of middle-management personnel, and decentralized operations, these neural networks have become a vital business tool that allows the organization to be even more efficient under these lean conditions.

In the field of scientific visualization, increasing computer power has allowed organizations, such as oil and gas companies, to gather huge databases of seismic records, geological satellite images, and the like. Huge is probably an understatement in this context because the database can easily reach into the Terabyte (1,000 Megabytes) ranges. For example, data can be accessed so that a view of the earth 100 feet down can be imaged and color-coded in whatever way the viewer desires. In the chemical and pharmaceutical industries, molecular arrangements can be visualized, rotated, and folded, yielding new insights into the possibilities of new materials, drugs, and genetics.

Virtual reality is one of the latest examples of powerful computer applications that were only dreamed of a few short years ago. Images of animals and jets in movies and on television appear to be real, but are not. *Star Wars* movies are shot entirely with digital cameras, often with nothing real on the scene except the people. Some even question the need for real actors at all, but the few films that have tried to portray "real" people digitally still produce characters that look fake.

But these are only films. Virtual reality extends this concept of fooling our senses to fooling ourselves into thinking that what we see and hear and sometimes even feel on the computer are as real as the world outside our window. Computer power today allows users to become as immersed in the virtual world as they wish, a computer-generated world of sight, sound, and mind. Virtual reality forms a real-time simulation that creates an artificial world where both objects and their environment have the illusion of being real. Naturally, the *Matrix* series has taken this concept to an extreme.

This list could be extended much further, but the point is already clear: Processing power is growing, and fast, while prices continue to fall. Figure 1-1 illustrates the increase

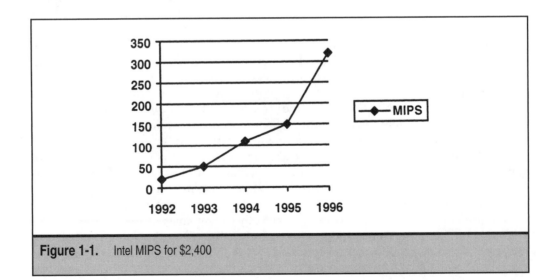

Figure 1-1.   Intel MIPS for $2,400

in millions of instructions per second (MIPS) that systems with Intel processors selling for about $2,400 were capable of from 1992 to 1996. Intel processors power about 85% of all desktop and laptop systems sold today. These figures, and all of the figures in this chapter, unless otherwise noted, come from advertisements appearing in Windows magazine, which published a retrospective on these trends in late 1996.

Rather than attempting to update all of these facts and figures and trying to hit a constantly changing target, the second edition preserved the information through 1996. It is all still valid. Suffice to say that the "average" PC in 2000 boasted between 128 and 256 Megabytes of RAM, a 16 or 20 Gigabyte hard drive, and about a 750 Megahertz Pentium III processor, all for less than $1,500. It should be kept in mind that prices seem to be stabilizing somewhat, especially memory prices. However, monitors still remain relatively pricey, the result of a high video chip rejection rate and the fact that monitors cannot shrink in size, but always grow. And new flat displays often cost more than the rest of the machine they are attached to.

A third edition update for 2002 is in order. Not so much to understand the invention or application of SONET/SDH, but just because everyone asks me about these trends and expects to find updates in new editions of this book. So here they are.

A new system purchased for $1400 in mid-2002 came with 512 MB of RAM, and 80 GB hard drive, a 1.8 GHz Pentium 4 processor, a 40-speed CD-ROM drive, a combined DVD-RW (4.7 GB) and CD/RW drive. Other goodies included a digital camera memory stick port, four USB ports, two FireWire ports, and a 10/100 Mbps Ethernet card. It should also be pointed out that some Apple computers include a Gigabit Ethernet (GE) adapter.

Figures 1-2a and 1-2b chart the same $2,400 system against the amount of random access memory (RAM in megabytes) and storage (in megabytes) installed for the same years. Although not as steep as the processor curve, the trend is just as pronounced.

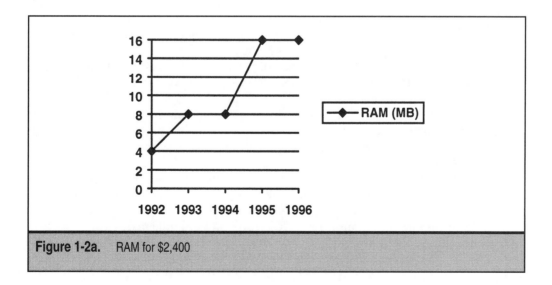

Figure 1-2a.   RAM for $2,400

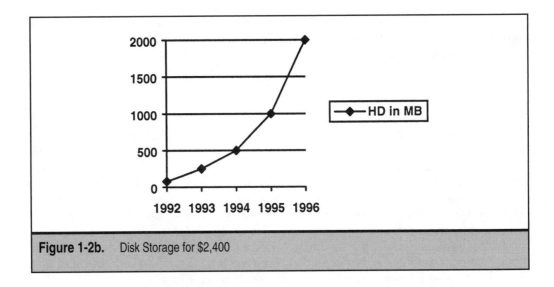

Figure 1-2b.   Disk Storage for $2,400

Systems based on the Intel processor continue to follow what is known as Moore's Law. Gordon Moore was one of Intel's founders who observed early on that doubling of semiconductor density and, therefore, processing power occurs about every eighteen months. Moore expected this trend to continue and, by 1996, it had resulted in Intel's introduction of five new Pentium versions and four new Pentium Pro chips within the previous twelve months. In spite of periodic negative statements about the limits of technology, such as semiconductors and masking techniques, the advance of integrated circuit and chip technology has continued without major difficulties for almost thirty years.

Not only is the hardware more powerful year by year, and sometimes month by month, but it is less expensive as well. Computer memory chips, in particular, have gone through a staggering decrease in price that has left users' heads spinning. Table 1-1 depicts the dramatic decrease in prices for PC dynamic RAM (DRAM) memory.

Most of the reduction in prices has come from the intense competition to supply memory-hungry Microsoft Windows 95 operating systems with enough memory to achieve the performance levels promised. When Windows 95 appeared, chip makers geared up production to supply the 16MB "recommended" (meaning "minimum") memory Microsoft mentioned frequently in documentation. DRAM pricing approached $5 per MB as manufacturers were accused of "dumping" (i.e., selling below manufacturing cost) to gain market share and reduce inventories.

In 2002, the Millennium editions of Windows, including XP, have kept up the trend to bigger memory sizes. There are more types of memory today as well, and prices vary based on type. For example, 512 MB is in the $110-156 range for a wide range of processors, and about $120 for the newer Rambus DRAM (RDRAM). This is a little more than $4 per MB, and no one is accused of dumping.

Size	1992	1993	1994	1995	1996	2002
1MB	$45	$65	$43	Not avail.	Not avail.	Not avail.
4MB	$143	$188	$156	$145	$37	Not avail.
16MB	Not avail.	Not avail.	$639	$514	$80	Not avail.
32MB	Not avail.	Not avail.	$1,194	$1,066	$282	$6
64MB	Not avail.	Not avail.	Not avail.	Not avail.	Not Avail.	$8
128MB	Not avail.	Not avail.	Not avail.	Not avail.	Not Avail.	$16-34
256MB	Not avail.	Not avail.	Not avail.	Not avail.	Not Avail.	$49-85
512MB	Not avail.	Not avail.	Not avail.	Not avail.	Not Avail.	$110-156

Table 1-1.   Memory prices

The reduction in memory prices has been mirrored by the reduction in the prices of hard drives for storage. Most PCs use internal device enhanced (IDE) hard drives, and most prices reflect *bare drive* configurations. Bare drives are sold without the controller boards or even screws needed to mount the drives in a PC. This fact, however, makes the price a good reflection of hard drive technology alone.

It is not uncommon for most popular $1,500 PCs to include a 40 GB hard drive. And it's a good thing, too. Usually, about 1 GB will be taken up by the software bundled with the new system. Fortunately, vendors are happy to sell users as much hard disk storage as they need. In another stroke of luck, the price for hard disk storage is decreasing rapidly as well. Seasoned computer users, who were used to the more leisurely pace of technology development in the 1980s, are sometimes shocked to find that their PC hard drives, which were never backed up but served them well for five years, cannot be replaced when they finally fail. The faithful 6 GB hard drive is gone and a replacement is nowhere to be found, except perhaps at a backyard garage sale. The more pressing problem, however, is what to do about the lack of a backup!

The largest hard drives readily available in 2002 are 160 GB monsters, costing about $200. The *smallest* hard drives commonly sold today are 20 GB, for about $40. Table 1-2 is essentially included just to put the drop in hard drive prices in perspective.

Size	1992	1993	1994	1995	1996	2002
200MB	$565	$301	$192	Not avail.	Not avail.	Not avail.
540MB	$1,589	$909	$369	$192	$160	Not avail.
1+GB	Not avail.	$1,208	$721	$320	$193	Not avail.
2+GB	Not avail.	$2,420	$1,519	$1,349	$285	Not avail.
40 GB	Not avail.	Not avail.	Not avail.	Not avail.	Not avail.	$58
80 GB	Not avail.	Not avail.	Not avail.	Not avail.	Not avail.	$84
120GB	Not avail.	Not avail.	Not avail.	Not avail.	Not avail.	$133

Table 1-2.   IDE hard drive prices

Strangely, computer monitors have been an apparent exception to this declining price rule. Monitor vendors have struggled to lower prices at all, because monitors have little to gain directly from improvements in semiconductor densities, and the monitor manufacturing process suffers from a high rejection rate, which raises prices for even perfect monitors. Nevertheless, a high-quality 14-inch VGA monitor costing about $1,000 in 1988 went for about $300 at the end of 1996 and was of much better quality as well. New monitors have crisper images, higher refresh rates, and improved brightness.

The newer LCD monitors often cost as much or more than the system they are attached to. Traditional tube-based monitors are still available in 2002. A 14-inch monitor costs about $70, but the 19-inch ($135) and 17-inch ($85) are more popular.

This list could be extended to include CD-ROM drives and printers. Interested readers are urged to consult the pages of any PC magazine for additional information. However, one more example may be in order: in 1992, a 300dpi, 8ppm laser printer cost nearly $1,500. Five years later, in 1996, the same device was selling for under $500. Five years after that, in 2001, a 600dpi laser printer could be had for less than $200.

What has all this to do with networks? Well, consider how much the end systems attached to these networks have changed in terms of power over the past five years. The obvious answer is radically. Now consider how much the wide area networks used to link these systems have changed in terms of speed over the past five years: hardly at all. Newer public network solutions exist, such as frame relay, but they often run at the same speeds as before.

A common guideline for networking is as follows: A computer can generate about 1 Mbps of network traffic for each MIPS that the computer CPU runs at. In the early 1980s, with only a couple of MIPS available on the most powerful desktops (and even minicomputers), it was not uncommon to see 200 to 300 PCs on a single Ethernet LAN. The 10 Mbps Ethernet handled it all with ease. In fact, the Ethernet design limit was 1,024 devices.

By 1987, the 386 chipset running at about 6 MIPS could push 6 Mbps out of the back of a PC onto a LAN. The only factor that kept Ethernet LANs from grinding to a halt was that traffic was extremely "bursty" (i.e., intermittent). As it was, the limit by the late 1980s became about 20 to 30 devices per Ethernet LAN. By the early 1990s, with speeds approaching 50 MIPS, a desktop could easily overwhelm a 10 Mbps Ethernet by itself. This need to give a server or even client a full 10 Mbps or faster led directly to the rise of switched Ethernet and 100 Mbps Ethernet on the LAN. However, as speeds go far beyond 250 MIPS, where is the bandwidth needed for the WAN?

Clearly, a need exists for new high-speed, wide-area networking methods to link these new desktops together. SONET/SDH fits this role admirably.

CLIENT/SERVER COMPUTING

The second development in network computing that became a driving force behind the bandwidth needs of networks is the popularity of client/server computing. The term *distributed computing* is closely linked to the concept of *client/server computing*. In fact, the use of client/server computing is sometimes seen as a reason for building the distributed network in the first place. If there seems to be no apparent reason why this should be so, it is undoubtedly true that until the rise of client/server computing there was little justification for an organization to go through the effort and expense of building a distributed network. This point needs further discussion.

One easy way to define a client is as follows: Any computing device on a desk with a person sitting in front of it doing work is a client. It makes no difference if the device is a terminal on an SNA network (SNA is IBM's System Network Architecture), a PC on a LAN, or a UNIX workstation on a TCP/IP network. These are all client devices.

Servers are easily definable too: Any computing device without a person sitting in front of it doing work is a server. Again, it makes no difference whether the device is an IBM mainframe, a DEC minicomputer, or a Novell Netware file server. These are all examples of servers. Servers do not have "workers" in front of them, but have at least one, and perhaps many, administrators who make the server available to a community of clients.

These definitions may seem simplistic, but they are extremely powerful in explaining the relationship of client/server computing and distributed networking, and the need for wide-area bandwidth as well. Once the basic definitions are understood, other terms fall into place. For example, the terms client/server computing or client/server architecture describes this overall practice of viewing all computing devices as either clients or servers (or even both at once in some cases). Client/server applications are specifically designed to run in a client/server environment with a client portion and a server portion to the application, and so on. The term *distributed processing* is sometimes used synonymously with this *client/server model* to acknowledge the usual geographic separation of clients and servers.

The relationship of the client/server model of computing and distributed networks is now easy to understand. Clients attach to servers over networks.Without a network, no easy way would exist for clients to reach the server they need. Therefore, the network is necessary to enable every client to access every server in the organization. The client/server concept is so important to networks that the key points are listed below:

1. Clients are desktop devices with workers.

2. Servers are devices with administrators.

3. Clients and servers are connected by networks.

4. Any client should be able to attach to any server.

5. Networks allow this to happen.

Of course, all of this ties in with the growth of powerful desktop devices discussed previously. The process began in the 1980s with the introduction of the PC into many organizations to replace the "dumb" terminal devices workers had used before to access applications on remote mainframes and minicomputers.

By 1986, the top-of-the-line PC was as powerful as the IBM mainframe in 1979. Many organizations were quick to take advantage of some of this computing power. Instead of having two or three terminals at each employee location, organizations could have a single PC running SNA terminal emulation when attaching to the corporate mainframe, VT100 emulation when attaching to the departmental minicomputer, and also when attaching directly to the local file server. In fact, with the rise of multi-processing operating systems, such as Microsoft's Windows and IBM's OS/2, these PCs were capable of doing all three at the same time.

It soon became common enough to link all of these PCs with LANs, and all of the LANs with routers, and all of the routers with point-to-point leased, private lines. The bandwidth between the routers was only a small fraction of the bandwidth on the LAN (64 Kbps on the WAN compared to 10 Mbps Ethernet) in most cases, but the bursty and intermittent nature of most data traffic made this a viable solution, at least for the short term.

In the long term, this was a problem. As both clients and servers grew more numerous and powerful, the available bandwidth between the routers became more scarce and pushed the network to the limit. Organizations with many PCs put everything on LANs with routers, including SNA devices and specialized terminals. This led to increased pressure to install more bandwidth between the routers to avoid bottlenecks. However, few service providers had much above 1.544 Mbps to sell to users, and nothing above 45 Mbps, even for those who could afford it. A few examples of typical client/server applications used by organizations today should emphasize the fact that the WAN bandwidth is not adequate for many applications today.

Sticking with the definition of client as a desktop computer and the server as a "back office" computer, there are numerous examples of client/server applications in common use today. Most database applications are now used in a distributed, client/server fashion. The database files can be huge and represent, in some cases, the entire record of an organization's business activities past and present; these files may reside on a number of servers. These database servers do not have to be concentrated in a single location, of course, but may be distributed among many locations that are geographically dispersed. The database client software is used to access the information stored on these servers.

Database applications now need to be installed and used in this client/server fashion. Installing such a database package defaults to this configuration: client and server. In most cases, the software is not only shrink-wrapped separately, but comes in different packing cases and boxes. Some configuration and installation parameters must be changed not to run the software in a distributed, client/server fashion, but to install and run the database and "query engine" (the client portion) on the same computer.

Another application that has become common in many organizations and that has pushed the capabilities of networks to the limit is simple electronic mail (e-mail) and the increased use of electronic data interchange (EDI) technology. E-mail has evolved from simplistic text to messages including voice and video portions. It is not unusual today for some large corporations to generate nearly 300 e-mail messages *per day* for key employees and positions. All of these messages must be distributed over networks, of course, and because e-mail is essentially a connectionless service—the intended recipient does not have to be accessing the network before someone can send them e-mail—most of these messages must be stored somewhere on the network in large mail servers (or "post offices") until users access, read, and dispose of them.

EDI involves the transfer of computer-readable data between organizations in a standard or agreed upon format. The partnered organizations may exchange EDI messages that represent an invoice, a packing list, or a purchase order. With EDI, instead of hand-processing multiple-copy forms and mailing or faxing them to the proper personnel for processing, an organization can automatically map the data into the correct electronic format and send it across a network. Upon arrival at the destination organization's computer system, the standard format can be translated into the organization's own internal format and processed automatically by the application software.

EDI, however, offers more than just a way to process batches of work orders. Graphics are increasingly being built into ordinary data transactions, and these graphics are often required in circumstances involving customized products. Merely ordering parts or materials from bland vendor listings is not sufficient to adequately convey the true nature of the products. Some graphic schematic or production drawing is usually necessary to complete the full business transaction. In some industries, notably the automobile industry in the United States, support for a host of standard EDI formats is required for all approved suppliers to the major automakers.

Another common use for networks today in many organizations is *distance learning*. As training (and re-training, for that matter) becomes more important, especially in high-tech industries, the pressure to cut back on things such as travel expenses and time away from the immediate work environment is also becoming greater. Instead of traditional instructor/classroom courses and materials, many organizations have chosen to explore the use of networks, usually both data and video networks, to deliver training more effectively right to the work location.

Sometimes seen in a form called *asynchronous learning,* the use of the organization's network to deliver training and instruction almost anywhere and anytime puts additional pressure on the network. Some organizations have gone so far as to commit to a given level of distance learning education ("20% of all course materials") at the end of a certain time period. Distance learning will only increase in popularity and sophistication in the future.

An ever-increasing number of people are working at home with either home-based small businesses or as part of a corporate job. Both employers and employees are finding this to be a viable alternative to long commutes and are finding increased productivity and job satisfaction. In some cases, telecommuting is an alternative forced on employees so that large companies comply with national and state environmental regulations. For example, the Federal Clean Air Act mandates a work-at-home option for qualified employees and job descriptions. A home office usually includes a computer, modem, fax, and answering machine hardware. In many cases, some form of screen sharing software could allow the sharing of data, images, and documents with colleagues and customers. E-mail software and a World Wide Web browser provide access to e-mail networks and the Internet, allowing communications with colleagues around the world. Other software allows access to electronic "card file" databases of business colleagues which can automatically display information about a business caller, making the fact that an individual works at home transparent to customers.

What's wrong with this picture? Nothing, except the twin issues of bandwidth limitations and network delay. As attractive as many of these client/server applications and their variations are, the universal use of all of them is always limited by the amount of network bandwidth available and end-to-end network delay users encounter when accessing any of these client/server variations. In the vast majority of these cases, the network bottleneck is not on the LAN side(s) of the network, but on the part of the WAN employed between them.

Many organizations employ client/server applications to allow distributed computing to take place over the mix of LANs and WANs used at various sites. In the vast majority of cases, the LANs are 10/100 Mbps Ethernet-type; more than 90% of all LANs built are one kind of Ethernet or another. The simple fact is that in spite of the availability for several years of technologies such as 155 Mbps ATM employed to the desktop in essentially a LAN configuration, 10/100 Mbps Ethernet continues to be the LAN technology of choice for organizations. And now there are Gigabit Ethernet (GE) and 10 Gigabit Ethernet (10GE) to think about.

Several reasons exist for the continued popularity of Ethernet LANs. First, the hardware and software components are extremely inexpensive and prices continue to drop. Next, several popular desktop architectures include built-in 10/100 Mbps Ethernet connectivity right on the computer motherboard. For example, a Sun Microsystems UNIX Workstation contains an integrated Ethernet connector. After all, what else would one do with a workstation besides connect it to an Ethernet LAN? Next, there is a vast pool of network managers, administrators, installers, and other miscellaneous personnel who know Ethernet well, and to the exclusion of familiarity with other LAN technologies. Colleges and universities around the world graduate more of these trained personnel every year. The pool is deep and wide and expertise for Ethernet is readily available. The increased use of high-speed Internet access to residences with multiple PCs has even led to the use of Ethernet right in the home.

The last reason for the popularity of Ethernet LANs is the one that is most relevant to this discussion. There is little incentive for organizations to increase the speed of their LANs, even for bandwidth-hungry client/server applications because the LAN is not the bottleneck in most cases. Although more powerful desktops have reduced the number of desktops linked by a single Ethernet from hundreds to tens to nearly one, the net result has been that the amount of traffic on a single Ethernet has remained fairly constant. The burden and bottleneck had now shifted to the interconnection devices between Ethernets and the bandwidth available on these links.

As common as the Ethernet is as a LAN type, so the router is as a LAN interconnection device. The roots of the popularity of the router go back to the days when LANs were almost universally linked by bridges. This is not really the place to debate the merits of bridging versus routing, but most industry observers agree that bridging makes sense in small, local LAN interconnection environments (such as when all of the LANs are in the same office park, or there are only a handful of dispersed LANs), while routing can handle anything bigger or more widely dispersed. Certainly the success of the Internet, which employs the router as the network node device of choice (currently), has also led to the increased popularity of the router itself as a network connectivity device.

The limiting factor in most client/server applications today, therefore, is the amount of bandwidth available between the organization's routers.

MULTIMEDIA

The third major development that has led to the need for more bandwidth is the increased use of multimedia in a number of different networked situations. Multimedia has already been mentioned in a variety of contexts previously in this chapter. Documents that run voice-overs while they display text, images that move, and graphics that evolve over time are all examples of multimedia. Multimedia is not just read, it is listened to and watched. Multimedia demands attention, it seems to many.

In spite of this almost obvious use of the term multimedia, multimedia remains a not-too-well-defined concept. Many people take the "I know it when I see it" approach and leave it at that. It seems clear that a "normal" database record will display on a computer screen while the user goes for coffee. But perhaps a multimedia database record will not. When the user returns refreshed, the multimedia aspect of the database record may have already "happened." The user has missed it. The event may be repeated, of course, but the point is the same. Multimedia happens: It is not static.

Multimedia has been variously described as either including, or even being characterized by, the inclusion of things such as video, three-dimensional graphics, animation, music, voice narration, voice recognition and synthesis, scanned documents, photographs, television clips, virtual reality, and even biofeedback-based input devices into otherwise data-only or text-only applications. Not all of these elements have the "evolving" characteristic emphasized above, but perhaps *multiple-media* use is just as good a definition. However it is defined, multimedia today seems to imply that networks must now support enough bandwidth to at least allow for the optional inclusion of voice, audio (quality sound), and video along with the data to be sent from server to client over the same physical network.

Standards such as MPEG and H.323 have been developed to define and support multimedia communications networks. Simple voice- and data-mixed applications could probably be adequately supported with existing WAN technologies; however, any applications involving mixed video, CD-quality stereo sound, and data will benefit by increased WAN bandwidths and lower delays. For example, experiments and trials are underway to allow the display of a Web site related to a sports event or even a night-time television drama right on the TV screen with the show.

PCs routinely play audio CDs in the same drive as that used for data-based CD-ROMs. Most PC systems come with better speakers (and features such as powered sub-woofers) than those found on stereo systems only a few years ago. PC users can easily listen to Mozart while writing a report for work or school. Cable TV connections can be hooked up to inexpensive PC boards to allow for the viewing of television shows on a monitor window. Strangely, this TV image can usually be iconized as would any other PC application window. But the TV icon contains a small, yet accurate, version of the television picture.

Multimedia is already pushing the limits of the capabilities of networks. The PC will probably become the focal point of home activity in the same way that the TV set is today. These combined TV/stereo/PC home units are already appearing and provide new multimedia interfaces between people and computers over networks. Instead of manipulating text and numbers, and later graphics and simple sound bites, a new breed of users will be manipulating moving images, changing the endings of movies or improvising a new climax to a popular piece of music. Photo-realistic graphics will blur the line between TV and PC.

A common thread runs through all of the above multimedia examples: The multimedia is either delivered locally to the user or accessed over a separate network built especially for this purpose. CD-ROMs with audio- and video-like multimedia encyclopedias are usually accessed directly from the local CD-ROM drive, not over a network, not even on a LAN. Audio CDs must be accessed and played in this fashion. The cable TV connection to the PC for video essentially makes a TV set out of the PC monitor. Any other network connectivity, for Internet access or telecommuting, must be maintained separately. The network, therefore, is a limiting factor here as well.

However, networks will have to handle a range of information: powerful PCs on LANs; client/server applications linking computers in diverse locations; and multimedia files and applications around the world. But what kind of network will be able to do it?

BROADBAND NETWORKS

The term *broadband* is used at least as frequently and probably as imprecisely as the term multimedia. Sometimes a distinction seems to be made between a *broadband network* and a *broadband application,* but this is of little help in defining the term broadband more precisely because these terms are usually defined circularly. That is, a broadband network is needed to support a broadband application, and a broadband application is one that requires a broadband network to function. A better definition must be devised.

Actually, there is a standard, and internationally sanctioned, definition of a broadband network; as defined by the International Telecommunication Union (ITU), a broadband network is characterized by "speeds higher than the Primary Rate." The Primary Rate refers to the ISDN primary rate interface (PRI) speed of about 1.5 Mbps in the United States and about 2 Mbps in most of the rest of the world. This would seem to imply that any network running faster than a few Megabits per second would qualify as a "broadband" network. Many networks do—LANs and WANs alike. The problem is that this definition was established by the ITU well before the increase of PC power, the popularity of client/server, and the promise of multimedia put so much stress on network capabilities. Therefore, in spite of this definition, the term broadband has come to mean much more than just a network built out of links that run faster that a couple of Megabits per second.

Because one of the main goals of this book is to explore the position that SONET/SDH holds in relationship to broadband networks, perhaps there is a better definition of broadband, one more appropriate to what SONET/SDH is and does. For the purposes of this book, the term *broadband* is defined as a network capable of supporting *interactive multimedia* applications. This does not imply that all broadband networks necessarily are used for interactive multimedia, only that they are capable of providing such support if needed. Nothing in this definition prevents such networks, for example, from supporting only high-speed file transfer or rapid client/server database access.

This definition rather nicely gives a more intuitive feel to many of the terms that include the term broadband. For example, a *broadband network* is now an *interactive multimedia network.* A broadband application is now one that includes support for interactive multimedia in its software code. In a sense, however, the new, admittedly subjective, definition is just as circular as the old one. What is *interactive multimedia*? No official definition of the term interactive multimedia exists. But perhaps that is not really a limitation, because official definitions from standards bodies, such as is the case with the term broadband itself, may be unenlightening. In this admittedly unofficial definition, the term interactive multimedia may be broken down into its two components: interactive and multimedia. Multimedia has already been defined as information delivered across a network that is not only read, but also viewed or heard.

Interactive may be defined as an application wherein user input action *now* affects application output *immediately.* User input action may be a click on a mouse button, the pressing of a function or enter key, or something else. Output is usually some monitor display, but could be some sound output or even printed images. Unfortunately, no absolute time intervals can be placed on the words *now* and *immediately* in this context. But again, this is not necessarily a bad thing. In fact, the only criteria are the ones users themselves apply to these terms. Not too long ago, an interactive (or real-time, or on-line, the terms varied) application was one in which user input resulted in output in about six seconds or less. Users today would never stand for such a sluggish system and network. Today, interactive users want sub-second response times from applications, systems, and networks.

The term, interactive multimedia, therefore, can be loosely defined as the delivery of multimedia content based on immediate user direction. When interactive multimedia is deployed according to the client/server model, the user is expected to be seated at a client computer, and the multimedia content resides on a remote server, or even several remote servers, because multimedia implies multiple information streams. Each stream may derive from an independent source. The user can immediately alter what is seen and heard through the touch of the mouse, or push of a button. A broadband network can be defined as a network capable of interactive multimedia delivery in a client/server architecture.

Few networks can effectively deliver interactive multimedia today. Even LANs struggle with this task, to the extent that much interactive multimedia content must be accessed from the CD-ROM directly attached to the desktop system. This struggle occurs because interactive multimedia applications make two very different, and yet related, demands on a network.

Multimedia applications, whether interactive or not, are characterized by a need for great amounts of bandwidth. This is usually measured in bits per second. A text-only application, such as accessing a World Wide Web page on a remote Web server with a Web browser (client), can usually make do with about 30 Kbps. Adding audio to the Web site will increase the loading time and make the higher bandwidth bonded ISDN channels, which run at 128 Kbps, more desirable. Viewing acceptable video from a remote Web site may force users to seek bandwidths in the megabit ranges, usually 6 Mbps, but as low as 1.5 Mbps for some video.

Obviously, SONET/SDH is a way to provide all the bandwidth that a multimedia application needs. But there is more to the broadband network story than increased bandwidth. This is where the *interactive* portion enters into the broadband equation.

Interactive applications that, of course, do not always involve multimedia content, are characterized by low and stable delays across a network. Both low delays and stable delays are necessary. When the delay across the network varies too much, unacceptable jitter is the end result. Sound is distorted by jitter, and video suffers from annoying speed-ups and slow-downs, resulting in jerky presentation. Low delays are needed to make interactivity possible in the first place. User input must travel across the network to the remote source, affect the output, and then the altered output must travel back across the network to appear at the client system.

If SONET/SDH were to only affect the bandwidth available to users on the broadband network, what would cause delays on the broadband network? Actually, bandwidth has an affect on network delay in general. It is not an obvious one, however, and requires a few more words about the effects of bandwidth on delay.

Bandwidth and Delay

The relationship between bandwidth and delay on any network is seldom acknowledged by end users. Much of the confusion arises because both terms use the unit "seconds" (bits per seconds, seconds of delay) in their measurements. Consider the network link shown in Figures 1-3a and 1-3b. The first figure illustrates how delay is measured in a network, while the second figure illustrates how bandwidth is measured in a network.

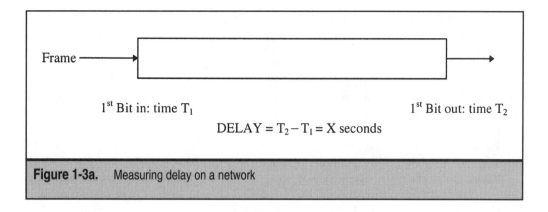

Figure 1-3a.   Measuring delay on a network

In the upper figure, a source sends a frame across a network link. In actual practice, this measurement does not need to be done over a single link. The entire network consisting of network nodes (the switches or routers) and trunks (links between network nodes) may occur between the end points of the measurements. The figure is only an illustration. The point is that the first bit of the frame leaves the source at some point in time (T1) and this first bit arrives at the destination at some later point in time (T2). The delay is just the time difference between these two time measurements of *first bit in* to *first bit out* of the network.

In the lower figure, the frame has arrived at the destination. There will also be a time interval for the entire frame to arrive off the network into the destination device. The number of bits in the frame (which varies) divided by the time interval between the first bit out (the same T2 as before) of the network and the last bit of the frame out of the network (T3) determines the *bandwidth* of the network available to the user.

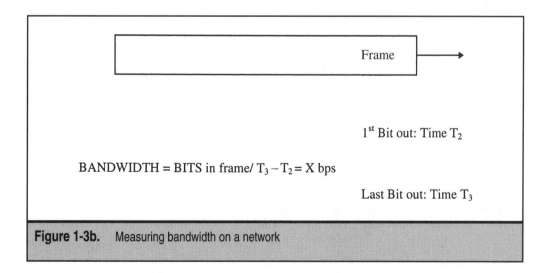

Figure 1-3b.   Measuring bandwidth on a network

Both the delay and the bandwidth will affect the *overall* response time that the user experiences from the network. To understand why this occurs, consider a network with *zero delay* (i.e., no delay at all) between end points. That is, the first bit of a frame appears at the destination as soon as it is sent (T1 = T2 in the figure). It is still possible for users to complain that "the network is too slow!" How can this be? Well, consider that the bandwidth available to users on this zero delay network is only 8 bps. Even a modestly sized 600 octet frame would take 600 seconds (10 minutes) to arrive. In the vast majority of cases, a computer cannot do anything with a frame until it is in memory completely because the error detection fields are usually located at the rear of the frame, and there is no sense in processing the beginning of a frame unless it is determined that no errors exist.

It such cases, the network delay is fine (zero), but there is not enough bandwidth on the network for the users. In other cases, the bandwidth may be adequate, but the delay is too high for the user's tastes. It is possible to attempt to classify network applications as "bandwidth bound" when they are constrained by the amount of bandwidth available, or "delay bound" when these applications are sensitive to overall network delay. For example, file transfers are typically bandwidth-bound applications. File transfers can be speeded up by increasing bandwidth or lowering delay, but increasing bandwidth will probably give better results. Users do not usually care how long it takes pieces of a file to arrive, as long as the whole arrives within a certain amount of time. Conversely, voice is typically a delay-bound application. Giving 64 Kbps digitized voice more bandwidth will not improve network performance. Only a suitable low and stable delay will make voice users happy.

It is important to appreciate the different effects that delay and bandwidth have on overall network speed. Users will always perceive both bandwidth-bound or delay-bound applications as caused by "slow" networks. When media, such as audio and data, are combined on multimedia networks, there must be both low and stable delays, as well as adequate bandwidth, to satisfy users. Interestingly, the International Telecommunication Union (ITU) defines the term *latency* on a network as the interval between *first bit in* and *last bit out* in several standards. This definition neatly combines the effects of bandwidth and delay. However, most network personnel loosely use the terms delay and latency interchangeably.

Bandwidth obviously has an effect on network delay, at least as perceived by the user; however, the more critical components of overall network delay are not dependent on bandwidth. There are two components of the network delays experienced by any user. These include the *propagation delay* of signals on the physical media and the *nodal processing delays* on each network node (switch or router) on the network. Only one of these elements can be speeded up in any realistic fashion.

The propagation delay on a network is a consequence of the physical fact that electricity, light, or any other form of electromagnetic signal can travel no faster than the speed of light from a source to a destination.

In fact, in most networks, regardless of the physical media, the signals travel much slower, at about 2/3 of the speed of light. The speed of light is almost exactly 300,000 kilometers per second, or about 186,000 miles per second. This means that most signals take about 10 millionths of a second or 10 microseconds to travel a mile. One microsecond (abbreviated as μsec) is equal to 1/1000th of a millisecond. A circuit 2,000 miles long would thus have a propagation delay of 2,000 ×10 μsec = 20,000 μsec = 20 msec, as mentioned above.

Nothing on a network will decrease the propagation delay of a signal from the source to a destination, short of moving the destination closer. This is not a viable option in most cases; therefore, the only component of network delay that a network service provider can improve upon is the nodal processing delay.

Nodal processing delays in a network, in turn, are dependent on two factors. The first is the overall speed of the node, whether central office switch, router, hub, or other more exotic device. It takes so long to move bits from an input port to an output port, even when no other traffic occurs on the switch. The second factor is the load on the network node at any point. The more traffic there is, the slower the node will operate. Eventually, operation may slow noticeably, and the node is said to be *congested*. This basically means that the network node is operating outside of its design parameters in terms of delay. The presence or absence of loads can make the end-to-end delay that an application manifests variable.

Figure 1-4 shows the relationship between all of the concepts discussed so far. Broadband is defined as support for interactive multimedia capability. The interactive portion depends on low and stable delays across the network. The delay, in turn, depends on the propagation delay and nodal processing speed. The bandwidth depends on the speed supported by the media available.

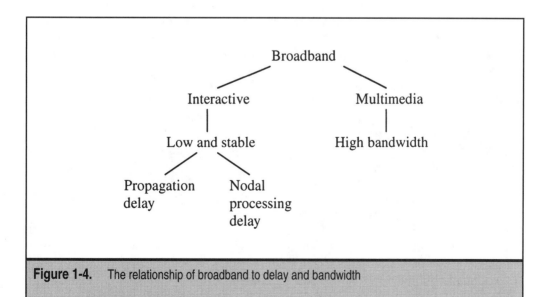

Figure 1-4.   The relationship of broadband to delay and bandwidth

CREATING BROADBAND NETWORKS

Two ways exist for equipment designers to deal with the two factors of basic delay and congestion. The first way is to make the network node as fast as technology allows. The second way is to make sure that the capacity of the network node is adequate for even the most extreme network loading conditions usually by adding more trunks and memory.

In the public switched telephone network primarily used for voice, the network nodes are central office or toll office switches and the bandwidth is provided by unshielded twisted pair copper local loops to the switching office and digital trunks on a variety of media between the switching offices. There are exceptions, but this is the general architecture.

Voice services work well on this architecture for two reasons. First, the bandwidth needs for voice are relatively modest. Digitized voice consumes only 64 Kbps, and compression can lower this even further. Second, international standards require voice switches to operate quickly, in less than 1/2 a millisecond. This means that voice services enjoy adequate bandwidth and delays on the public switched telephone networks. Ten switches on a voice circuit add less than 10 milliseconds to the end-to-end delay.

Networks that support interactive multimedia and many other networks today benefit from the deployment of fiber optic cable and SONET/SDH. SONET/SDH can easily provide all the bandwidth needed for any application, from interactive multimedia and imaging to medical applications. The trickier part to address is the nodal processing delay.

Early attempts to combine voice and data applications go back to early days of X.25 packet-switched networks in the early 1980s. X.25 proved not to be adequate for the delivery of packetized voice services for two reasons. The older X.25 switch was not nearly as fast as a telephone central office switch, taking anywhere from 5 to 10 msec to process a packet. This was 10 to 20 times slower than the voice network. Ten switches on an X.25 network added from 50 to 100 msec of delay to the end-to-end delay experienced by the user or application. This was far above the propagation delay.

The second reason was that the delay in any particular X.25 switch was extremely dependent on the traffic load of the switch at any point as are all network devices. This made the delay encountered by a particular packet containing digitized voice highly variable. The service was usually unacceptable because of great voice distortion. Vendors applied more inventive techniques to their X.25 products to compensate for these twin problems, but with modest success. (Of course, X.25 was never *designed* for voice—but some still tried.) Of course, the answer to building network nodes with low and stable processing delays was simple to describe. The answer, however, was difficult to implement until computer architectures ran fast enough to perform the needed tasks. In a nutshell, the answer is to make the nodal processing delay, even in a worst-case scenario in terms of traffic load, an insignificant fraction of the propagation delay end-to-end through the network. All applications—voice, video, or whatever else—*had* to work within a given propagation delay, or they would not work at all.

For example, consider a circuit with an end-to-end propagation delay of 20 msec. Suppose that there are 10 network nodes along the circuit and that these contribute only 5 μsec ±1 μsec of delay regardless of network load. Then the total nodal processing delay would be about 50 μsec, or only about 1/400th of the propagation delay. The variability would be only ±10 μsec, or only 1/2000th of the overall propagation delay. These nodal processing delays of a fraction of a percent of the propagation delay would be indistinguishable to any end user, regardless of content. This simple example ignores finer statistical points as the additive effects of the variances, not the standard deviations (we should only see ±3 μsec delay variation), but the point is still valid.

Once this realization is made, the only question remaining is to decide which nodal architecture is the best one available to offer the low and stable processing delays required. Basically, any "fast-packet" technology will do. Fast-packet technology delivers packets of information with a low and stable enough delay to allow for multimedia operation across the network. The international standard for broadband services built on these fast packet networks is B-ISDN. The reference to ISDN is not accidental. B-ISDN, as conceived and standardized by the ITU in 1988 (the "Blue Books"), was intended as a logical extension of ISDN technology and services. The B-ISDN standard settled on the cell relaying ATM technology, for a variety of reasons. ATM switches provide the nodal processing delays that were low enough and stable enough for all services. SONET links provided the huge bandwidths needed to effectively combine voice and video and data networks on the same physical infrastructure.

In a sense, B-ISDN is a framework and an evolutionary standard. Work on SONET/SDH, ATM, and especially B-ISDN services, was not completed in 1988. Indeed, the work had barely begun. The ITU is not the only standards organization working on completing the B-ISDN vision. Other organizations are working on various aspects of broadband networking. The Institute of Electrical and Electronics Engineers (IEEE) has extended broadband work into the LAN arena. In the United States especially, the American National Standards Institute (ANSI) and Bellcore (now Telcordia) have adapted ITU standards for use within the United States. Other organizations, such as the Electronic Industries Association (EIA) have contributed as well.

Figure 1-5 extends the previous picture to include the present broadband or fast-packet technologies. SONET/SDH is the choice to supply the bandwidth. ATM switches provide the network nodes. B-ISDN services may be offered based on this network architecture. Now, broadband networks may be built without SONET/SDH links and ATM switches. After all, customers buy services, not technology. But only a network based on SONET/SDH and ATM may offer ITU-compliant B-ISDN services to customers.

B-ISDN

To their enduring credit, the ITU recognized the limitations of ISDN in terms of raw speed and multimedia capabilities by the middle of the 1980s. The increased power of desktop computing devices and the widespread use of client/server architectures made it apparent that the 64 Kbps to 2 Mbps (1.5 Mbps in United States) speeds of ISDN would not be able to keep pace with the rapid evolution of these devices for long. In 1988, the ITU produced a "blueprint" for taking telecommunications into the twenty-first century. This blueprint was B-ISDN.

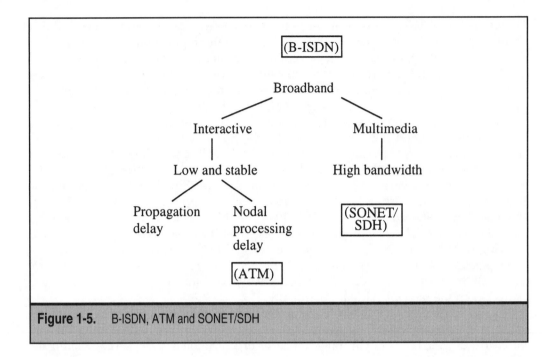

Figure 1-5.   B-ISDN, ATM and SONET/SDH

B-ISDN was a vision of unified services, all delivered on the same unchannelized physical network with fast switching, state-of-the-art multiplexing, and very fast and yet low-bit-error-rate links tying all of the components together. Perhaps overly optimistically, some observers hailed B-ISDN as a network architecture for the next 30 years. No matter how long it takes to implement B-ISDN, or how long the B-ISDN vision remains viable, the whole concept remains an impressive achievement as an attempt to address network limitations.

The services that B-ISDN encompasses are all-inclusive. They amount to no more or less than nearly everything that has been done on a network before, everything that can be done now, and anything that can be done in the future. After all, once a network has all the bandwidth technology allows, operates fast enough for any interactive application, and virtually has no errors, what else could users possible need that a B-ISDN network could not deliver? The fast switching and state-of-the-art multiplexing are provided in the B-ISDN scheme by ATM network nodes. The very fast and yet low bit- error-rate links that tie everything together are SONET/SDH fiber links.

The whole network has a public portion in the form of large, service provider-based ATM switches, which mostly switch and do some multiplexing, and a private component consisting of smaller (usually), customer-premises-based ATM switches, which mostly multiplex and do some switching within the customer site. This entire "hybrid" private/public ATM on a SONET/SDH network delivers services defined by B-ISDN to the users.

The B-ISDN network concept is illustrated in Figure 1-6.

A partial list of B-ISDN services is listed in Table 1-3.

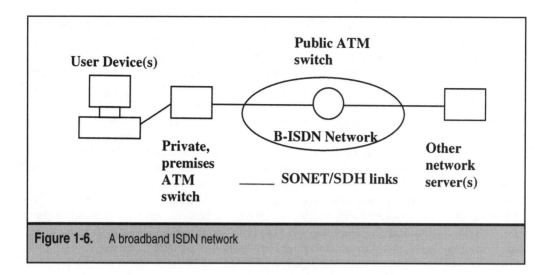

Figure 1-6.   A broadband ISDN network

Network Connections

The list of services in Table 1-3 naturally tend to fall into one of a number of categories. There are voice and data services, of course, but also imaging (fax) and video services. Few of these depend on B-ISDN to allow them to be available to users who want them badly enough. Currently, for example, many organizations buy equipment to supply voice, fax, or data services for their users without a thought for B-ISDN; however, this means that all of these various types of user equipment and the network nodes that support the service must be linked together separately with some form of network connections to supply the needed connectivity.

Services		
Voice	CAD/CAM	Telemetry and security
LAN interconnectivity	Telex/Teletex	Home consumer services
Fast-packet service	E-mail	Electronic banking
High-definition TV	Fax	Teleshopping
Videoconferencing	Videotex	Telecommuting
Private line emulation	Electronic document interchange	

Table 1-3.   B-ISDN services

In the United States especially, network connections between routers, LANs, and other network service delivery devices tend to take the form of leased private lines. Also called point-to-point links, dedicated circuits, or even some other terms, these leased private lines are basically the bandwidth on digital links bought from the telephone company (not always, but usually) on a long-term lease. They are considered "private" due to the fact that only the organization that pays the bill has the right to put bits on the link. Even when the link is idle for days, the service provider cannot reclaim the bandwidth for its own purposes. The bandwidth belongs to the customer. It is, however, still a leased or rental situation. If the link were to fail, the owner (the service provider) must fix it. If the lease were to expire or were not paid, the bandwidth would revert to the service provider, which would then be free to lease the bandwidth to another customer.

There are a number of reasons why the process of building network connections became one of leasing private lines in the United States. First, when they began building computer networks in the United States in the 1960s, organizations had plenty of money to spend on whatever they needed to make the networks possible. Second, service providers were only too happy to expand their facilities to meet the demand for private lines that the new data networks consumed.

Most organizations had enough money to afford their own network nodes. Whatever switches or other devices they needed could be purchased from the manufacturers. The leased private lines then became the method of choice for linking these network nodes together. This solution was neither inexpensive nor painless. Leased private lines were purchased by the mile, and more bandwidth cost more money. In an attempt to cut down on the increasing burden of leased line networking costs, many organizations built a bare minimum of private lines to link their sites. This also resulted in unintentional bottlenecks as traffic to and from remote sites funneled through intermediate sites. Because more bandwidth cost more money, organizations also attempted to provide connectivity with the absolute minimum amount of bandwidth required, usually the digitized voice bandwidth of 64 Kbps.

This is not to say that organizations employed individual 64 Kbps digital links on their own physical facilities. It was common for many organizations, especially larger ones, to lease "bundles" of 64 Kbps private lines. These were known as channelized DS-1s or T-1s in the United States. The term, *channelized* refers to the fact that the bandwidth on these circuits was divided into 24 circuits, or channels, of 64 Kbps each, called DS-0 channels. The whole practice became so common in the 1980s that the term *channelized networking* began to be applied to this configuration. This process was accomplished by time-division multiplexing, but it was the ends that were important, not the means. The whole idea of channelized T-1 networking is shown in Figure 1-7.

Organizations would divide the use of the channels on the T-1 according to their needs. It was not unusual to find an organization assigning six DS-0 channels to connect their corporate routers on various LANs, using six other DS-0s to link the corporate PBXs with private voice lines (called *tie-lines*), and using six more DS-0s to link videoconferencing equipment from one site to another. The other six DS-0s were usually reserved as spares for growth. This mixed data, voice, and video network achieved this support at the price of flexibility.

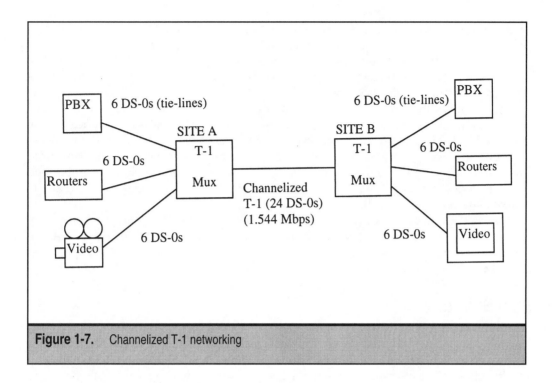

Figure 1-7.   Channelized T-1 networking

The bandwidth that the channels represented could not be easily reassigned or shared by other channels. For example, even if no one were on the phone at 3:00 A.M., the 384 Kbps bandwidth associated with the tielines could not easily be used by the data channels. Each channel was a private line by itself, even though it was bundled and delivered on a T-1 (DS-1). In many cases, even the use of these T-1 circuits running at an aggregate bit rate of 1.544 Mbps did not help to alleviate the need for more bandwidth within most organizations due to the use of channelized networking.

In the rest of the world, the situation was different than in the United States. Other countries, instead of prospering in the 1950s, were attempting to recover from the devastation of World War II. Still other countries, although relatively unharmed by the nearly global conflict, nevertheless lacked the internal funds and resources to build the telecommunications infrastructure a leased-line market required. When computer networks began to be built around the world in the 1960s and 1970s, the plentiful and inexpensive leased-line environment found in the United States was absent in the rest of the world, with very few exceptions.

To address this need for networks without private lines, other countries created public data networks. In a public network environment, the service provider, usually a government entity, such as the national telephone company, bore the expense of buying and installing the network nodes and the links connecting them. Users then bought the service that the node type delivered, but not the links or nodes themselves. It was a simple and elegant solution. Eventually, the ITU adapted the X.25 Packet Switch Data international standard to ensure compatibility and interoperability between these public data networks.

Just as the leased-line networks in the United States had limitations, so did the public data network have limits as well. In the case of public data networks, the problems were slightly different, but just as serious.

The bandwidths available over X.25 public packet switched networks did not exceed 64 Kbps, and in many cases did not exceed even 9.6 Kbps. There was no technical reason for this. The X.25 standard simply peaked at 64 Kbps. Delays were highly variable, just as in all packet-switched networks of the time, precluding the use of X.25 for packetized voice and video. Because much of the customer traffic was carried on the same public network facilities also meant that performance was highly variable, and security trickier to implement than on private lines.

None of these factors prevented X.25 from becoming the data network solution of choice in most countries around the world. X.25 was used for everything from SNA networks to the emerging client/server applications. Frame relay extended X.25 capabilities into the multi-megabit ranges.

B-ISDN Splinters

By the beginning of the 1990s, all of the trends discussed to this point seemed to be coming to a crisis point. The power of computers threatened to overwhelm the networks they were increasingly dependent on to function within the client/server model. Demands for multimedia only worsened the situation. The practice of building leased-line networks with channelized private lines led to the splintering of the available bandwidth among many network connections, each with idle bandwidth. Even the use of public data networks did not improve matters greatly.

The need for increased networking bandwidth and faster network nodes eventually resulted in the creation of B-ISDN. B-ISDN used the incredible bandwidths available on fiber optic links and combined this with the unparalleled speed of ATM switching. A sophisticated multiplexing technique used with ATM made it possible to create unchannelized networks which allowed all of the available bandwidth to be used by whatever application (i.e., voice, video, or data) needed bandwidth. The invention of flexible (or dynamic) bandwidth allocation became popularized as *bandwidth on demand,* although this is a less precise term. Bandwidth on demand encourages the belief that bandwidth can be created out of nothing, while flexible bandwidth allocation correctly focuses on the dynamic reaction of the network to users' changing bandwidth needs.

However, even though B-ISDN offered a potential solution to all these problems, not all of the pieces of B-ISDN were equally mature and ready for implementation in hardware, software, and network protocols. The three components of B-ISDN (i.e., SONET/SDH, ATM, and B-ISDN services) were not pillars, but a tower. Supporting the whole network structure were the fiber SONET/SDH links, coincidentally the most mature part of the plan. On top of the SONET/SDH foundation were the ATM switches.

Although ATM was emerging from laboratories and test networks by 1990, there is a lot more to networking than just sending bits from place to place, especially in a public network. Networks must be managed and controlled. It was in just these two areas that early ATM products proved most lacking. Without the firm support of stable and robust ATM products, the B-ISDN services, perched atop the B-ISDN tower, seemed ready to come tumbling down.

What actually happened in the 1990s was not as bad as that, but it was something that may not be fixed for some time to come. The different stages of the components of B-ISDN soon led to the breakdown of the whole structure, a splintering that may take years to put back together, if it happens at all. SONET/SDH was soon deployed independent of ATM and/or B-ISDN services; after all, bandwidth was just bandwidth. SONET/SDH neither required nor cared whether the bits carried over the link came from or went to ATM switches or equipment. SONET/SDH grew up in the 1990s as an orphan in some respects, but an orphan with a very bright future.

ATM suffered a somewhat different fate. ATM was effectively split in half by two related, but independent, camps. LAN equipment vendors seized the fast switching aspects of ATM and immediately made plans to use ATM to allow for unprecedented speeds to the desktop. Because network management and control are much simpler in a LAN environment, these ATM shortcomings were much less obvious in this application.

At the same time, WAN equipment vendors embraced unchannelized support for voice, video, and data that was inherent in ATM and immediately made plans to use ATM internally with their network products. Service providers liked the idea of merging separate channelized facilities onto ATM networks, but needed the network management and control components desperately.

B-ISDN suffered the worst fate of all. All of the attention to SONET/SDH and ATM left little time or effort to devote to B-ISDN services. If SONET/SDH was just another private-line solution, and ATM was being buried inside LAN hubs or telephone company toll office switching networks, what was the sense of moving forward with B-ISDN service descriptions? Each of the two lower pieces of B-ISDN—SONET/SDH and ATM—have taken on their own identities independent of B-ISDN. The question is now becoming one of whether SONET/SDH and ATM should be reunited with B-ISDN, no matter how far into the future. Of course, networks and the technologies that make them possible exist to serve users. If users are happy with SONET/SDH and ATM the way they are, why not leave well enough alone? The answer to this question is complex. A full answer would point out the advantages of having a global standard for interoperable and compatible services. However, as time goes on, the success of the pieces of B-ISDN, especially SONET/SDH, makes it apparent that a reunification of SONET/SDH, ATM, and B-ISDN may not happen for years to come, if at all.

In the past few years, the rise of the Internet and the Web have all but laid to rest grand plans for a global, unified network based on B-ISDN applications, ATM switches, and SONET/SDH links. The brave new world of Web-based television and telephony, IP routers, and DWDM links has fundamentally changed the future of networking. Networks today grow from the ground up, like the Web, from instance to standard, rather than from standard to instance as outlined by the vision of a standards body, such as X.25.

Oddly, the only widespread use of H.323, the most mature piece of B-ISDN yet defined, is on the Internet in the form of IP telephony. But certainly people will not wait for

B-ISDN applications while they buy and sell, work and play, and even live their lives on the Internet and the Web. The only role for ATM today appears to be as a high-speed backbone technology to connect IP routers, and even that role has been challenged. Even SONET and SDH are challenged by leaner, more IP-friendly techniques within the DWDM framework.

A word here is appropriate about adding quality of service (QOS) to IP networks like the global public Internet. Like all technologies designed to meet mainly data needs, IP networks struggle to provide the low level of delays, stable delays, and reliability that are needed to support interactive multimedia applications such as voice and newer forms of video. These QOS parameters must be added either to the network or to the IP application, or to both, if the application is to become acceptable and common. There are currently more than a dozen schemes for adding this QOS to IP networks, including DiffServ, RSVP, and IntServ. It is not necessary to discuss these schemes further, only to point out that as long as there are multiple ways to add QOS to an IP network, the pressure to add bandwidth to an IP network will only increase.

Why should this be so? Because the benefits of adding bandwidth to a network are well known, studied, and accepted. The benefits of adding DiffServ, RSVP, or something else to a network to improve QOS are at the present unknown. The situation is like the early days of PCs: with limited cash on hand, and faced with a slow PC, the smartest thing to do was to add memory—as much as affordable. The benefits of the added memory were immediate and well documented. The reasons that bandwidth is the network equivalent of memory have been discussed already. So bandwidth will be the network QOS resource of choice for some time to come, until one IP QOS scheme or another emerges that is as reliable as adding bandwidth to improve overall network QOS.

SONET AS PRIVATE-LINE SERVICE

In the United States, where facilities were plentiful and organizations had money to spend in relative abundance, most networks were built as private-line networks. Telephone companies were almost always the source of these private lines. Private lines amounted to the selling (actually, long-term leasing) of bandwidth on telephone facilities to customers. The telephone companies gratefully saw the increase in private-line sales as another revenue stream. It may seem strange that the selling of pieces of the telephone companies' lifeblood was considered a sound business practice, because the bandwidth leased to the customer could not be used internally by the service provider. The practice of leasing private lines, however, was embraced because a private line service is barely a service. The risk and exposure of the telephone company as a service provider (or *carrier* of the service) was minimal.

The risk and exposure were small because a private-line service is used to create a private network. In a private network, the service provider's main duty is to furnish the basic transport service, in which the service provider merely accepts bits from a piece of compatible customer premises equipment (CPE) and delivers them to a compatible CPE at the other end of the point-to-point private line. The service provider neither looks at the bits (voice, video, or data), stores the bits (e.g., messaging system), nor alters the bits (e.g., to ensure compatibility). In fact, telephone companies were forbidden from providing any more than basic transport services or selling the CPE to the customer. Anything beyond the basic transport of bits qualified as enhanced services. Enhanced services were strictly controlled by regulating agencies such as the Federal Communications Commission (FCC) before 1996.

Service providers may not be overjoyed at losing large chunks of bandwidth that could otherwise be used to deliver dial tone and other types of service; however, the tradeoff was still acceptable because the only thing the service provider was responsible for was the transport of bits from point A to point B. It made no difference whether the bits made sense or not because the CPE and not the service provider generated them. All that mattered was whether the bits arrived, and all that could really go wrong was that the link stopped relaying the supplied bits, which was fairly easy to fix. Coupled with the fact that the private line had to be paid for, month after month, whether it was used 100% of the time, 50% of the time, or not at all, these factors made private-line services quite popular in the United States.

What has all this to do with SONET/SDH? SONET/SDH started out as an international standard for fiber optic transmission and was quickly included in the B-ISDN plan by the late 1980s; however, B-ISDN is not a private networking solution. B-ISDN specifies the services on a public network. Yet, service providers in the United States quickly embraced SONET/SDH and deployed it as a private-line service independent of B-ISDN. Why did they do this?

There are two reasons that SONET/SDH is not tied to B-ISDN, especially in the United States. First, early SONET (not including SDH, just SONET) standards proposals predated the work on B-ISDN; therefore, SONET began independent of B-ISDN. Second, service providers (i.e., mainly telephone companies) in the United States needed SONET to keep up with the demands for private-line service. Each point will need a little more elaboration.

Although it is true that it forms the foundation of a B-ISDN and ATM network, SONET/SDH did not begin this way. The idea of creating a standard digital network scheme for fiber was around before the idea of combining SONET/SDH with high-speed switching and multiplexing in the form of ATM eventually created B-ISDN services and networks. Although SONET/SDH came to be *intended* for use on public B-ISDN networks, there is nothing to *prohibit* its use as a private-line solution.

Private-line services sell off "pieces of network" in the form of bandwidth to customers. Now, how much bandwidth any one customer can buy is limited only by the total amount of bandwidth available, which only makes sense. But as multimedia applications and powerful desktop computers began to proliferate in many organizations, the demand for increased bandwidth speeds could not easily be met with existing telephone company networks and facilities.

For example, early LANs and router networks could be linked productively at 64 Kbps. To supply this bandwidth, the telephone company usually installed 24 channels of 64 Kbps each, providing 1.5 Mbps combined. If the customer were really pressed for speed, the whole bandwidth of 1.5 Mbps could be linked to a single router. However, above 1.5 Mbps, things were expensive and difficult to provision. The only realistic solution above 1.5 Mbps was 45 Mbps, some 30 times faster. In many cases, this was overkill and, in most cases, the price was too high.

Several lower speed 1.5 Mbps would sometimes do for multimedia, but not always. It is the same in many other industries: If a customer were to need a race car that goes 200 miles per hour, selling the customer two that would go 100 miles per hour each would not be the same thing.

The highest standard speed used internally in the telephone company network was 45 Mbps. Ironically, higher speeds were possible, but were never standardized and remained proprietary. Equipment could only be purchased from the one vendor that made it. This clearly would not work in an environment where the CPE remained beyond the control of the service provider.

All of these factors combined to keep the cost of private-line bandwidth above 1.5 Mbps too high to be easily affordable by most organizations. Lack of facilities kept the cost high on the service provider side, and proprietary equipment kept the cost high on the CPE side. After all, if the piece of network being sold is 45 Mbps, there must be higher speed standard interfaces on the service provider's network backbone to deliver these speeds. Unfortunately, there really was not.

SONET/SDH private-line services will break the high bandwidth logjam. SONET/SDH runs at many times the 45 Mbps bandwidth customers frequently require for interactive multimedia services. SONET/SDH will effectively lower the price for previously expensive links running at 45 Mbps and even higher (of course, SONET/SDH may require changes to the wiring inside a building as well).

CHAPTER 2

A Fiber Optic Tutorial

The synchronous optical network, or SONET, is the North American version of an international standard method (SDH) for a family of high-speed transmission links on fiber optic cable. SONET takes advantage of the many attractive features available on today's fiber optic facilities. This chapter explores these features and advantages in more detail, and shows why SONET is built on fiber optic technology. This chapter assumes no special knowledge of fiber optic cable, or anything more than a passing knowledge of digital transmission systems. At the same time, this chapter attempts to detail the reasons why SONET and fiber optic transmission systems are invariably intertwined today.

This chapter and the next few, concerned as they are with the roots of SONET and SDH, consider only SONET and mention SDH only in passing. As will become apparent later on in this book, the International Telecommunication Union (ITU) essentially took work already begun in the United States on SONET and created SDH for the global community. Today, it is proper to treat the two technologies as essentially one: SONET/SDH. But when speaking historically, especially before 1990 or so, SONET was the only real game in town. As recently as 1998, a major vendor of SONET/SDH test equipment was teaching SDH alarm (error) conditions internally using SONET documentation. There was little available for SDH specifically. Today, of course, SONET and SDH are more like two sides of the same coin.

The rapid deployment of fiber optic transmission systems allowed service providers to deploy equipment that could transmit at very high data rates. This process had been going on since the 1970s, and (in some areas like Chicago) even in the late 1960s. But, with a lack of standards above 45 Mbps (the DS-3 rate), this equipment utilized vendor-proprietary multiplexing techniques to transmit data at rates of up to 1.7 Gbps. Before SONET came along, identical vendor equipment was needed at each end of a fiber span, and the addition or drop off (usually just called *add/drop*) of various data streams for switching (usually at the DS-1 level) necessitated demultiplexing the entire data stream. When identical equipment was not available at both ends of a fiber span, such as in a handoff between a local carrier (LEC) and an interexchange carrier (IXC), handoff occurred at the lower "standard" data rates of DS-3 and below.

The previous chapter explored the origins of private-line networks and why SONET forms such a key piece in these networks today. This chapter explores the reasons that SONET (and soon after, SDH) came to be based on fiber. Today, the answer seems obvious, because fiber optic networks have become common and accepted not only on the service provider's backbone network, but also on the desktop. But it was not too long ago that fiber optic networks seemed much too fragile and expensive to become the basis for a new generation of digital networking standards. Therefore, it is worthwhile to examine why fiber exploded in the networking arena.

It sometimes seems that fiber is the media of choice for an increasing number of cabling situations. It is not only the service providers that have installed miles and miles of fiber, about half of which is installed privately in a campus or building environment. Fiber optic cable installations will continue to increase both on the customer premises and on the service provider's network. Fiber is the media of choice in increasing telecommunications situations, from voice, to video, to data.

FIBER OPTIC TRANSMISSION

The transmission of digital signals on fiber optic cable is simple to explain and understand. It is really as simple as shining a light down a strand of fiber. Even a hand-held flashlight can do it, but not very well. It is quite common to simply use a brief flash of light (a *very* brief flash of light: the faster the link, the briefer the flash) to indicate a "1" bit in a time slot and the absence of a flash in a time slot, to indicate a "0" bit. This simple binary on/off technique is the easiest and cheapest to implement. Many will recognize this method as just another form of amplitude modulation. This method is assumed in this tutorial. However, it should be pointed out and always kept in mind that there are other modulation techniques used in many configurations. Newer fiber modulation techniques are very sophisticated and are nearly as elaborate as the complex modulation techniques used in high-speed modems. The idea is the same: increase the bit rate available on a given bandwidth.

Fiber optic cable is made of very pure glass. Actually, there are other constructions, including very pure plastic, but these fibers have more restricted applications. The manufacturing process is fascinating, but of limited interest here. In one very popular method, fiber optic cable is made by heating and drawing out a large chunk of very pure glass (called a "preform") into a strand of fiber thinner than a human hair (a typical fiber is 120 microns (millionths of a meter) in diameter, and a human hair is about 140 microns in diameter). Various layers of similar materials are added (called "buffers"), along with strengthening materials (usually Kevlar fibers). The composite fiber optic cable, which may contain numerous fiber strands, is jacketed, labeled, and taken up on reels. Early fibers had a reputation for fragility, and often broke due to bending stresses at installation time. Fiber is actually stronger than steel when it comes to pure tensile strength (pulling). Bending fiber causes stress fractures (cracks), however, and the strengthening material is present more to prevent kinking and resultant fracturing than to prevent excessive pulling forces.

For commercial applications, whether within a building or over a long distance, a laser or light-emitting diode (LED) is used, depending on the speed and distance the application must cover. LEDs have a more "spread out" light flash, or pulse. Lasers generate much sharper pulses, which make them useful for higher speeds and longer distances, as shown in Figure 2-1. Although an LED may only be able to generate a light pulse with useful power about 40 nanometers (nm, billionths of a meter) in width, a laser pulse can

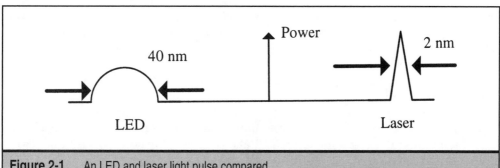

Figure 2-1. An LED and laser light pulse compared

be as narrow as 2 nm. Narrower pulses translate to higher bit rates in even the simplest modulation schemes.

Of course, many good safety rules exist that make the use of the lasers previously used for fiber optic communication difficult in general office space. Older lasers were usually not found on someone's desktop. The laser was usually down the hall in a locked communications closet with a warning sign on the door.

Perhaps a word should be said about lasers and the human eye, since safety is always a concern. Although it is true that the power of most commercial laser products have very low power rates (thousandths of a watt, or milliwatts), the human eye is a very effective lens. When looking directly at even the lowest power laser, the eye will focus all of the laser's power on the retina. In a relatively short amount of time, permanent damage can be done to a person's eyesight. This is mentioned not to frighten people, but only to remind them that a healthy respect is needed when lasers are in use nearby.

Lasers used to be very much more expensive than LEDs, another factor to consider in fiber optic networks design and implementation. But today, mass production of chip-based lasers has narrowed the gap considerably. Most lasers today for SONET and SDH come in *intermediate range* (IR) and *long range* (LR) versions. LR lasers are appropriate for fiber spans of 10 to 40 kilometers (about 6 to 25 miles), and IR for anything shorter than that. LEDs are still used in campus-wide fiber optic networks.

One of the major reasons that fiber optic cable is popular and effective is that the loss of signal strength over a given distance is much less than with other transmission media. That is not to say that fiber has no signal loss, or attenuation. Much of the loss of signal strength on a fiber link happens right at the interface between the light source and fiber. This is called the "injection loss" and is a function of what is known as the "numeric aperture" of the particular fiber. It is important to note that the interface between the light source and fiber does not have to be a bonded, soldered, or direct physical interface. If anything, the connection is essentially glued using a strong epoxy resin. Typically, the light source and fiber are just very close together. As a result, the light source can become misaligned with the fiber and needs periodic "tuning" to stay within specifications.

On the fiber optic cable itself, the light pulse travels as a coherent package of light waves. The light waves travel because of "total internal reflection" (it isn't really total) at the core/cladding interface of the fiber. The difference in the *index of refraction* (the same effect that makes a spoon appear bent in a cup of water) between the core and cladding bends the light and confines the light-wave package to the core itself. The core is simply very pure glass, about 5-50 micrometers (millionths of a meter) in diameter. The cladding can be glass or plastic, and is about 120 micrometers in diameter. However, the fiber optic cable itself is typically heavy, due to the presence of multiple fiber strands, strengthening materials, and jacketing.

It is important to realize that fiber optic cable shares an important characteristic with other digital transmission systems. Fiber optic cables are almost universally deployed in pairs. One fiber is used for outbound digital signals and the other fiber is used for inbound digital signals between the same endpoint transmitters and receivers. Fiber can be made to run full-duplex on a single strand, but this tends to be more expensive and is not a common practice. Therefore, most diagrams of fiber networks in this book will show two strands of fiber connecting any two pieces of equipment.

At the receiver end of the link, which may be literally thousands of miles long, an avalanche photo detector (APD) or positive-intrinsic-negative (PIN) device detects the light. It is the nature of light to travel in a wave and be detected as particles called photons. The receiver counts the photons in the corresponding time slot. When the number of photons is above some threshold, a 1-bit was sent, anything else is a 0-bit. In theory, a 1-bit represented by 75 photons worth of light at the sender can be detected by 21 photons in the same time slot at the receiver. However, the number is usually about 38 (about half of those sent), which gives a characteristic bit error rate (BER) of 10^{-9} on the link. Again, this modulation technique is only intended for illustration purposes; actual techniques may be much more sophisticated.

Another aspect of the quantum physics "weirdness" of fiber and light transmission, in general, is that some of the photons will be detected *before* they are sent. A light pulse can "spread out" from the time slot in which it was actually sent, spreading forward as well as backwards. Fiber optic cable signals also do not travel at the speed of light through the fiber; light travels at 300,000 kilometers per second only in a vacuum. The index of refraction of the fiber slows the light down somewhat. Usually the propagation speed of light in a fiber optic cable is about 2/3 of the speed of light in a vacuum, or some 200,000 kilometers per second (about 125,000 miles per second).

Even from this simple description of fiber transmission systems, some major points are obvious: fiber links are sensitive to configuration differences and must be maintained; timing is obviously quite important on fiber links; and fiber cable is quite small. A simple picture of a fiber optic transmission system is shown in Figure 2-2. Although the numeric aperture (na) is represented as a "device," the actual na is just a theoretical construction of the fiber itself.

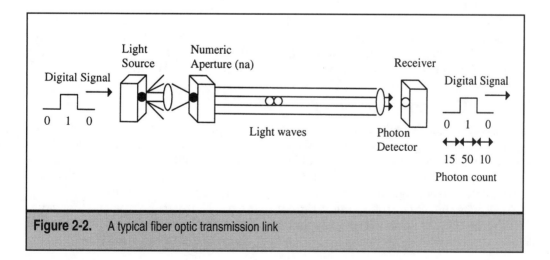

Figure 2-2.   A typical fiber optic transmission link

FIBER OPTIC CABLE TYPES

A few words should be said about the various types of physical structure that fiber optic cables can have. Two basic classifications of optical fiber are defined here:

▼ Multimode fiber (mm fiber), which allows light to take many paths (modes) as it travels through the fiber.

▲ Single mode fiber (sm fiber), which has a core so small that only one path (mode) is available for light to travel.

Multimode Step-Index Fibers

The simplest type of fiber is the multimode step-index fiber. This fiber type typically has a core diameter ranging from 125–400+ microns (.005 inch to over .016 inch). This relatively large core allows many modes, or ways for light propagation down the fiber. The larger the core diameter, the more modes it supports. Because light reflects at a different angle for each mode, some waves follow longer paths than others. A wave that travels straight down the core without any reflecting arrives at the other end sooner than other waves. Other waves arrive later in the receiving sample period, and the more times a wave is reflected, the later it arrives.

Therefore, light entering the fiber at the same time and representing the same bit or bits may arrive at the other end at slightly different times. Although it is not physically accurate, most people would say that the light has "spread out" (in time). This process of light-spreading is more accurately called *modal dispersion*. Modal dispersion limits both the speeds and distances that signals can be sent through a fiber.

Dispersion, along with some other complicating factors, makes multimode step-index fiber the least efficient of the fiber types. It is used for short runs, such as linking a computer mainframe to peripheral equipment and other LAN applications. Conversely, this type of fiber's relatively large size and simple construction do offer advantages.

This fiber gathers light well because of the large numerical aperture (the light-gathering cross-section of the fiber). Next, it is the easiest to install. Finally, it is quite inexpensive compared to other types. The disadvantages are the large dispersion of the light signals and the relatively low bandwidth, which limit both speed and distance.

Multimode Graded-Index Fiber

Obviously, modal dispersion can be a limiting factor in long-haul, fastfiber networks, such as SONET. One way to reduce modal dispersion is to use multimode graded-index fiber, sometimes called GRINs. With this type of fiber, the index of refraction is highest at the center of the core and gradually tapers off toward the edges. Because light travels faster in a lower index of refraction medium, the light furthest from the center axis travels faster.

Because the waves following different paths travel at various speeds, the waves will reach the same point at roughly the same time, which is desired. Note that modal dispersion is still present, but now it is much lower than in step-index fiber. Light waves are now no longer sharply reflected, they are gently bent.

Multimode graded-index fibers are generally small, typically 125 microns. This type of fiber's size still makes it fairly easy to install, and the use of the graded index makes it more efficient. However, the more complex structure of the core makes it more expensive than the multimode step-index fibers, but it is very popular.

Overall, multimode graded-index fiber offers a good intermediate choice between the other two types (multimode step-index and single mode step-index), and the advantages and disadvantages of it are in between these two types. Multimode graded-index fibers are good for medium distances and medium speeds. Usually, this type of fiber is used for more than merely a single floor of a building.

Single-Mode Step-Index Fibers

Of course, the best way to remove the effects of modal dispersion is to reduce the fiber's core diameter until the fiber only propagates one mode. This method is used in the single-mode step-index fiber. This type of fiber has a core diameter of only 2–8 microns (.00008 to .0002 of an inch). These fibers are by far the most efficient, but their small size can make them difficult to handle for many applications. Use of this fiber is usually limited to very demanding, high-speed, long-distance applications, such as SONET. The advantages of single-mode fiber include minimal dispersion, higher efficiency, larger bandwidths, and thus longer distances and higher operating speeds. Conversely, these fibers are more expensive, more difficult to splice and connectorize, and require a laser as the light source because of their small numerical aperture.

Single-mode step-index fibers are used for long distances for telecommunications networks. Generally, SONET/SDH spans can be run up to 30 miles (about 48 kilometers) without the need for repeaters, although the state-of-the-art is much higher today. They can carry signals at high speeds, up to 40 Gbits. Naturally, most SONET/SDH networks use single-mode fiber.

FIBER OPTIC ADVANTAGES

What makes fiber so attractive for so many uses? Actually, there are eight distinct advantages that fiber has over almost any other medium. Some are simple and obvious, while others are more subtle. In fact, fiber offers more pluses and fewer minuses than any other medium in use today. The price of fiber optic links, which must include the cost of not only the fiber itself but also the transmitter, receivers, and any repeaters necessary, is now comparable to other, copper-based media, such as coaxial cable or unshielded twisted pair (UTP). Certainly, the installation cost per bit/second is the least of any cable or wireless medium. Beyond the fact that the bandwidth of fiber is practically unlimited, it is also true that no other cable medium is even being researched today.

Moreover, in the fiber industry, speeds achievable on fiber quadruple every two years, and the distance achievable without digital repeaters also quadruples every two years. The eight fiber advantages are listed below:

▼   Lower BERs. Fiber offers lower bit error rates (BERs) than any other transmission media. In some cases, it can take three years of running a fiber link to generate the same number of errors as on a copper (coax, twisted pair) link (1,000 to 1 ratio).

■   Higher bandwidths. Speeds achievable on fiber are unsurpassed. Unlike copper media, fiber speeds are increasing constantly, in some cases by orders of magnitude.

■   Longer distances without repeaters. Digital repeaters, which detect and repeat the string of 0's and 1's onto another fiber link, require power and add potential points of failure. A given length of fiber will need fewer repeaters than the same length of cable composed of another medium.

■   Immunity from interference. Most interference in the form of noise comes from the fact that copper cables are long antennas. Fiber will not pick up electromagnetic noise from the environment.

■   Security. Fiber networks can be constantly monitored for signal loss increases, which may indicate the presence of taps.

■   Less maintenance costs. Solid-state fiber optic components need much less maintenance than do devices used in copper networks.

■   Small size and weight. The same size crew can install much more fiber in one day than any other cable media.

▲   Bandwidth upgrades. When designed with this in mind, it is possible to increase the bandwidth available on the fiber link simply by changing the sender and receiver components.

Because the success of SONET is intimately linked with the advantages of fiber optic systems, let's take a closer look at each of these advantages.

Bit Error Time Intervals

It is not just the lower BERs that make fiber so attractive for transmission links and standards. Copper networks from T-1 to X.25 have always been engineered for BERs of about 10^{-6}. This indicates one bit in error for every million bits sent (106 is 1 million) which is not bad, but these links are used for very high bit rate transmission. For instance, with a BER of 10^{-6} at the DS-3 rate of 45 Mbps, a bit error will occur every 22.0 milliseconds on average, or about 45 bits in error every second. Actually, copper errors are usually in *bursts*, but the average BER is still a valid concept.

The significance of the BER for high-speed networks is twofold. First, at higher speeds, such as 135 Mbps, only fiber can give acceptable average bit error intervals. Second, at lower bit rates, fiber can give astonishingly low bit error intervals, leading to the common observation that "with fiber, bit errors just go away." This is not really an exaggeration. For example, some service providers offer 64 Kbps links on fiber with a BER of 10^{-13} (1 per 10 *trillion* bits). At this rate, the average bit error interval at 64 Kbps will be one bit error every five *years*. A decrease in the BER from 10^{-6} to 10^{-9} is a thousand fold decrease. In other words, the number of bit errors a network has in one day at 10^{-6} on copper is the same as the number of errors it has in three years at 10^{-9} on fiber!

Fiber links today run at 10^{-10} or better; therefore, the number of errors that a copper network has in one day, takes 30 years to occur on fiber. This fact fuels the perception that errors are eliminated with fiber. Table 2-1 shows some average bit-time intervals (i.e., time between bits in error) for a variety of line speeds.

Obviously, as networks' speeds climb, lower BERs become not a luxury, but an absolute necessity.

Typical Media Bandwidths

Nothing can match fiber for high bandwidth and speeds. In fact, no digital signal has ever been generated that fiber cannot carry. The bandwidth of a transmission medium is measured in terms of the analog frequency range (bandwidth is more properly an analog term) measured in hertz (cycles per second, or Hz). The symbol for cycles per second is f, and the rule for determining the bandwidth of a medium is given by the formula $f = c/\lambda$, where c is the speed of light and λ is the characteristic wavelength of the medium.

For fiber, c is 186,000 miles per second or 300,000 kilometers per second, and λ is about 1 micron or 1,000 nanometers (billionths of a meter) for visible light. Using these numbers gives a characteristic bandwidth for fiber optic cable of about 300 terahertz, or 300,000 gigahertz. This is only a naive computation, in the sense that it concerns only analog bandwidth and does not take into account digital modulation techniques for sending digital signals on the fiber. However, it is representative even assuming only 1 bit for each hertz of bandwidth.

In contrast to fiber, Figure 2-3 shows the typical bandwidth of other common media. The scale is logarithmic, which means that each step on the vertical axis is actually 1,000 times the previous one.

Bit Error Rate	64 Kbps	256 Kbps	1.5 Mbps	45 Mbps	135 Mbps
10^{-6} BER	16.0 seconds	3.9 seconds	0.7 seconds	22.0 millisec.	7.4 millisec.
10^{-9} BER	4.3 hours	65.0 minutes	11.0 seconds	22.0 seconds	7.4 seconds
10^{-12} BER	6.0 months	1.5 months	7.7 days	6.2 hours	21 hours

Table 2-1.   Average bit error time intervals

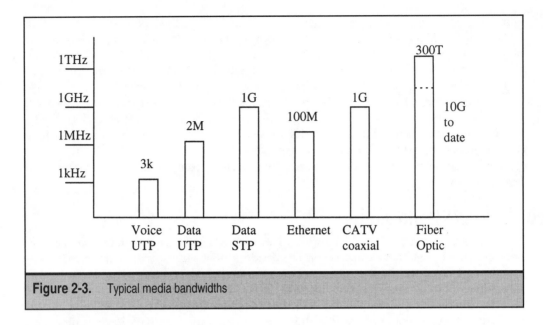

Figure 2-3. Typical media bandwidths

Long Distances

Fiber offers the ability to traverse long distances without repeaters. Repeaters "clean up" and re-send the signal on long fiber spans. The distance between repeaters on a digital link is purely a function of the signal loss over the distance. Greater signal loss means increased numbers of repeaters are needed. Figure 2-4 illustrates the typical signal losses

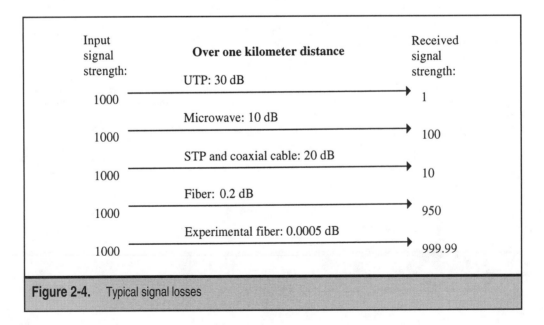

Figure 2-4. Typical signal losses

over one kilometer (about 5/8 of a mile) for various transmission media. In each case, the signal strength is rated at an arbitrary 1,000 "units."

The ratio of transmitted-to-received signal strength is known as attenuation. Engineers measure attenuation in decibels or dB, which are easier to work with mathematically than are simple ratios. The formula is dB = 10 log SNR, where SNR is the signal-to-noise ratio. By this scheme, a send/receive ratio of 1,000 to 1 causes a 30 dB signal loss.

Typical attenuations in terms of dB per kilometer are shown in Figure 2-4. However, it is more understandable to translate this into received signal strength. For example, coaxial cable has a typical attenuation of 20 dB per kilometer and a received signal strength of 10 out of 1,000. Fiber comes in at an eye-catching 950 out of 1,000.

Incredibly, experimental fiber receives 999.99 of the 1,000 transmitted signal strength over a kilometer. Such "lossless" transmission has only previously been an engineer's dream.

IMMUNITY FROM INTERFERENCE

Copper media, from the twisted pair copper wires used in telephone company local loops to the coaxial cables used in cable TV systems, are basically long antennas. Because these all operate in the region of the electromagnetic spectrum used for wireless and other forms of transmission, copper wires are liable to pick up noise. Transmission lines can be shielded from noise, but this adds expense to the cable and must be maintained and installed properly to achieve maximum benefits.

Noise is just an unwanted signal that must be distinguished from the sent signal at the receiver. Noise interferes with the intended signal and must be addressed. Impulse noise can come from electrical motors and lighting strikes nearby. Other environmental noise can come from the normal operation of household appliances and TV sets. Even PCs can interfere with TV reception in another room or adjacent apartment.

On copper transport networks, interference is a nagging problem. Telephone companys' local loops share utility poles with power companies. These telephone cables must be carefully installed to avoid annoying hums from the power lines. When used for digital data, these hums are more than annoying, they are fatal to the bits, causing numerous errors.

Coaxial cable has a *self-shielding* property and, therefore, has a reputation for being more noise-resistant than unshielded wires. Although this is true, many sources of electromagnetic noise make their way onto coaxial cable as well. For instance, when cable TV companies began to experiment with two-way communication over the cable TV coaxial network, they quickly discovered that many common household devices generated noise that was inadvertently picked up by coaxial cables. These sources of noise and, thus, interference included microwave ovens, TV sets, and PCs. This is one reason why every PC has an FCC sticker. The data bus on every PC radiates electromagnetic signals into the surrounding area and may interfere with other electronic devices. Digital cell phones can even interfere, in some models, with heart pacemakers and hearing aids.

Fiber is not only noise resistant, but it is also basically impervious to noise because fiber operates in the optical range of the electromagnetic spectrum. Although stray electrical signals may be transmitted by copper media, no "stray light" effect occurs with fiber.

Put more simply, although they overlap with those of copper media, "bandwidths" of storms, power supplies, and many wireless transmitters do not overlap with optical bandwidths.

Fiber optic cable has been used for years where noise is a factor in transmission systems. For example, many factories with heavy power equipment cannot even use coaxial cable for LANs on the factory floor. Many power companies that run communications and control lines along high-voltage power transmission lines cannot use it either. The 400,000 volts of alternating current will knock out the signal of many coaxial cables. Fiber, in many cases SONET fiber, is used for this purpose instead.

Security

One of the most persistent myths regarding fiber is the idea that fiber is more secure than copper because fiber is difficult to tap into. In fact, this is not true. Fiber used to have a reputation for being difficult to handle; this security myth was a consequence of that reputation. Today, all that is needed is a simple optical coupler. The fiber is then tapped and any signal that travels down the fiber is picked up by the eavesdropping device.

Now, the fact that fiber is guided and not a broadcast medium (e.g., satellite, wireless, or microwave) affords it some measure of security over these media. However, any fiber link can be tapped by bending the fiber, scraping off the outer layers of cladding, and strapping a simple optical coupler device to the cable at that point.

Why does fiber have a reputation for being secure? Because when this coupler is inserted, it is very easy to detect. The inserted optical coupler will add at least 0.5 dB of signal loss (attenuation) to the fiber link. In normal situations, the attenuation on any link, even a fiber one, may increase with time—especially when the link is not maintained properly. However, a jump of 0.5 dB very quickly is not normal under any circumstances.

Security is handled on fiber networks by closely monitoring the endto-end attenuation (signal loss) of a link. If this were to rapidly increase by a half dB (the dB is a standard measure of signal loss) or so, then some physical damage would have occurred. It could be innocent (e.g., environmental damage), or it could be an intruder. There are even ways to determine precisely how far along the fiber the loss was injected, focusing detection efforts at that area.

All of this indicates that fiber is attractive for secure and sensitive applications. Needless to say, the federal government and military prefer fiber to any other type of link in many cases. Even banks and many businesses have come to recognize the enhanced security that fiber optic networks can provide.

Less Maintenance Costs

Less maintenance costs means both fewer and lower in this case. All network links, whether fiber or not, need to be maintained. The cost factor is a function of the number of devices on the network that may fail, and the raw number of these devices throughout the entire network.

Fiber optic links and networks require fewer active devices that are less expensive to maintain. This translates to a cost-effective solution because there are lower costs involved in maintaining the fiber network than in any other media.

Even the fiber itself has lower maintenance costs. For instance, fiber does not corrode as copper media does. Furthermore, squirrels constantly chew aerial copper cables, especially in the spring, but have demonstrated less of an appetite for fiber. Fiber, however, does have its problems: It must be carefully protected from water. Even in the ground below the water table level, microscopic air bubbles in the fiber can fill with water and cause the fiber to fail. Ultraviolet radiation can destroy the glass in fiber, so fiber must be carefully deployed in aerial fashion. Fiber works best and lasts longest when buried underground, and this is how most fiber is deployed today.

The greatest risk to underground fiber today in North America is from gophers. There is actually a standard test placing fiber cable in a gopher enclosure to measure the risk (the gophers are not harmed). Squirrels will chew on anything, including aerial fiber. In other parts of the world, termites cause damage to fiber cables. However, reports of sharks chomping on undersea fiber are greatly exaggerated.

Low attenuation on fiber links means a smaller number of repeaters. Fiber links need much less time and energy devoted to maintenance because of this factor. The smaller number of active components needed on fiber links is a considerable cost savings in terms of installation costs. Solid-state electronics also draw less power and have a mean time to failure of about 20 years, just as a modern stereo system.

Once the fiber optic cable is in place, little can go wrong with it outside of simple cable cuts. Early fiber links were said to either work or not. Some never quite worked right, no matter how much post-installation tinkering occurred. However, if installed correctly, fiber links worked well. Some service providers even buried the fiber under poured concrete, which helped to prevent backhoes and other excavation equipment from damaging the fiber (known as "backhoe fade"). Other types of cable, however, needs constant access for maintenance purposes and cannot be buried under concrete.

Small Size and Weight

Fiber also offers numerous advantages in terms of size and weight. Even 10,000-foot fiber reels weigh only a couple hundred pounds. Smaller reels can be carried under a person's arm and weigh about 60 pounds. Fiber optic cable can be installed above hung ceilings without worry, and can be attached to a building slab without concern about an engineering report to determine if the building floor load factor may be exceeded. Fiber easily snakes its way through crowded conduits. In fact, one of the earliest incentives to run fiber optic telephone cable in metropolitan areas was to relieve the stress on crowded trunk cable conduits.

In contrast, 2,700 pair copper wire is 3.38 inches in diameter and weighs almost 7 pounds per foot. Installing such cable in hung ceilings or attached to floor slabs is usually out of the question due to the weight. The problems are just as bad outdoors. A 1,000-foot reel weighs close to 4 tons, and a four-person crew would struggle to install a fraction of a mile per day. Simply setting up the enormous trucks and pulleys consumes the better part of a day. Fiber crews work fast and lean, and the increased productivity is an added benefit not to be ignored today.

Fiber represents many times the bandwidth of the 2,700 pair cable. Is it any wonder that a service provider like a telephone company can install more bandwidth on fiber in a week than their entire network had on copper previously?

There is even an added benefit to fiber optic cable's small size and weight that is not obvious, but has come to be quite important: 2,700 pair cable (there is even 3,600 pair cable) is difficult to handle because of its weight and much of it is buried below grade. When installed across bridges, the cable is particularly vulnerable, in spite of warning signs. A jackhammer can rip through the cable in a few seconds, seeming to the construction workers to be at first just a particularly tough patch of road.

This is just what happened in the summer of 1992 in White Plains, New York, when a bridge was being resurfaced. Immediately, a tent was erected on the site, traffic diverted, floodlights installed, and a crew worked around the clock for three days until all service was restored. It was not just a matter of "I found a purple and slate. Who has a purple and slate over there?" Documentation was delivered to the site and consulted, because service must be restored according to a particular customer hierarchy. First hospitals get service back, then police, then doctors, then pharmacies, and so on (details vary state by state), down to the ordinary residential customers. This task proved as time consuming as the actual splicing of the copper pairs themselves.

Now consider the same outage on fiber optic cable. A fiber optic cable may contain 12 to 36 strands of fiber. Each can be spliced in about 10 to 20 minutes, depending on the type of fiber and immediate conditions.

Documentation is usually not needed, because thousands of users may be on the same fiber optic cable pair. Service is typically restored to all affected customers in a few hours, not a few days (although severe outages may take longer).

Bandwidth Upgrades

Another advantage of fiber is that when the link is configured with upgrading in mind, the same fiber link can be run at a higher speed just by exchanging the transmitter and receiver. This is a huge cost advantage for fiber networks.

Upgrading a digital link with copper networks has been difficult. For instance, converting from a 1.5 Mbps link (a DS-1 or T-1) to a 45 Mbps link (a DS-3 or T-3) requires the installation of new media in most cases. T-1 is defined on unshielded twisted pair (UTP) copper wire and T-3 is defined on coaxial cable, fiber, or microwave. These are pre-SONET definitions. Commercial buildings and even other types of buildings have plenty of UTP from the local telephone companies for supporting T-1s, but few have the media needed for T-3s.

Many organizations are asking for 45 Mbps links to support new interactive multimedia or other high bandwidth, broadband applications. The price to the customer for the T-3 must offset the cost of running the new media to the service provider in a reasonable payback period. This fact has kept the cost of T-3 links high, and forms yet another incentive for installing fiber optic cable initially.

Once installed, the fiber is not a limiting factor. In most designs, fiber installed to support 45 Mbps service can easily be upgraded to support 155 Mbps service, or even 622 Mbps service. The fiber link grows with customer expansion and is not obsolete in a short time frame (often before the payback period!). In these cases, it is only necessary to change the transmitters and receivers to make the fiber optic link function at the desired higher speed. Because most fiber is buried below grade—the perils faced by submersed and aerial fiber have already been mentioned—cost of burial is the most significant cost factor in fiber optic installations. In most parts of the country, it costs tens of thousands of dollars per mile simply to dig a hole in the ground. Obviously, the cost component is to be avoided when possible.

Again, it is important to note that this upgrade capability is not inherent to fiber optic networks. It must be carefully planned. But it can be done, and is a vast improvement over older digital media.

This chapter explored some of the operational characteristics and benefits of fiber optic cable. These are fiber benefits and not direct SONET/SDH benefits. But because SONET/SDH is an international standard for fiber optic transmission, SONET/SDH shares all of the advantages of fiber optic networks, with the added benefit of being an open standard tied to no single vendor or company.

This brief initiation into the world of fiber optic transmission has introduced the key concepts of fiber optic cable, transmitters, and receivers. But each of these three basic components of fiber optic networks such as SONET/SDH has undergone a revolution in the past few years. These developments have lead to the feasibility of advanced optical networks, independent of SONET/SDH, based on WDM and, more importantly, DWDM. This new world of optical networking deserves a chapter of its own.

CHAPTER 3

Optical Networking, WDM, and DWDM

The previous chapter introduced some of the basic concepts of fiber optic transmission. This chapter uses the basic terms and ideas examined in the previous chapter and builds on that foundation. This chapter explores some of the most interesting and vital ideas in networking today. There are three major areas: state-of-the-art optical networking, combining multiple wavelengths of light in the same fiber to produce wavelength division multiplexing (WDM), and, finally, literally jamming these wavelengths as close together as technically possible to produce a denser version of WDM known simply as DWDM. All three concepts are related and have implications for legacy SONET/SDH systems.

The very term *SONET* contains the key concept of an *optical network* as opposed to a network where 0's and 1's are represented by electricity. When applied to SONET, oddly enough, the term *optical network* is somewhat misleading. The fact of the matter is that SONET is *not* an optical network at all. The same is true of SDH, of course, although at least SDH makes no claims in its acronym to be any form of optical networking. SONET and SDH are still very much electrical transmission networks that just happen to consist of fiber optic links.

This just-optical-links aspect of SONET/SDH might not seem like much of a big deal, but it makes all the difference. In SONET/SDH the bits only shuttle back and forth between *network elements* (NEs) in the form of lightwaves on the optical links. In order to do anything at all interesting with these optical bits in a network element—switch them, route them, multiplex them, even simply check them for errors—it is first necessary to change the optical bits back into electrical bits. Naturally, this need for constant and frequent electro-optical (E/O) conversion is itself expensive and error-prone, and adds considerable complexity to the network. Even simple regenerators along a SONET/SDH link (traditionally about 40 km apart) must first convert the optical bits to electrical bits in order to send the bits over the next section of the line. But as long as the network elements are incapable of acting directly on the lightwaves, SONET/SDH has little choice but to deal with the electrical bits to do more than shuttle bits between two points over a fiber optic link.

The whole point of this chapter is that networks no longer have to function in this way. The term *optical networking* can be defined as *a network in which at least some of the network elements are capable of dealing with bits in an optical form*. This might seem like a broad definition, and it is. The network need not be all optical, and not all network elements need to be optically enabled. The key is that bits might be switched, routed, multiplexed, amplified, checked for errors, or all of the preceding, while still in their optical form without the need for E/O conversion and the reverse.

Once this definition of an optical network is established, more elaborate plans and variations of the basic concept are easily understood. For example, an *all-optical network* is an optical network in which all multiplexing, switching, routing, and so on, is done directly on the 0's and 1's while they are still in their optical form. A *passive optical network* is an optical network in which all of the components from sender to receiver do not require electrical power to operate. And so on.

Multiplexing has proven to be the simplest network function to make into an optical operation. This leads directly to the creation of WDM and DWDM. Optical cross-connects are right behind, and optical switches might follow soon. These more advanced, but related, network elements will be discussed in Chapter 19, where the impact of optical cross-connects and switches on SONET/SDH is assessed.

OPTICAL NETWORK BREAKTHROUGHS

Optical networks, and to a lesser extent WDM and DWDM, depend on the recent development of four key optical components:

▼ Special fibers with nonzero dispersion characteristics

■ The in-fiber amplifier, such as the Erbium-Doped Fiber Amplifier (EDFA)

■ The tunable laser diode operating in the "third window" at around 1550 nm

▲ The in-fiber Bragg grating to isolate individual wavelengths at the receiver

Figure 3-1 shows each one of these breakthroughs in use in a WDM system. A full discussion of WDM and DWDM appears in a later chapter. The emphasis here is on each of the optical networking breakthroughs that pose somewhat of a challenge to SONET/SDH and lead directly to WDM and DWDM systems.

Each one of these breakthroughs is important enough to deserve considerable discussion. In this chapter, there will be no specific mention of SONET/SDH itself. However, optical networking concepts apply equally to SONET/SDH and all other networking architectures that include fiber optic links, such as Gigabit Ethernet.

SPECIAL FIBERS

The previous chapter outlined the main differences between multimode and single-mode fibers, the major forms of fiber optic cable. All of the advantages of fiber optic cable in general were detailed in that chapter, but mention should also be made of the fact that the connectivity in an optical network is optical, not electrical. This is not a trivial difference. Electrical communications systems always have the possibility of ground loops. Grounding is an important part of electrical networks and basically prevents static electricity from building up on the cable and preventing signals from being sent. But if there is an appreciable voltage potential difference between two grounding points, the resulting current flowing over the cable can easily swamp the actual signal sent. Furthermore, electrical connections can be hazardous since high voltages can harm people touching the wire, as anyone touching a "wet" T-1 can assert. Finally, electrical communications cables can be struck by lightning, even when buried underground. Lightning strikes constitute a hazard in many tropical areas around the world. Optical networks suffer from none of these drawbacks, except when electrical power is distributed to powered fiber components along the link. But this is just an argument for as many passive optical network elements as possible.

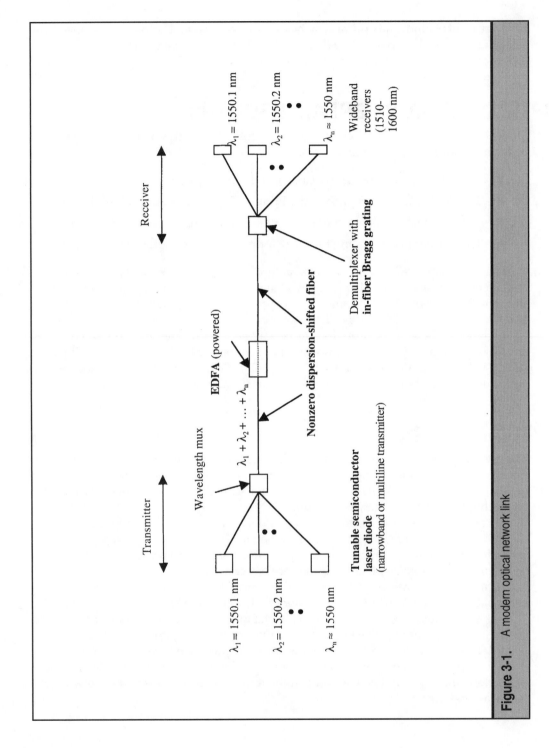

Figure 3-1. A modern optical network link

Other potential drawbacks of fiber optic networks include joining of the fiber cables, the risk of bending or kinking the cable, and the effects of gamma radiation and very high voltage electrical fields. The best way to join fiber is by fusion splicing: melting the glass so the ends fuse to one another. This is a problem outdoors during harsh weather. Connectors can be used, but add considerable loss to the link. The attenuation can be about 0.3 dB for multimode fiber and higher for single-mode, due to the smaller core diameter and the increased difficulty in aligning the two ends of the splice.

If fiber is bent too much, the light is no longer reflected into the core properly, and it escapes. The allowable bend radius depends on the actual difference between the refractive index (RI) between the cladding (lower RI) and core (higher RI). The RI is just the ratio of the speed of light in a vacuum to the speed of light in the material, which is glass in the case of fiber optic cable. Confusingly, in formulas relating to fiber optics, RI is represented by the letter n. If $n = 1.5$, for example, the speed of light in the fiber is about 200,000 km/sec, not 300,000 km/sec as in a vacuum (300,000/200,000 = 1.5).

The bigger the difference in core and cladding RI, the tighter the allowable bend radius. Unfortunately, the special fibers discussed in this section all tend to minimize the RI difference between core and cladding, leading to a larger allowable bend radius. Kinks in fiber are tight, permanent bends caused by physical damage during installation, such as dropping a heavy weight on a fiber cable. A more correct term for a kink is a microbend.

Finally, gamma radiation causes some types of glass to emit light (which interferes with the signal) and permanently darkens the glass as well. But this level of gamma radiation is normally found only inside a nuclear power plant. And when used in environments where voltages exceed 30,000 V (not all that uncommon today), fiber glass also emits light and discolors. Fiber needs to be shielded in high-voltage environments.

LIGHT IN GLASS

The traditional picture of light traveling down a fiber optic cable as a little arrow is somewhat misleading. This arrow representation is known as the *ray theory of light* and was common right up until the last part of the nineteenth century. The current theory of light acknowledges that, in accordance with quantum physics, light is both a wave and a particle. Light normally travels as an electromagnetic wave and is detected as a particle. Both forms can be called *photons,* although it is more common to call a particle of light a photon and refer to light in its wave form as just a *light wave.*

This is important because light does not really travel down a fiber as a ray in the form of little arrows, bouncing off the walls between core and cladding from transmitter to receiver. The truth is much more complex. Light consists of fluctuating electric and magnetic waves, orthogonal (90 degrees apart) to each other (so they do not interfere). As light propagates, the whole structure normally rotates. When the light travels down a fiber, a whole array of effects takes over. Light may be elliptically polarized, as when the period of rotation of the fields is not the same as the wavelength of the light. The main point is that this complex light structure actually snakes its way down a fiber in a spiral form known as a *helical pattern.*

Interference is another important issue when trying to understand how modern fibers function. Light interferes even with itself in ways that can be described perfectly with mathematics, but are hard to explain or understand in an everyday sense. Even a single photon of light can interfere in odd ways with itself when a system is set up just right. It is one thing to think of light as a wave phenomenon and to understand interference as similar to water waves on a pond interfering with each other. It is quite another to think that a single photon can somehow interfere with itself, when obviously there is only one of them to begin with.

Interference is important in all aspects of optical networking, even in the simple act of injecting light from air onto the fiber itself. Typically, about 4% of the incident light is reflected, and the higher the angle, the greater the reflection. Differences in RI always cause reflection—4% into the fiber at the sending end and 4% into the air at the other end of the fiber link. But when monochromatic light (one wavelength) shines on a thin sheet of glass, the total reflection is not 8%, as might be expected. There could be anywhere from 16% of the light reflected to no light reflected at all. The result depends on the thickness of the glass in terms of the wavelength of the light. The maximum of 16% reflection occurs when the round-trip thickness of the glass is exactly a multiple of the wavelength of the light. This is constructive interference. The minimum of zero reflection occurs when the round-trip thickness of the glass is an odd number of half wavelengths. This is an example of destructive interference.

It is important to point out that interference does not destroy the energy present on the fiber. Energy is always conserved, so the lost energy just shows up in some other part of the optical system, probably as heat or noise. With this in mind, modern fibers include antireflection coatings of special thickness and RIs to keep reflections to a minimum. This is not to say that interference is always a negative. In optical networks, filters, gratings, and other related components all depend on light interference to function.

Wavelengths are always important in optical networking. But today many wavelengths and ranges of wavelengths are used for different purposes. This is a good place to take a quick look at the wavelengths that will be discussed in the rest of this chapter. Both of these are shown, and explained, in Figure 3-2.

The best perspective on the wavelengths used for optical networks is linked to history. The range from 800 to 900 nm can be called the *1970s window* or *first window* (this is the term used by Harry Dutton of IBM). All early optical links in the 1970s used this range. The band around 1310 nm is the *1980s window* (*second window*), and this is where SONET/SDH, which matured in the early 1980s, operates. Today, the range from 1510 to 1600 nm is the *1990s window* (*third window*) and is used because it is at the lowest point of attenuation on the fiber. This window is used in all modern WDM and DWDM systems.

However, the opening of the third window was not easy. Light sources are expensive in this range, and there is more to the fiber story than simple attenuation. More was needed to enable fibers to function well in the 1990s window. This work concerned the degradation effects due mainly to *dispersion*.

There are many reasons why a pulse of light sent along a fiber will *degrade*, which is just another way of saying *change*. The light will be weakened, a pulse lengthened in time, and so on. There are five major reasons for the degrading of a pulse of light sent along a fiber optic cable.

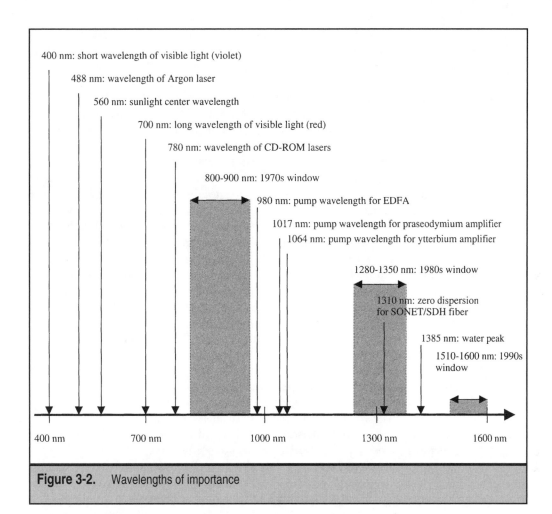

400 nm: short wavelength of visible light (violet)

488 nm: wavelength of Argon laser

560 nm: sunlight center wavelength

700 nm: long wavelength of visible light (red)

780 nm: wavelength of CD-ROM lasers

800-900 nm: 1970s window

980 nm: pump wavelength for EDFA

1017 nm: pump wavelength for praseodymium amplifier

1064 nm: pump wavelength for ytterbium amplifier

1280-1350 nm: 1980s window

1310 nm: zero dispersion for SONET/SDH fiber

1385 nm: water peak

1510-1600 nm: 1990s window

| 400 nm | 700 nm | 1000 nm | 1300 nm | 1600 nm |

Figure 3-2.   Wavelengths of importance

ATTENUATION This is always the major concern. A light pulse will always weaken because the fiber absorbs light. The glass itself does not absorb the light, but any and all impurities in the glass will. And if the fiber is not manufactured with precise uniformity in diameter, and so on, these flaws will cause scattering of the light. Scattering effects are the biggest cause of light loss in fibers today.

DISPERSION This is exactly what it sounds like: a pulse of light becomes spread out (disperses) in time as it travels down a fiber. Over long enough distances, all of the pulses eventually merge together and the peaks needed at the receiver to distinguish 1s all disappear. There are many different types of dispersion, some of which can even be used to advantage in optical networking. Three types of dispersion are of primary concern. *Material dispersion* (also called *chromatic dispersion*) results from the fact that fibers present slightly different RIs at different wavelengths and so each wavelength travels at a slightly different speed. Even the best lasers produce more than one wavelength of light (although in a very narrow range). *Waveguide dispersion* is caused by the shape of the core and RI profile across the core. Waveguide dispersion can actually cause the light pulse to bunch up in time and disappear. This also means that waveguide dispersion, if carefully planned, can counteract the effects of material dispersion. (It should be noted that in some references chromatic dispersion is defined as the sum of material dispersion and waveguide dispersion.) This will be explained more fully later on in this section. *Modal dispersion* occurs in multimode fibers because the distances traveled, and hence the arrival times, are slightly different for each mode.

POWER LIMITS   One way of overcoming attenuation effects is just to pump up the input power. But the amount of optical power coupled onto a fiber is limited by nonlinear effects. This just means that doubling the input does not double the output and, in fact, might decrease the output to lower levels than before. Optical networks must function in the linear power region to be useful. Most single-mode fiber systems are limited to about a half a watt of input power to remain in the linear region. The nonlinear effects in a fiber are caused mainly by the intense electromagnetic fields in the core when light is present.

POLARIZATION EFFECTS   The core of fiber optic cable should be perfectly round and symmetrical. But there are always imperfections, either in manufacture or during installation, which distort the core in places. As a result, light in the fiber is changed in polarization. Currently, few if any optical networks rely on polarization of the light to carry information. But in systems running at 40 Gbps and above, polarization effects will be a concern.

NOISE   It might seem odd to speak of noise in an optical network, since immunity from electrical interference is one of the big advantages of fiber. It is true that optical networks will not pick up noise from *outside* the network. But noise is just an unwanted signal detected by the receiver. In optical networks, *modal noise* is a complex effect present on multimode fibers, and *mode partition noise* is a problem with single-mode fibers as power hops around (is partitioned) among many close wavelengths.

None of these optical impairments are fatal, fortunately, and most can be overcome by a variety of methods. Generally, it is desirable to make the fiber core as narrow as possible and allow only one mode of propagation at a particular wavelength. This is the idea behind singlemode fiber, of course. Next, the wavelength(s) chosen for operation must be selected based on the characteristics of the fiber itself. Cost is a factor as well, since historically the shorter the wavelength used in terms of nanometers (nm), the lower the cost. Also, dispersion effects that depend on wavelength can be minimized by narrowing the *spectral width* of the light source. The spectral width is a measure of the range of wavelengths generated by the light source. Even lasers do not produce all of their power at one particular wavelength, although lasers are much better than LEDs. Modern semiconductor lasers can have spectral widths of around 1 to 5 nm, while some more sophisticated (and expensive) lasers can have spectral widths as low as 0.01 nm. LEDs on the other hand will have spectral widths between 30 and 150 nm. Finally, material and waveguide dispersions act to spread out and narrow waves, respectively, and so can be used together to counteract each other, to the benefit of the optical system.

All in all, modern optical network design depends on the interplay of two factors, with a third factor added in some cases. The two most important considerations are signal strength and dispersion effects. The third factor is noise, but noise is rarely a problem in simpler optical systems.

Fiber as Glass

Optical fibers are not made out of the same kind of glass as is used in windows. In scientific terms, a *glass* is any substance (mostly containing large portions of silicon) that does not transition to a solid state when cooled from a liquid. Glass acts as a solid but is still technically a liquid. However, even the oldest window panes are still just as thick at the top as they are at the bottom (reports of "flowing" medieval cathedral windows have proven to be greatly exaggerated).

Pure fused silica, or silicon dioxide, is the primary ingredient in optical fiber cables. But using silica requires high temperatures and the RI of silica is not particularly well suited for fiber optic systems. So *doping* is used to change the RI of the silica core or cladding, or both. Doping is just the process of intentionally adding impurities to a substance during fabrication. Even chandeliers can have lead oxide added in to produce the sparkle of "lead crystal" glass. In optical fiber, 4% to 10% of germanium oxide is added to the silica to increase the RI. This is a huge amount of dopant, considering that semiconductor fabrication processes dope materials with only 1 part in 10 million or so. Boron trioxide will decrease the RI. Other substances are also used, but they all tend to also increase the attenuation of the fiber.

When all is said and done, the typical attenuation characteristics of modern fibers are fairly consistent. It should be noted that the infrared range starts at about 730 nm, so none of these wavelengths are visible. Also, the typical fiber attenuation in 1970 was about 20 dB/km, which is quite high for any medium. By the 1980s, this was reduced to about 1 dB/km, a figure widely used by vendors and writers. In the mid-1990s, attenuation on fiber of around 0.2 dB/km was common.

Fiber attenuation varies dramatically with wavelength. Figure 3-3 shows the variation in some detail. There is a "water peak" at around 1400 nm caused by water ions that are always present in the fiber.

So the battle against fiber attenuation can be fought and won by opening the 1990s window wavelengths for operation. However, dispersion effects have to be taken into consideration as well. One of the reasons that the 1970s window was not the best window was that the fiber *zero dispersion point*, where material and waveguide dispersions exactly canceled each other out, was around 1310 nm. So, pulses at 1310 nm stayed nice and tight as they made their way along many kilometers of single-mode fiber, although they still weakened from attenuation, of course.

The standard single-mode fibers used in SONET/SDH had their zero dispersion point at 1310 nm. By using changed fiber fabrication methods (doping) and manipulating the RI profiles across the fiber core, a whole generation of *dispersion-shifted fibers* appeared in the 1990s that made operation in the even lower attenuation 1990s window attractive.

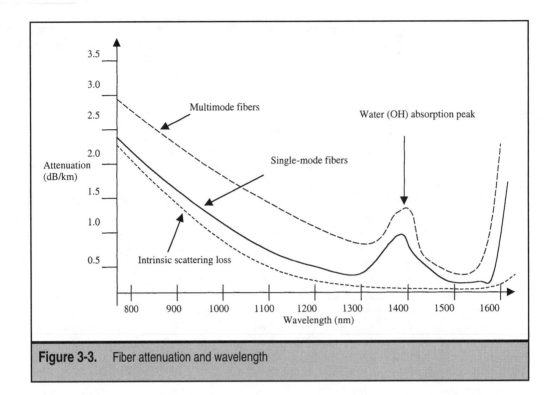

Figure 3-3.   Fiber attenuation and wavelength

So in the 1990s newer fibers moved the zero dispersion point into the 1990s window, centered around 1550 nm. This interplay between material dispersion and waveguide dispersion is shown in Figure 3-4.

The story of special fibers for optical networks might stop here, if not for a few other important points. First, SONET/SDH still is defined to operate at 1310 nm, yet obviously the current way of building optical networks is firmly in the 1510- to 1600-nm range. Second, this is not to say that wavelengths in the 1310-nm range cannot be used with dispersion-shifted fibers (sometimes called DSF). However, the dispersion (and attenuation) will be higher at 1310 nm (the characteristic SONET/SDH wavelength) than at, say, 1550 nm. Third, optical networks today rely more on WDM and DWDM than a single serial stream of bits at a single wavelength. So a single zero dispersion point right in the middle of the transmission band is not always desirable. For WDM and DWDM, it would be better to have a range between 1510 and 1600 nm where dispersion is low, but still nonzero. Again, clever doping and RI profiles made the manufacture of such *nonzero dispersion-shifted fiber* (NZ-DSF) possible.

The RI profile of NZ-DSF is shown in Figure 3-5. These fibers are recommended for WDM and DWDM systems, although regular 1310-nm zero dispersion-shifted fibers can be used as well (as long as the higher dispersion is accounted for during system design).

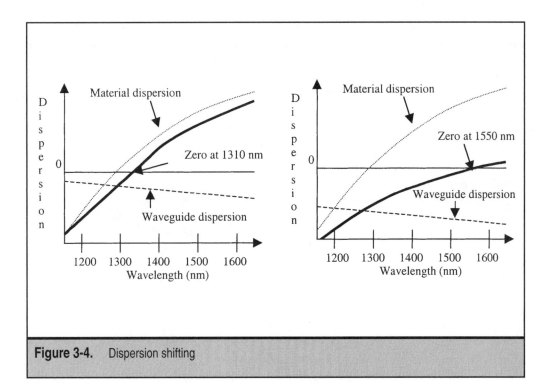

Figure 3-4.   Dispersion shifting

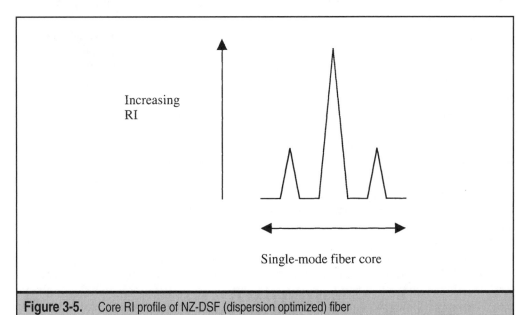

Figure 3-5.   Core RI profile of NZ-DSF (dispersion optimized) fiber

There are many other interesting developments in the field of fiber optic cables. These can only be mentioned here. There are large effective-area fibers (LEAF) that will accept more optical power, dispersion flattened fibers with low dispersion in the 1990s window (but high attenuation), and dispersion compensating fibers that can be spliced in periodically to combat dispersion effects on long NZ-DSF runs.

The Erbium-Doped Fiber Amplifier (EDFA)

The second key piece of the optical networking puzzle, in addition to the creation of fibers with special characteristics in the 1500- to 1600-nm range suitable for WDM and DWDM, was the invention of the in-fiber optical amplifier. There are also semiconductor optical amplifiers (SOAs), but these are not considered here. The most well known and most widely deployed in-fiber optical amplifier is the optical amplifier based on the rare earth element erbium. When erbium is used, the amplifier is called an EDFA, or Erbium-Doped Fiber Amplifier. As might be imagined, an EDFA is created by doping an otherwise ordinary single-mode fiber with erbium during fabrication.

The use of EDFAs has become so widespread that the acronym EDFA is in danger of becoming a synonym for fiber amplifier in general. This is unfortunate, as other elements can be just as effective as erbium when it comes to amplifying optical signals. For example, praseodymium and ytterbium, which are also rare earth elements, can be used. Sometimes, this whole class of optical devices is called *rare earth doped optical amplifiers* (REDFAs), but this terminology is far from standard.

SONET/SDH does not—and right now cannot—use EDFAs or other forms of optical amplifiers. This is because SONET/SDH, as a firmly entrenched 1980s technology, uses the 1980s optical communications window, usually at 1310 nm. EDFAs, on the other hand, were developed in the 1990s and so function in the 1990s optical communications window from 1500 to 1600 nm. The essential characteristic of any optical amplifier is that these devices operate across a wide range of wavelengths, independently and effectively. The electro-optical repeaters used in SONET/SDH typically function only at the SONET/SDH wavelength of 1310 nm. If SONET/SDH links are to be mapped onto WDM or DWDM systems, it is necessary to use a transponder to shift the native SONET/SDH wavelength into the 1500- to 1600-nm range.

It is somewhat ironic that amplifiers were used in analog transmission systems but abandoned in favor of digital regeneration and repeater techniques, culminating in SONET/SDH electro-optical repeaters, due to the cumulative noise effects when cascading amplifiers. In other words, in an analog system, noise is added by each amplifier, since analog noise is indistinguishable from the analog signal. In a digital system, noise can be removed at each regenerator as the signal is strengthened and cleaned up. Why then do optical networks emphasizing WDM and DWDM rely on in-fiber optical amplifiers instead of digital repeaters? Does not noise also accumulate in optical amplifiers? Of course it does! What saves optical amplifiers and allows them to be used in WDM and DWDM systems is the fact that the degree of noise in optical systems is very low to begin with. However, given enough EDFAs, low level of optical noise or not, eventually a long enough cascade will result in more noise than signal (this is the case with 80 or more EDFAs on a link several thousand kilometers long).

The attractions of EDFAs and all optical amplifiers in general are many. Electro-optical repeaters are complex and can fail. EDFAs are built into the fiber itself. Repeaters are tuned to a specific wavelength and bit rate. Amplifiers just amplify anything on the link. With WDM, each wavelength would have to be demultiplexed, converted to electricity, regenerated, converted back to optical signals, and then remultiplexed. In-fiber amplifiers make WDM technically feasible. Finally, repeaters cost much more than optical amplifiers. The differences between electro-optical repeaters and in-fiber optical amplifiers are shown in Figure 3-6.

The operation of an EDFA is simple to describe, in principle. A relatively high-powered (compared to the weakened input signal) laser source at 980 nm is mixed with the arriving input signal on a special section of erbium-doped fiber, typically about 10 m of fiber coiled to reduce the form factor of the device. This is called the pump signal (there is also a second pump window at 1490 nm). The erbium ions present in the EDFA fiber section are excited to a higher energy state by the pump laser. When photons from the input signal in the range from about 1530 to 1620 nm (the longer wavelengths are not as effectively amplified) strike the excited erbium ions, some of the erbium ions' energy is transferred to the signal and the erbium ions return to a lower energy state (until they are pumped again). The nice thing is that the photons have exactly the same phase and direction as the incoming photons. There are just more of them, so the input signal is amplified. That is all there is to it.

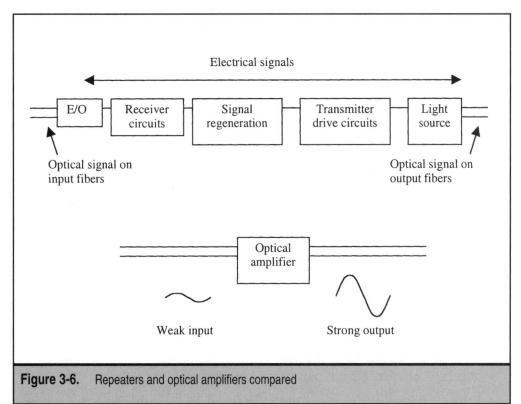

Figure 3-6. Repeaters and optical amplifiers compared

This does not mean that there is no more to be said about EDFAs. For example, reflections at the end of the EDFA section can cause the EDFA fiber itself to become a laser, so an optical isolator is usually employed to prevent these reflections. Also, noise in an EDFA manifests itself as amplified spontaneous emissions (ASEs), which can emerge from both ends of the fiber link and must be dealt with. ASEs in the same direction as the signal become a major source of cumulative noise, while backward-propagating ASEs can harm source lasers if not filtered out.

Oddly, the pump signal can actually be injected into the EDFA section in the opposite direction from the input signal! In fact, there are firm technical reasons for doing so. The erbium ions do not go anywhere, of course. EFDAs can also be pumped from *both* directions. EDFAs can even be pumped remotely, from the transmitter or receiver side. In some configurations, an EDFA can be pumped *through* another EDFA section, as long as the attenuation at 980 nm is not too bad. Usually the pump laser is right at the EDFA section it pumps, however—although the electrical power needed to operate the pump laser can be delivered locally or remotely (sometimes in the metallic fiber protection sheath). Obviously, EDFA design is a very active area of research. EDFAs have been used not only as line amplifiers, but also as preamplifiers directly in front of a receiver and as power amplifiers placed directly off the transmitter. A more detailed schematic picture of EDFA operation is shown in Figure 3-7.

In summary, optical amplifiers have numerous advantages over repeaters. Optical amplifiers are simpler and smaller, and they cost much less than repeaters and yet they have much longer mean times to failure. They operate over a range of speeds and wave-

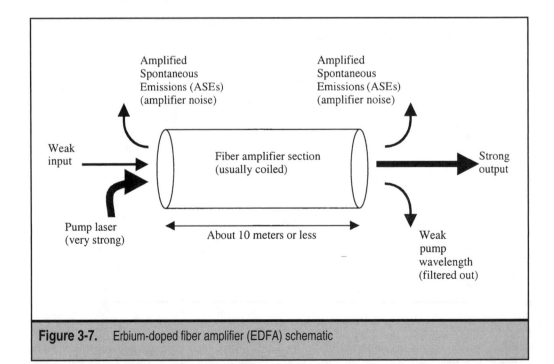

Figure 3-7. Erbium-doped fiber amplifier (EDFA) schematic

lengths at the same time and produce a gain in the weakened optical signal of anywhere from 25 to 50 dB. They amplify all forms of optical signals, whether digitally encoded or analog, and in fact the coding on the fiber can be changed without changing the in-fiber amplifier. They add no appreciable delay to the fiber run and, if designed properly, can allow the operation of diagnostic tools such as optical time domain reflectometers (OTDRs) right through the amplifier.

Naturally, there are drawbacks to optical amplifiers as well, especially since they are relatively new optical networking devices. The biggest problem is that they add noise to the system that can accumulate, as in any cascade of amplifiers, until the noise swamps the signal. They do not yet operate over the full WDM and DWDM wavelength window, and the gain produced is not the same at all wavelengths of interest (that is, the gain spectrum is not flat). They do not perform any clean up of the optical signal as regenerators do, such as reshaping pulses. This means that if an incoming optical signal is dispersed (distorted), the output of the optical amplifier will be stronger but still distorted. Finally, in many cases the optical amplifiers must be placed more closely than electro-optical repeaters.

But when all is said and done, the in-fiber optical amplifier will be a key piece of any optical network.

The Tunable Laser Diode Operating at 1550 nm

The third key piece of the optical networking puzzle, in addition to the creation of special fibers and in-fiber amplifiers for WDM and DWDM, was the creation of the tunable laser diode operating at around 1550 nm, right in the middle of the 1990s optical communications window. There are actually three pieces to this invention. The first part is the *laser diode* itself. The second part is the *tunable* nature of the laser, which enables the same physical device to be used to generate different wavelengths of light. Finally, the third part is the operation around 1550 nm. All are equally important for optical networking today, and each aspect will be discussed in some detail.

In the early days of fiber optic systems, lasers were bulky affairs that were very expensive and hard to work with in spite of their numerous advantages over LED-driven fiber systems. Not only was the cost often prohibitive, but lasers often had to be locked in a closet with a warning sign on the door and anyone who entered had to put on protective eyewear. Today, both LEDs and lasers are semiconductor devices mounted on a simple computer card assembly. LEDs remain simpler than lasers when implemented in a chipset, but they also have a lot of features in common. So it makes sense to start with a quick look at semiconductor LEDs.

An LED is just a *forward-biased p-n junction*, but this is not the place for a full discussion of semiconductor electronics. All that is needed is an appreciation of the fact that when the proper electric potential (voltage) is applied to a wafer of p-semiconductor material and n-semiconductor material, light will be emitted at the junction between the materials. That's the easy part. The hard part is getting the LED light at the desired wavelength out of the junction and onto the fiber. In practical LEDs, sometimes the light emerges from the surface of the device (SLEDs) or from the edges of the device (ELEDs). In most SLEDs, the n-semiconductor has a hole in the middle to allow the light to emerge from the junction. A very simple diagram of a SLED (sometimes called a "Burrus LED") is shown in Figure 3-8.

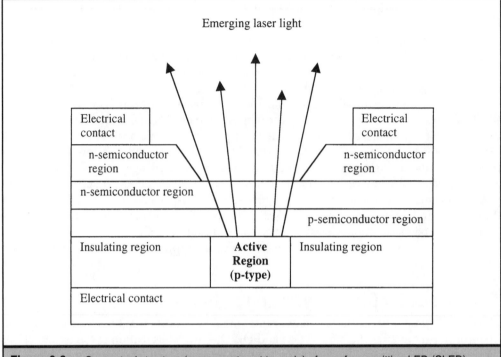

Figure 3-8. Conceptual structure (components not to scale) of a surface-emitting LED (SLED)

The characteristics of LEDs as light sources (transmitters) in optical networks compared to lasers are important. LEDs have been lower in cost than lasers, although this price differential has all but disappeared in recent years. Much of the price differential was based on production volume. Certainly the use of lasers in simple CD-ROM data and audio devices has leveled the cost playing field drastically. LEDs have had lower power outputs than lasers (100 microwatts or so), and produce light not at a single wavelength, but in a large band or spectral width of about 5% of the wavelength (50 to 100 nm). The spectral width can be reduced through the use of filters, but of course this also filters out part of the signal power. LED light is not a highly collimated (parallel) beam like a laser, but LEDs produce a large, divergent cone of light. So, the LED light must be focused with a lens onto the fiber (this is why LEDs are not used with single-mode fibers: it is too hard to get much light into the small core). LEDs cannot be turned on and off fast enough to drive gigabit-per-second systems as lasers can. Most LEDs top out at about 300 Mbps. On the other hand, LEDs can easily be used with analog modulation techniques. This can be done with lasers, but only with much more difficulty.

In contrast to LEDs, lasers are much better to use in optical networks as light sources. Laser light in theory is coherent and consists of a single wavelength. This is not strictly true in practice, as all lasers used for communications exhibit a characteristic *linewidth*.

But lasers can be turned on and off very quickly, in some cases on the order of femto-seconds (one millionth of a nanosecond, or 10^{-15} seconds), and they produce higher powers than LEDs (about 20 milliwatts for communications systems, and up to 250 milliwatts for optical amplifier pump lasers). Laser light, which emerges highly parallel from the device, can inject 50% to 80% of its power onto a fiber. These characteristics are related to, but not strictly dependent upon, a laser's coherent light emission.

But however close lasers and LEDs come in price, the fact remains that temperature control and power control are needed in lasers. These controls require electronic circuits to function. And lasers are generally nonlinear with respect to input power.

In spite of some claims to the contrary, most semiconductor lasers used for communications do *not* produce a single wavelength of light. Just like LEDs, a range of wavelengths is produced, and the spectral width is the measurement of this. This turns out to be important if a *tunable* laser is desired. Usually, the spectral width of the laser is about 8 nm, scattered across eight peaks or *modes* (not the same as fiber modes). Not all of these wavelength peaks are of equal strength. The laser first outputs power at a *dominant mode*, then at another mode and another, switching back and forth unpredictably. Thus power is distributed in a bell-shaped curve, with ill-defined edges. So it is common to quote spectral width as the distance between the points on the curve where the power drops to one-half the maximum mode power. This is called the *full width half maximum* (FWHM) and is a valid measure for LED spectral widths as well.

Spectral width is a crucial laser parameter for a number of reasons. First, the wider the spectrum, the greater the dispersion on the fiber. When using WDM or DWDM, the smaller the spectral width, the closer the wavelengths can be packed and the more channels that can be used on a single fiber. Also, sophisticated modulation techniques such as phase modulation cannot be used unless the spectral width is very narrow. Finally, very narrow spectral widths, usually desirable, can actually cause unwanted nonlinear effects in the fiber, such as Stimulated Brillouin Scattering (SBS), discussed more fully later in this chapter.

In addition to spectral width, semiconductor laser output has a characteristic *linewidth*. So not only are the peaks spread across the spectral width, but each individual peak is slightly different in width. The relationship between spectral width and linewidth is shown in Figure 3-9.

There are many other characteristics of semiconductor lasers, such as coherence length and time, power, operational range in terms of wavelength, and wavelength stability. However, these are beyond the scope of this discussion.

Oddly enough, it is conceptually easy to make a semiconductor laser out of a semiconductor LED. Whenever light is confined in a proper material of the proper shape and size, the material will "lase." Sometimes this can happen even when not desired, as with reflections in a fiber amplifier section. So to make a semiconductor laser out of a semiconductor LED, just add mirrors at the ends! In practice it is a little more complicated, but not by much. The end result is a Fabry-Perot laser, or FP laser. Essentially, the FP laser is an edge-emitting LED (ELED) with mirrors on the ends. The only problem is that simple FP lasers have very wide spectral widths, so a lot of time and effort are spent repackaging FP lasers to get just the right operational characteristics.

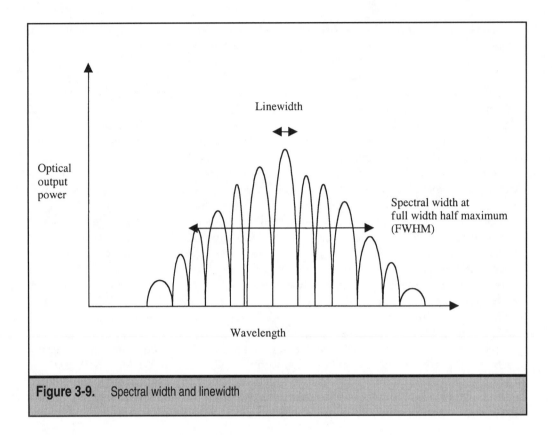

Figure 3-9. Spectral width and linewidth

Usually, some form of diffraction (Bragg) grating is placed within the lasing cavity between the mirrors to narrow the spectral width, typically to the 0.2- to 0.3-nm range. A diffraction grating is just a series of etched lines (discussed more fully in the next section) which reflects the light in a certain way. There are two major classes of this type of laser: the distributed feedback laser (DFB) and the distributed Bragg reflector laser (DBR). Of the two, the DFB laser is of the most interest in optical networking, because it leads directly to the tunable laser. A simple diagram of the active region of a DFB laser is shown in Figure 3-10.

The quarter wavelength phase shift in the middle of the grating is optional, but it narrows the linewidth of the laser output significantly. The main purpose of the grating is to reduce the spectral width of the basic FP laser. The grating is just a small variation in the RI of the material with a characteristic period (wavelength). In fact, since much of the action of the bending of the light below the lasing cavity (where the grating is usually located today) is reflection, the DFB laser does not even need end mirrors, but they are usually present. The form and position of the grating can be manipulated to yield a tunable laser as well.

The major drawbacks of DFB lasers, besides expense, are that they are sensitive to reflections (light entering the device from the attached fiber will eventually destroy the DFB

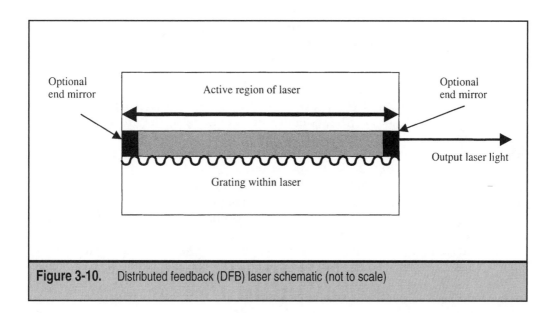

Figure 3-10.   Distributed feedback (DFB) laser schematic (not to scale)

laser) and that they are sensitive to temperature variations (not only environmentally, but also due to the *internal* heat generated by sending a long string of 1 bits). So DFB laser bit streams are typically balanced to guarantee an equal number of 0 and 1 bits over time.

The inclusion of a Bragg grating within the structure of the DFB laser leads directly to consideration of tunable lasers. In WDM and DWDM systems, having physically identical light sources that can be tuned to slightly different wavelengths is highly desirable. In one sense, tunable lasers make a virtue of necessity. When attempting to construct simple WDM systems, it proved difficult to fix the wavelengths of the transmitters exactly. This was because slight variations in the semiconductor laser fabrication process meant that the lasers produced varied slightly in wavelength anyway. Whenever a precise wavelength was needed, many lasers had to be made, tested, and selected, and the rest discarded. This was wasteful and expensive. Also, the materials that make up the laser diode deteriorate over time and change the wavelength, leading to a risk of the wavelengths overlapping. If the laser could be tuned, this would correct both situations.

But the most important reason for using tunable lasers in optical networks is the interest in *optical wavelength routing* (more precisely, direct optical cross-connecting). Tuning lasers hooked into single strands of fiber allows optical network devices to quickly accept a signal arriving on one wavelength from an input fiber and route the signal onto an output fiber at another wavelength. The arriving and departing wavelengths can be changed quickly to establish new routes or paths through the optical network. It is even possible to set up wavelength paths, just like telephone calls, across such an optical network.

Typically, tunable lasers can be tuned directly over a range of about 10 nm, mostly centered around 1550 nm, which is right in the middle of the third optical communications window. This remains a very active area of research, naturally. To extend the tunable range even further, a *sampled grating* can be added to the basic DFB laser design. A sampled grating is just a short, but otherwise normal, defraction grating with periodic *blanked sections* between the grating sections. In the tunable laser, two sampled gratings are made to interact with each other so that a small, induced change in the RI of the device causes a change in the wavelength output.

The gratings produce reflection peaks at slightly different wavelengths. Most of the peaks from the two gratings will be at slightly different wavelengths, but one set of peaks can be made to coincide, reinforcing that wavelength and effectively tuning the laser to that wavelength in particular. When a tuning current is applied to one grating, the "comb" of reflection peaks shifts slightly, reinforcing another set of peaks at another wavelength. Tuning range can be extended in this fashion to 100 nm. The basic principle of operation of the tunable laser diode is shown in Figure 3-11.

The two gratings essentially produce two sets of output wavelengths with slightly different spectral widths. By "tuning" these sets of wavelengths, one pair of peaks can be made to coincide and reinforce each other, while the other peaks are made to interfere. This is somewhat of an oversimplification of what occurs, but it is close enough for this level of detail.

There are other forms of tunable laser currently under research, such as the external cavity DFB laser that employs a tunable grating on a crystal structure. There is more than enough interest in optical routing to make tunable lasers a highly active field for some time to come.

Recently, DWDM systems with hundreds of channels have appeared. Rather than use many lasers, *multiline* lasers have appeared to drive these systems. These encourage and equalize the multiple wavelengths for external modulation. However, the most common use of SLEDs remains in the 1970s window for Gigabit Ethernet (GE), ELEDs (FP) lasers for SONET/SDH around 1300 nm, and DFB lasers for DWDM.

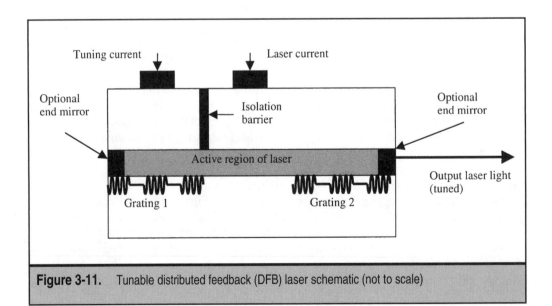

Figure 3-11. Tunable distributed feedback (DFB) laser schematic (not to scale)

In-Fiber Bragg Grating

The fourth and final key piece of the optical networking puzzle, in addition to the creation of special fibers, in-fiber amplifiers, and tunable lasers for WDM and DWDM, was the invention of the in-fiber Bragg grating. The in-fiber Bragg grating is used to isolate individual wavelengths at the receiver of the WDM or DWDM system. Some form of isolation is needed because the general-purpose receivers used in most optical systems today are wideband receivers, meaning they will register photons not only from one specific wavelength, but across a wide range of wavelengths. This problem is shown in Figure 3-12.

Of course, narrowband receivers tuned to a specific wavelength can be developed and used, but this approach is an expensive proposition and an increasingly difficult task for DWDM systems where wavelengths are jammed closer and closer together all the time. There are other ways to filter out the unwanted wavelengths at the receiver besides the in-fiber grating, naturally. The attraction of the in-fiber Bragg grating, like the in-fiber optical amplifier, is that these optical devices are part of the actual fiber itself and allow the continued use of a standard, wideband optical receiver.

Some observers of the optical networking scene rank the invention of the in-fiber Bragg grating (sometimes called an FBG) as second only to the invention of the laser itself

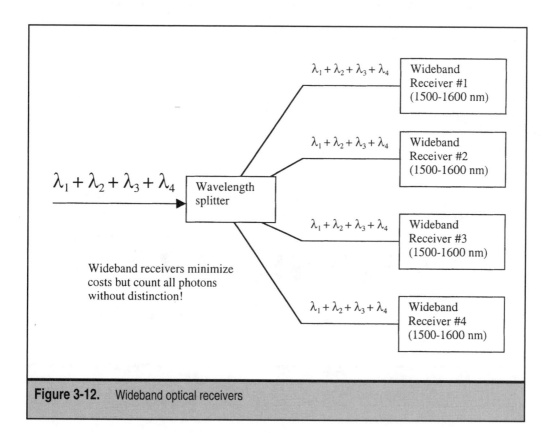

Figure 3-12.   Wideband optical receivers

as a milestone in optical networking. The FBG is simple in the extreme to make, very inexpensive, and highly reliable. Not much more could be asked for. The FBG is just a short length (only a few centimeters) of ordinary single-mode fiber into which variations in the core RI have been etched using beams of ultraviolet radiation. When etched with just the right spacing, the in-fiber Bragg grating will reflect (select) one precise wavelength back in the direction from which the light entered the fiber and pass the other (non-selected) wavelengths through the fiber unchanged (there is some small attenuation effect on the other wavelengths). The general idea behind the FBG is shown in Figure 3-13.

The small changes in RI lengthwise in the fiber core of an FBG section will reflect a small amount of light at each discontinuity. If the wavelength of the light and the spacing (period) of the grating are the same, then there is positive reinforcement in the backward direction. In technical terms, the grating forms an *electromagnetic resonant circuit*.

There are many variations on this basic in-fiber Bragg grating design. Some of these can be used to deal with the issue that the wavelength selected is reflected *back* toward the source. It would be more useful to pass the desired wavelength *through* the fiber and eliminate the others, as a true filter should. And so it turns out that it is possible to etch many different FBGs into the same section of fiber. Then a *blazed grating* can be added which is built at an angle to the fiber core and reflects light out of the core. So it is possible to combine many FBGs and blazed gratings to filter out all but the desired wavelength.

The use of FBGs as filters at the receiving end of a simple WDM link allowing the use of identical wideband receivers is shown in Figure 3-14.

The only problems with FBGs have been some sensitivity to temperature variations and stretching of the fiber section with the gratings. Both effects have been relatively easy to control in practice. FBGs can be used not only as filters in WDM and DWDM systems, but also to control dispersion in the 1550 nm band and to stabilize the output of a laser, tunable or otherwise.

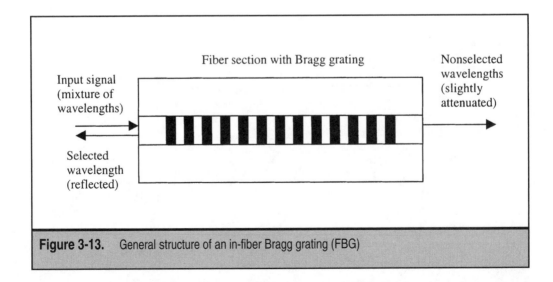

Figure 3-13.   General structure of an in-fiber Bragg grating (FBG)

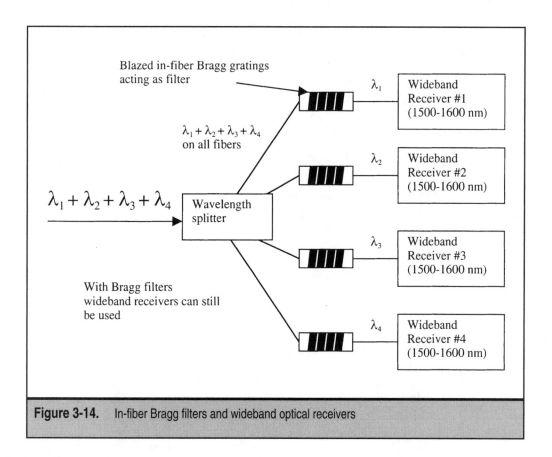

Figure 3-14.   In-fiber Bragg filters and wideband optical receivers

WDM

The first edition of this book did include a consideration of WDM systems. However, the WDM section appeared in a chapter on the future of SONET. So in just a few short years the practice of WDM has moved along to the point where the topic deserves to be treated right up front. This section basically reproduces the original WDM section, with minor updates along the way.

Strangely enough, there is really no such thing as WDM. Actually, it would be more accurate to say that the technology known as WDM is basically familiar analog frequency division multiplexing (FDM) applied to fiber optic transmission. This is because the frequency of any electromagnetic waveform (and light is just another form of electromagnetic wave phenomenon) is directly related to its wavelength by the formula $c = \lambda * f$. That is, the wavelength of any electromagnetic wave multiplied by its frequency is always equal to the speed of light (c).

Analog systems, which multiplex signals together in different frequency channels (e.g., broadcast TV channels set 6 MHz apart), use FDM. So did older L-carrier networks using coaxial cable. Each channel uses its assigned bandwidth all the time, but no channel gets the entire bandwidth. The formula $c = \lambda * f$ could easily be used to convert the megahertz or kilohertz channels to wavelengths, and then the method could technically be called a form of WDM. However, there is a little more to WDM than just FDM.

Digital transmission systems, in contrast to analog systems, usually use time division multiplexing (TDM) to combine signals onto a higher speed medium. This is the multiplexing method used in both T-carrier and SONET, of course. In TDM, the entire bandwidth (frequency range) of the transmission medium is broken up into time slots, and each input uses the entire bandwidth, but only for its allotted time interval. For example, T-carrier divides the entire 1.544-Mbps bandwidth (less overhead) into 24 time slots of 64 Kbps each.

The key to understanding WDM is to realize that the enormous bandwidth of fiber optic cable, which spans the whole visible light spectrum and beyond, makes it possible to combine FDM and TDM on the same path through the same fiber. The differences between FDM, TDM, and WDM are shown in Figure 3-15.

Figure 3-15 shows that both FDM and TDM techniques can be combined on fiber optic links to create WDM. There is no reason that different "colors" of light cannot be used on the same fibers, as long as transmitter and receiver are tuned to function that way. In the same fashion, voice nursery intercoms can function over home electrical wiring without interference.

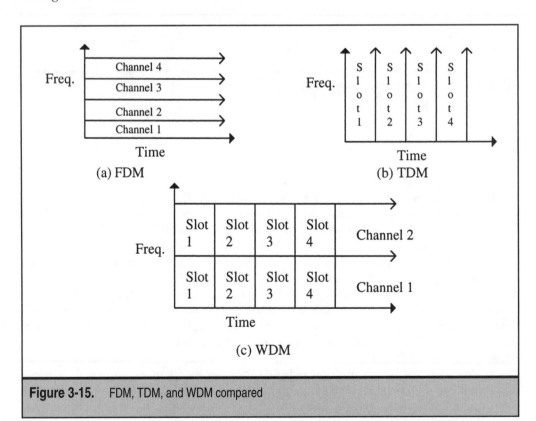

Figure 3-15.   FDM, TDM, and WDM compared

WDM and SONET/SDH

In conventional fiber optic transmission systems like SONET/SDH, information is transmitted in the form of light pulses. Pulses of light represent individual bits: zeros and ones. The electrical bits are converted to optical signals, transmitted over a fiber optic cable, and converted back into electrical signals at their destination.

WDM involves the simultaneous transmission of light from multiple lasers with different wavelengths over a single fiber optic line. Sources send these wavelengths of light to a multiplexer, which consolidates them for transmission over a single WDM fiber link. WDM breaks the optical spectrum into channels, each with a different wavelength. Time slots, as in SONET/SDH, can be used with each of these channels.

Special optical amplifiers, usually spaced tens of kilometers apart, amplify all of the wavelengths simultaneously. Finally, at the far end of the WDM link, the signals arrive at a demultiplexer, where they are separated and sent to receivers at the destination points. WDM equipment also includes optical fiber filters, which selectively transmit or block a particular range of wavelengths, and transponders, which function as the transmit/receive systems.

WDM is useful whenever congestion occurs on fiber networks, SONET/SDH or otherwise. Greater capacity can be created without the need to install more fiber along the right-of-way. The trade-off occurs in the end-equipment expense, but electronics usually beat the backhoe anytime. WDM can be used to expand to greater capacities over the same number of fibers. WDM also gives a service provider one way of achieving 10 Gbps and even higher transmission rates.

Early WDM devices and links involved two-channel multiplexing. These used the 1310- and 1550-nm (nanometer) laser operating regions, or windows, in short-span applications, or two widely spaced channels in the 1550-nm region (e.g., 1538 and 1558 nm) in long-span applications.

WDM systems may be unidirectional or bidirectional. A bidirectional WDM device will pass a number of multiplexed frequencies in both directions on one fiber. For example, in a four-channel bidirectional WDM span, signals at 1549 and 1557 nm are multiplexed for transmission in one direction, and 1533- and 1541-nm signals are multiplexed for transmission in the other direction, all on the same working fiber. This is shown in Figure 3-16.

Conversely, a unidirectional WDM device multiplexes a number of different frequencies for transmission in one direction on one fiber. For example, in a four-channel *unidirectional* WDM span, signals at 1557, 1549, 1541, and 1533 nm are multiplexed for transmission in one direction on the fiber. This type of WDM is shown in Figure 3-17.

Both of these techniques allow for such fiber configurations as combining four OC-48s on a single fiber span for an aggregate signal speed of OC-192. This newer WDM technique can use a previously installed fiber base to support extremely narrowly spaced channels. This newer WDM system does have a number of requirements. For instance, using WDM to combine four OC-48s requires maximum reuse of any existing OC-48 equipment: compatibility with current regenerator spacing, a minimum of 4-wavelength operation, the ability to upgrade to 8 or even 16 different wavelengths (or more) in the future, the capability to add or subtract wavelengths (no mean feat), and unidirectional and/or bidirectional operation.

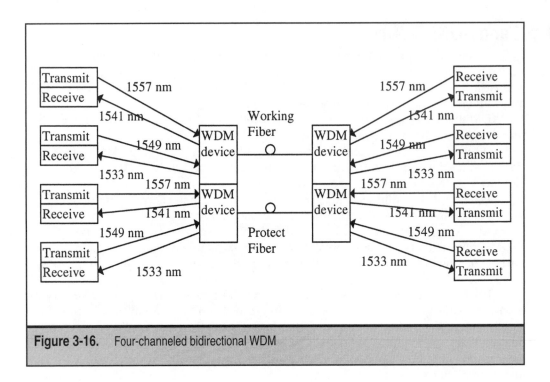

Figure 3-16.   Four-channeled bidirectional WDM

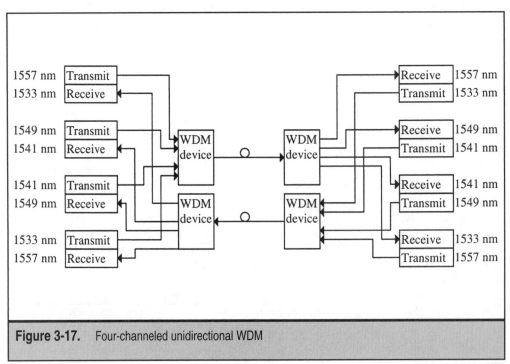

Figure 3-17.   Four-channeled unidirectional WDM

Sometimes there are references to something called *dense WDM* (DWDM). There are varying interpretations of the difference between standard WDM and DWDM. Some would define DWDM as the multiplexing of two or more channels on a single fiber, with the channel spaced 2 nm or less from each other. Others basically consider anything involving four or more channels to be DWDM. The International Telecommunication Union (ITU) is busily working on a complete, standard set of WDM guidelines.

The next generation of WDM is waiting in the wings. This type of WDM, which is known as *wide-spectrum multiwavelength multiplexing*, is ideal for SONET and all long-haul fiber applications because it allows for economical incremental growth. Such wide-spectrum WDM is needed to realize the full benefits of future fiber optic technologies, such as optical cross-connects.

In wide-spectrum multiwavelength multiplexing, the wavelengths needed for the transmission system can be produced by wavelength-specific transmitters, or *wavelength adapters*. Wavelength adapters allow transmission equipment to have any wavelength output at the transmitter. For example, a lower-cost 1310-nm laser could be used. This is good for multivendor environments or for backward-compatibility, but wavelength adapters are quite expensive and the high cost is accrued with every added wavelength.

Before implementing wide-spectrum WDM, however, a whole new set of challenges must be addressed. One such challenge is finding the proper spacing of transmission frequencies to allow a large number of channels without crosstalk or expensive filters. The evolving ITU-T standards have set the channel spacing for wide-spectrum WDM on 100 GHz intervals with 193 100-GHz as a reference frequency.

DWDM

Not only is WDM with wavelength spacings of about 10 nm an accepted optical technology today, but WDM has moved along to the point where it is *dense* WDM (DWDM) systems that are the major topics of discussion and consideration. The major difference between WDM systems and DWDM systems is that in DWDM the wavelengths used are packed within 1 nm or even 0.1 nm of each other, or about 10 to 100 times denser than in WDM. The current record for wavelength packing on a single fiber is 1,022 channels, although the total throughput is only about 40 Gbps. However, since the current serial bit transmission record on fiber is 160 Gbps, DWDM systems with thoughputs in excess of 160 Tbps (160,000 Gbps or 160,000,000 Mbps) are potentially possible. So it would seem that there is very little real difference between WDM and DWDM. One has closer spacing of wavelength channels, true enough, but essentially WDM and DWDM are the same technologies. However, this is not really true. The close spacing of wavelengths in DWDM systems translates to a whole new set of concerns.

This is because in practice the closer spacing of wavelengths in DWDM makes a big difference. If the packing of wavelengths to within 0.1 nm (or less!) over a single fiber core were trivial, DWDM would merit no more than a footnote in WDM discussions. But there is a world of difference between 10-nm spacing and 0.1-nm spacing of light waves. For example, using 0.1-nm spacing allows 100 light sources to concentrate their energy over a 10-nm bandwidth into a single fiber core 9 microns or so in diameter. This is a huge energy flux concentrated in a very small space, although in terms of raw power the total amounts are still relatively small. But confining this electromagnetic energy in the fiber core for long distances without debilitating attenuation, dispersion, and other fiber effects is a challenge that deserves a section all its own.

Everything is magnified in DWDM, so that even small effects in WDM become crippling in DWDM unless steps are taken to compensate for these impairments. Fortunately, there are methods available that make DWDM systems not only practical, but attractive.

To take a simple example, in DWDM systems the center wavelength of a channel must be tuned carefully to make sure it is exactly where it is supposed to be. This is because even simple modulation of the carrier signal will essentially double the bandwidth required (in precise accordance with Nyquist) and might otherwise overlap and interfere with wavelengths still within their band, but not properly centered. Naturally, as the wavelengths move to within a fraction of a nanometer of each other, these *band edges* become more and more important.

Before getting into a full discussion of DWDM system impairments and what must be done to control their effects, this is the place to say a few words about two often neglected components of optical networks. These are the *combiners* and *splitters* that mix wavelengths onto a single fiber at the sender and split the mixed wavelengths onto separate fibers at the receivers. These components have come a long way as well. In many cases, the same device can serve both functions. Used in one direction, it is a splitter. Turn it around and it is a combiner or coupler.

Simple couplers make use of a phenomenon called *resonant coupling*. This occurs when two fiber cores are brought into close proximity. If the lengths are just right, half of the input power ends up in each fiber core. This is called a *3-dB coupler*. The basic building block can be employed to make (for example) eight-way combiners or splitters. However, each stage causes some signal loss and attenuation. This is a physics phenomenon, and even two light streams directly coupled onto one fiber *must* lose half their power.

When operating at different wavelengths, DWDM couplers and splitters are more efficient. Modern systems can use optical technology such as *beam splitter prisms and/or birefringent materials*. Many splitters make use of diffraction gratings of one form or another, including planar diffraction gratings.

At the source, some lasers are now *multiline lasers*. Instead of using the comb of spectral width wavelengths to tune the laser, mutliline lasers attempt to even out linewidths and equalize the power of the several wavelengths produced. This is probably a good idea in the long run. If DWDM systems evolve to 1,000 inputs or more, 1,000-input lasers might be needed. But if multiline lasers can produce 10 usable wavelengths, only 100 input lasers would be needed. Of course, the multiline laser must be modulated externally at each wavelength to send bits, since there is only one drive device, but this is simple enough to do.

DWDM Impairments

It has been established that jamming many wavelengths onto a single fiber core is difficult. This is due to a number of fiber impairments that are annoying in any optical system, but absolutely crippling to a DWDM system. Oddly, some effects can be controlled by jamming the wavelengths even closer together, and other impairments require keeping the wavelengths well apart. Some effects can be dealt with by boosting the transmitter power, while others require power limits. But just what are they, and why do they interact this way?

All fiber impairments can be roughly characterized as dispersion effects. In this book, they are further divided into modal dispersion effects, scattering effects, and miscellaneous effects, but they all manifest themselves as dispersion. Table 3-1 lists the effect, the

Modal Effects	Cause	DWDM Workaround
Intermodal dispersion	Multiple modes interact	Single-mode fibers
Intra modal dispersion	Various:	Fiber construction methods:
Material (chromatic)	RI varies with wavelength	Effects cancel at 1300 nm
Waveguide	20% light travels in cladding	(Can be shifted)
Polarization mode	RI varies with polarization state	Perfectly round core (difficult)
Scattering Effects		
Rayleigh	RI varies with fiber density	Perfectly uniform density
Raman (SRS)	Low wavelength pumps high, produced "power tilt"	Lower power, limit span, pack wavelengths close together
Brillouin (SBS)	RI varies with acoustic waves	Lower power, limit number of channels, avoid two-way systems
Miscellaneous Effects		
Linear crosstalk	Mostly result of SRS	Use filters, separate wavelengths
Four-wave mixing (FWM)	Additive wave response	Watch channel wavelengths, avoid 10+ Gbps systems
Cross-phase modulation	RI varies with intensity (called *optical Kerr effect*)	Balance power in system
Self-phase modulation		Limit power in system

Table 3-1.   Fiber impairments

cause, and the workaround in DWDM systems. Keep in mind that this table is a distillation of dispersion effects that most impair DWDM systems. There are many other types of dispersion and fiber impairments.

It should be noted that many of these effects also vary with the serial bit speed used in the fiber channel of a DWDM system. As a general rule, doubling the speed from 5 to 10 Gbps will quadruple the dispersion on the same length of fiber. This only reinforces a truism in networking: to go farther, go slower. The following sections take a quick look at each effect in some detail.

Modal Effects

The first modal effect is *intermodal dispersion* and is included here really just for the sake of comparison. Intermodal dispersion is the result of the interplay between the various physical paths which a beam of light can make its way down a fiber core. The workaround is to use single mode fiber (SMF), and SMFs are used in all practical DWDM systems. Even older single-mode fibers designed for 1300-nm use can be employed, but the attenuation is high and runs must be short.

But there is also *intramodal* dispersion that affects even a single propagated mode. The causes of this impairment are varied according to type, and all the workarounds concern fiber construction methods. *Material (chromatic) dispersion* is a result of the RI of the fiber varying with wavelength. *Waveguide dispersion*, which has an opposite sign to material dispersion and can act to counteract the effects of material dispersion, is the result of 20% of the light being carried not in the fiber core, but in the cladding. The difference in RI in the cladding causes dispersion. These effects usually cancel out at around 1300 nm, but dispersion-shifted fibers move this point to around 1550 nm. So, most DWDM systems will use dispersion-shifted single mode fibers. The final modal effect is *polarization mode dispersion*. This is the result of the RI of the fiber varying with the polarization state (polarity is not the same thing) of the signal. It is generally only of concern at serial bit speeds of 10 Gbps and higher. The workaround is to use fiber with a perfectly round core. (Surprisingly, there are times that perfectly round cores are not desired and polarization mode dispersion is actually encouraged!) Now, most fiber fabrication techniques do their best to produce round cores. However, it is easy for the fiber to be stepped on or otherwise deformed during installation.

Scattering Effects

The *scattering effects* are an interesting group. These are named, curiously enough, after physicists and investigators into optics. *Rayleigh scattering*, common to all fibers, is unavoidable and results when light is scattered, or deflected, by particles in the fiber. It is a concern in DWDM systems because variations in the density of the fiber make the dispersion worse. Again, the only way to compensate is to construct the fiber to very exacting standards.

Of more interest is *stimulated Raman scattering* (SRS). SRS is caused when multiple wavelengths interact on one fiber through the crystal lattice of the fiber core and so became a concern only in WDM systems. With SRS, shorter wavelengths will "pump" longer wavelengths, resulting in a power loss at shorter wavelengths and a power gain at longer wavelengths. This causes a pronounced *power tilt* in the received power of the wavelengths, even though all might be sent with the same power. SRS is not always bad: optical amplifiers are nothing more than fibers in which the pumping from the lower wavelength (the pump laser's wavelength) to all the higher ones is encouraged to the extreme. SRS can be controlled by lowering the input power, limiting the span distance, and packing the wavelengths (generally, the farther apart the wavelengths, the more pronounced the SRS between them).

Finally, there is *stimulated Brillouin scattering* (SBS). SBS occurs even at low power levels and is the result of acoustic vibrations in the fiber. That is, the atoms of the fiber literally move back and forth, creating an acoustic (pressure) wave in the fiber. The pressure change affects the RI, which scatters the light, producing SBS. Both SRS and SBS are *stimulated* in the sense that the light that is itself scattered produces the effect. Oddly, SBS is less severe at higher data rates as the bit time shrinks. However, SBS is much worse when light signals propagate in both directions at the same time in bidirectional WDM systems (because of the complex acoustic waves generated). In fact, SBS is one of the main reasons that two fibers are used, one for each direction, in most optical systems. But if runs are short enough, even DWDM can be used as long as the number of wavelengths used is limited. So the workarounds are limited power (which limits runs), limited wavelengths, and avoiding two-way transmission on a single fiber.

Miscellaneous Effects

The final category of *miscellaneous effects* is also an interesting group. The first member, *linear crosstalk*, is not really a separate impairment at all. Linear crosstalk is caused by SRS, but many sources treat linear crosstalk separately. Crosstalk is just the interaction or interference between two wavelengths. The workaround is to keep the wavelengths well separated and filter out extraneous wavelengths not used for information carrying.

The name *four-wave mixing* (FWM) is an unfortunate one. This impairment was first noticed in early WDM systems with four channels, hence the name. But it really should be *N-wave mixing*, since the interactions of any number of wavelengths can lead to this phenomenon. FWM is just an additive response to multiple wavelengths, the optical equivalent of *beats* in electrical systems. Three wavelengths will combine to yield a fourth wavelength at the receiver, according to a formula (so FWM is appropriate in this sense). The formula, in frequencies used, is $f_{new} = f_1 + f_2 - f_3$. Naturally, if the new wavelength is also used in the DWDM system, crosstalk results. The workaround is to watch the wavelengths used and space them out unequally (the ITU grid, discussed later, allows for this). FWM is also worse at 10 Gbps and higher.

Finally, there are two impairments known as *phase modulation* effects. As the result of the RI of the fiber varying with the intensity of the light, a phenomenon called the Kerr effect occurs. Cross-phase modulation (CPM) occurs between the wavelengths and self-phase modulation (SPM) occurs within even a single wavelength of light. CPM is compensated for by carefully balancing the power between the channels in a DWDM system, and SPM is dealt with by limiting the power used (which also limits distance, of course).

The ITU DWDM Grid

The ITU-T has been active in exploring international standards for DWDM. One specification is G.692, entitled "Optical Interfaces for Multichannel Systems with Optical Amplifiers," which contains a number of interesting points about DWDM. Unfortunately, DWDM developments have been proceeding so fast that the ITU-T and other standards organizations have been hard-pressed to keep up with system capabilities. These words might soon be obsolete, if they are not already.

The ITU-T specifications mainly address point-to-point DWDM systems using 4, 8, 16, or 32 channels. The maximum specified span without amplifiers for a DWDM link is 160 km (100 mi) or up to 640 km (400 mi) when amplifiers are used.

A standard wavelength grid is established based on multiples of 50 GHz spacing between the wavelengths used. There is also a center reference frequency ("wavelength") at 193.1 terahertz (THz: 1,000 GHz). The whole grid is sometimes called the "ITU 193 THz grid." In keeping with the tradition of citing frequency in ITU-T specifications, all of the wavelengths are specified in Hertz (frequency), not nanometers! But conversion is easy enough to do. The reference frequency of 193.1 THz equates to 1553.5 nm, right in the middle of the 1990s window, as might be expected. The 50 GHz spacing works out to about 0.4 nm between the wavelengths. The wavelengths used are in the range from 1528.77 nm to 1560.61 nm.

The G.692 draft proposal allows:

▼   4 channels with 3.2-nm (400 GHz) spacing

■   8 channels with 1.6-nm (200 GHz) spacing

■   16 channels with 1.6-nm (200 GHz) spacing

▲   32 channels with 0.8-nm (100 GHz) spacing

Supervisory wavelengths are allowed at 1310, 1480, 1510, and 1532 nm. In the draft, users are allowed to use any wavelength on the grid, but only those wavelengths on the grid. However, working DWDM systems have progressed to 128 channels or more, and spans in excess of 640 km are possible, challenging the ITU-T plans. Allowable speeds for each channel are 622.08 Mbps, 2.488 Gbps, and 9.9 Gbps. These correspond to SONET/SDH speeds of OC-12/STM-4, OC-48/STM-16, and OC-192/STM-64. Optical amplifiers of the EDFA type are expected to be used, operating in the 1550 nm region. In addition to the basic 50 GHz spacing, systems are allowed to use 100 GHz (8 or more channels), 200 GHz (4 or more channels), 400 GHz (4 or 8 channels only), 500 GHz (8 channels only), 600 GHz, (4 channels only), and even 1,000 GHz (4 channels only) spacings.

Potentially, the ITU-T grid spacing could be firmly allocated on a national basis, with each country limited to using the wavelengths assigned to it. This is actually a good idea, since the day may come (and many feel this day is closer than anyone thinks) when a single strand of fiber will literally circle the globe carrying hundreds or even thousands of wavelengths with in-fiber optical amplifiers. Each country would just attach to this global fiber ring and never fear interfering with some other countries' wavelengths. To communicate with any other country, just send on a channel allocated by the receiving country. To receive, just listen on the channel assigned for the return path.

It might seem odd to spend so much time in a book on SONET/SDH discussing optical networks, WDM, and DWDM. None of these concepts are dependent on SONET/SDH in and of itself. And SONET/SDH is defined to run at 1310 nm, in the 1980s window, reflecting the time frame for SONET/SDH origins. Nevertheless, it is quite common to find SONET/SDH running somewhere in even a DWDM system. This is because SONET/SDH is a much more mature technology and includes consideration in a standard manner for essential network operational features such as network management and protection switching. In networking, bit transfer is really the easy part. Controlling all aspects of the process is much more difficult.

There are a number of ways to use SONET/SDH as a *management channel* in a DWDM system. One way is just to couple in a SONET/SDH channel at 1310 nm with the DWDM channels at 1550 nm. After all, there is only one fiber. If the SONET/SDH management channel indicates a problem, there is obviously a problem in the whole fiber. Alternatively, a transponder can be used to shift the 1310-nm management channel into the band centered around 1550 nm. Sometimes the SONET/SDH management channel is used to carry additional network management information about the other DWDM channels, such as error-correcting codes. Additional issues concerning the future of SONET/SDH in a DWDM world will be considered in a later chapter.

CHAPTER 4

The Digital Hierarchy

SONET/SDH is an international standard for very high-speed digital transmission over fiber optic cable. SONET is an abbreviation for Synchronous Optical NETwork and was first considered by Bell Telephone Laboratories (Bell Labs, now Lucent Technologies) in the early 1980s to solve a series of nagging problems with existing digital transmission systems. This chapter is not intended to emphasize these problems and limitations (all technologies always have problems and limitations, even SONET/SDH). In a sense, SONET/SDH represents as natural an evolution of the former digital hierarchy as DWDM systems represent the evolution of SONET/SDH. DWDM "solves" a series of SONET/SDH issues, but the rise of DWDM does not necessarily mean the end of SONET/SDH. In the same way, the rise of SONET/SDH did not doom the former digital hierarchy. SONET/SDH just made the problems and issues of the existing digital hierarchy less limiting.

Digital transmission systems are characterized by the fact that these communication links only carry information in the form of binary digits (universally known as bits). Binary digits can only represent a "0" or a "1." Strings of 0's and 1's can be constructed to represent almost anything from computer-based data to digitized voice to stereo audio on a music CD (compact disk) to the soundtrack of a movie (in several varieties), even potentially to the movie itself.

It may seem that we are now living in a "digital world" where nearly everything from music to movies comes out of a computer. This impression is not far from true. Music can be generated much more precisely and accurately (not to mention much less expensively) from a computer than from a human. Movies are so laden with digital special effects that it sometimes seems silly to even generate the images on film for projection onto a screen. Why not just watch a big computer monitor?

How has this transition to the digital representation of almost everything come about? What has all this to do with SONET/SDH? In order to understand the position of SONET/SDH in this digital world, it is necessary to understand the problems with digital transmission that SONET/SDH was designed to overcome.

ANALOG AND DIGITAL

The alternative method of transmitting voice, video, or data from one place to another is to employ analog transmission systems. Analog signals can take on more values than merely 0 or 1—for instance, a signal of "0.5" is perfectly fine in an analog system, but meaningless in a digital transmission system. In fact, analog systems allow signals to have any value between some maximum and minimum value. Typically, analog signals will vary smoothly over time in an unpredictable pattern, wandering between maximum and minimum as the input signal (voice or otherwise) varies. In voice transmission, the input signal is an electrical representation of the human voice's acoustic pressure waves (sound). This analog signal is sent over the telephone wire, which has a number of physical forms, as an analog electrical signal and reconverted to an acoustic pressure wave at the receiver.

These analog systems evolved to deliver voice over the national telephone network, or public switched telephone network (PSTN) as it is more properly called. Because people cannot talk in 0's and 1's, the builders of the PSTN had little choice but to initially deploy the national voice network as a purely analog network. Sometimes it is said that the PSTN is "optimized for voice," meaning that nothing can really be sent over it unless it is an analog sound. Even office and home computer users with personal computers (PCs) on their desks are usually familiar with the use of modems (which is a contraction of MOdulator/ DEModulator) to communicate over analog voice telephone lines. A modem changes the 0 and 1 digital signals that computers generate into "sounds" (actually, the electrical representation of that "sound") which can be carried over the PSTN. Simply put, a "0" bit "sounds" differently than a "1" bit to the receiver.

The differences between analog and digital signals are shown in Figure 4-1. Note that the signals vary by amplitude (signal strength), but other signal parameters, such as frequency, are possible and even quite common. Also note that digital signals are not limited to basic 0's and 1's, but may include more possible values. Even when this is the case, however, all signal values are interpreted as strings of 0's and 1's, and the signal is considered to be undefined when not expressed as one of this limited set of values. For example, four level digital signals would express the bit strings 00, 01, 10, and 11.

It is somewhat ironic that the analog voice telephone network largely replaced a perfectly functional digital network that was some thirty years older than the telephone, which was invented in 1876. This was the national telegraph network. After all, Morse

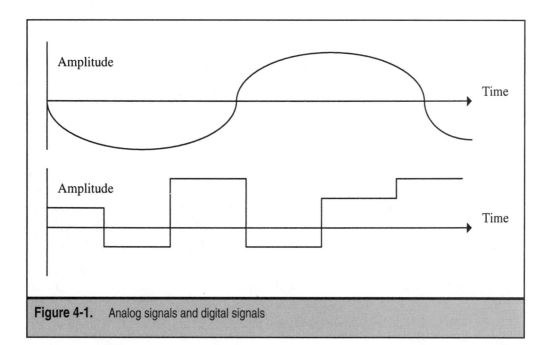

Figure 4-1.   Analog signals and digital signals

Code is easily understood as representing information as a string of 0's ("dots") and 1's ("dashes"). The letter "a" is just a dot followed by a dash in Morse Code (01), the letter "b" is dash-dot-dot-dot (1000), and so on. Morse—it was really his assistant, Theodore Vail, who invented "Morse" Code—gave common letters shorter strings and less common letters longer strings out of consideration that Morse Code was keyed by a human being.

Alexander Graham Bell invented the telephone while trying to develop a way to *multiplex* (combine) telegraph signals on a single wire. The concept of multiplexing has enormous significance in SONET/SDH, but it obviously is an idea that has been common for many years. Without multiplexing, many wires must be installed to increase the total message carrying capacity of the network. Multiplexing makes more efficient use of existing facilities, minimizes expense in terms of physical wiring, and is mostly limited only by the degree of ingenuity of electrical engineers operating within the limits of a few basic physical laws, such as the fact that signals cannot travel faster than the speed of light.

Bell's background in education of the hearing impaired uniquely qualified him to hear in the tones of multiplexed telegraph signals not the mixture of 0's and 1's or dots and dashes, but the sound of the human voice. Time was needed to work out the details, but the telephone was the first major network technology that was analog instead of digital. All previous methods, from smoke signals to war drums to "one if by land and two if by sea" were inherently digital in nature. The analog telephone required no special knowledge or training to use.

It soon became apparent to early telephone users, however, that the blessing of analog had a dark side as well: Analog signals are much more susceptible to *noise* than are digital signals. In any transmission system, noise is simply a spurious and unwanted signal that interferes with the signal sent into the system. To understand why, consider the fact that in Morse Code, a dash was simply defined as being at least three times longer than a dot. Early telegraph operators all had a distinctive "hand" because their basic dots and dashes all varied somewhat in duration (the phrase "heavy-handed" comes from this era). When a noise on the telegraph line (usually a nearby lightning strike) induced a spurious electrical signal into the line that somehow stretched a dot to twice its normal duration, the receiving operators were still likely to say to themselves "that wasn't a dash, so it must have been a dot."

Analog signals can vary smoothly from maximum to minimum. A signal twice as long as a dot was just as likely as one three times as long, or one and a half times as long, or nearly any duration at all. In analog systems, it quickly becomes impossible for receivers to distinguish noise (the unwanted part) from the signal (the wanted part). In the telephone network, this noise was simply *noise* which swamped the sender's speech in a constant background hiss of clicks, pops, and buzzes, mostly due to adjacent digital telegraph lines.

Early telephone pioneers quickly invented ways to minimize the analog noise on telephone lines. But it was never eliminated and limited the quality of voice that could be delivered, especially over longer and longer distances, to a barely acceptable level. Clearly, there was a reason to try to invent a more effective way to minimize noise effects on the analog network. The answer, when it came, was to digitize analog voice signals.

THE DIGITIZATION OF VOICE

Many people in the telecommunications field today associate the digitization of voice with such modern developments as integrated services digital network (ISDN) and digital T-carrier networks. However, efforts to digitize voice began around World War II mainly as an attempt to eliminate nagging noise problems on analog voice circuits. These early efforts were interrupted by the war, but work continued in the United States after the war. By the 1960s, the phenomenal growth of the United States economy after World War II was the added incentive needed to complete this work.

Before the war, only about 40% of households in the United States found it economically feasible and socially desirable to have a telephone. Letters, backyard fences, and other social activities filled the hours people spent not otherwise engaged in work and related activities. But after the war, when the United States was essentially the only developed nation not devastated by the conflict, the economy expanded rapidly as other countries understandably turned to the United States to purchase goods and raw materials to rebuild their shattered cities. The United States was only too glad to oblige, and the immediate result was a kind of "Golden Age" in the 1950s, when companies rapidly expanded to meet the growing demand for goods not only overseas, but for the newly prosperous citizens in the United States.

One of the luxury items at the top of nearly everyone's list was a telephone. The telephone companies, especially the AT&T Bell System (as the self-declared "nation's telephone company"), were hard pressed to keep up with the demand for local loops, switching offices, and trunking networks that the swelling system now required. Local loops were needed to provide a pair of wires for every new telephone. Switching offices (or central offices, as they were frequently called) were needed to switch the calls to their destinations. Trunking networks were needed to link these offices together; trunks are telephone lines that link not individual telephones to switching offices, but switching offices to each other. These trunks became the focal point of digitization efforts in the 1960s.

Many switching offices employed an analog trunking system that multiplexed, or combined, 12 voice channels carrying telephone calls onto two pairs of copper wires. Each voice channel was assigned a separate frequency band for this purpose. This technique was known as *frequency division multiplexing* (FDM) because each voice channel had its own frequency band assigned to it regardless of whether it was in use or not; therefore, idle channels existed. When someone made a telephone call that terminated at a different switch than the one that the telephone's local loop was attached to, an idle trunk channel frequency band was assigned to that person's conversation for the duration of the call. A typical switching office could have up to 10,000 local loops, while trunks between switching offices typically numbered only about hundred. This basic telephone network structure of analog loop, switch, and trunk is illustrated in Figure 4-2.

As network use grew due to the number of new telephones and telephone calls, this trunking network threatened to be overwhelmed. If no trunks were available between switches, no more calls could be completed, and the telephone company would lose potential revenue. Obviously, adequate trunk capacity was a key parameter in telephone

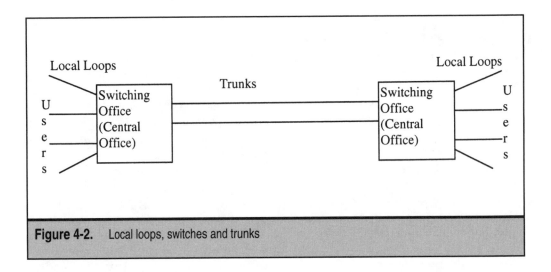

Figure 4-2. Local loops, switches and trunks

network design and operation. The problem was that adding trunk capacity was neither easy nor quick. Funds had to be allocated, routes planned, materials purchased, and the work done. The task was made even more difficult by the fact that the switching offices usually were many miles apart.

The search for a more efficient way to quickly add capacity to a trunking network led several groups of engineers back to digitization. It was already understood that digitized voice was less noisy than analog circuits. However, the end equipment (transmitters and receivers) needed to convert analog voice to digital voice and back again was quite expensive. An added economic motive was sought to justify the expense of the equipment: Additional trunking capacity provided that added motive.

A digital system was designed that could double the capacity of 12-channel analog voice trunks to 24-voice channels. Multiplexing trunk networks were known as *carrier systems* to the telephone companies. This allowed telephone companies to cost-justify the new digital equipment needed by doubling capacity on the trunking network without having to go through the expensive process of laying new cable and had the additional benefit of improving voice quality by reducing noise. The new digital multiplexing trunk network was designated by the engineers at Bell Telephone Laboratories as *T-carrier*. The "T" had no significance other than a letter designation. There were also N- and L-carrier systems, among others, mostly analog methods. But it was the digital T-carrier that fueled a revolution in digital transmission that led directly to SONET.

T-Carrier

With the advent of T-carrier trunking networks in the early 1960s, telephone companies addressed one of a number of problems brought on by the unprecedented growth demands of the rapidly expanding economy. At about the same time, electronic components and processors were introduced into the switching offices. These looked similar to

1960s-style computers, except that the Bell System was reluctant to call them what they were, because an agreement in 1955 (the "Final Judgment" of an anti-trust suit) with the federal government essentially banned AT&T from developing and marketing anything that could be construed as a computer. Therefore, under the name "Electronic Switching Systems" (ESS) computers slowly made their way into the AT&T network. The introduction of area codes (called "Numbering Plan Areas," or NPAs by AT&T) helped to complete long-distance calls without requiring the wholesale hiring of more long-distance operators. Understanding these related developments helps to place the T-carrier within the proper deployment context.

T-carrier was the basis of a whole family of digital trunking methods. The basic unit was the digitized voice channel, which produced a stream of bits at the constant rate of 64,000 bits per second, or 64 Kbps. The 64 Kbps digitized voice channel was designated Digital Signal-0 (DS-0) in the T-carrier system. The "0" indicated the lowest level of what was to become the T-carrier hierarchy. There was no carrier involved in DS-0, because this basic signal was not multiplexed with anything else, but merely served as the lowest level T-carrier input and output.

Twenty-four DS-0 signals were combined to yield a DS-1. The "1" indicated the first level of the T-carrier hierarchy. The DS-1 signal was transmitted on a T-1 physical network. Although DS-1 specified a certain structure of 0's and 1's from 24 input DS-0s, T-1 specified the transmitters, receivers, and wiring requirements for transporting the DS-1. Most people use the terms "DS-1" and "T-1" interchangeably, although this is technically imprecise.

A DS-1 performed its multiplexing task in a fundamentally different fashion than did analog carrier systems using FDM. The DS-0 produced 64 Kbps by generating 8 bits 8,000 times per second. There were many technical reasons that 8 bits and 8,000 analog "samples" were taken, but these are beyond the scope of this discussion. It is only important to realize that each digitized voice channel (DS-0) produced 8 bits every 1/8,000th of a second, or 125 microseconds (μsec). Obviously, the combined output of 24 DS-0s would require a way of sending at least 24×8 bits = 192 bits every 125 microseconds. To accomplish this, a DS-1 was organized into *frames.* Each frame was sent and received 8,000 times per second. An additional bit was added to help the receivers distinguish the beginning of one DS-1 frame from the end of the previous DS-1 frame. So the aggregate bit rate of a DS-1 was 193 bits/frame $\times$ 8,000 frame/second = 1,544,000 bits per second or 1.544 Mbps. The structure of a DS-1 frame is shown in Figure 4-3.

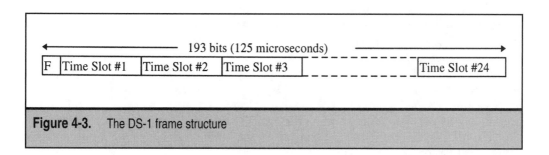

Figure 4-3.   The DS-1 frame structure

Within the DS-1 frame, a DS-0 input channel always occupied the same position in the frame. For instance, a particular DS-0 channel may be the first 8 bits after the framing bit, or the next 8 bits after that, and so on. Because the DS-1 frame was thus divided by *time* and not by frequency, this method was referred to as time division multiplexing (TDM). Although it is not impossible to FDM digital signals, or to TDM analog signals, the equipment needed to do so is complex and expensive. It is easier and cheaper to multiplex analog signals with FDM and multiplex digital signals with TDM, and this approach is nearly universally observed today.

It is important to note that there is some *overhead* associated the DS-1 frame structure. Overhead are bits that are needed, but do not represent the information sent. Because the 24 DS-0s only generate 24×8 bits/frame $\times 8,000$ frames/second = 1,536,000 bits/second, a difference of 1.544 Mbps - 1.536 Mbps = .008 Mbps = 8 Kbps occurs between the aggregate DS-1 bit rate and the sum of the inputs. This "extra" 8,000 bits per second of course is due to the presence of the framing bit in every DS-1 frame. Telephone companies use this overhead for a variety of purposes today, such as error control, but 8,000 bits per second is not much in a world where even 30,000 bits per second or so is considered barely adequate for checking e-mail from a home computer.

Other information needed to be conveyed over a T-carrier link as well. For example, both ends needed to know which trunks were idle, which were in use, which were now being seized from one end or the other for a voice channel. No place existed in the T-1 frame structure to carry this signaling and supervisory information; therefore, the T-carrier equipment routinely "robbed" bits from each individual voice channel for signaling and supervisory and maintenance purposes. This *robbed bit signaling* became a distinguishing characteristic of the T-1 frame structure.

It quickly became apparent that robbing a bit from each voice channel 8,000 times a second was overkill for signaling purposes. It was simply not necessary to know to the nearest 125 microseconds when an idle trunk channel was about to be seized. When not used for signaling purposes, the robbed bit in each voice channel (DS-0) came to be used for other things as well. Because T-1s indicated 0 bits by a lack of voltage on the line, too many consecutive 0's in a row caused the receivers to go out of synchronization with the senders. It was then impossible to determine whether 20 or 21 consecutive 0's were actually sent. Initially, T-1 multiplexers set the maximum number of consecutive 0's sent to 15. If a user were to generate more than 15 consecutive 0 bits on adjacent DS-0 channels, the robbed bit would be used to enforce this minimal "1's density" requirement.

The robbed bit was also used to convey certain alarm conditions (specifically, *yellow alarm*) and to carry special maintenance messages between the T-1 multiplexers. These four uses of the robbed bit—signaling, 1's density, yellow alarm, and maintenance—effectively limited the useable bandwidth of a DS-0 channel in a T-carrier network to 7 bits/frame $\times 8,000$ frames/second = 56 Kbps, especially in data applications. The quality of normal voice conversations were unaffected by this wholesale bit robbery.

T-carrier grew to include many hierarchical levels. That is, DS-1s were combined in a variety of ways to yield higher and higher capacity trunks. The goal was to aggregate as much traffic as possible on a single trunk T-carrier, and it was enormously successful in doing this in a cost effective fashion.

Other Digital Carrier Schemes

While all of this development was going on in the United States, the rest of the world, mainly meaning Europe, was not idle. The Europeans were busily completing their own hierarchy of digital carrier systems, which differed from the United States version in a number of significant ways. First, even though the European systems also produced 8 bits 8,000 times per second, the 8 bits were coded differently than in the United States version, and actually produced even less noise than the United States digital encoding method; however, the differences were great enough to prevent a digital voice channel from being directly connected to an international circuit in most cases. Confusingly, both are often designated DS-0, which means the exact coding method must be determined by context.

Next, the Europeans also improved on the United States DS-1 frame structure. The 8 Kbps DS-1 frame overhead that was barely adequate for management purposes, and totally inadequate for signaling in switched environments, such as normal voice operation, was replaced by a more advanced structure. In the United States, many functions had to be done by "robbing" a bit from each voice channel, limiting many applications to 56 Kbps instead of 64 Kbps at the DS-0 level.

The Europeans eliminated bit robbing by adding a full 16 bits of overhead to each frame, rather than just a single bit, as in the United States T-carrier system. A group of 8 bits usually is referred to as an *octet*, but in the United States this group of 8 bits is more commonly, but less correctly, called a *byte*. There were, and perhaps still are, bytes that are not always 8 bits long. But all octets are exactly 8 bits long, by definition. The 16 bits of overhead in the European digital carrier standard was organized into 2 octets. The first octet was located at the beginning of the frame and allowed sending and receiving equipment to stay in timing synchronization for the entire duration of the frame, regardless of how many consecutive 0's were encountered in adjacent DS-0s. This framing "word" (yet another term for a group of 8 bits) was followed by the octets from 15 DS-0 digital inputs. Then a group of 8 bits used for signaling, alarms, and maintenance was inserted in the frame. The frame concluded with another series of octets from another 15 DS-0 digital inputs. The whole frame structure yielded 30 DS-0 64 Kbps channels with 2 overhead channels, also functioning at 64 Kbps. But this allowed the European digital carrier system to offer the full 64 Kbps channel to any application.

This frame structure gave an aggregate bit rate of $32 \times 64,000$ bits/second = 2,048,000 bps = 2048 Mbps. This became the Conference of European Post and Telecommunications (CEPT) administrations standard. This was designated E-1 to distinguish it from the T-1 frame structure of 1.544 Mbps; the structure of the E-1 frame is shown in Figure 4-4.

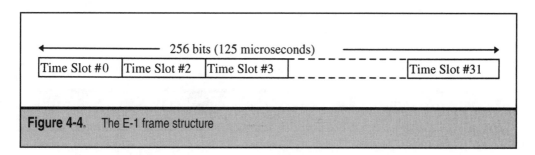

Figure 4-4. The E-1 frame structure

DIGITAL HIERARCHIES

T-1 was just the beginning of a whole hierarchy of multiplexing schemes. It made sense (and saved money) to aggregate larger and larger groups of digital channels onto a single trunk or transmission channel. After all, the channels in the trunk were all going between the same two points anyway. This hierarchical structure of course was not unique to digital multiplexing. Analog multiplexing hierarchies based on FDM (such as L-carrier) had existed, and continued to exist, side by side with digital multiplexing based on TDM. Unfortunately, the differences between T-1 and E-1 frame structures translated to differences in their higher levels of multiplexing as well. Even more confusing, other nations, which had adopted the basic T-1 frame structure (e.g., Japan), organized higher levels differently.

To sort all this out, it became common to refer to the first-level digital carrier hierarchy as T-1 in the United States (technically North America), E-1 in Europe, and J-1 in Japan. (There are some other differences between North American and Japanese carrier systems beyond the scope of the present discussion.) This terminology extended to all levels of the hierarchy as well. Table 4-1 shows all of the defined levels of digital TDM multiplexing in use around the world. (This table also appears at the beginning of this book for ease of reference.)

Digital Multiplexing Level	Number of Equivalent Voice Channels	Bit Rate (Mbps) N. America	Bit Rate (Mbps) Europe	Bit Rate (Mbps) Japan
DS-0/E0/J0	1	0.064	0.064	0.064
DS-1/J-1	24	1.544	---	1.544
E1	30	---	2.048	---
DS-1C/J1C	48*	3.152	---	3.152
DS-2/J2	96	6.312	---	6.312
E2	120	---	8.448	---
E3/J3	480	---	34.368	32.064
DS-3	672	44.736	---	---
DS-3J	1344*	91.053	---	---
J3C	1440*	---	---	97.728
E4	1920	---	139.264	---
DS-4	4032	274.176	---	---
J4	5760	---	---	397.200
5	7680	---	565.148	---

(*) Intermediate multiplexing rates.

Table 4-1. Digital TDM hierarchy used in North America, Europe and Japan

The use of the asterisk to indicate "intermediate multiplexing rates" simply means that the indicated level of the hierarchy was formed primarily to act as a intermediate stage on the way to some other level. For instance, the North American DS-1C was formed primarily to act as an intermediate stage when going from a DS-1 to a DS-2. Until relatively recently, these intermediate levels were not seen as user interfaces or even on the transmission network itself.

Of course, it is nice to have all of the world's digital hierarchies in one place for reference purposes. However, SONET was designed and developed primarily to replace the existing T-carrier hierarchy in North America. This is not to say that SONET is not an important part of the full international standard. The fact that SONET is primarily intended for use in North America merely acknowledges SONET's origin and main sphere of interest.

It may be instructive to describe the levels of the T-carrier portion of the hierarchy used in North America. These are all digital multiplexing techniques using time-division multiplexing (TDM). The T-carrier hierarchy is detailed in Table 4-2.

The term *intermediate multiplexing* rates again refers to the same practice as before. DS-3Cs were primarily formed to carry digital signals on digitized microwave links and to carry North American television signals on older fiber optic networks. For many years, events like Monday Night Football carried the video portion of their signal on a nation-wide DS-3C network from interexchange carriers such as AT&T or MCI.

The hierarchy begins with the basic building block of a single voice channel digitized at the international standard rate of 64 Kbps (which is 0.064 Mbps). No transmission facility is associated with this rate, which is known as DS-0 (digital signal level 0). Twenty-four of these DS-0s are multiplexed together with the TDM method to become a DS-1. In this case, the transmission facility is known as the T-1 (T-carrier Level 1). T-carrier describes certain factors, such as the copper wire pairs or coax cable characteristics and the transmitter and receiver specifications, for the whole hierarchy. The DS part describes the signal format on these physical links.

Digital Multiplexing Level	Number of Equivalent Voice Channels	Bit Rate (Mbps)
DS-0/E0/J0	1	0.064
DS-1/J-1	24	1.544
DS-1C/J1C	48*	3.152
DS-2/J2	96	6.312
DS-3	672	44.736
DS-3J	1344*	91.053
DS-4	4032	274.176

(*) Intermediate multiplexing rates.

Table 4-2.   Digital TDM hierarchy used in North America

Most private networks in organizations today are built from leased T-1s and T-3s, or DS-1s and DS-3s. In either case, thousands of organizations now have at least one T-1, and many have 10 or more. T-3s are less common, but are selling well today.

T-2 has begun to be sold, mostly for LAN connectivity. T-3C exists partly because it fits nicely into a microwave tower. Note that the "C" is always capitalized and stands for "concatenation." A "concatenated T-1" (T-1C) is two T-1s "pasted" together for transmission purposes. No one has ever seen a standard DS-4 or T-4. Although DS-4/T-4s with various proprietary elements did see limited deployment, the standard was never developed and most service providers have tried to fill in the gap with various proprietary ways to aggregate multiple DS-3 signals onto a fiber optic link.

T-carrier links do not have to be channelized into a given number of voice channels. It is possible to have an *unchannelized* T-1, which offers an unstructured, raw bit rate of 1.536 Mbps (and, in a few cases, even the full unframed 1.544 Mbps).

Proprietary T-3 and Above

Because leased line T-carrier networks give a user "all of the bandwidth all of the time," the philosophy of service providers has always been to sell "pieces" of bandwidth delivered on higher-bandwidth links farther up the digital multiplexing hierarchy. For instance, in the early 1980s, when T-1 was the most common digital link used internally between switching offices on the service provider network, the highest commonly available rate was the basic DS-0 line running at 64 Kbps.

By the mid 1980s, service providers (then almost exclusively the local and long-distance telephone companies) began to offer full T-1 rates transmitting at 1.5 Mbps. These T-1 connections could either be provided channelized into twenty-four 64 Kbps DS-0s or unchannelized as an unstructured data steam transmitting at 1.5 Mbps. It made little difference to the service providers because the entire 1.5 Mbps could no longer be used by the service provider for anything other than the transport of the customer's bits.

Of course, by the late 1980s, users began to experience pressure to be able to go even faster. Multimedia, networked applications, imaging, and the raw horsepower of LAN-attached desktop computers began to overwhelm even a full T-1 in many cases. The next logical step beyond T-1 in the digital hierarchy was T-3; T-2 equipment existed, but mainly as an intermediate step to T-3. However, if a service provider were going to offer T-3 speeds to customers, for technical and economic reasons the service would need to be based on something that operated even faster than T-3. The problem was that no standard really existed beyond T-3, in spite of the existence of T-4 on paper, and a standard fiber optic interface for T-3 was never fully completed and standardized.

This did not stop many service providers from more or less inventing their own ways of carrying and delivering bit rates above T-3 over their networks. All of these methods were nonstandard and effectively proprietary, meaning that a piece of sending equipment from one vendor had to be matched by a piece of receiving gear from the same vendor at the other end of the link. Because the user was not buying the "T-4," but rather the T-3s (or multiple T-1s) that it carried, this was not a bad solution.

Even the T-3s became the target of much activity. Because no standard fiber optic interface existed for T-3, many service providers became creative with how a T-3 signal operating at 45 Mbps was picked up from a customer site or switching office and carried internally on the service provider's network. There was nothing wrong with this approach. After all, users did not care much about how their bits got from place to place on the networks, only that they got there.

Many proprietary T-3 and above digital multiplexing schemes were in use by 1990 in North America; these are listed in Table 4-3.

Most of the levels listed in Table 4-3 were implemented on fiber optic cable, and all are tagged with a nonstandard designation. The terms applied to the levels in the digital hierarchy are in quotes because of the lack of a standard with these carrier systems.

As long as these digital systems were used internally, no real problem with the proprietary nature of the technology existed. However, if any of these systems were to be offered to customers as leased private lines (and expensive ones at that), the equipment located at the customer site would be considered as customer premises equipment (CPE). As such, the service provider had no control over the customer's vendor selection for this equipment. However, if the proper vendor were not chosen—and in many cases, a low bid process or other open supplier procedure had to be followed—then the link would not work. Thus, these proprietary levels of the digital hierarchy were never offered for customer leasing.

The lack of a fully standardized T-3 rate and above method for digital multiplexing and transporting high-bit rate data streams was one of the problems that ultimately led to the development of SONET. Nevertheless, the use of all levels of the T-carrier hierarchy has become the accepted way to build private networks.

Carrier Multiplexing Level	Number of Equivalent T-3s	Bit Rate (Mbps)
N.S. T-3	1	48-50
N.S. T-3	2	90
N.S. T-3	3	135
N.S. "T-4"	9	405 or 432
N.S. "T-4"	12	560 or 565
N.S. "T-4C"	18	810
N.S. "T-4C"	24	1,100 (1.1 Gbps)
N.S. "T-4C"	36	1,100 (1.1 Gpbs)
(N.S. = Nonstandard)		

Table 4-3. Propriety T-3 and above links in North America

T-Carrier and Private Networks

As strange as it may seem at first, most major corporations in the United States today (and in many others around the world) are little telephone companies. People do not often think of the corporate networking department in this way, but corporations in the last 20 years have not looked to public telephone companies for networking solutions to corporate networking problems. Rather, they have built their own private networks to provide custom networking solutions.

The difference between a "public" and "private" network is an important one, especially when it comes to a discussion of SONET. All networks consist of a number of simple elements. One of these elements is the "end system" itself: the source and/or destination of the information traveling across the network. Another element includes the "network nodes" that reside within the network and forward the traffic from an input port to an output port. Information makes its way across the network in a series of "hops" from one network node to another from the source end system to the destination end system.

In a public network, the service provider owns and operates the network nodes. These network nodes may be switches or routers or even some more exotic network device. The key is that the user/customer need not own or operate any of the network nodes. The service provider takes care of all of this on behalf of the customers who are usually called subscribers on a public network (such as a frame relay network).

In a private network, conversely, the customer owns and operates the network nodes. If any role exists for the public service provider, it would be to supply the bandwidth needed to link the customer-owned-and-operated network nodes together. These "private leased lines" are really just rented pieces of the telephone company network that supply the bandwidth needed for this private network node connectivity.

Although many telephone carriers in the United States have had success as Internet service providers (ISPs), they have more or less accepted their role as simple bandwidth providers for their corporate customers, at least for the present. Of course, corporate customers have had to assume a larger role in building, operating, and managing private networks. In this sense, corporations have had little choice but to become small telephone companies on behalf of their employees.

The origins of this trend are no mystery. Even before the rise of large corporate data networks, corporate voice networks evolved to the point where it made sense to have a small voice switch on the corporate premises. This was the origin of the "private branch exchange" (PBX). The term "private" accurately reflected the migration of the "exchange" (voice switching) function from being the responsibility of the public telephone company to being the responsibility of the private corporation. The resulting voice network was small (a "branch" office of the larger public telephone switch to which it was still attached), but the message was clear: A service that was usually provided by a public service provider was now supplied to customers (i.e., employees of the corporation) by a private network. Another arrangement, called *outsourcing*, treats another's private network similarly to a public one.

When data networks became common in the corporate environment, which usually meant the deployment of an IBM SNA network, the trend continued. There was no longer any question of obtaining network services for SNA networks from a public network service provider because IBM sold front end processors (FEPs) directly to corporate customers. These FEPs formed the network nodes of the IBM SNA network. Corporate data managers liked the control they enjoyed over the corporate network, and because the bandwidth had to be purchased (i.e., leased) from the telephone company anyway, the public telephone company profited without needing to understand SNA data networking.

Outside of the United States, there were more determined efforts to provide public data networking services. The X.25 international standard for public packet-switched data networks, explained in a little more detail later in this work, was a major step. Bandwidth in the form of leased private lines was much more expensive and facilities scarce outside of the United States for economic and policy reasons, and X.25 enjoyed mild success, especially in Europe. Within the United States, the practice of building private corporate networks had become such a part of the corporate SNA networking environment that efforts on the part of telephone companies to promote public X.25 services largely failed. IBM helped by pricing X.25 support on SNA network so high that it was much cheaper to lease point-to-point private lines to link FEPs together.

Linking LANs with Private Lines

Before the mid 1980s, corporate networks usually consisted of a hierarchical SNA network which employed low-speed, analog and digital- leased private lines. In many cases, the SNA network spanned the entire organization, coast-to-coast, and the cost of leasing these private lines by the mile from telephone companies was controlled by using a *multidrop* arrangement. With a multidrop circuit, many remote SNA cluster controllers could share a single port on the mainframe's FEP. Because the remote cluster controllers in an SNA network could not communicate directly, all traffic to and from a particular site had to pass through the mainframe location. The use of multidrop lines was an elegant solution for the modest amounts of text-based information sent over an SNA network from terminal device to host application and back again.

By the mid 1980s, many companies were replacing desktop terminals with PCs, which could do more for an individual employee than provide access to the corporate mainframe. To be sure, mainframe access was still an important part of the employee's job function. In fact, most workers simply used the desktop PC to access the corporate mainframe applications they needed to perform their jobs. The PC did not magically make applications and data jump off of the mainframe onto the desktop PC. A whole class of PC applications known as *terminal emulation programs* evolved to meet this continued need for central computer access. With terminal emulation, a PC user could still be wired to the cluster controller for mainframe access.

The corporate networking environment situation changed in the 1980s, quickly and dramatically. Most private corporate networks had been built to accommodate the IBM SNA environment. Even when other computer vendors' networking products were used (e.g., DECnet and HP networking products), the resulting networks looked much the same as SNA networks: hierarchical, star-shaped, wide area networks (WANs).

These networks typically linked a remote corporate site to a central location where the IBM mainframe or other vendor's minicomputer essentially ran the whole network. In this sense, the networks were hierarchical, with the central computer forming the top of a pyramid and the remote sites only communicating, if at all, through the central computer. Building private networks for this environment was easy and relatively inexpensive. Private lines were leased from the telephone companies to link all the remote sites to the central site. Although private lines were leased by the mile (i.e., a 1,000-mile link cost more than a 100-mile link), there were various ways around letting this variable expense limitation impose too much of a burden on the private network.

This situation is illustrated in Figure 4-5; five sites are networked together with point-to-point leased lines. The connectivity needed is simple: hook all four of the remote sites to the remaining central site where the main corporate computer was located.

A quick look at the figure makes it easy to see how many links are needed to create this network. Four private leased lines are depicted; fewer cannot be used to link each remote site directly to the central location. These links are the main expense when building and operating such private corporate networks. These links, however, are not the only expense; a communications port must be maintained on each computer in the network. This expense is minimal at each remote site. A glance at the figure shows that each remote site needs only one communications port and related hardware. The situation is different at the central computer site where the communications port and related hardware needs to link to each of the remote sites. Of course, the central computer must be configured with the total number of ports needed for remote site connectivity. This last requirement usually was not a problem, especially in the IBM mainframe environment.

However, what if the number of remote sites grew to ten? Twenty? One hundred? How many links and ports would then be needed to deploy such a hierarchical network? As corporations—and corporate networks—merged, expanded, and otherwise grew

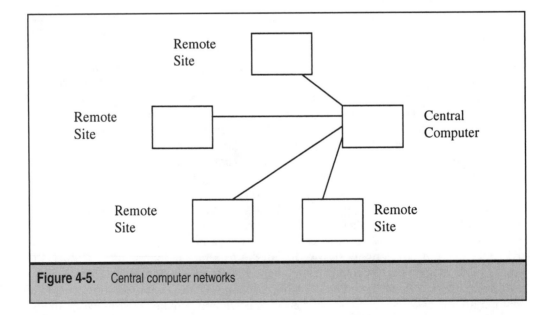

Figure 4-5.   Central computer networks

throughout the 1970s and into the 1980s, this issue became a problem for corporate data network designers.

Fortunately, it is not necessary to draw detailed pictures of these larger networks to determine how many links and ports are needed. A simple mathematical relationship can be used to determine how many links and ports are needed to link any number of sites into a hierarchical, star-shaped network. When the number of sites is designated by the letter "N" (including the central site), then the number of links needed would be N-1. For instance, in the above figure, N = 5 and the number of links needed is N-1 = 4. The number of communication ports needed throughout the network is given by 2(N-1) (read as "2 times N-1"). When N = 5, the number of ports is 2(N-1) = 8. Fully half of these ports (given by N-1 = 4) would be needed at the central site.

It is now easy to figure out that when the number of sites is 20 (N = 20), the number of links would be 19 (N-1 = 19) and the number of ports would be 38 (2(N-1) = 38), with 19 of them (N-1) at the central site. If N = 100 (100 locations on the corporate network, a figure easily approached in larger corporations and governmental agencies as well), the number of links would be 99 and the number of communications ports would be 198, with 99 at the central site. These networks were large and expensive, but not prohibitive.

What has all of this to do with T-carrier? Simply this: The rise of corporate LANs and client/server computing in the 1980s has meant that building private corporate networks in hierarchical stars is no longer adequate for private corporate networks. This is not the place to discuss the evolution of LANs and client/server in detail. It is enough to understand that LANs and PCs running client/server software (e.g., database client package to a database server or even corporate e-mail applications) are best served by networks with more than just a single link to some central location.

In a client/server LAN environment, it can no longer be assumed that all communications occur between a remote site and a central location, as hierarchical networks assumed. In a client/server environment with LANs connected by WANs, any client may need to access any server, no matter where the client- or server-PC happened to be located. Client/server LANs at corporate sites that need to be connected are better served by peer, mesh-connected networks. Today, SONET/SDH has new life as a technology to connect an organization's or service provider's *routers*, not older networking devices. Routers also work better the denser the mesh of connecting links becomes.

This need for a different type of private corporate network created problems. The number of T-carrier links and ports needed for peer, mesh-connected private LAN networks were much higher than the modest link and port needs in the older hierarchical, star environment.

To see why this is true, Figure 4-6 illustrates the problem. A mesh network consists of direct point-to-point links between every pair of locations (LANs) on the corporate network. This way, no client and server is separated by more than one direct link; however, the associated numbers for the required links and ports have exploded.

The figure demonstrates that the peer network requires ten links and four communications ports at *each* location to establish the required mesh connectivity. The total number of ports needed is now twenty. The formulas, again based on well-understood

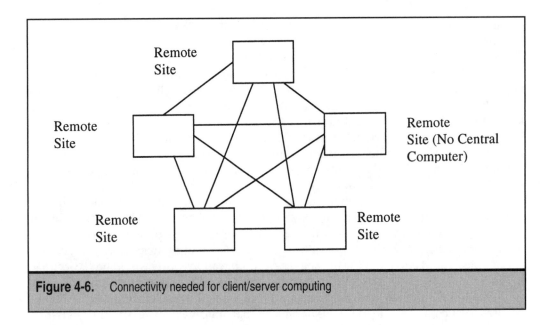

Figure 4-6. Connectivity needed for client/server computing

mathematical principles, are now N(N-1)/2 = (5 × 4)/2 = 10 for the number of point-to-point links needed to mesh connect N = 5 sites and N(N-1) = (5 × 4) = 20 for the total number of communications ports needed (four at each site).

For twenty sites (N = 20), the numbers would include (20 × 19)/2 = 190 for the links and 20 × 19 = 380 for the ports (19 at each site). Although it would not be impossible to configure 19 ports for each site, the hardware expense alone would be enormously high, even prohibitive. Most network managers would balk at providing 190 WAN leased lines paid for by the mile to link the sites together. For 100 sites (N = 100), the numbers would require an astonishing 4,950 links ((100 × 99)/2) and 9,900 communications ports (100 × 99). Each site would need hardware to support 99 links to other sites, an impossible task for any common communications device architecture.

Again, various strategies were employed by corporations to keep LAN (and router) connectivity networking expenses in line. Partial meshes were deployed in varying "backbone" and "access network" configurations. These measures were quite successful, in the sense that these arrangements satisfied connectivity requirements at a minimal cost.

T-Carrier and Network Applications

The above section mentioned some of the "workarounds" that corporate data communications managers have employed to try to avoid the expense involved with creating mesh-connected LAN internetworks. These strategies all involved paring down the number of links and ports needed so that even if all the sites were not point-to-point directly connected, at least each site was reachable from any other site. In most cases, this meant that any client could still attach to and access data on any server, but not always directly. This put more of a traffic load on some network links and network nodes. Higher speed T-carrier links offered relief from this situation. Instead of a few 64 Kbps links between sites, a full, unchannelized T-1 running 1.5 Mbps was normally employed.

There is another reason that T-carriers, especially full T-1 links operating at 1.5 Mbps, have been employed for many of the applications that users run on LANs in a client/server environment. The reason is referred to here informally as "application overload," which means that LAN speeds and application data exchange needs have made the WAN portion of the corporate data network an automatic bottleneck that slows these networks to a crawl when the link bandwidth is not adequate.

Initially, the new and powerful PCs in an organization were used mainly to access applications and data on the mainframe running terminal emulation. This worked, but hardly took advantage of the computing power now available on the desktop. Fortunately, this computing power did not go untapped for long. Organizations began to develop—or even purchase off-the-shelf—applications that operated and accessed data available locally, within the department in which the employee was located. This movement toward *departmental computing* led to a separate need to network these PCs not throughout the organization, but within individual departments. The LAN was a perfect solution. LANs provided networks for many intelligent devices (PCs) at a single location, rather than the WAN connectivity provided by the SNA network.

By about 1990, a whole new way of looking at corporate applications and the networks they ran on had evolved. The need to share information and resources among users led to the development of the idea of distributed processing, most often along the lines of the client/server model. Although some of these terms have no precise definitions, it is generally agreed that sharing "information" refers to the need for PC-based applications to sometimes access data stored on another computer, whether local or not. Shared "resources" refers to the expensive hardware that PC users sometimes need to complete their jobs, hardware that could not effectively or affordably be provided to each individual PC user. Not long ago, both modems and laser printers fell into this "shared resource" category. Falling hardware prices have served to make both of these pieces of computing equipment common on most desktops, but other devices that are new and expensive fit into the sharing category, such as color laser printers and DVD-RW writing devices.

Distributed processing acknowledges the fact that desktop PCs are powerful enough to perform work-related functions that were previously reserved for the central computing site. There are various forms that any movement toward distributed processing or computing can take. One of the most common is the employment of the client/server model for distributed processing. In the client/server model, most desktop PCs are classified as clients, meaning these PCs run some form a LAN Network Operating System (NOS) software. The NOS (for example, Novell's once highly successful NetWare product) allows the client PC to attach to another kind of PC device known as a server over the LAN to which both devices must be attached. The server runs the bulk of the NOS software, and clients can only communicate among themselves by first sending data or messages to the server, which forwards the data or messages across the LAN to the correct destination client PC.

The whole scheme is quite elegant and takes advantage of the tremendous processing power available on each desktop in a modern corporation. A problem arose with this idea when departments with similar functions had a need to share information and resources not over an individual LAN, but with other remote sites in the same company. For example, a national company often had a need to share information between a client on the west coast and a server on the east coast. This commonly occurred when a corporate customer located on the east coast was traveling on business to the west coast. An order placed to the west coast sales office meant that the sales representative had to access the customer's sales account information on the east coast server.

The most efficient and effective way of doing this was to allow the west coast client to access the east coast server. This meant that the LANs in both locations had to be networked together over a WAN. Because the applications that had been developed in the client/server environment typically used a Windows-based graphical user interface (GUI) rather than the older text-based terminal interface, the amounts of data exchanged by a relatively simple sales transaction were much larger than anything that the corporate SNA network had been designed to handle.

As a result, in most cases separate WAN links were leased and used to connect special LAN internetworking devices together at the various locations needing such connectivity. These special network devices could be LAN bridges or routers, but as time progressed the LAN internetworking device of choice became the router, for a variety of reasons beyond the scope of the present discussion. A new philosophy and terminology of corporate networking arose in this environment, which is most commonly known as "router-based networking."

The most critical networking consideration in router-based networking is the need for efficient router connectivity across the corporate WAN. Only then could the logical conclusion of the client/server model ("any client can access any server") be realized. But this led to a new series of challenges for the builder and managers of this new breed of networking.

The number of PCs attached to corporate LANs increased rapidly. The number of LANs needing connectivity and the associated corporate support across a WAN increased rapidly as well. It quickly became apparent that client/server LAN traffic was not the same as SNA traffic. LAN traffic is much more diverse, of a much higher volume, and much harder to predict in terms of traffic patterns than is the traffic on SNA networks.

As the day-to-day operational applications in corporations became more dependent on the LAN-based client/server applications, LAN internetwork traffic grew dramatically. The level of network complexity grew rapidly as well because client/server router-based networks are not well served by either low-speed multidrop lines or hierarchical, star-shaped WANs. The needs of LAN applications led to attempts to connect each router with a direct link to every other router. As the router network complexity grew, it became impossible to keep up with client/server demands, especially as more SNA devices came to be attached to LANs as well.

A newer 1990s corporate trend developed Internet-based corporate services. These newer applications often needed the services of a more powerful UNIX-based workstation to provide the company with a "web site" presence on the logical subset of the public Internet known as the World Wide Web. Other companies relied on UNIX-powered workstations for more processor-intensive applications, such as computer-aided design/computer-aided manufacturing (CAD/CAM) that remained beyond the abilities of the desktop PC. New multimedia applications requiring a network handling not only data, but audio and video were developed for both PC and UNIX computing platforms. The TCP/IP protocol played a large part in this as well.

In this environment, routers were best served by "mesh connectivity" with direct point-to-point leased lines to every other router site. The numbers of communication ports needed on each device grew as well. A compromise solution deployed "partial mesh" and "router backbone" networks, but the amount of LAN traffic quickly exhausted the capabilities of these compromises to deliver the speed and bandwidth that LAN-based client/server user applications demanded.

Older LANs built in the early 1980s operated at the familiar Ethernet and Token Ring speeds of 10 megabits (millions of bits) and 4 megabits per second (Mbps), respectively. By the mid 1980s, Token Ring LANs operated at 16 Mbps. Of course, these LANs had to be connected over a private corporate WAN with leased lines from the telephone companies. The most common affordable speed for these links was 64 kilobits per second (Kbps, thousands of bits per second). One million is 1,000 times larger than one thousand; therefore, trying to let LANs running at millions of bits per second communicate over WAN links running at only 64 Kbps created a severe congestion problem on the corporate network, with the associated delays that exasperated users.

Even when a corporation employed a full T-1 operating at 1.5 Mbps between LAN sites, the bottleneck created trying to link two Ethernet LANs running at 10 Mbps over the T-1 was still about 6 to 1 (6 times 1.5 Mbps = 9.0 Mbps). There was only one sure way to make this type of network functioned even marginally acceptably: reduce the amount of information that needed to be sent over the WAN portion of the network.

Unfortunately, this philosophy went against the trend of the applications the users were running that led to the need for LAN internetworking in the first place. By the early 1990s, these applications were no longer just sending modest amounts of data bits from client to server and back across the WAN. Company employees were looking at including multimedia in their LAN applications. Multimedia combined images, graphics, audio, and video with the traditional data into new and exciting applications for sales and marketing, research and development, and even information systems departments.

The distinguishing characteristics of multimedia applications in a client/server environment are two-fold. First, multimedia applications need plenty of horsepower on every client desktop to deliver realistic audio and full-motion video to the user. The rapidly falling cost of PC hardware guaranteed that this would not be an insurmountable problem for most organizations, either for the clients or the servers. Second, multimedia applications need plenty of bandwidth to deliver the much greater amount of bits these applications required, and within a smaller network delay time frame. Users who would wait patiently for about 10 seconds for a database record or Web page would not stand for 10 "freezes" of video or 10 seconds of silence during an audio playback.

Furthermore, the whole philosophy of distributed, client/server applications prevented the organization from simply duplicating information at every LAN site in the organization. The resulting complexity of propagating updates everywhere, and the sheer volume of the updates sent over the network in this case, easily offset any gain realized from cutting back on transporting the multimedia information across the overburdened WAN links. By the mid 1990s, the problem had only gotten worse.

Even LANs running at 10 and 16 Mbps were being swamped by many users all running the newer applications over shared-media LANs which were, after all, based on early 1980s technology. Even when multimedia applications were not an issue, newer PCs were so fast that they could easily fill up a 10 or 16 Mbps LANs with traffic from only a handful of users.

LAN network managers responded with one of two solutions, and sometimes even employed both in many cases. First, so-called "LAN switches" appeared, giving each user attached to the LAN essentially a dedicated 10 or 16 Mbps right to each PC. The result was that there were even more bits waiting to travel across the WAN link to remote locations, in some cases 10 times as many bits or more. Second, newer LAN technologies appeared that ran at 100 Mbps, turning a barely acceptable LAN-to-WAN 6 to 1 bottleneck into an unbearable 60 to 1 bottleneck at the LAN-WAN interface.

These network devices employed at the LAN-WAN interface were usually routers, which were capable of linking two separate LANs over a WAN link. Sometimes bridges were used for the same purpose, and although the two devices behaved differently, both enabled clients on one LAN to access a server or servers on another LAN at a remote location.

A much better fit for 10 Mbps Ethernet and 16 Mbps Token Ring LAN connectivity would be T-3 running at 45 Mbps. Many organizations have begun to use these higher speeds for this purpose. Until the Web explosion and the concerns with heavily-visited web site congestion, T-3s were relatively rare and expensive (in part because of their scarcity) and so were not easily used to link LANs together.

Many modern applications need more bandwidth than is easily affordable or available to build the types of LAN-WAN-LAN connectivity required for distributed, client/server environments, especially when faster LANs are factored in. Additional problems arose with the T-carrier structure that became more apparent in the late 1980s.

The Trouble with T-Carrier

By the 1980s it became obvious that drawbacks existed with the digital hierarchy that were neither debilitating nor trivial, yet serious enough to require action. T-carrier worked well within certain limitations, but given the accelerated pace of desktop device development and LAN speeds and capabilities, problems needed to be addressed quickly. Accentuating the problem was the fact that more often, T-carrier was used to connect LANs in an organization.

T-carrier was basically a single vendor world at the time. The Western Electric Company (WECO), the manufacturing arm of the Bell System, made T-carrier equipment for all components of the Bell System. It was difficult to obtain accurate performance statistics for these T-carrier links, and they were awkward to troubleshoot.

The Mid-Span Meet and Missing Pieces

The strange fact about the T-carrier hierarchy is that it was never completed. Annoying gaps in the original Bell Labs specifications existed, some of which have been mentioned above. For instance, no DS-3 method of sending a T-3 signal over a fiber optic link was fully specified and deployed. Therefore, even a simple thing, such as getting a DS-3 from New York to New Jersey, was often an adventure and costly. DS-4 was in even poorer shape, but even less of it existed at the time. For these reasons, service providers that wanted to deploy the T-carrier hierarchy—which was all of them—had to fill in the blanks in the specifications with their own proprietary methods. As long as the Bell System remained intact and all of these carriers bought WECO, a problem did not exist. After all, how many T-1s would start at Illinois Bell and end up at GTE? Not very many. If worse came to worse, the signal could be "re-analoged" for the interface. This lack of ability to have one vendor's equipment or telco on one side of the link and another vendor's equipment on the other side of the link is known as the lack of the *mid-span meet*. In a mid-span meet, one carrier is responsible for one end of the link and another carrier is responsible for the other side of the link.

However, after 1984, there were a lot of situations where the two ends of a T-carrier link were under the control of two different organizations. With the "equal access" rules of divestiture, local callers had the right to use their long-distance carrier of choice, on a per call basis. This required the use of digital T-carrier trunks from the local exchange carrier's (LEC) central offices to the long-distance carrier's points of presence (POP). "No problem," said the new AT&T (which still included WECO), "Everybody just buy WECO." This philosophy was not well received by anyone.

The question then became one of completing the T-carrier standard or trying to invent something new in the world of divestiture.

The Need for Operational Procedures

Network management, troubleshooting, and other administrative networking functions are needed on T-carrier networks as well as any other type of network. Digital T-carrier links need to be monitored, problems detected and fixed, and so forth. Because the customer has basically leased only the bandwidth that the T-carrier link represents, and not the transmission facilities themselves, this network management function remains the responsibility of the service provider. The customer may initiate the process (usually the customer is the first to know that a problem exists on the link), but the service provider carries out the tasks needed to maintain the network link up to the standard set for the leased line service (errors, availability, and so on).

On a T-carrier network, this network management function is known as Operations, Administration, Maintenance, and Provisioning (OAM&P); outside of North America, OAM&P is usually known as OAM, especially in Europe. The process of OAM&P on a T-carrier network is rather primitive. All OAM&P measures in T-carrier, from alarms to maintenance signals, robbed bits from the user channels. The service providers really had no choice because the overhead in a T-carrier transmission frame (1 bit out of the total 193 bits in the transmission frame, or 0.518%) was not adequate for even the simplest OAM&P tasks.

One of the side effects of this was to limit the effective bandwidth of a DS-0 channel to 56 Kbps because one in eight bits in every channel had to be "robbed" for OAM&P (the term "robbed" may be unfortunate, but this is what it was usually called). There were other functions for this "robbed bit," such as signaling and maintaining "1's density," but only OAM&P is important in this limited discussion of the robbed bit. Even the simple act of measuring the Bit Error Rate (BER) on a T-1 link (usually because the customer had complained about it) involved taking the link out of service, performing a BER test (BERT), and then returning it to service (many times with a "no trouble found" designation because the problem was frequently in the CPE).

The overhead in the T-carrier network for OAM&P, even at the higher levels of the hierarchy, was totally inadequate for both the customer and service provider needs.

Byte and Bit Interleaved Multiplexing

The problem of bit interleaved multiplexing is fairly complex. However, because it is an important topic in T-carrier, especially when contrasted with SONET, some time will be spent explaining the related concepts of bit and byte-interleaved multiplexing.

Whenever digital data streams are combined to form a higher-rate carrier system (so called because they "carry" many low-speed links in one higher-speed package), the source digital data streams are almost invariably organized as bytes. Bytes are groups of 8 bits (0's and/or 1's), and officially called octets in international standards. This once was not universally true, but in a world where almost everything from data, to voice samples, to video conferencing comes from a computer, the organization of digital information into bytes (or octets) of exactly 8 bits is a standard. Whether the bits come from a desktop PC, or a digital PBX, or a digitized video stream, a computer chip pumps out those bits somewhere.

Now combining these individual source streams involves the actual multiplexing of them onto a higher rate link and of course de-multiplexing them into the composite signals (if desired) at the other end of the link. There are two main ways to accomplish this multiplexing of bits, and both of them are used in the T-carrier hierarchy. Both involve *interleaving*, or merging, because it would hardly be possible to take more than a few bits from one input port and forget about the others, even for a short period of time. The receiver would notice a "gap" in the bits arriving immediately and fail. All inputs must be interleaved, even idle ones, because absence of bits always is interpreted as a network failure by the T-carrier receivers. The question is whether to take the input data stream and interleave by *bytes* or by *bits*.

In byte-interleaved multiplexing, input digital streams are multiplexed by interleaving 8 bits (the byte) from each input port with the bytes from the other input ports. The result from one round of servicing all input ports (active or not) is usually organized into a *transmission frame* that is sent on the link. The term *transmission frame* is preferred because there is less confusion when this term is used with other types of frames generally seen on networks, such as Ethernet frames or Token Ring frames.

In a T-1, the multiplexing is byte-interleaved multiplexing. That is, 8 bits are taken from the 24 input ports on the *T-1 mux* (short for multiplexer) CPE. A T-1 mux is also called a *channel bank* when voice is the primary focus of the equipment. The T-1 mux/channel bank terminology came about when channel banks were positioned as voice-only T-1 muxes and the term *T-1 mux* was reserved for data muxes, a distinction based on technology. It is a frustrating fact of life in both the computer and networking industries that distinctions based on technology never last, but terminology based on technology never dies. So the term channel bank is still heard occasionally. Whatever the device, these input bits are taken from each input port and combined to form a DS-1 frame which is sent over the link. This process is shown in Figure 4-7. The trans-

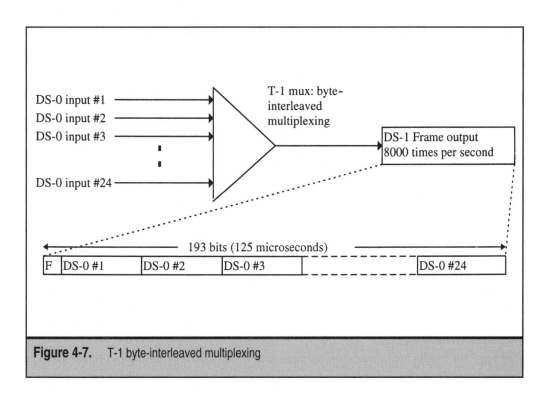

Figure 4-7. T-1 byte-interleaved multiplexing

mission frame has the familiar time slot arrangement characteristic of all time-division multiplexing techniques.

Many nice things happen when byte-interleaved multiplexing is used. First, it is simple and efficient because the processor chips used in the T-1 mux itself move bits around 8 bits at a time, in byte form, the same way as a computer would. Second, it is efficient to store bytes in the memory of the T-1 mux and form the frames with a minimum of effort and does not require much memory (only a byte or so on the input side, only a frame or so on the output side). Third, it is easy for equipment located within the network itself to find an individual DS-0 inside the DS-1 frame. There are several reasons to do this, such as in equipment known as an add/drop multiplexer (ADM).

ADMs are common pieces of equipment in T-carrier networks. Of course, they are not really called ADMs; the most common term is to call these devices digital cross-connect systems (DCS) because these devices do essentially the same thing as a cross-connect panel does in an analog network. When used to rearrange digital customer circuits for higher-level multiplexing, a process known as *grooming*, a DCS is often called a digital *access* cross-connect systems (DACS). But DCS is the more generic term. A cross-connect panel used to be a piece of plywood nailed to the wall that enabled technicians to manually connect an input link to an output link, usually by "punching down" the wire pairs associated with the signal. Today cross-connects are rack-mounted, hightech units in many cases, but manual cross-connect panels still exist. A DCS is just a digitized, computerized, manageable, high-tech version of the manual cross-connect that does electronically what technicians ("craft" people in the telephone industry) had to do manually in the past.

A DCS enables a service provider—or customer when the DCS is configured as a CPE device—to *drop* one or more DS-0 data streams and *add* one or more DS-0 data streams to replace them. Another name for this process is *drop and insert*, but the process is the same. This gives more flexibility to the user of the link.

For example, consider an organization with offices in San Francisco, Atlanta, and New York. Like most organizations, the network linking these sites is a T-carrier network made up of leased private lines. Some DS-0 channels handle voice channels from PBXs, others handle data between routers at these locations, but all are multiplexed onto a T-1 (DS-1). The question is how can these sites be linked most efficiently?

The answer varies based on actual user needs; however, it is not unusual to find the organization employing T-1 muxes in San Francisco and New York, and an ADM in Atlanta. Here is why: When a single T-1 is used between San Francisco and New York, then obviously all 24 channels that start on the west coast end up on the east coast. But what about Atlanta? The organization could lease another two T-1s to link both San Francisco and New York to Atlanta, which gives the network the appearance shown in Figure 4-8. This configuration requires three links and, because these links are paid by the mile, this could be an expensive solution, especially when only 10 or so of the DS-0 links are needed in each location. The usual pricing for T-1s from most service providers means that a T-1 is typically less expensive than as few as six DS-0s.

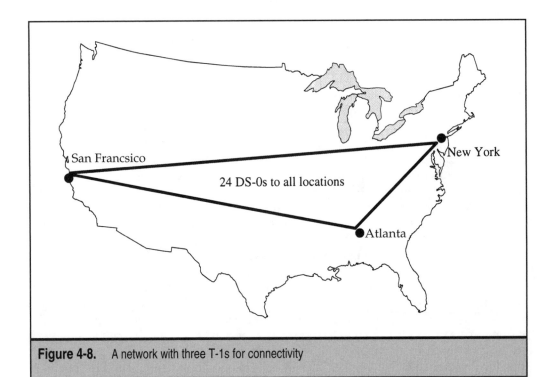

Figure 4-8. A network with three T-1s for connectivity

A better solution does exist. Again, the details would depend on the actual network and usage patterns. The organization would need to employ an ADM as a CPE device in Atlanta. The ADM could *drop* 12 (for example) DS-0s from San Francisco in Atlanta and *insert* (or *add*) 12 DS-0 in Atlanta on the way to New York. The same thing would happen to digital signals sent in the other direction. As a result, 12 DS-0s occur between all three locations, but are delivered over only two T-1s. The basic operation of an ADM is shown in Figure 4-9 showing the adding and dropping of only two DS-0s for simplicity.

It is important to note that in the figure, the DCS shares many of the features of the basic T-1 mux. The DCS both demultiplexes and multiplexes, creates DS-1 transmission frames, has memory buffers, and so on. A buffer is merely a special memory area dedicated to an input or output port on a network device. The input and output DS-1s do not need the DCS. It makes no difference to the links whether two of the DS-0 channels going into the device are not the same as the two input channels. "Twenty-four-in" is still equal to "twenty-four-out." Of course, the input frame must be processed to yield the 24 time slots, and rebuilt on the way out of the DCS; however, this happens quickly. Although the remaining 22 DS-0s in the figure are shown passively shuttled through the DCS, a few configuration commands could allow any of these other channels to be cross-connected as well.

Technically, the term "add-drop" or "drop and insert" should be used only when DS-0s are terminated at a site. The term "cross-connect" should be reserved for the process of rearranging or aggregating DS-0s (a process sometimes called "grooming"). However, the terms are used loosely in many instances today.

The network now looks as it does in Figure 4-10. The network is still private, the links are still leased, but the cost to the organization is much less. DCS and its associated adding and dropping of constituent channels is much easier when byte-interleaved multiplexing is used.

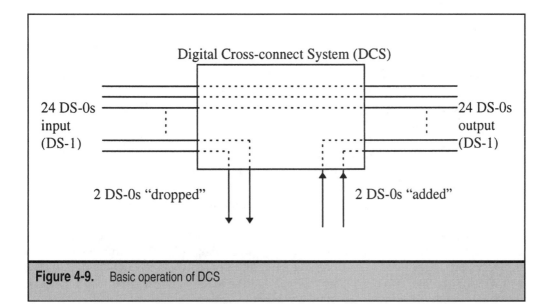

Figure 4-9. Basic operation of DCS

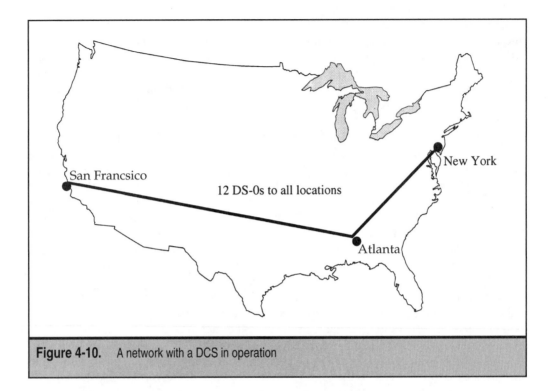

Figure 4-10.   A network with a DCS in operation

Byte-interleaved multiplexing is often called *synchronous* multiplexing, but the term is not universal. The use of the term *synchronous* is unfortunate because it also appears in the acronym SONET. The unpleasant fact is that words such as *synchronous* and *asynchronous* are overused in the telecommunications and networking industry, and often the precise meaning of the term is unclear. In the context of byte-interleaved multiplexing, the term synchronous refers to the fact that the input clocks must be running at the same speed as the output clock.

Not all multiplexing in the T-carrier hierarchy is byte-interleaved multiplexing. In fact, byte-interleaved multiplexing is the exception rather than the rule. With the exception of a special kind of T-3, explored more fully in Chapter 5, the remaining digital T-carrier hierarchy forms higher bit rate frames, and employs *bit*-interleaved multiplexing rather than byte-interleaved multiplexing used when going from 24 DS-0s at 64 Kbps to 1 DS-1 frame at 1.5 Mbps. As may be imagined, the term, *asynchronous multiplexing* is often used as a synonym for bit-interleaved multiplexing.

In bit-interleaved multiplexing, the input digital data streams (i.e., voice, video, or data) are not taken one byte (8 bit octets) at a time to form a transmission frame, but one bit at a time. This leads to some interesting—that is, complex and error-prone—situations on the network link. With byte-interleaved multiplexing, bits are accumulated into a buffer (usually only two bytes or so long, but not necessarily) on the input side. When a bit arrives a microsecond or so "late" or "early" (due to network timing inaccuracies) the input buffer can mask these timing differences from the actual multiplexing portion of the equipment that builds the output transmission frame. It makes no difference if 9 bits or 10 bits were actually in the input buffer, as long as 8 bits (the byte) were ready to go into the output frame. T-1 muxes are specially designed so that this will always occur.

Because bit-interleaved multiplexing relies on smaller buffers and feeds only one bit at a time from each input buffer to be sent out with the output transmission frame, there may be times, for example, when a bit is needed from input port 22 and there is no bit in the buffer. Obviously, the position must be occupied by something, or two things will happen, neither of them particularly good from a networking standpoint. First, the receiving multiplexer will not be able to find the bit in the arriving transmission frame to be demultiplexing to port 22 on the receiving side of the link. Second, the total number of bits sent per unit time would fall below the accepted line rate of the link (such as 44.735 Mbps instead of 44.736 Mbps). In either case, the equipment and link will fail. This makes bit-interleaved multiplexing more awkward to handle than is byte interleaved multiplexing. The reasons for this will be explored in more detail a little later in this chapter. The concept of bit-interleaved multiplexing is illustrated in Figure 4-11, which may be compared to Figure 4-7.

Although Figure 4-11 is accurate in concept, some details are worth noting. First, the framing and overhead bits for a DS-3 frame are much more complex than is the simple framing that the byte-interleaved DS-1 requires. If a bit were not ready to be sent from an

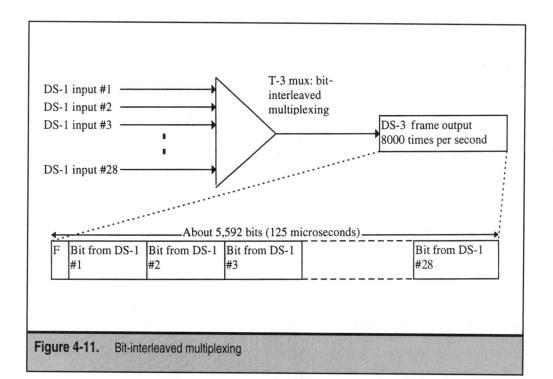

Figure 4-11. Bit-interleaved multiplexing

input buffer (i.e., the input DS-1 clock is running slower than the output DS-3 clock), then a bit must be "invented" by the T-3 mux in order to maintain proper frame structure and line speed. These *stuffed bits* must be removed by the receiving equipment because this is not a "live" data bit on one of the T-1s. Much of the added overhead bits in the full DS-3 frame structure are dedicated to allow receiver equipment to find and eliminate these stuffed bits, if employed. Another important point is that the output data rate is about 5,592 bits every 125 microseconds (8,000th of a second) because when the whole T-carrier structure was set up in the 1960s and 1970s, clocks in processors had difficulty in dealing with such small time slots. After all, 5,592 bits in an 8,000th of a second is only about 0.022 microseconds per bit. At these speeds, the multiplexing equipment had to settle for approximations. There may have been 5,596 bits actually sent in the 8,000th of a second, or 5,588. This variation in timing on a digital link and, thus, the number of bits actually sent or received per unit time, is known as *jitter*. Another term for the same effect is *wander*.

Byte-interleaved multiplexing leads to easy byte (or 8 bit octet) handling in equipment memory and results in easy add/drop and cross-connect equipment arrangements. However, the use of bit-interleaved multiplexing is more complex, error-prone, and awkward both for memory buffer handling and the adding and dropping of both DS-1s and DS-0s—how do you easily find the 8-bit bytes when spread around the arriving frame as individual bits? With this known, then why does the rest of the digital T-carrier hierarchy employ bit-interleaved multiplexing? The answer is that, at the time, no other way existed to do multiplexing at these speeds except using bit-interleaved multiplexing. Also, the cost of the memory used in the buffer was much more expensive than memory costs today. (In 1961, a Meg of memory cost about $4 million.) A T-1 must only buffer 1.5 Megabits to buffer a complete frame. A T-3 mux must buffer 45 Megabits to accomplish the same task, some 30 times more. Also, the simple fact is that jitter is easier to deal with when bit-interleaved multiplexing is used, and dealt with more quickly with fewer buffers. This lack of fast processors and inexpensive memory led to the use of bit-interleaved multiplexing at levels above T-1 in the North American digital hierarchy.

DS-1 TO DS-3 MULTIPLEXING

A good deal of the initial work done that led to the development of SONET was done to address issues of DS-1 to DS-3 multiplexing. Therefore, a more detailed look at this process is in order, above and beyond the brief conceptual introduction above.

The biggest liability in the T-carrier hierarchy had to do with the way that TDM worked in T-carrier. The device used in a T-carrier network that went from 28 DS-1s (each with 24 DS-0s in most cases) to a single DS-3 is known as an M13 multiplexer (pronounced "M-one-three"). The M13 multiplexing process involves the combination of 28 DS-1 (1.544 Mbps) signals into a single DS-3 signal (44.736 Mbps). This is done in a two-stage process, the first stage of which is diagrammed in Figure 4-12. Two stages were needed because of the relatively rudimentary electronics used at the time.

The construction of the DS-1 signal is performed in a synchronous (or *isochronous*) fashion by byte-interleaving single-byte voice samples or other input bits from 24 DS-0 (64 Kbps) signals into an aggregate signal of 1.536 Mbps (without the 8 Kbps framing overhead). In this

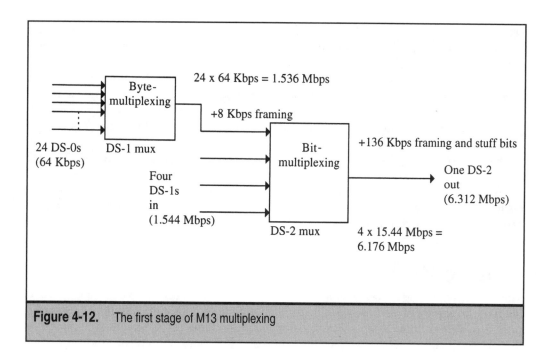

Figure 4-12.   The first stage of M13 multiplexing

process, bytes are grouped into transmission frames of 192 bits each (24 channels × 1 byte per channel). These frames are repeated 8,000 times per second. At the beginning of each 192-bit frame, a single framing bit is added bringing the aggregate DS-1 rate up to 1.544 Mbps.

In Figure 4-12, Stage 1 of the M13 multiplexing process involves the combination of four DS-1 (1.544 Mbps) signals into a single DS-2 signal (6.312 Mbps). In this process, 136 Kbps of overhead is added to the DS-1 signals to produce the DS-2 signal. Part of that overhead is for DS-2 framing purposes. The remainder consists of stuff bits that are added to the DS-1 data to make up for differences in the timing rate (jitter) of the incoming DS-1 signals. Jitter also affects the "phase" of the arriving frame. Technically, the term for two or more signaling events that occur at *nominally* the same rate (within certain defined limits), but that may occur with any phase difference, is *plesiochronous*. Thus, the DS-1 to DS-2 multiplexing process is said to be plesiochronous. In fact, because the only isochronous piece of the T-carrier digital hierarchy is DS-1, the whole T-carrier hierarchy is often called the plesiochronous digital hierarchy (PDH). This term is much more often used outside of North America, where E-carrier is used.

The important point is that the step from 24 DS-0s to one DS-1 is *byte* interleaved, while the step from four DS-1s to a DS-2 is *bit*-interleaved. It is easy to find a DS-0 in a DS-1 because all of the bits are grouped together. But it is next to impossible to find a DS-0 (or even a whole DS-1) inside a DS-2 because the bits are scattered throughout the frame and include stuff bits as well.

The second stage of the M13 multiplexing process is shown in Figure 4-13. This involves the combination of seven DS-2 (6.312 Mbps) signals into a single DS-3 signal

(44.736 Mbps). This is also a plesiochronous process. Of the 552 Kbps of overhead that is added to the DS-2 signals to produce the DS-3 aggregate, much of it consists of stuff bits, which ensure rate and phase alignment of the incoming DS-2s.

When the M13 multiplexing is complete, T-3 data streams are often multiplexed with additional T-3s and transported over a single fiber optic link, but this is a nonstandard process. Therefore, this process is proprietary (i.e., different vendors' equipment complete the process in various fashions) and necessitates the need for matching equipment at both ends of the fiber span.

Figure 4-13 also shows an electro-optical conversion process that combines 12 DS-3 signals (metallic) into a single 565 Mbps optical carrier signal.

Although the example shown on the visual is not hypothetical (i.e., equipment that performs this process exists), many other variants of this electro-optical conversion process also exist. Unfortunately, the example shown, along with the aforementioned variants, is proprietary to each particular equipment manufacturer.

Problems with M13 Multiplexing

The M13 multiplexing process has several inherent disadvantages. From a user perspective, the most important of these disadvantages is that no fiber optic carrier standard existed at rates above DS-3. In today's environment of bandwidth-hungry, distributed applications (e.g., video, imaging), 44.736 Mbps simply falls short of user requirements and provides a strong impetus for development of standards above the DS-3 rate, such as SONET.

From a carrier perspective, the lack of electro-optical standards is equally problematic, especially in the United States. With no optical transmission standard, carriers that

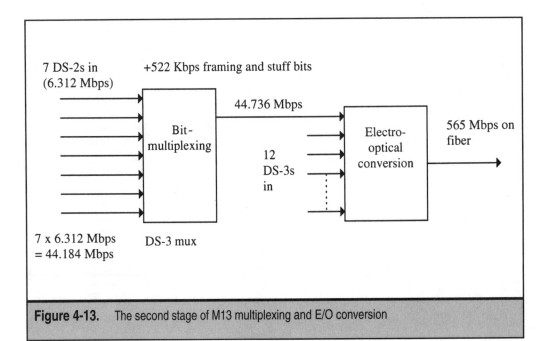

Figure 4-13.   The second stage of M13 multiplexing and E/O conversion

purchase equipment from different vendors are unable to interoperate at interface points. This lack of a "mid-span meet" standard was one of the factors that prompted the development of the SONET standard.

Due to the plesiochronous nature of the M13 multiplexing process, full demultiplexing is required to access an individual DS-1 stream within the DS-3 aggregate. Bits that were stuffed in the multiplexing process must be destuffed in the demultiplexing process. In the M13 process, two stages of destuffing/demultiplexing are required. Provision of enhanced services, such as DS-1 cross-connection and DS-1 add/drop, are thus rendered difficult and expensive in the M13 environment.

Finally, OAM&P interfaces and systems in the M13 environment are typically vendor-specific. Craftspeople need to be trained on a number of such systems to become proficient in fault diagnosis and repair operations. In addition, even when equipment is purchased from a single vendor, OAM&P systems may not be fully integrated; that is, a craftsperson performing a diagnostic operation on an M13 multiplexer may not be able to "see through" its associated electro-optical converter to another M13 multiplexer.

WHAT A CONCEPT!

This chapter concludes by tying the advantages of fiber optic network links discussed in the previous chapter to the nagging and persistent shortcomings of T-carrier links. Why shouldn't service providers struggling with the limitations of T-carrier use the advantages of fiber optics to offset the drawbacks of T-carrier networks?

The advantages of fiber could be employed as a basis for a family of scaleable, standard transmission systems and equipment. It would be more flexible, and easier and cheaper to install, run, and maintain. Such a network would use optical transmission with synchronous multiplexing in a fully standardized fashion. This network would be a synchronous optical network—SONET!

IMPORTANCE OF SONET

The drive toward higher capacity transmission systems with low-deployment costs helped lead to the development of the SONET (and later SDH) standard. This standard has several key features. First, the SONET standard defines a transmission hierarchy that begins where the current digital hierarchy ends; in T-carrier systems (in North America) that rate is 45 Mbps, in ETSI systems (in Europe) that rate is 139 Mbps. The second feature of the SONET standard is the synchronous nature of its multiplexing system. Lower bit-rate channels (called "subrate channels" in SONET) can be identified, extracted, or inserted by a single device without the need to demultiplex the higher rate SONET signal. This allows simplified add/drop capability. The third feature is SONET's support for OAM&P systems. Bandwidth is allocated for use as a common OAM&P transmission path. This allows each element in the transmission system to communicate with a centralized management system, or to distribute the management responsibility throughout the elements. This extensive OAM&P support could conceivably redefine the way communication networks are managed and the way these networks provide services to customers.

PART II

How SONET/SDH Works

art I of this book introduced how fiber optic transmission links and the T-carrier digital hierarchy used in North America function in a private line network. These chapters concluded with the thought that the standardization of fiber optic transmission would both take advantage of all of the positive features of fiber optic systems and overcome all of the drawbacks that were becoming more obvious in the T-carrier world. However, an idea is not reality. Much hard work remained to be done before SONET/SDH was created as a standard fiber optic transmission system.

This section of the book explores how the standardized fiber optic transmission system eventually gave birth to SONET, and details some of SONET's operation. Due to the mainly historical nature of this section, the emphasis will be on SONET rather than SDH, although everything that applies to SONET ultimately applies to SDH as well. Where appropriate, the plain term SONET is used to indicate a particular feature that was developed for SONET alone and eventually made its way into SDH as well. SONET/SDH has a very detailed architecture and series of protocols to implement this architecture. Key aspects of this architecture include considerations of network timing and synchronization. Some of this information was introduced briefly during the survey of T-carrier in Part I of this book. This is the time to explore this architecture and timing structure in more detail, at the risk of introducing more technical terms and relatively advanced concepts. These terms and concepts are discussed on a nonmathematical basis, however, because details about things like network synchronization—while crucial to network operation (especially SONET/SDH operation)—remain somewhat arcane to users and even technicians.

This part of the book is organized into five chapters. Chapter 5 forms the basis for a high-level description of SONET/SDH functioning. It also investigates the formation of the original SONET standard from important forerunners, such as SYNTRAN. SYNTRAN is discussed and contrasted with T-3 (really M13) multiplexing as a first step on the road to SONET. An important section of this chapter explains why some differences still exist between North American SONET and synchronous digital hierarchy (SDH) intended for use in the rest of the world. The important point is that both are standard, both are compatible in the vast majority of cases, yet both will remain different (for very good reasons).

Chapter 6, which is of necessity quite long, forms a complete exposition of the SONET/SDH architecture and operational protocols. All aspects of SONET/SDH are explored in this chapter, from the basic SONET STS-1 and SDH STM-0 frame structures, to sub-rate payloads maintained for backward compatibility with existing T-carrier and E-carrier services, to concatenated SONET payloads, such as STS-3c (OC-3c to some) and the basic SDH STM-1 payload, which are intended for use with asynchronous transfer mode (ATM) and B-ISDN networks.

Chapter 7 explores all of the overhead bytes defined and used in SONET and SDH. SONET section, line, path, and virtual tributary overhead byte structures and use are investigated in full. At the same time, SDH regenerator section, multiplex section, higher order (also called high order) path, and lower order (also called low order) path overhead are given the same extensive treatment.

Chapter 8 was new to the second edition of this book and is expanded here. This chapter explores the details of four possible SONET/SDH payload types. There are other payload types defined, but these four constitute the bulk of all SONET/SDH uses today. The chapter begins by considering SONET/SDH for 64 Kbps DS-0 voice, which is still the main content of many SONET/SDH networks. This third edition of the book adds some details on this use of channelized SONET/SDH. Then the chapter moves on to investigate asynchronous DS-3 and other E-carrier multiplexing level support, another key SONET/SDH application. Next, the chapter explores one of the original goals for SONET/SDH transport: the filling of payloads with ATM cells. The chapter closes with a discussion of probably the most interesting aspect of SONET/SDH today: using SONET/SDH to transport IP packets directly instead of inside ATM cells.

Chapter 8 also investigates two of the more interesting applications of SONET/SDH for the networks of today: the Generic Framing Procedure (GFP) and the concept of Virtual Concatenation. Both show the extensible nature of SONET/SDH, its ability to adapt to new situations, and at the same time show that reports of a mass exodus to direct IP over DWDM systems might be premature. Both GFP and Virtual Concatenation allow SONET/SDH more flexibility in carrying all types of data over a network made up mainly of routers and SONET/SDH links between them, rather than more rigid digital access link to multiplexer to backbone arrangement.

Chapter 9 contains details about the operation of two key elements and features of SONET/SDH: synchronization and timing. The chapter describes how SONET performs all aspects of network synchronization and timing, and then treats the same material from an SDH perspective. The emphasis is not a theoretical discussion of these concepts, but rather describes their real world application in SONET/SDH.

CHAPTER 5

The Evolution of SONET/SDH

The first part of this book detailed the need for higher bandwidth links for private-line networks and the lack of adequate standards or even technologies to provide them. Higher-speed equipment like desktop computers needs higher-speed links to connect them over a wide area. These links were traditional private lines, but no such private-line standard or technology existed in a service provider's network to accommodate the customer's needs.

A real force behind this drive to create SONET (SDH was not yet in the picture) was a significant lack of hierarchical transmission standards greater than 45 Mbps. However, prior to SONET, a technology known as SYNTRAN (SYNchronous TRANsmission), operating at the DS-3 rate of 45 Mbps represented the upper end of the bandwidth. Only highly specialized, and generally proprietary, applications operated at speeds in excess of 45 Mbps. As a result, little standards work had been done for rates greater than DS-3.

The importance of SYNTRAN to the ultimate evolution of SONET cannot be overemphasized. Because SYNTRAN shared (and still shares) many of the basic architectural features of SONET, especially with regard to synchronization and timing (as the name implies), a brief look at SYNTRAN is in order.

The second driving force behind the efforts to develop the SONET standard was the need for what was called the *mid-span meet*. This was an open and standard interface so that different service providers, such as local and long-distance telephone companies, could establish links between their switching offices. These twin forces—high-bandwidth synchronous needs and mid-span meets—combined to produce SONET.

However, the time has also come to introduce the main players in the creation of the SONET standard, namely the international and national standards organizations themselves. Because these organizations figure prominently in the development of SYNTRAN and mid-span meets, this chapter will explore them first.

MEET THE PLAYERS

Many organizations play a role when it comes to developing and overseeing SONET standards. Most of the main players are shown in Figure 5-1. The role of each is described below, with emphasis on their contributions to SONET standardization. Once SONET and SDH came together, the main organization of current concern has become the ITU, of course. In the United States, ANSI still releases SONET-specific documentation.

American National Standards Institute

American National Standards Institute (ANSI) was created in 1918, making it one of the first standards bodies to be formed in the United States. Oddly, ANSI develops no standards of its own. Rather, ANSI's role is to play an important part in standards coordination and approval. ANSI's main function is procedural; that is, ANSI coordinates and blends private sector standards development. ANSI is supposed to be an impartial mediator, resolving conflicts and duplications of effort among various private groups working on technology implementation. ANSI is also the United States representative to the International Standards Organization (ISO).

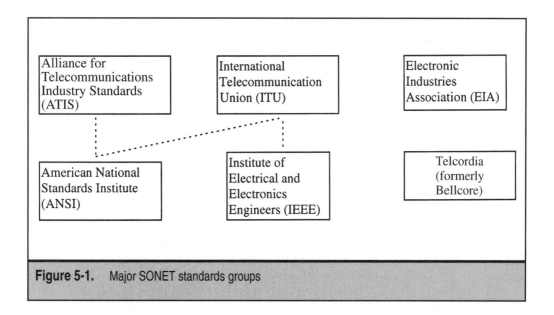

Figure 5-1.   Major SONET standards groups

ANSI has over 300 different standards committees, as well as associated groups like the ATIS. Specific procedures must be followed for standards development. Under the various groupings, such as ATIS, are various technical committees and task groups. These subcommittees and working groups perform the actual standards research in a variety of computer and communications areas.

Most relevant here are the T1 subcommittees:

▼ T1E1—Carrier to customer installation interfaces

■ T1M1—Internetwork operations, administration, maintenance, and provisioning (OAM&P)

■ T1S1—Services, architecture, and signaling

■ T1Q1—Performance

■ T1X1—Digital hierarchy and synchronization

▲ T1Y1—Special subjects, such as video and data

The T1 subcommittees that are especially active in SONET standards include the T1X1 and T1M1 subcommittees. The T1X1 group played a key role in organizing the effort that led to the specification of the SONET rates and format, as well as the optical interface for the single-mode fiber. The T1M1 group has responsibility for guiding the development of the SONET standard for OAM&P. These efforts, crucial in any networking scheme, focus on such things as alarm surveillance, performance monitoring, fault-location, and link management.

OAM&P will be a matter of ongoing standards activity for years to come. Even after they are implemented, the initial standards for OAM&P will not cover all issues for all

SONET products operating under every circumstance. This fact can be frustrating to users and service providers, but it is a fact of life.

Alliance for Telecommunications Industry Standards

ATIS was organized after divestiture (discussed in more detail later) to represent the interests of inter-exchange carriers. After the breakup of the Bell System, some degree of cooperation between these competitors was still needed. Formed in February, 1984 as the Exchange Carriers Standards Association (ECSA), ATIS's Committee Tl-Telecommunications is a direct result of the divestiture process. ATIS sponsors the operation of and provides administrative assistance to this standards-making group. The Committee T1 has more than 100 member companies and operates under the formal rules and procedures of ANSI.

More specifically, the Committee T1 is focused on the functionality and characteristics associated with the interconnection and interoperability of telecommunications networks. This, in turn, concerns various interfaces with end-user systems, other carriers, Internet service providers (ISPs), and other types of enhanced service providers (ESPs).

The Committee T1 does not have wide-ranging responsibility for its own set of standards. However, through the combined processes of direct liaison, cross-membership, and coordination with other national standards bodies (much of it through ANSI's Joint Telecommunications Standards Coordinating Committee), ATIS still has much to say about standards. The Committee T1 standardization process is becoming more efficient at addressing a wide range of connectivity issues, including those related to the development of SONET.

ATIS standards usually make their way to other standards bodies in the United States and eventually become ANSI standards. It was ANSI that adopted the SONET Phase 1 specification for SONET implementation (discussed in greater detail later) in September of 1988; in November of that same year, the International Telecommunication Union (ITU; then the CCITT) also approved it. Phase I of SONET contained over 90% of the SONET standard.

Telcordia (Bellcore)

Bell Communications Research (Bellcore), now Telcordia, was founded as the result of the divestiture process splitting off the operating companies from American Telephone and Telegraph (AT&T) (the Bell System) in 1984. Once the former Bell Operating Companies (or Regional BOCs, the "RBOCs") were separated from AT&T, the new localized telephone companies did not have easy and convenient access to AT&T Bell Laboratories, which was retained by AT&T under the divestiture decree. The seven Regional Bell Holding Companies (RBHCs, the "owners" of the RBOCs) established and supported (by way of "contributions") Telcordia as their research and development arm. Telcordia has played a key role in developing industry standards, including those for SONET, mainly because of the need for equal access capability, which is detailed later. Bellcore is now known as Telcordia and is totally independent of the RBOCs. Although many SONET documents now say "Telcordia," the work embodied in them was mostly Bellcore's.

Telcordia issues a number of documents that pertain to telecommunications in general and SONET specifically.

Telcordia issues technical advisories (TAs) which inform the industry about standards issues. These roughly correspond to draft requests for comments (RFCs) in the Internet world. Interested parties can provide input to Telcordia before the TAs become finalized into technical requirements (TR) documents. The TRs provide the information necessary for vendors to begin developing equipment that complies with the technical requirements of the RBOCs, which naturally constitute a huge market for these vendors.

The major ANSI standards for SONET are largely the result of a close cooperative effort between ATIS and Bellcore. The main impact of the Telcordia TRs was on equipment, while the interface definitions were the primary contribution of ATIS. For example, the fiber optical line signal in the ATIS interface standard specifies the use of bit-interleaved parity bytes to perform error checking. The Bellcore TRs specify that the receiving equipment monitor the parity bytes. If a disagreement were to occur between the transmitted and received parity byte values, a transmission error would have resulted, of course. SONET equipment determines the number of error-free seconds, erred seconds, severely erred seconds, and unavailable seconds. This information is stored and may be relayed to the network management monitoring station in the form of instantaneous alarms, performance statistics, or both. Thanks to the presence of the relevant Bellcore TRs, all vendors can develop SONET-compliant equipment in this respect.

Telcordia also issues standards in the area of SONET test equipment. Test equipment standards allow service providers to implement sophisticated diagnostic routines to identify problems with regard to such SONET features as virtual tributaries, fiber optic regenerators, and other network equipment. The test procedures developed by Telcordia will probably penetrate to the user community through the Electronic Industries Association (EIA).

Electronic Industries Association

The EIA was founded in 1924 as the Radio Manufacturers Association. The EIA is a trade group, meaning a group whose members all engage in a like activity or sell a similar product. EIA members represent manufacturers of electronics devices in the United States. Several thousand industry representatives participate in over 200 EIA technical committees to develop mainly hardware standards. An EIA committee known as TR-30 developed the famous RS-232 (now officially the EIA-232) and RS-449 physical interface standards. Through the TR-30 committee, EIA interacts with both ITU and ANSI. In fact, EIA sponsors many of the ANSI committees.

With SONET, the EIA role is to assist in the development of optical interfaces for customer premises equipment (CPE). Another EIA committee, the FO2.1.4 Committee, along with Telcordia, will continue to develop testing procedures. Why are both Telcordia and EIA involved? Because Telcordia develops SONET test procedures for service providers, while EIA develops SONET test procedures that will be used by end users employing SONET CPE and buyers of private SONET equipment. There will be a lot of joint activity between Telcordia and the EIA to achieve the highest degree of compatibility in this key area.

Institute of Electrical and Electronics Engineers

The Institute of Electrical and Electronics Engineers (IEEE) is an international professional society for electrical engineers. The major contribution of the IEEE to telecommunications has been the ongoing development of Local Area Network (LAN) standards. The IEEE has been only peripherally involved with the SONET (and SDH) standardization process, but the IEEE Technical Activities Board will have a great impact on SONET/SDH, due to the IEEE's LAN involvement.

The IEEE 802.x series of standards specify the characteristics of Ethernet (IEEE 802.3), Token Ring (IEEE 802.5), and various other LAN architectures. The LAN work of the IEEE 802 committee is currently organized into the following subcommittees:

- ▼ 802.1 High Level Interface
- 802.2 Logical Link Control
- 802.3 CSMA/CD Networks
- 802.4 Token Bus Networks
- 802.5 Token Ring Networks
- 802.6 Metropolitan Area Networks—Distributed Queue Dual Bus (DQDB)
- 802.7 Broadband Technical Advisory Group
- 802.8 Fiber Optic Technical Advisory Group
- 802.9 Integrated Data and Voice LANs
- 802.10 Network Security
- 802.11 Wireless Networks
- 802.12 Demand Priority Networks
- 802.14 Cable TV Based Broadband Communications Network
- 802.15 Wireless Personal Area Networks
- 802.16 Broadband Wireless Access
- 802.17 Resilient Packet Ring
- ▲ 802.18 Radio Regulatory Technical Advisory Group

The IEEE works in cooperation with other standards bodies. IEEE committees routinely pass their recommendations to ANSI for approval as national standards, and on to the ITU for approval as international standards.

International Telecommunication Union

Technically, it is the International Telecommunication Union-Telecommunication Standardization Sector (ITU-T) that is concerned with SONET/SDH. The ITU-T was formerly known as the CCITT. The ITU is supported by the United Nations and is the body that makes recommendations and coordinates the development of telecommunications standards for the entire world, including North America. The official United States representative to the ITU is the Department of State. The ATIS Committee T1 provides much input to the United States ITU-T Study Groups, such as Study Group C. They, in turn, support the Department of State in establishing positions of the United States at ITU meetings, which is a curious blend of technical talk and political infighting.

Study groups in the ITU have specific technical responsibilities for developing international standards, in much the same way that technical subcommittees of the ATIS Committee T1 have specific responsibilities in national standards development. The ITU is comprised of 15 study groups, each of which has specific responsibilities in the field of data communications. The study group that has made the most contributions toward the development of an international SONET standard is Study Group 15—formerly Study Group XVIII, but the study groups lost their Roman numerals (and were renumbered) when the CCITT changed its name. This study group focuses on digital networks, including B-ISDN.

The study groups work on technical issues assigned to them by the Plenary Assembly of the ITU, which meets every four years at various sites around the world to approve the work of the study groups and to issue new assignments for the next development cycle. The study groups hold meetings six to nine months apart. It is important to note that although they are only "recommended," the standards approved by the ITU do carry the weight of law in many European countries. The ITU has clout: Service providers can be sanctioned for violating ITU "recommendations" in many countries. (The ITU specifications are called "Recommendations" only because the ITU has no enforcement agency. The ITU "recommends" that each nation enforce ITU rules on its own.)

Today, of course, it is the ITU that determines what SDH is and should be, and that, in turn, drives SONET in North America. In other words, the current environment is more like SDH/SONET than SONET/SDH.

SONET Interoperability Forum

The SONET Interoperability Forum (SIF) was formed in 1994 to address the interoperability issues in SONET. Particular attention was focused on the methods needed to ensure that equipment from different vendors worked together. The forum emerged mainly because of the shortcomings in the original SONET standards, as defined by the ANSI T1X1 Committee. These standards, helpful as they were, gave vendors far too much latitude for developing variations of messages contained in the built-in SONET data communications channel. This made interoperability difficult, if not impossible.

SIF was essentially absorbed into ATIS in 2000 and its name was changed to the Network and Service Integration Forum (NSIF). But NSIF documents and activities should be closely monitored by interested parties.

FORGING STANDARDS

The organizations introduced above do not follow explicit rules when developing standards. All things considered, developing communication standards remains an art rather than a science. Although many industry observers treat standards as explicit rules, standards are most often open to considerable interpretation by vendors, software developers, and others. Although mostly perceived to be driven by advances in technology, standards are often the result of severe political infighting between corporations, governments, standards developers, and even influential individuals. Therefore, in the real world, all standards—SONET/SDH included—are documents that are continually evolving.

Standards are not static and do not spring fully grown from standards bodies like Athena from the brow of Zeus. Even a firmly established standard like X.25, for instance, which defines the facilities offered by the worldwide public packet networks, has changed over the years. X.25 was developed in stages and has remained open to amendment since the first draft was issued in 1971. This was the document known as the "gray book." The X.25 gray book was followed by a series of amendment documents in 1976, 1980, 1984, 1988, 1992, and 1996, all at four-year intervals. All of these were issued by the International Telegraph and Telephone Consultative Committee (CCITT, now the ITU-T), the international standards body; even though X.25-based public packet networks enjoy worldwide acceptance, the ITU will continue to improve X.25, if need be, into the future.

Generally, the main function of all standards bodies is to coordinate activities or directly engage in activities that lead to a consensus on how best to implement a particular technology. This process can eliminate needless duplication of effort, waste of often limited resources, and alleviate the confusion that inevitably results when too many organizations produce their own (proprietary) solutions.

A good example of the standardization process is from the early days of telephony. Each telephone company—hundreds had emerged after the original Bell telephone patents expired about the turn of the century—set up its network differently from all the others. As a result, users could only communicate with each other if they were served by the same telephone company. If a subscriber wanted to call someone served by another telephone company, connections to both networks were required, which often meant having more than one telephone.

Although the eventual solution to this problem perhaps should not be repeated (i.e., AT&T systematically bought up all the independent companies and put them on the "Bell System Standard"), this standardization process did lead to the explosion in popularity of this new communications medium. Therefore, AT&T came to occupy its position in the telephony standards arena that it sustained through the breakup of the Bell System.

SYNTRAN

A full recapitulation of how M13 multiplexing operates in the T-carrier environment is unnecessary here. But it is important to note that the M13 multiplexing process that formed one DS-3 out of 28 input DS-1s was a plesiochronous process that involves the bit-interleaved multiplexing of the input DS-1s. As a result of this plesiochronous operation, the formation of the DS-3 transmission frame also involved using many overhead bits to determine the presence (or absence) of stuff bits based on the differences between input timing (called *clocking*, in most cases), and output timing (clocking).

Assuring the same rate of ticking (which also included phase of ticking because ticks could occur at the same rate, but not at the same instant) between input clock and output clock on the network would allow the multiplexing equipment to dispense with the need to stuff bits into the DS-3 transmission frame. This would, in turn, allow byte interleaved multiplexing in an isochronous fashion to occur at the T-3 level of 45 Mbps. As has been shown, a real incentive exists to perform byte-interleaved multiplexing at any level. Byte-interleaved multiplexing allows for the easy *recovery* of constituent DS-1s from a DS-3, and the recovery of any DS-0s directly from the DS-3 frame.

This recovery of a DS-0 (64 Kbps) from a DS-3 frame (45 Mbps), of course, could be done without SYNTRAN or SONET. However, it was tedious and messy because the DS-1s with their embedded DS-0s were scattered bit-wise throughout the DS-3 frame. The DS-0 channels were needed in many cases to add and drop individual 64 Kbps channels to make more efficient and cost-effective use of the links. In T-carrier networks operating at the T-3 rate, it was necessary to fully recover each of the twenty-eight DS-1s in order to determine the proper DS-0 channel needed. This resulted in what came to be known as *back-to-back multiplexing* in networks using M13 equipment.

With back-to-back multiplexing, an incoming DS-3 frame was broken down into its constituent DS-1s, the relevant DS-0s were cross-connected (or dropped and inserted) as required, and then the entire DS-3 frame was reassembled for the outbound link. Naturally, if the T-3 were delivered on fiber optic links, the proper fiber optic transmission system (FOTS) equipment would also be needed. This whole operation is shown in somewhat simplified form in Figure 5-2.

In actual practice, the process of back-to-back multiplexing was much more complex than is shown in Figure 5-2. The short lines in the figure labeled as "DSX-3" were typically coaxial cables that snaked around the site to a cross-connect panel, where they extended back to the M13 mux itself. Of course, the same thing happened on the other side.

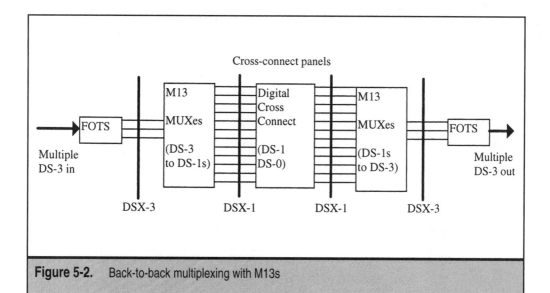

Figure 5-2. Back-to-back multiplexing with M13s

Naturally, the lines labeled "DSX-1" were typically masses of twisted pair copper wires extending across or under the floor as well. The appearance of the site was usually a nightmare of four telephone-booth-sized equipment boxes (the FOTS and M13 muxes) and tangled masses of coaxial and copper cables. All of the various DSX cross-connect electrical interfaces had maximum cable distances associated with them. Otherwise timing problems and their related alarms occurred. These distances for T-carrier DSX interfaces are shown in Table 5-1. Plus, all of these devices had problems: they needed power supplies, generated heat, and were prone to failures.

One of the main reasons that SYNTRAN was developed was to try to reduce the complexity of this arrangement needed to get at a poor 64 Kbps DS-0. Work on SYNTRAN began in the mid 1980s and was approved as an ANSI standard in 1987 as ANSI T1.103. SYNTRAN's intent was to "augment the asynchronous, bit-stuffing multiplexing design currently in use." SYNTRAN allowed "direct" DS-0 and DS-1 "observability" in the 45 Mbps bit stream. It would be more common today to use the terms, *plesiochronous* and *recovery*, instead of *asynchronous* and *direct observability*, but the meaning is identical. The document also described the DS-1s inside the DS-3 as *tributaries*, a term that SONET eventually used in the same context.

SYNTRAN allowed the new, "synchronous" DS-3 frame format to be transmitted over the same link with the same transmitters and receivers as before. After all, 45 Mbps is 45 Mbps. Only the format of the bits changed from bit-stuffed to byte-interleaved. The equipment changed, the transmission frame structure changed, but not the bit rate.

The fact that much of the DS-3 frame overhead bits were no longer needed in SYNTRAN to determine whether stuff bits were employed allowed these bits to be used for other purposes. In an obvious moment of triumph, the ANSI standard announced that "the stuffing and control bits, which have been liberated by the synchronous format, are assigned new functions."

These new functions included an embedded 64 Kbps data link just for network OAM&P, an identification field, and so on. One new function was the inclusion of some error checking on the entire transmission frame. This was not for retransmission pur-

Cross-connect Type	Maximum Cable Footage
DSX-1	655 Feet
DSX-1C	655 Feet
DSX-2	1000 Feet
DSX-3	450 Feet
DSX-4 (nonstandard)	225 Feet

Table 5-1. Cross-connect interface distances

poses as with data frames, which would have been impossible for voice anyway, but rather to allow for the gathering of error statistics either on the link or even end to end. Such error statistics could be used to verify performance levels or to detect problems.

There is no need to discuss the details of the new SYNTRAN frame format any more than there was a need to discuss the older DS-3 frame in a SONET book. It should only be noted that SYNTRAN still struggled with timing variations (i.e., jitter). SYNTRAN introduced a process called *controlled slips* to prevent the accumulation of timing variations on the link that would throw the framing process off.

SYNTRAN was good, as far as it went. In many cases, however, SYNTRAN was too little, too late; events of the 1980s in the telecommunications industry in the United States overshadowed the accomplishments of SYNTRAN, as formidable as they had seemed. The main event was *divestiture.*

QUEST FOR THE MID-SPAN MEET

The SONET concept was born in the early 1980s, before the signal event in the telecommunications field in the United States known as *divestiture.* Before 1984, the bulk of the telephone system in the United States was owned and operated by AT&T. AT&T Bell Laboratories was the research and development arm of the organization, and had invented T-carrier and SYNTRAN. AT&T Long Lines handled the vast majority of "long distance" telephone calls, those made beyond a certain regional area. AT&T also handled local telephone calls through wholly owned subsidiaries called the Bell Operating Companies (BOCs). In fact, because people made mostly local calls, the whole conglomerate was usually referred to as the Bell System.

Other independent local telephone companies existed mostly in rural areas at first. Some grew quite large in their own right, such as GTE around the country and Southern New England Telephone (SNET) in Connecticut. Others remained quite small, such as the Sylvan Lake Telephone Company in rural New York, with only a few hundred subscribers. All of these local companies more or less were forced to use AT&T Long Lines so that their subscribers could make long distance calls. In order to do so, a link (trunk) was needed to connect a local switching office to the AT&T Long Lines switching office. In these cases, one end of the trunk used was owned and operated by the local company and the other end was owned and operated by AT&T. Because the responsibility for the link was assumed to change from one company to the other in the middle of the link, this arrangement was known as the *mid-span meet.* Naturally, in order to work, the equipment at both ends had to be compatible.

This was not much of a problem as long as AT&T was the only game in town. AT&T also had a manufacturing branch, called Western Electric (WECO), that made equipment used throughout the Bell System. As long as AT&T Long Lines was the only long distance carrier needed, any local company could just buy WECO equipment and be sure that the long-distance trunks to the AT&T Long Lines office would work.

In 1984, the process of divestiture, however, broke up the Bell System into several large, regional local telephone companies (Regional Bell Operating Companies or RBOCs). The other pieces of AT&T, Long Lines, WECO, and Bell Labs, remained more or less intact, although the new RBOCs founded their own research and development company, Telcordia, with a lot of personnel from Bell Labs. A key piece of the divestiture process was called *equal access*. Equal access meant that a local company, and any subscriber to any local company's telephone service, was free to use other long-distance carriers besides AT&T. These carriers, called Inter-exchange Carriers (IXCs), were represented by a new breed of competitive long-distance service providers. The most notable were MCI (which originally meant Microwave Communications Incorporated), whose competitive long distance efforts had largely brought about divestiture in the first place, and Sprint (originally part of the Southern Pacific Railroad network).

Under the new divestiture rules, a local company could just as easily connect to an MCI or Sprint switching office (called *point of presence* or POP) for long-distance service as to AT&T. In fact, in most cases local companies had installed trunks to MCI and Sprint POPs so that their customers could enjoy the equal access that divestiture promised. The question was, what type of link should these be? It seemed obvious that the equipment at each end still had to be compatible.

But because neither end of the link was AT&T in many cases, it was no longer a given that WECO T-carrier equipment would be used in all cases. Certainly, none of the other IXCs, including MCI and Sprint, were interested in using WECO equipment. However, the T-carrier hierarchy was never fully completed at the T-3 level and above. Mid-span meets at these levels required both end companies to purchase the same vendor's equipment. It hardly seemed possible (or probably violated the spirit of the equal access provisions) to require the newly competitive telephone systems to all buy the same equipment. Therefore, how could the anticipated flood of calls be handled by the new long-distance companies with a variety of vendor's equipment when no universal, scaleable, fiber-based digital trunking standard existed for the mid-span meet?

SONET AND SDH

MCI issued a request to the ICCF, set up for this very purpose, that they go to ANSI's T1 Committee and ask them to develop standards for interconnection of multi-vendor and multiple-owner fiber optic transmission systems. This request, also known as the mid-span meet capability request, was one of several that the ICCF was working on with ANSI, and one of two submitted to the T1 Committee.

Early in 1985, Bellcore proposed to the ANSI T1X1 group a network solution to fiber system standardization. Their proposal was designed to allow for the interconnection of multi-vendor, multi-owner fiber optic terminals. It also offered interconnection of functionally different network hardware. For example, it would allow, for the first time, line-terminating multiplexers from different vendors to be connected simultaneously to a digital cross-connect system. Further, Bellcore proposed a digital signal/transmission rate hierarchy, wherein transmission speeds were integer multiples of some predefined base rate. Finally, the proposal offered a bit-interleaved multiplexing scheme that would allow for optimal use of available bandwidth.

Both of these offerings, SYNTRAN and the mid-span meet initiative taken together, resulted in the creation of the Synchronous Optical Network, or SONET standard.

This was not an easy task. Until 1988, no universal fiber-optic system standards existed. Digital standards had developed differently around the world, to the point that the multiplex hierarchies that govern the public networks of major commerce centers, such as North America, Europe, and Japan, supported different transmission rates.

Therefore, at the T-3 equivalent rate and above, each manufacturer's product was designed according to a proprietary set of specifications. As a result, no two manufacturer's fiber optic products were compatible with each other. This meant that an installation at one end of a connection between networks had to be exactly matched at the other end with a corresponding product from the same manufacturer.

The situation led to a strong tendency among carriers to choose a single supplier of fiber systems, which resulted in a loss of some potential economic and technical benefits of a competitive multi-vendor market. To ensure interconnectivity across network boundaries, either the network operators had to agree to use the exact same fiber products, or they had to confine themselves to electrical interfaces that limited them to transmission rates far below the actual bandwidth capacity of the fiber optic systems being used. This was usually done using multiple T-1s instead of the higher capacity T-3s.

Moreover, based on the technological promise of the synchronous standard, many manufacturers were already looking at synchronous multiplexing, and a network approach to bandwidth offerings. Some of them were already planning product rollouts based on minimal knowledge of what was coming. Naturally, this complicated the standards-development process. Manufacturers were so anxious to get on the SONET bandwagon that they did not wait for all of the standards to appear. Telephone companies needed and wanted SONET badly. As a result, the momentum for its design and standardization grew quickly. The SONET standard originated from a cooperative effort between various standards bodies and private industry.

As mentioned above, in 1984, MCI proposed an interconnecting fabric of multiple carrier and manufacturer fiber optic transmission connections to the ICCF. The ICCF, in turn, requested Committee Tl, the carrier-to-carrier interfaces subcommittee of ECSA, working under ANSI T1X1, to develop standards for an optical interface.

By February, 1985, Bellcore had proposed to T1Xl the SONET concept, which went beyond the MCI request to allow the interconnection of all network elements with fiber terminations. The Bellcore proposal was based on the following assumptions about the future direction of telecommunications networks:

▼ Fiber optic media is the transmission medium of choice

■ Customers will require high-bandwidth services (broadband services)

■ Remote switches connected by fiber optic media are highly desirable

▲ Switching equipment will eventually accommodate a direct optical interface

Using this proposal from Bellcore as a foundation, the T1Xl committee's efforts eventually involved more than 400 detailed technical proposals from 120 individuals representing more than 50 corporations. That year, a Bellcore Technical Advisory (TA) was issued at the Technology Requirements Industry Forum (TRIF).

In August of 1985, the T1X1 Committee rolled out a proposal document on the SONET principle. Of course, the design was not strictly theirs. It was based on input from a variety of manufacturers, local and long-distance carriers, ad hoc user groups, and others. One of the greatest points of contention existed between Bellcore and AT&T: Bellcore wanted the base-line speed to be 50.688 Mbps; AT&T wanted it to be 146.432 Mbps. After much debate, a compromise rate of 49.94 Mbps was reached and the "virtual tributary" concept was introduced as a way to transport DS-1 based services (backward compatibility).

By January of 1987, widespread agreement had been reached on SONET design, and a draft document was prepared for a vote by the T1X1 Committee. Just as the committee was to vote in session, the ITU-T (then the CCITT) threw a wrench into the works. In 1986, the international community had taken notice of the standardization effort in the United States, but these initiatives addressed the T-carrier hierarchy in the United States, not the one used in Europe. Therefore, ITU-T began an effort of its own that became known as the Synchronous Digital Hierarchy (SDH).

Throughout the SONET design process, the needs of domestic consumers in the United States had driven the efforts of T1X1, with little regard for the needs of those who may want to use SONET as a standard outside of the United States. T1X1 saw SONET as the answer to many problems, most notably a way to stop the virtually uncontrolled growth of incompatible fiber optic terminals in the manufacturing sector to make mid-span meets easy. In Europe, however, where the manufacturing arm of the telecommunications business was still completely regulated, no such urgent need was perceived. As a result, the ITU was less willing to assign SONET the priority that delegates from the United States wanted. Little incentive existed in completing ANSI work when the ITU was dallying.

Just before things came to a complete halt, the problem was somewhat resolved when delegates from Japan and Great Britain began to attend the United States T1X1 Committee meetings. This resulted in two benefits. First, European interests began to be fairly represented in T1X1 and, second, a communication channel of sorts was created between the European standards group CEPT (the acronym is not easily rendered in English), and the T1X1 Committee.

In February of 1987, after a great deal of development and deliberation, the United States formally proposed SONET to the ITU as an answer to the network node interface (NNI) study, begun in 1986. NNI was a study established to find a non-media-specific interface, different from that specified for B-ISDN. In the eyes of T1X1, SONET seemed an ideal NNI solution; however, it was not ideal for those outside the United States. The proposal specified a base rate of 50 Mbps. European transmission systems, which have no benchmark level anywhere near 50 Mbps, asked that the new hierarchy be based on 150 Mbps, because this would be closer to their 139.264 Mbps basic signal rate. This was a substantial request. They executed the necessary modifications and, within three months, had an acceptable design, based on what is called the STS-3 frame structure.

In July of 1987, the ITU asked all its subscribing administrations to consider two proposals for a 150 Mbps NNI specification, one from CEPT, and one from the United States. For technical reasons, the United States proposal fell out of favor. Instead, CEPT asked the United States to make a few simple modifications to their proposal that would bring it in line with European needs. These involved changing the basic bit rate from 49.92 Mbps to 51.84 Mbps and employing a byte-interleaved multiplexed frame structure, as SYNTRAN did. Oddly, initial SONET proposals still used bit-interleaved multiplexing. In any case, the United States agreed to the changes. Following some intense work, a draft proposal was finally created.

In early 1988, T1X1 accepted the changes, and SONET was on its way to becoming a true international standard. In February, 1988, ANSI TlX1.4 submitted for ballot two documents: TlX1.4/87-505R4, specifying fiber optic rates and format; and TlX1.4/87-014R4, specifying the optical interface for a single-mode fiber.

All of the comments about the ballot were resolved by a new Optical Hierarchical Interfaces working group TlX1.5. Final approval for SONET was achieved in June, 1988, in the "blue books." The SDH specifications were known as G.707, G.708, and G.709 (now everything is bundled back into G.707 and its additions).

In the United States, equivalent specifications, ANSI T1.105-1988 and ANSI T1.106-1988, published later that year are now commonly referred to as the SONET Phase I specifications. This provided a North American Standard. The ITU-T also recommended an international standard based on OC-3 which would allow Europeans to multiplex a 34 Mbps signal as readily as Americans could multiplex a DS-3 signal.

Refinements to all of these resulting documents continued into 1989. Most of the activity centering around the details for full product internetworking and network management. Because it is an internationally recognized standard, SONET has been the target of considerable interest among organizations that operate international networks. With SONET deployed around the world, users will not only realize cost-savings in transport networks and be able to expand the range of useful applications, but enjoy seamless interconnectivity across international boundaries.

It is important to state that SONET *is* in many ways an international standard, although the SDH rules take precedence on international links.

All SONET specifications are in compliance (despite some differences in implementation) with the ITU's Synchronous Digital Hierarchy (SDH) recommendations. In the United States, service providers and equipment vendors will follow the ANSI and Telcordia specifications. For international links and SONET use, the ITU specifications are followed.

SONET is both a United States and, as SDH, an international standard. SONET is being implemented in phases due to its complexity, worldwide scope, and required collaboration on the part of implementers. Because of this phased SONET standardization, vendors are all adopting different strategies to design equipment that can be readily upgraded to conform to emerging standards as they become available.

Because all vendors followed the phased approach to SONET rather closely, it may be helpful to explore the SONET phases in a little more detail, but mostly for historical reasons.

SONET STANDARDS: THREE PHASES

The standards development work was divided into three separate efforts.

The first phase (Phase I) of this activity set the stage for interoperability between different vendors' equipment, regardless of transmission rate. This stage defined things such as signal construction (i.e., attributes that define the frame structure and the assignment of overhead versus user payload), multiplexing parameters (i.e., definition of the method used to achieve higher rates in the SONET hierarchy), optical requirement (i.e., specification of the pulse characteristics and optical parameters necessary for light-based transmission systems), and payload mapping (i.e., methodologies used to carry user data in the SONET).

Phase II addressed the problems associated with OAM&P functions. Because SONET has its own embedded operations channel, called the data communications channel, a new protocol would have to be designed to allow operations messages to be swapped between dissimilar network elements (i.e., hardware components). Phase II activities also defined an electrical interface specification (for optical-to-electrical conversions). The work associated with Phase II took considerably longer than expected. A benchmark model, based on the OSI Reference Model, was still not ready to serve as a basis for OAM&P design work when the T1X1 Committee met in February of 1990.

The solution to the overburdened Phase II took the form of a third phase, which further defined specific message formats and message sets for performing the OAM&P functions. Phase III was completed with the adoption of the OSI network management protocol message set.

With the completion of Phase III, the SONET standards became effectively complete. Three standards organizations now have specifications for defining SONET. The three sets of specifications—ITU-T's G.707-G.708, ANSI's T1.105, and Telcordia's TR-NWT-000253—define a common architecture and common interfaces, and employ a common implementation for SONET. Equipment designed to one of the three specifications will interoperate with equipment designed to another.

Today, SONET phases are only of historical interest, but it may still be informative to detail the major parts of the SONET Phases.

Phase I

Phase I of SONET standards was issued in 1988 as the initial ANSI and related Telcordia documents. Phase I has two components, namely T1.105 and T1.106. The T1.105 defines the following:

- ▼ Byte-interleaved multiplexing format
- ■ Line rates for STS-1, 3, 9, 12, 18, 24, 36, and 48
- ■ Mappings for DS0, DS1, DS2, DS3, and SDS3 (SYNTRAN)
- ■ Monitoring mechanisms for Section, Line 1E Path structures
- ▲ A 192 Kbps and 576 Kbps Data Communications Channels (DCC)

All of these concepts are discussed in detail in Chapter 6. T1.106 specifies the optical parameters for the longer distances with single-mode fiber cable systems.

Phase II

Phase II standards, released during 1990/1991, had three components. These included Tl.105Rl, T1.117, and T1.102-199X. Tl.105Rl is a revision of T1.105 issued earlier and specifies a number of key SONET points.

1. SONET format clarifications and enhancements. Timing and synchronization enhancements are addressed in Section 7. The synchronization issues are still pending completion as is the jitter caused by the SONET pointer actions (described in more detail later). Due to its serious nature, jitter is addressed in Phase III standards as well.

2. Automatic Protection Switching or APS (Section 13).

3. The full seven-layer protocol stack for DCC and Embedded Operations channel (Section 14).

4. Mapping of North American DS-4 (139 Mb/s) signal into STS-3c.

These concepts are fully discussed in Chapters 7 and 13. T1.117 specifies the optical parameters for short haul (less than 2 kilometers), multimode fiber cable systems. T1.102-199X gives electrical specifications for STS-1 and STS-3 signals.

Phase III

Phase III of SONET addressed the crucial, but still controversial, issue of communication over the embedded operations channel for OAM&P. The controversy involved continued use of non-standard features in operations packages. Without this aspect of the SONET standards, equipment of different vendors cannot exchange control information and be managed by other than a specific vendor's network management software packaging. As recently as 1990, traffic could be successfully passed at the OC-1 and OC-3 rates in a multi-vendor SONET environment, but network management commands could not be exchanged.

Without this resolution of the OAM&P message format (which still continues to some extent, including debates about TCP/IP inclusion), existing RBOC operational support systems (OSS) cannot work with SONET's network management schemes. The SONET OSS is still being defined for operations, administration, maintenance, and provisioning of network services. Present RBOC OSS's are embedded in an ASCII-based TL1 or Transaction Language 1. This works fairly well for smaller SONET links and networks, but TL1 is still a "man-to-machine" language. This means that a human being must issue the TL1 commands at one end of the link. At SONET speeds, automation is preferred.

SONET network applications with elaborate meshed and multi-vendor links will require a machine-to-machine language, such as one based on the OSI-RM's network management standards. Resolution of this OSS issue is one of the sticky items in the completion of the SONET standards. These issues will be discussed in more detail later in Chapter 13.

Phase III consisted of four components. These included T1.105.01, T1.105.05, T1.105.03 and Tl.l l9. All of these documents are now available in updated versions. The current specifications are listed in the Bibliography.

ANSI T1.105.01-1994 provided the requirements for two-fiber and four-fiber bidirectional, line-switched SONET rings. These rings allow for failure restoration within 50 milliseconds and are an important part of many network recovery plans.

ANSI T1.105.05-1994 described the Tandem Connection Overhead layer for SONET. This allows carriers to maintain series of STS-1 signals as a single unit and offers users improved monitoring and service as well.

ANSI T1.105.03-1994 provided the jitter requirements at SONET interfaces and at interfaces between SONET and asynchronous (or plesiochronous) networks. The Phase III SONET jitter specifications are important for efficient SONET performance.

ANSI T1.119-1994 allowed for the management of SONET network equipment using an OSI-Reference-Model-based (OSI-RM) interface. The interface is based on the OSI-RM Common Management Information Protocol (CMIP). CMIP, in turn, is based on the OSI Common Management Information Service Element (CMISE), which provides a standard set of application program interfaces (APIs) for OSS, and the Abstract Syntax Notation version 1 (ASN.1) standard, which provides a common format for internal data representation. The use of CMIP will provide a standard for SONET OAM&P.

The significance of these SONET phases is as follows: Phase I enabled SONET equipment to pass bits back and forth; Phase II enables SONET equipment from different vendors to pass bits back and forth; and Phase III enables SONET equipment from different vendors to pass bits back and forth and to be managed by service provider OSS software regardless of vendor.

CHAPTER 6

SONET/SDH Architecture and Frame Structures

ONET/SDH is a set of coordinated ITU, ETSI, ANSI, and Telcordia (formerly Bellcore) standards and specifications that defines a hierarchical set of transmission rates and transmission formats. These are to be used by equipment vendors and service providers at or above a basic optical transmission rate of 51.840 Mbps, which forms the base transmission rate for SONET, and is a special form of SDH called STM-0. SDH multiplexing speeds begin at 155.52 Mbps, called an STM-1. Today, SONET/SDH addresses transmission speeds as high as 40 Gbps.

This chapter describes the fundamental architectural layers of SONET/SDH and the frame structures used by SONET/SDH at these various layers. The basic SONET/SDH frame formats are introduced, as well as the important concepts of virtual tributaries (SDH calls them virtual containers) and some of the basics of the protocols used with SONET/SDH. Virtual tributaries and virtual containers are used for backward compatibility with existing digital hierarchies, not only the North American T-carrier hierarchy, but also E-carrier. Concatenated (formerly called super-rate) frame structures were originally to be used for the transport of B-ISDN traffic, most notably the cells used in Asynchronous Transfer Mode (ATM) networks. Now concatenated frames are used mainly to transport IP packets over national or global Internet backbones. All of the latest frame structures are detailed in this chapter, but the details of the SONET/SDH overhead, virtual tributaries, containers, and other concepts such as SONET/SDH rings are discussed in detail in later chapters of this book.

Because this chapter mentions ATM cell transport inside SONET/SDH, an important distinction must be made. The term *asynchronous* is part of the ATM acronym and the term *synchronous* is part of the SONET/SDH acronym. So how can asynchronous ATM cells be transported inside synchronous SONET/SDH transmission frames? The answer lies in the overuse of the terms asynchronous and synchronous in the telecommunications industry. Each of the terms may have a variety of meanings depending on context. In the case of ATM, asynchronous refers to the way ATM performs multiplexing in an asynchronous, or non-time-division-multiplexed fashion. In the case of SONET/SDH, synchronous refers to the way the SONET/SDH performs multiplexing in a synchronous, or purely time-division-multiplexed fashion, with input clocks synchronized to output clocks. ATM and SONET/SDH can still work together because they function at different levels of the overall B-ISDN architecture that brought them together.

A more complete discussion of key SONET/SDH concepts, such as clocking and timing, jitter, and network synchronization, is deferred until later chapters. Some of the mechanisms for dealing with the effects of these timing variations, however, are examined in this chapter. This should not be a great inconvenience, however, since the exact details of factors like jitter will not prevent a basic understanding of SONET/SDH itself.

SONET/SDH SPEEDS

Table 6-1 shows the speeds of the SONET/SDH hierarchy, which consists of a few basic building blocks of terms and speeds. For SONET, each level has an optical carrier (OC) level and an electrical level transmission frame structure to go with it, called the synchro-

Optical Level	Electrical Level	Line Rate (Mbps)	Payload Rate (Mbps)	Overhead Rate (Mbps)	SDH Equivalence
OC-1	STS-1	51.840	50.112	1.728	STM-0
OC-3	STS-3	155.520	150.336	5.184	STM-1
OC-12	STS-12	622.080	601.344	20.736	STM-4
OC-24	STS-24	1244.160	1202.688	41.472	N.A.
OC-48	STS-48	2488.320	2405.376	82.944	STM-16
OC-192	STS-192	9953.280	9621.504	331.776	STM-64
OC-768	STS-768	39813.120	38486.016	1327.104	STM-256

Table 6-1.   SONET/SDH digital hierarchy

nous transport signal (STS). It makes sense that an STS-3 frame is sent on an OC-3 fiber optic link, just as a T-3 frame is sent on a T-3 link in the T-carrier hierarchy.

SDH does not use this awkward STS/OC distinction, and almost always just uses a simpler STM-N notation, as shown in the table. However, SDH also must reference optical and electrical network portions and equipment. Usually, the optical or electrical meaning for the term STM-N is determined by context. Whenever confusion is possible, sometimes the letters "O" and "E" are added as in STMO-4 or STME-16 to make the meaning absolutely clear.

Table 6-1 also details the physical line bit rate, the payload bit rate, and the overhead bit rate for each SONET/SDH level shown. The payload is merely the bit rate remaining after the overhead bits, which cannot be used for customer data, are subtracted (actually, overhead calculation is a little more complicated than that). The payload is carried inside an envelope, a special part of the transmission frame, in SONET/SDH. Finally, the table gives the synchronous digital hierarchy (SDH) designations for each of the SONET levels. Obviously, the STS level divided by three (except for STM-1) gives the synchronous transport module (STM) level. In other words, SDH counts by threes, while SONET counts by ones. The SONET counting units are the basic 51.84 Mbps bit rate of OC-1. STM levels increase by units of 155.52 Mbps (3 × 51.84 Mbps).

Every currently defined SONET level has an STM equivalent except for OC-24. This SONET level is still defined by ANSI documentation, but seldom seen today. This brings up an interesting point about SONET. Originally, the OC-N notation covered N = 1 to 256. There was even an 8 bit field in the SONET (and SDH) overhead to enumerate this value (this point will be discussed further in the overhead chapter following this one). What happened?

Note that not all possible values of N are represented in SONET/SDH. For example, no OC-5 or STM-7 exists. Although the value of N can technically take on any value from

1 to some maximum, currently 768, SONET/SDH standards define only a few of the levels. Otherwise, there would potentially be hundreds of different types of SONET/SDH equipment, making a joke of interoperability in spite of the presence of standards.

Sometimes there is a technical reason behind this preference for some levels and not others. For instance, early equipment vendors found that it was actually more cost effective to build SONET gear operating at the OC-12 level than at the OC-9 level because of laser chip costs and economies of scale. Little OC-9 SONET equipment was ever built, while the OC-12 equipment market has flourished.

It is absolutely crucial to remember that unless other arrangements are made in equipment configurations, *SONET/SDH will remain as channelized as a T-carrier or E-carrier link.* That is, an STS-3 contains three STS-1s. An STS-12 contains twelve STS-1s, and so on. By default, all of the STS-1s inside an STS-48 on an OC-48 fiber optic link run at 51.84 Mbps, the STS-1 rate. In the same fashion, in T-carrier, a DS-3 frame contains 28 T-1s, all operating at 1.544 Mbps. Ironically, SONET/SDH, despite the incredible speeds available at the higher ends of the hierarchy, would seem to limit networking to channels operating at 51.84 Mbps or 155.52 Mbps from any individual user.

However, just as DS-3 (or DS-1 for that matter) is available as an *unchannelized* transport without any structure (save the DS-3 frame overhead bits), so is SONET (SDH has its own channel structure). An unchannelized DS-3 merely offers a raw bit rate (no more 28 DS-1s) at 45 Mbps. In the same way, an unchannelized STS-12 would offer the user a raw bit rate (no more 12 STS-1s) at about 622 Mbps. The same applies to SDH channelization. More details on "unchannelized" SONET/SDH levels—they are not called unchannelized frames in SONET/SDH, but concatenated frames—are detailed later in this chapter and in the next one.

A nice trick for converting a given OC/STM level to Gbps is to divide the OC number by two and then again by ten: The result is the line rate in Gbps. Of course, this works by dividing by ten and then two, or even directly by twenty. The outcome is the same. For example, the speed of an OC-48/STM-16 is $48 \div 2 = 24 \div 10 =$ about 2.4 Gbps. The speed of an OC-12/STM-4 is $12 \div 2 = 6 \div 10 =$ about 0.6 Gbps or 600 Mbps.

The terms used in Table 6-1 appear frequently in the rest of this book. Because these form the basis of the SONET/SDH architecture, a review of these terms will aid in the understanding of the architecture.

OC-N This notation refers to the SONET transmission characteristics of an Nth level transmission link. This notation is analogous to the T-1 notation in today's T-carrier transmission network. Within SONET, an OC-3 transmission link is assumed to have an STS-3 frame structure. With concatenation, which can be contiguous or virtual, the frame becomes an STS-3c frame.

STS-N In this notation, STS refers to the SONET frame structure of an Nth level transmission link. This notation is analogous to the DS-1 notation in today's transmission network. Although most SONET links should properly be referred to as STS-N, it is much more common (and just as correct) to speak of OC-N. T-carrier labors under the same burden, but has not suffered for it because most people say "T-1" when they really mean "DS-1."

Payload The term used to indicate user data within a SONET/SDH frame. For SONET, payloads can be sub-STS-1 level virtual tributaries (e.g., T-1s, E-1s, DS-2s), STS-1 level payloads (e.g., DS-3s), or concatenated payloads (e.g., IP packets, ATM cells). SDH has different payload structures for sub-STM-1 payload, called virtual containers, but can also carry concatenated payloads for packets and cells.

Envelope   The portion of the STS-1 frame used to carry payload and end system overhead. SDH sometimes calls this an Administrative Unit. In general, SDH terms and structures for frame contents are very complex, even compared to SONET, and will be discussed in full in Chapter 8.

Overhead   The portion of the STS-N/STM-N frame used to carry management data for OAM&P.

Concatenation   A term that refers to the linking together of multiple STS-1/STM-1 frames to form an envelope capable of carrying higher bit-rate payloads, such as when SONET or SDH is used to carry ATM cells or IP packets. A concatenated SONET link is referred to as an STS-3c or an OC-3c, or whatever the SONET level happens to be. Note that in T-carrier, a form of concatenation existed as well and was also referred to by adding a letter "C," as in DS-3C. But also note that in T-carrier the concatenation indicator was always a capital "C", while in SONET the concatenation indicator is always a lower case "c". This is not an example of change for the sake of change. The concatenation in SONET is fundamentally different than the concatenation done in T-carrier. In SONET, the "c" really stands for "unchannelized," as will become obvious later on in this chapter.

SDH has its own terminology for concatenation in the form STM-Nc. There are two types of concatenation in SONET/SDH. Concatenation can be contiguous (the concatenated STS-1s/STM-1s are closely associated and related in the frame structure) or virtual (the concatenated STS-1s/STM-1s are unrelated in the frame structure in any way and only logically associated). Both contiguous and virtual concatenation will be discussed in full in this book.

A unique aspect of the SONET/SDH hierarchy, as opposed to the T-carrier hierarchy, is the relationship between the speeds of the different levels within the hierarchy. In SONET/SDH, a simple multiplication by 51.84 Mbps (or 155.52. Mbps for SDH) yields the correct speed at any level of the hierarchy. STS-3 operates three times faster than an STS-1, and $N \times 51.84$ Mbps = 155.52 Mbps. This simple multiplication is an aspect of SONET's synchronous, byte-interleaved multiplexing, and cannot be done with other carrier systems like T-carrier.

For example, a DS-3 frame contains the combined bits of 28 DS-1 frames, scattered throughout the DS-3 frame bit-by-bit. There is a lot of additional overhead in the DS-3 frame needed to locate these scattered bits properly at the receiving end of the T-3 link. The line rate of 28 combined DS-1 frames is just 28×1.544 Mbps = 43.232 Mbps; however, a T-3 operates at 44.736 Mbps. The extra bits are the added overhead from the DS-3 frame structure (e.g., C-bits, M-bits). This amounts to 44.736 Mbps – 43.232 Mbps = 1.504 Mbps, which is almost an added DS-1 all by itself.

SONET/SDH, conversely, adds no additional overhead at all at higher levels of the multiplexing hierarchy. All the overhead needed for synchronous, byte-interleaved multiplexing is present even at the lowest level of the hierarchy, at the STS-1 level. Of course, the trade-off is higher levels of overhead at the lower levels for no additional overhead at the higher levels. But this is not an insignificant advantage, especially when the need for higher and higher rate transports is considered, and today there are things that can be done with this "unused" overhead at higher levels of the hierarchy.

The overhead percentage in SONET/SDH is fixed, while the overhead percentage continually rises in other, asynchronous multiplexing schemes. In SONET/SDH, the ratio of overhead to payload (1.728 Mbps to 50.112 Mbps) remains constant at 3.45%, regardless of the value of N.

Figure 6-1 shows the percentage of overhead-to-payloads in SONET/SDH and several common T-carrier-based line signaling rates. The burden of asynchronous, plesiochronous multiplexing becomes obvious at higher line rates. The advantage of SONET/SDH becomes more pronounced at these levels, the very levels needed for modern network links.

STS/OC DISTINCTION

A few final words are in order about the distinction between the STS and OC designations. The remainder of the SONET frame after the overhead bits is the SONET payload itself, more commonly (but less accurately) known as *user data*. This payload may be information from the existing T-carrier hierarchy (e.g., DS-1, and so on), ATM cells or IP packets (if the SONET link were concatenated, like an OC-3c), or even something entirely different (e.g., MPEG video).

The designation STS refers to the electrical side of SONET. This seems odd at first, because SONET is supposed to be an optical network. Although modern network equipment excels at the sending and receiving of serial bits over a fiber optic link, the same is not true of any other network task. All SONET (and SDH) frames must be created, multiplexed, managed, and so on. This must be done in the electrical world, rather than in the optical world. The outbound user to the network side of this process is illustrated in Figure 6-2.

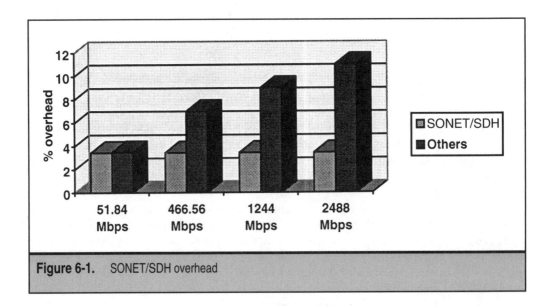

Figure 6-1.   SONET/SDH overhead

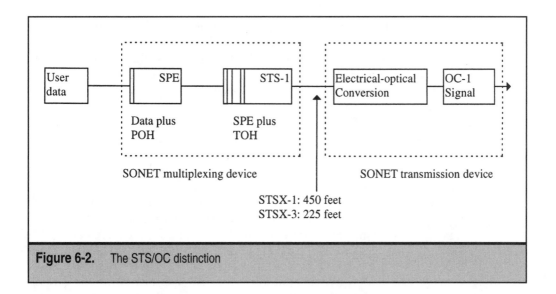

Figure 6-2.   The STS/OC distinction

In Figure 6-2, user data is sent to a SONET STS-1 multiplexing device. This device adds the path overhead (POH) to yield a synchronous payload envelope (SPE), and then adds the transport overhead (TOH) to yield the entire SONET STS-1 frame. Multiple user data streams can be combined as well. The SONET multiplexing device next sends the STS-1 frame to a SONET transmission device, which can be up to 450 feet away (an STSX-1 interface on copper for inter-equipment links for an STS-1) or 225 feet away (an STSX-3 interface on copper for inter-equipment links for STS-3). (It should be noted that in most SONET/SDH equipment packages today, these multiplexing steps are most often accomplished in the same device.) Multiple SONET multiplexers cannot share a transmitter, although several user data streams can share a SONET multiplexer, as in the case when three STS-1s are using a single STS-3.

Therefore, in SONET, the multiplexing hierarchy, framing formats, overhead functions, and so on are properly referenced by STS terminology. An STS-1 frame can carry a clear-channel DS-3 signal (44.736 Mbps) or a combination of sub-DS-3-rate signals, such as DS-1 or DS-0. Other possibilities are also mentioned above. All of these signals are electrical in origin.

Note that distance limits are strictly imposed by various standards on the distance that SONET signals can travel electrically. The serial bit durations must be kept long enough so that electrical noise and high frequency attenuation remain at a minimum. Otherwise, timing is lost and the frame structure degrades to the point where it can no longer be found on the receiving equipment. The distances are comparable to DSX- 3 and DSX-4 distance limits and should not be a burden to users or equipment vendors.

For transmission on fiber optic links, the STS-1 structure is converted and an optical carrier level 1 (OC-1) signal results. The OC signal is sent after most, but not all, of the electrical STS-1 frame is scrambled to avoid long strings of zeros, and so enable receiver clock recovery from transitions in the incoming data stream. (More details on scrambling follow later in this chapter.) The OC terminology is typically used when the carrier system itself is discussed, but SONET customers seldom seem to use the STS terminology. Most users and service providers alike typically refer to SONET speeds by OC references.

Purists still argue that since an OC-3, for example, is just an unstructured serial stream of bits, the concatenation term should properly be applied only to the STS-3c frame that rides the OC-3. Yet most service providers and equipment vendors happily speak and write about "OC-3c" or even "OC-3C" just because the terminology is easier. (One SONET vendor admitted labeling equipment "OC-3C" only because the die printer had no lowercase letters!) Users of SDH tend to look at this SONET controversy and shrug. This is because SDH uses the term STM for *both* electrical and optical signals, as mentioned. If ever any confusion were to arise, the terms STME-1 and STMO-1 can be used to indicate the electrical and optical aspects of STM, respectively.

Before looking at the details of the SONET and SDH frame structures used at the different levels of the hierarchy, it would be a good idea to examine the overall architecture of SONET/SDH links and systems. This is because while SONET/SDH frame structures for different speeds vary widely in detail (although they still consist of the same general structure), the overall SONET/SDH architectures are essentially the same no matter what speed the links actually run.

SONET/SDH ARCHITECTURE LAYERS

SONET/SDH, like almost everything else in computers and telecommunications today, has a layered architecture. Layers help to break up the communications tasks into more manageable pieces, and can furnish a certain amount of flexibility for implementers by making the layers more or less functionally independent of each other. Thus, vendors have a fair amount of latitude when implementing layer functions internally, as long as the interface between layers is maintained in a standard fashion.

Although SONET/SDH has a layered architecture, it is important to note that SONET/SDH mainly deals with the physical layer of the OSI Reference Model (OSI-RM), similar to T-1 or T-3 services. In other words, no relationship exists between the SONET/SDH layers and the layers of the OSI reference model, except for the existence of SONET/SDH at the physical layer of the OSI-RM. A "Layer 3" in SONET/SDH has no relationship to the Layer 3 (network layer) of the OSI-RM.

There are four layers in the SONET/SDH architecture, but of course the names are slightly different. A simple picture of the SONET layers is shown in Figure 6-3. The photonic layer refers to the optical properties of the transmission path, which involves the sending of the serial 0's and 1's from a sender to a receiver. SONET/SDH lasers and LEDs and their paired receiving devices operate at this layer. The photonic layer deals with the transport of bits across this physical medium. No overhead is associated with this layer—it is only a stream of 0's and 1's. The main function of this layer is the conversion of STS/STM electrical frames into optical OC bit signals.

Associated with each of the remaining SONET levels is a set of overhead bytes that govern the function of each layer. These overhead bytes are logically known as section overhead (SOH), line overhead (LOH), and path overhead (POH). The SOH and LOH

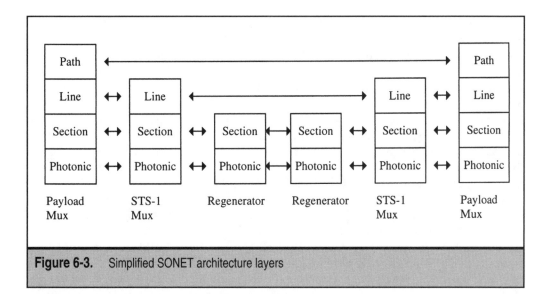

Figure 6-3.   Simplified SONET architecture layers

make up the transport overhead (TOH) located in the first columns of the STS frame. The POH bits are included as the first column of the synchronous payload envelope (SPE).

The section layer refers to the *regenerator section* of the transmission link. In fiber optic transmission, as in most digital schemes, 0's and 1's must be *regenerated* at fairly regular intervals. Of course, the signal does not necessarily terminate at this point; therefore, a single SONET link may have regenerators spaced every 10 kilometers or so on a 100 kilometer link. In this case, every segment between regenerators forms a section in SONET.

The section layer manages the transport of STS frames across the physical path, using the photonic layer. The functions of this layer include section error monitoring, framing, signal scrambling, and transport of section layer overhead. Section layer overhead consists of nine bytes of information that contain the information required for the section layer to perform its functions. This overhead is created or used by what is known in SONET as section terminating equipment (STE).

The line layer refers to the *maintenance span.* A maintenance span forms a segment between two SONET devices, excluding the lower layer regenerators. A single SONET link from one user site to another may consist of many such spans. In fact, much of the terminology in this chapter that refers to SONET "links" would be more properly expressed in terms of SONET "spans." However, outside of SONET network maintenance and engineering groups, the term "span" is rarely heard. To most, a SONET network consists of many SONET "links" between sites. However, the term mid-span meet is a constant reminder of the importance of the span concept in SONET networks, especially when maintenance and protection is considered.

The line layer manages the transport of entire SONET payloads, which are embedded in a sequence of STS frames, across the physical medium. The line layer functions include multiplexing and synchronization, both required for creating and maintaining SONET payloads. Line layer overhead consists of 18 bytes of information (twice as much as the SOH) that provide the line layer with its ability to perform its functions, its ability to communicate with the layers that surround it, and to provide certain protection and maintenance features. This overhead is used and created by SONET line-terminating equipment (LTE).

Finally, the path layer covers end-to-end transmission. In this case, end-to-end refers to customer-to-customer transmission. One end of the path is always where the bits in the SONET payload (i.e., SPE) originate and the other end of the path is always where the bits in the SPE are terminated. This may not be the actual end user device (such as a desktop workstation or server), but usually refers to some kind of premises SONET multiplexing device. Such devices can combine the bit streams from many end user devices and send them on to another site, where the data streams are broken out in the companion SONET CPE. The POH associated with the SONET path is considered to be part of the SPE for this very reason. The POH is passed unchanged through the SONET line, section, and photonic layers and is thus indistinguishable from user data to the SONET equipment (but the POH still must be there).

The path layer transports actual network services between SONET multiplexing equipment. These services would include the transport of customer DS-1s, DS-3s, ATM cells, and so on. The path layer maps these service components into a format that the line layer requires and then communicates end-to-end, using functions made possible by the POH bytes (nine in all) to ensure overall transmission integrity. The POH remains with the data (payload) until it reaches the other end of the SONET link.

Conceptually, each SONET layer requires the services of the layers below it to perform its functions properly. If two path layer processes were exchanging a DS-3 data steam from one customer site to another over an OC-3, the three layers would complement each other in the following fashion: The DS-3 signal and the associated POH bits would be encapsulated into a SONET payload (SPE) and handed off to the line layer. The line layer would then multiplex several signals (but only the link is larger than an OC-1, which can only fit one DS-3) from the path layer and add line overhead (LOH) to the mixture. It would then hand the signal element off to the section layer, which would add its own overhead, perform framing and scrambling, and then hand the whole frame off to the photonic layer.

The overhead in SONET/SDH will be totally unfamiliar to those experienced with T-carrier. SONET/SDH overheads obey a strict hierarchy in SONET/SDH equipment types; therefore, it may be a good idea to explore these relationships before examining the actual overhead byte structure in detail.

SONET/SDH Overhead Hierarchy

Many types of SONET/SDH equipment and the ways that they may be combined to provide SONET/SDH services are explored later in this book. For now, it is enough to realize that some types of SONET/SDH equipment are designed for CPE use, and so must generate the proper path overhead bits and SPE packaging. These equipment types are path terminating equipment (PTE) in SONET talk, and higher order path terminating equipment in SDH.

Consider SONET alone for the moment. It is important also to point out that the SONET CPE must also form a termination point for a SONET span (LOH) and even the section (SOH). It is common in SONET, as in T-carrier, that the CPE performs a regenerator function as well. Obviously, the SONET CPE must form and process *all* of the overhead bytes of whatever function.

Other pieces of SONET equipment have more specialized functions with regard to overhead. For example, a user's SONET signal at the OC-1 level may be combined with other users, perhaps from other sites, to form a single OC-3 in the same way that 28 T-1s from various users may be combined by the service provider to form a single T-3. The individual user's POH will not change, but this LTE will generate its own LOH and SOH from the new STS-3 frame structure. So an STE performs both LOH and SOH processing on the maintenance span.

At the lowest level, each optical segment between any sender and receiver at all in SONET forms a section. Regenerators all examine, process, and change the SOH, but do not touch the LOH or POH. This SONET overhead hierarchy, with different pieces of equipment performing different overhead functions and terminating sections, lines, and paths as required, is shown in Figure 6-4.

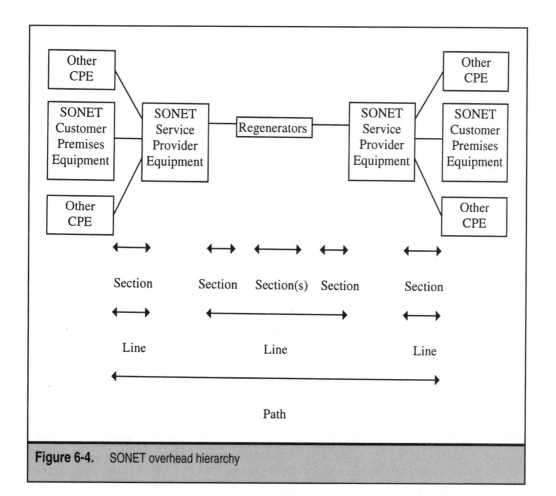

Figure 6-4.   SONET overhead hierarchy

In Figure 6-4, two PTE devices communicate by exchanging SONET payloads over a SONET link. Not only do these CPE devices terminate the path, but the line and section as well. Once the user's payload arrives at the service provider's network, the payload is typically combined with the payloads of other users in the same geographical area. Therefore, this piece of SONET LTE terminates the line (maintenance span) and the section, but not the users' paths. Finally, each regenerator along the way terminates the section, but neither the line nor the path.

Between each component, down to the regenerator, which is no longer the relatively passive device it once was in T-carrier, SONET SOH is terminated, examined, and modified. The LOH passes unchanged through the regenerator, but is terminated, examined, and modified by what SONET calls the *major components* of the link, which generally translates to everything else but the regenerators. The SOH is processed by these major components as well.

Of course, the POH passes unchanged across the entire link, from CPE to CPE. All POH generated by one CPE arrives unprocessed and unchanged at the other CPE. Thus, the line and POH is just more bits (data) to the sections and POH is just more bits (data) to the lines (that is, the maintenance spans).

Due to the hierarchical nature of the SONET overhead(s), each line endpoint is also a section endpoint and each path endpoint is also a line and section endpoint. However, the converse is not true: Each section endpoint is not necessarily a line or path endpoint and each line endpoint is not necessarily a path endpoint. Therefore, the required overhead processing varies at each component of a SONET network. It all depends on exact function.

A simple way to refer to SONET devices without reference to LTE or PTE terminology is just to refer to them as *SONET network elements* (NEs).

SONET Layers in Detail

There are many types of SONET/SDH network elements or NEs. All of these types will be discussed in full in Chapter 11. But even now it would be helpful to put the ideas of SONET PTE and LTE equipment into perspective. After all, no one buys or installs a SONET PTE. Path or line or section termination are functions related to overhead processing, not equipment packaging.

The rest of this chapter makes no attempt to discuss SONET and SDH more or less at the same time. This quickly becomes very awkward and it is hard to separate SONET and SDH details when encountering them for the first time. It is usually better to deal with one or the other first and completely with regard to architecture, overhead structure, and frame structure before turning to the next. So this chapter considers SONET first, then SDH. After examining the SONET architecture in detail, then the SONET path, line, and section overhead in general, and the various SONET frame structures in full, the same treatment is applied to SDH. Along the way, especially in the SDH section, SONET/SDH differences are noted.

The SONET architecture introduced so far is very general. In the real world of SONET, the PTE is usually a SONET device known as a terminal multiplexer or TM. The TM is the start or end of the SONET world, because beyond the TMs at each end of the SONET link there is no SONET at all, only non-SONET tributaries (such as voice) or payloads (such as IP packets). The TM at the sender loads the payload into the SONET system and adds the path, line, and section overhead. The values of the overhead bytes are based on configured parameters and current link conditions. At the receiver, the TM processes the overhead, removes it, and passes the payload content on to the non-SONET portion of the network. Now, a pair of routers with SONET (or SDH) interfaces can be TMs. In this case, wherever the IP packets came from or are going to forms the non-SONET portion of the network. In voice applications, the non-SONET payloads are typically virtual tributary voice channels carrying "voice," which can be two people talking, a fax transmission, analog modem data, or telephone touch tones.

The SONET LTE is usually some form of add-drop multiplexer (ADM). The role of the ADM as the key SONET/SDH equipment and NE is discussed more fully in Chapter 11 and other chapters. For now, the ADM's main role is as a SONET LTE device. That is, the ADM never examines or processes the path overhead, with the exception of the optional tandem connection byte in the path overhead, discussed later on in this chapter. Between ADMs, a physical fiber link in each direction carries the frames across a series of regenerators if the link is long enough. Regenerator spacing varies from ADM vendor to ADM vendor, but the usual maximum SONET span between ADMs or LTEs is 40 km (25 miles).

This more realistic look at the overall SONET architecture is shown in Figure 6-5.

In the figure, physical links, pairs of unidirectional fiber optical cables, connect the TMs (PTE), ADMs (LTE), and regenerators (R). These five links form the photonic layer (not shown) and section layers in SONET. Line layers are found between "major" SONET NEs, the TMs and ADMs in the figure, or everything but the regenerators. The ADMs usually multiplex other STS-Ns onto the OC-N fiber. The STS path layer is between the TMs, where the contents of the SONET payload is created and delivered as tributaries. Each layer or level, section, line, and path, has its own overhead bytes in the SONET frame. Finally, the tributaries themselves have their own overhead, detailed in Chapters 7 and 8, and the virtual tributary (VT) path layer also runs between TMs.

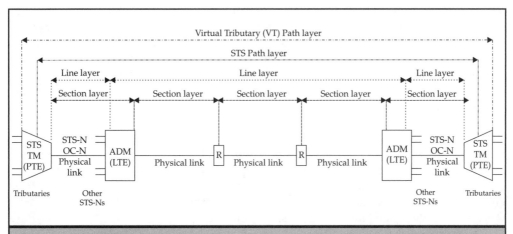

Figure 6-5. SONET architectural layers and equipment

Tandem Connection and Tandem Bundles

The basic SONET architecture of path and line and section is augmented today by another, optional layer. This is the *tandem connection* layer and essentially sits between the line and path layers of SONET (tandem connections are defined in SDH also). What is a tandem and why is a tandem connection a useful concept in SONET (and SDH)? If lines are defined between major pieces of SONET equipment and the path layer is end-to-end, from customer site to customer site, then what else is needed? How is a tandem bundle used in a SONET network?

Good questions, and they all have good answers. Along the way to understanding what a tandem connection is for, we'll also take a first look at how individual STS-1s become higher rate STS-Ns. Without an appreciation of looking at an STS-3 (for example) and seeing three STS-1s, the concept of a tandem bundle is difficult if not impossible to grasp.

PTEs for customers work in pairs to send information from one customer site to another. SONET/SDH is not a switched service like ATM, or a routed service based on IP routers. SONET/SDH is just a trunk circuit, purely and simply, as point-to-point as a T-1 or E-1. In many cases the customer does not even have any SONET/SDH gear on site, but just passes voice or video or data traffic off to the SONET/SDH service provider in some other format, from T-1 or E-1 voice frame to serial line protocol frames. The service provider's SONET/SDH equipment is typically in an equipment room in the building, or central to the office park, or housed somewhere in a campus environment. The point is that the PTE is where the SONET/SDH frames come from and where they go to.

But what kind of service provider is supplying the SONET/SDH service? SONET services in the United States, as will be pointed out in Chapter 18, can come from local telephone companies, a local ISP (often a reseller of wholesale capacity obtained from another service provider), an alternate provider of metropolitan area network (MAN) services, or one of many different entities (power and cable TV companies sometimes offer SONET/SDH services as well). These local service providers usually have limited service areas. In the United States, in many cases, regulation prevents local telephone companies from offering any type of service at all outside of their local service areas.

When SONET end points are far apart, perhaps spanning the United States, SONET services can be obtained from a telephone inter-exchange (long distance) carrier, a national ISP, an optical backbone provider, and others. But in many cases, these long haul providers do not have local networks to reach every possible customer site, and the local providers do not have networks to span the country (and they might be prohibited by regulation from doing so anyway).

In these cases, even though the customer contracts with either a local company or long-haul provider, there are really *three* service providers involved. There are usually two local networks, plus the nationwide network in between. All three have their own operations, repair, and other maintenance departments, all independent of the others. Yet there is only one path as far as the customer is concerned. When things go wrong from end-to-end, all that the customer knows is that something in between is broken. This is where the SONET/SDH tandem connection comes in.

The term *tandem* is borrowed from the U.S. telephony world. A tandem office has both local switching facilities (the central office switch) and links to inter-exchange carriers in *tandem* (both together). A tandem forms the bridge between local and long haul service providers. The SONET/SDH tandem connection is used to break up the natural SONET/SDH path layer into logical pieces, each of which is assigned to a different service provider. This situation is shown in Figure 6-6.

In SONET/SDH, the logical tandem paths are maintained by tandem connection terminating equipment (TCTE). The presence of this equipment is completely optional and depends on the cooperation of the service providers involved. The tandem connection overhead and protocol allows one of the three service providers to pinpoint exactly where an elevated bit error rate is occurring along the end-to-end customer path. This helps to prevent finger-pointing and assists efforts to isolate and fix the problem more quickly.

The figure also shows a related tandem concept. This is the idea of a *tandem bundle*. Tandem offices with TCTE devices can logically associate many STS or STM links into a bundle, treating all of them the same as far as the tandem path is concerned. In the figure, the long haul provider is associating many local SONET/SDH links with the long haul link through their network. This bundling and unbundling of customer links is accomplished by the TCTE devices in the tandem offices, as shown in the figure.

Tandem connections and TCTE devices are totally optional in both SONET and SDH. When a tandem connection is added to a SONET link, the entire architecture takes on the structure shown in Figure 6-7.

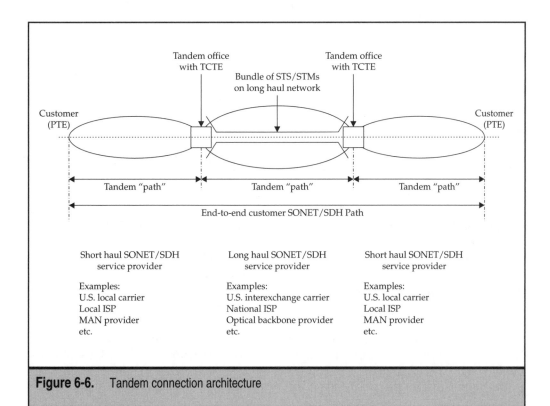

Figure 6-6.   Tandem connection architecture

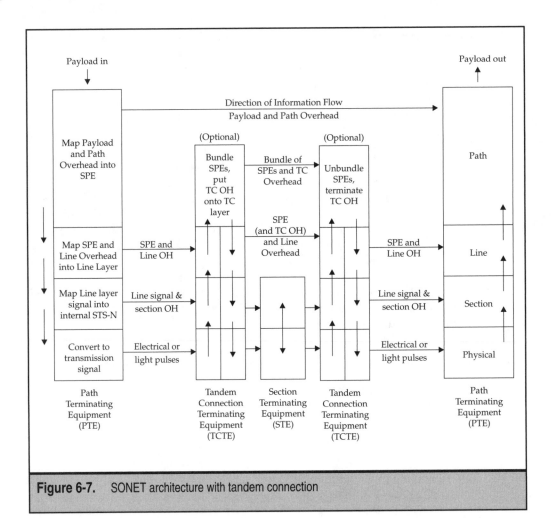

Figure 6-7. SONET architecture with tandem connection

This figure might seem complex, but it is not. At the upper left, a non-SONET payload (voice or video or data) enters the SONET network system at the PTE. The payload is *mapped* and path overhead is added to create a SONET synchronous payload envelope (SPE) at the virtual tributary and STS path level. The SPE and line overhead is mapped into a line layer unit, then this signal is mapped into an internal STS frame with section overhead attached. The PTE could be a telephone company central office, or a router with SONET interface. Note than the PTE forms the endpoint of a SONET section and line as well as the path.

The whole frame is converted to an electrical or optical signal depending on distance to the tandem equipment (TCTE). Once at the TCTE, which is entirely optional, the SPEs are bundled with special tandem connection (TC) overhead and then the entire STS-N frame is re-built. After this frame is sent through one or more section equipment (STE) devices, which process only the section overhead, the matching TCTE device unbundles the collection of STS-1 frames and processes (terminates) the tandem connection overhead. The SPEs are then re-packaged as separate STS-1s and passed along to the receiving PTE. After receiving the frame (physical), the receiver processes in turn the section, line, and path overhead, eventually restoring the payload to exactly what entered the SONET network.

SONET Multiplexing and Tandem Bundles

It has already been pointed out that unless other configuration steps are taken, the basic, default structure for an STS-N link is as N STS-1s (the same is true in principle for SDH). So a basic STS-3 has 3 STS-1s, an STS-12 has 12 STS-1s, and so on.

But there is an *order* implied in the sequencing of STS-1s inside an STS-N. Details of SONET frames will be discussed shortly. Until then, it is enough to realize that STS-1s will be loaded inside an STS-N in certain positions. The positions must be well-known and fixed so that other multiplexing equipment can easily find them.

The positions of the three STS-1s inside an STS-3 are easy to understand. The bytes from the three STS-1s are just multiplexed together in the sequence 1-2-3-1-2-3-1-2-3... and so on. This is called byte-interleaved multiplexing. However, in an STS-12, the twelve STS-1s must be ordered as if they were first made into an STS-3, and then four STS-3s were byte-interleaved multiplexed into an STS-12. The resulting STS-12 structure in terms of STS-1s is therefore 1-4-7-10-2-5-8-11-3-6-9-12... This sequence must be followed even if the STS-1 to STS-12 multiplexing is done in one device, in one stage, or even in one chipset. This same type of byte-interleaved pattern is seen over and over again in SONET frame structures, so it helps to see it here first without the complications of overhead to consider.

In the same way, an STS-48 with forty-eight STS-1s must look like it came from 16 STS-3 multiplexers all running at the same time. The resulting STS-1 sequence, along with the STS-12 sequence, is shown in Figure 6-8. In each case the intermediate multiplexing stages are shown as well as the final result.

In the upper part of the figure, twelve STS-1s create an STS-12. The two stages of multiplexing are shown, first a 3:1 stage producing 1-2-3, 4-5-6, 7-8-9, and 10-11-12 units. In other words, these are normal STS-1 to STS-3 multiplexers. Then a second 4:1 stage takes the input bytes one by one and produces the 1-4-7...6-9-12 multiplexing pattern.

The lower part of the figure shows how 48 STS-1s become an STS-48. Now there are a total of sixteen 3:1 multiplexers feeding a 16:1 multiplexer producing the extended 1-4-7-10-13-16...46-2-5-8-11-14-17...47-3-6-8...45-48 pattern of the STS-48 frame.

What has all this to do with tandem connection STS-1 bundling? Everything. When STS-1s are bundled and unbundled inside TCTE devices, *the STS-1s to be bundled must be contiguous*. This means that the tandem connection bundle could consist of an STS-3 inside an STS-12, an STS-12 inside an STS-48, or even other combinations. But when a tandem bundle is created, the STS-1s forming the bundle must have the pattern shown in Figure 6-9. The pattern is that the STS-1s must all come from the same 3:1 multiplexer or adjacent sets of 3:1 multiplexers.

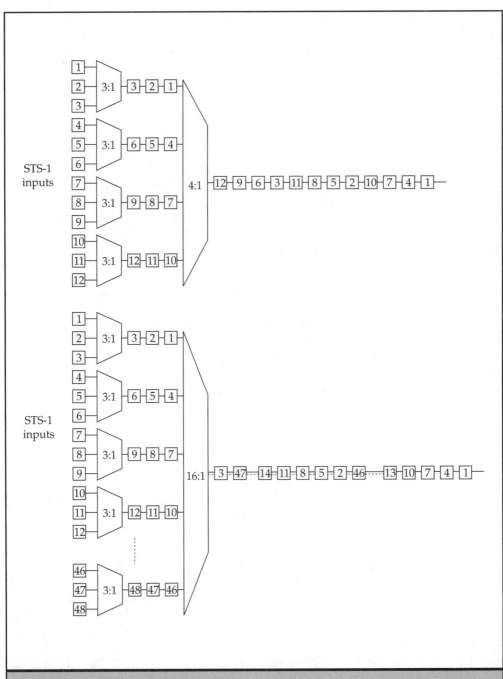

Figure 6-8. Multiplexing STS-12 and STS-48

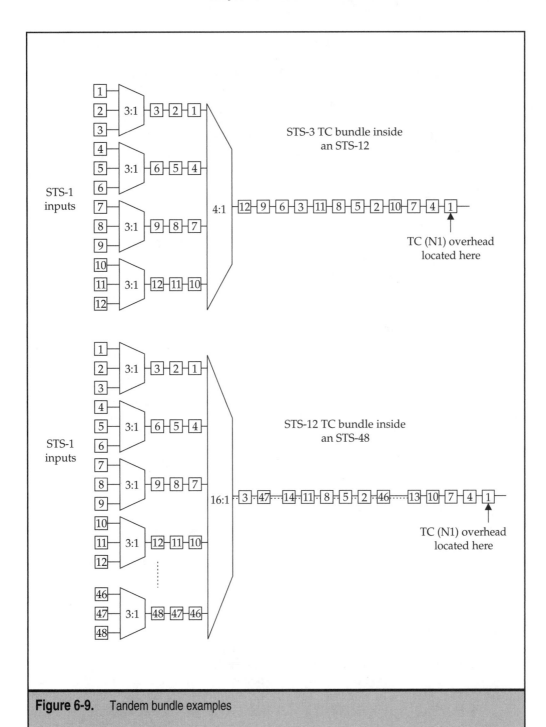

Figure 6-9. Tandem bundle examples

These are only examples of tandem bundles and are not to suggest that tandem bundling possibilities are limited to these configurations. One new detail has been added. In each case, the special TC overhead used by the bundle is located in the first STS-1 of the sequence. This TC overhead byte is called the N1 byte and is discussed in the next chapter, Chapter 7.

Tandem connections are defined for the SDH architecture. But the mention of STS-N multiplexing patterns and overhead bytes make it more important to consider SONET frame structures and overhead locations in more detail before moving on to SDH.

SONET FRAME STRUCTURES

Before looking at the details of the SDH architecture and terminology, this is a good place to finish up the SONET section of this chapter with a look at the SONET frame structures. We'll start by exploring the STS-1 frame in considerable detail, including overhead byte locations, but only so that as the other, high-rate frame structures appear, they will seem less complicated than might otherwise be the case.

STS-1 Frame Structure

The basic building block of the SONET digital transmission hierarchy is called the STS level one, or STS-1 frame. The SONET frame is a transmission frame used for the transport of a package of bits over a physical link; therefore, a SONET frame exists at a lower level on a network than the more well-known Ethernet or Token Ring LAN frames. The SONET frame is the exact analogy of the T-carrier level transmission frames in function, such as the T-1 frame structure.

In SDH, the SONET STS-1 frame is called the STM-0 frame. This sounds odd, but there is a good reason for the STM-0 designation. SDH originally contained no STS-1 equivalent frame. But since an *STM-1* frame is quite complex, it was common even in SDH environments to speak of a "conceptual STM-0" frame identical to an STS-1 purely for educational purposes. Today, of course, the STM-0 frame is the same as the STS-1 frame.

The basic STS-1 SONET and STM-0 SDH frame consists of 810 bytes, transmitted 8,000 times per second (or once every 125 microseconds) to form a 51.840 Mbps signal rate. This basic rate, the fundamental building block of SONET (but not SDH), is derived as follows:

810 bytes/frame × 8000 frames/sec × 8 bits/byte = 51.840 Mbps line rate

In other words, the 810 bytes of the basic SONET frame structure are sent 8,000 times per second, and because each byte consists of 8 bits, the signaling rate on the link is 51.84 Mbps.

Figure 6-10 shows the basic structure of the STS-1 SONET (and STM-0) frame in visual format. The STS-1 frame is 9 rows of 90 columns; it is always shown in this format, so that the overhead bytes will line up properly at the beginning of the frame. The STS-1 frame is transmitted one row at a time, from top to bottom, and from left to right within each row. Therefore, the byte in row 1, column 1 is sent first, and the byte in row 9, column 90 is sent

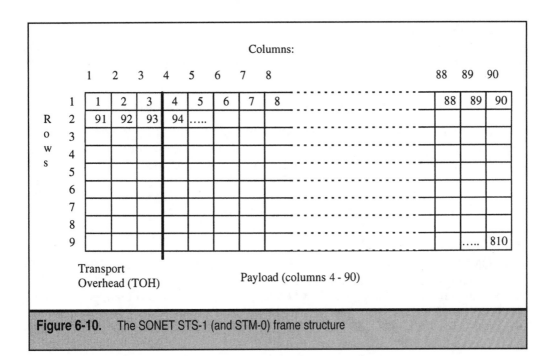

Figure 6-10.   The SONET STS-1 (and STM-0) frame structure

last. After the 90th byte is sent at the end of row 1, the next byte sent is the first byte in row 2, the byte in column 1.

One frame is sent every 125 microseconds; therefore, SONET/SDH can maintain time-slot synchronization that is required for delivery of normal, uncompressed PCM voice (8 bits 8,000 times per second or 64 Kbps). SONET/SDH also adheres to frame synchronization timing with existing asynchronous network standards (e.g., DS-1, E-1, and DS-3). More details on this timing synchronization are examined in the next chapter.

An STS frame is composed of two main sections, each with their own structures. The first three columns of the STS-1 frame form the TOH for the entire frame. All of the SONET overhead information (divided into section, line, and path) that is used to manage defined parts of the SONET network and transported data (called a payload), is in the first three columns of the frame. This overhead section, therefore, consists of 27 bytes (9 rows × 3 bytes/row) sent as part of each and every SONET frame. This overhead cannot be eliminated or converted for user data.

It is important to note that this means that the overhead in SONET is sent three bytes at a time throughout the entire SONET frame. That is, when a SONET frame is sent, three bytes of overhead begin the frame, and then the payload follows with 87 bytes of user data. These payload bytes comprise the remainder of the first row of the frame. The 91st through 93rd bytes are again overhead bytes, in this case the first three bytes of row 2 of the SONET frame. Then 87 bytes of payload follow, and so on throughout the entire 125 microsecond frame time.

The SONET payload is carried in the synchronous payload envelope (SPE). The capacity of the SPE is 9 rows of 87 columns (since the first 3 columns in each row are for overhead). This comes to 783 bytes (9 rows × 87 columns) of payload in each frame, giving a total user data rate of:

783 bytes/frame × 8000 frames/sec × 8 bits/byte = 50.112 Mbps payload rate

Actually, there is a little more to the SPE structure than just this. For example, more overhead, called path overhead (POH), is contained in the SPE but considered part of the user data. However, this should be enough to understand the basic SONET STS-1 frame structure. The SPE, just as the entire SONET frame that contains it, is sent 8,000 times per second. Finer points about the SPE and POH will be explored a little later.

The basic structures of the STS-1 frame are described below.

Section and Line Overhead

The SONET STS-1 frame transports a considerable amount of overhead associated with operational functions within the SONET network. The first three columns of the STS-1 frame are dedicated to section overhead (SOH) and line overhead (LOH), performing a wide variety of functions. Together they are considered the TOH. The first three rows of the TOH are the section overhead (SOH; nine bytes). The last six rows of the TOH are the line overhead (18 bytes).

Synchronous Payload Envelope Capacity

The remaining 87 columns of the STS constitute the SPE capacity. It is more precise to say that these 87 columns represent the SONET information payload and that the SPE is the structure of the information payload, but usually only the SPE itself is singled out for attention.

The SPE contains both the path overhead, called POH, and the actual user data. The POH, combined with the user data (payload) that follows, constitutes an SPE. The POH can begin in any byte position within the SPE capacity. This typically results in the SPE overlapping into the next frame.

It should be noted that the extensive amount of overhead in SONET is useful for surveillance and network management, as well as other activities. Although criticized for excessive overhead, the fact that SONET overhead does not increase proportionately at higher levels of the hierarchy is a real advantage.

It is easy to construct the overall structure of a frame at any level of the SONET hierarchy once the basic STS-1 format is understood. All SONET frames are sent 8,000 times per second. All SONET frames have exactly nine rows. The only variable is the number of columns. This may sound simplistic, but it is true and results in all of the different speeds at which SONET operates.

For example, an STS-3 frame (which is the same as an STM-1 frame) consists of 9 rows and is sent 8,000 times per second; however, an STS-3 frame is not 90 columns wide. The STS-3 frame is three times wider (N=3, after all). Therefore, the STS-3 frame is 270 columns wide, and that is not all. The STS-3 overhead columns are multiplied by three as well, as are the SPE capacity columns. An STS-3 frame then is 270 columns wide (3 × 90), of which the first 9 columns are TOH (3 × 3) and the remaining 261 (3 × 87) are payload capacity. The whole STS-3 frame is 2,430 bytes (9 rows × 270 columns). The line rate for an OC-3, therefore, must be:

2430bytes/frame × 8000 frames/sec × 8 bits/byte = 155.52 Mbps line rate

The same result can be derived simply by multiplying the OC-1 line rate of 51.84 Mbps by 3. Thus, the "trick" of SONET frame structures and line rates is that each frame is N times larger in terms of columns than an STS-1 frame, so the line must signal N times faster than the OC-1 line rate to transmit the same 8,000 frames per second. After all, the STS-3 frames may be carrying hundreds of digital voice samples, all of which must arrive every 8,000th of second.

Figure 6-11 shows some basic structures of other common SONET/SDH frame types. These frame structures will be examined in more detail shortly.

SPE Mapping into the STS-1 Frame

It may seem obvious that the first byte of the payload should be the 4th byte of the SONET STS-1 or SDH STM-0 frame. After all, if the SONET SPE contains the user data and the SPE is sent 8,000 times per second, why shouldn't the SPE be aligned with the overall SONET frame structure? In a perfect world, there would be no answer to this question. But this is not a perfect world, and the clocks used in networks to time bit streams do not always co-operate and remain synchronized. The whole question of timing and network synchronization is explored more fully in Chapter 9. For now, it is enough to point out that jitter and phase differences make it difficult to fix the SPE inside the SONET frame in all cases and at all times. In fact, an insistence on fixing the SPE may actually be foolish and ultimately counter-productive.

The SPE, therefore, does not have to be aligned with the first row and fourth column of the SONET frame. It all depends on the clock source and SONET network configuration. Figure 6-12 depicts how an SPE would be mapped into a SONET frame when there are phase differences occurring between the incoming and outgoing payloads. The same logic applies to SDH, as will be explained later in this chapter. In plain English, this determines how the customer's data is carried on the SONET system.

Study a few items in Figure 6-12. First, the SPE is not necessarily frame aligned with the beginning of the SONET frame. This is due to the fact that the SPE must be frame synchronous with the *customer's* payload (which may be a DS-3 CSU/DSU, or other CPE device) and not the service provider's SONET system. This only makes sense. A SONET link does not generate any payload of its own to transport, but delivers customer payloads from site to site over the SONET link. Phase differences also occur in the multiplexing of various customer payloads onto a higher speed SONET link as well, such as when three STS-1 payloads are combined into a single STS-3 link.

The customer's payload then is mapped into a payload envelope and the POH is added, forming an SPE. This is done by the CPE device at the customer's site. The SPE is then mapped into the SONET frame. A good analogy is to compare the SPE construction to the functioning of a T-1 CSU (a portion of the T-carrier CPE) and the final STS-1 frame construction to the functioning of a T-1 DSU (another portion of the T-carrier CPE).

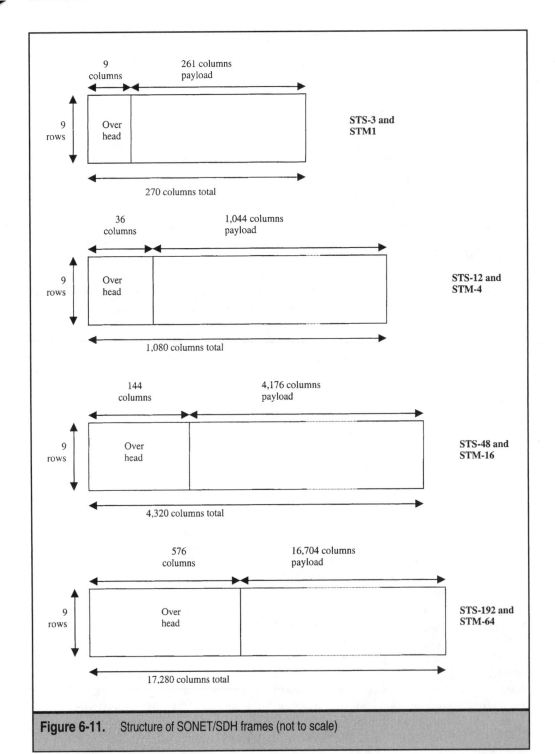

Figure 6-11. Structure of SONET/SDH frames (not to scale)

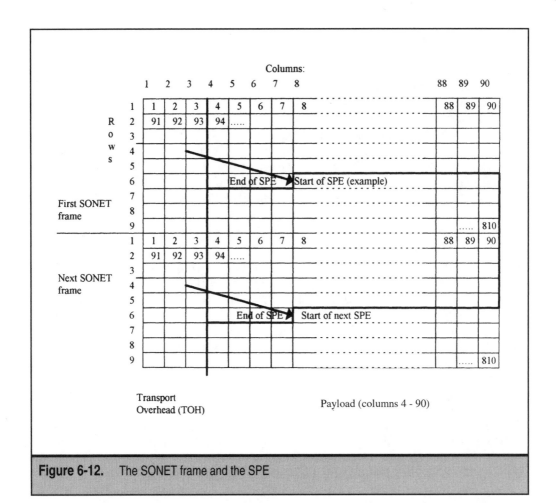

Figure 6-12. The SONET frame and the SPE

The second item to notice in Figure 6-11 is that a mechanism must exist to find the SPE at the receiving side of the SONET link because the SPE is separated from the TOH by its position within the frame. That is, the SPE is no longer fixed in the fourth column of the SONET frame but can now begin almost anywhere within the information payload section of the SONET frame. How can the SPE be found by the receiving SONET equipment?

The answer is that components of the TOH (namely "pointer bytes") identify the position of the start of the SPE within the SONET frame. This allows the receiving SONET transmission equipment to align with the incoming transmission stream and to identify the start of the SPE. Otherwise, the buffers would be needed to delay the sending of the SPE until the start of the next SONET frame. This does not sound like much of a penalty, but the buffers would be quite large at SONET speeds. If an SPE were ready to be sent halfway through a SONET frame on average, then the buffer needed would be half the size of the SONET frame. This is about 1,200 bytes at the OC-3 level, but a huge 80,000 bytes or so at OC-192. STS-192 frames are $9 \times 90 \times 192 = 155,520$ bytes long. Larger buffers add expense and complexity in the form of buffer management to SONET equipment. Today, buffers (memory) are quite inexpensive, but this was not the case when SONET/SDH was defined.

In practice, the delay needed to align the incoming SPE with the outgoing SONET frame is usually achieved once at the ingress point of the SONET network. Then for the rest of the links until the SPE emerges at the other customer site, the SPE is fixed at column 4 of the SONET frame on the service provider's SONET network.

The last point in the Figure 6-12 is the concept that the SONET frames and the SPEs within them are continuous. Therefore, a realistic drawing of the SONET transmission systems would show SONET frames and SPEs before and after the sequence shown in the figure. That is, the last byte of an SPE is immediately followed by the first byte of the next SPE, regardless of where the SPE may appear in the SONET frame.

STS-1 Overhead

So far, three different types of SONET overhead have been mentioned. There is section overhead in the first three rows and columns of the SONET STS-1 frame, line overhead in the last six rows of the first three columns, and path overhead, always pointed to by the line level pointer bytes and forming the first column of the SPE.

The next chapter, Chapter 7, will investigate all aspects of the SONET/SDH overhead bytes. But this is the time to at least introduce them by their proper names and positions. The structure of all other SONET/SDH frames is incomprehensible without appreciating the role that SONET/SDH overhead plays in the overall frame structure.

Figure 6-13 shows the positions and names of the bytes defined as SONET overhead. SDH is a little different and will be shown later. At this point, all that is needed is an understanding that some of the overhead bytes are needed for each STS-1 inside an STS-N. So if there are 48 STS-1s inside an STS-48, 48 separate A1/A2 framing bytes (for instance) are needed. Other overhead bytes are needed only once for the entire STS-N frame. After all, there is really only one link and one fiber carrying the N STS-1s.

In some cases, the path overhead is "locked" as a fourth column inside the STS-1 frame, but in most cases it does not. But all SONET equipment must be able to allow the SPE to "float" inside the STS-1 frame, as already pointed out.

SONET FRAMES BEYOND STS-1

In addition to the basic STS-1 frame, SONET specifications currently define STS-3, STS-12, STS-48, STS-192, and STS-768 frame structures. Although STS-24/OC-24 is still defined as a SONET level, no frame structure details about STS-24 are present in current ANSI specifications and so need not be discussed here.

In keeping with the philosophy of "N STS-1s makes an STS-N," the basic frame format is just N times bigger and the link N times faster to maintain the 8,000 frames per second relationship. But the transport overhead complicates this simple formula. Path overhead will be considered later, but for now only transport overhead is important. Let's start with the STS-3 frame.

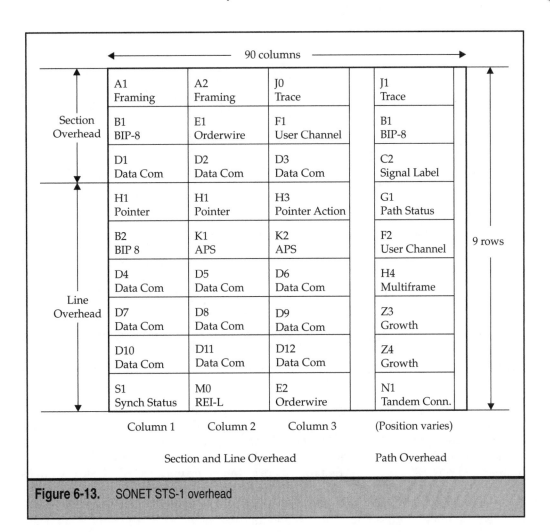

Figure 6-13.   SONET STS-1 overhead

The STS-3 Frame Structure

An STS-3 frame has 9 rows and 270 columns. This is three times the number of columns in an STS-1, as expected. The overhead is three times larger as well. So there are now nine columns of overhead leading off the frame. The nine overhead columns have the three overhead columns of the three STS-1s in the familiar 1-2-3 byte-interleaved multiplexing pattern. That is, column 1 consists of the first column of overhead from the first STS-1, column 2 consists of overhead from the second STS-1, and column three of the STS-3 frame consists of overhead from the third STS-1.

The transport overhead of the STS-3 frame is shown in Figure 6-14. The overhead columns from the first STS-1 appear in columns 1, 4, and 7 of the STS-3 frame.

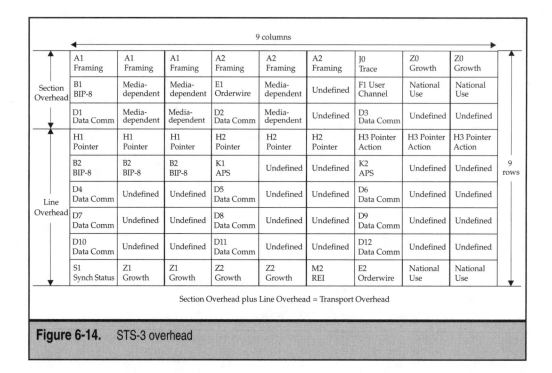

Figure 6-14.   STS-3 overhead

But something is not quite right. The three sets of A1/A2 framing bytes are there, one set each for the three STS-1s inside the STS-3. The set of A1 and A2 framing bytes for the first STS-1 are in columns 1 and 4, the set of A1 and A2 framing bytes for the second STS-1 are in columns 2 and 5, and the set of A1 and A2 framing bytes for the third STS-1 are in columns 3 and 6. But already in the first row, there is something unexpected. The J0 trace byte appears only once, in column 7. The other two columns, columns 8 and 9, have Z0 growth bytes in place of the J0 trace bytes from the second and third STS-1 inside the STS-3.

Because some overhead bytes are needed for each and every STS-1, while other overhead bytes are needed only for the whole STS-3 frame, many overhead positions become undefined or available for other functions. An STS-3 frame is still only a single, unified frame traveling on a single fiber, even though there are three STS-1s inside. For example, there need only be one data communication channel (D1-D12) for the STS-3 equipment to use at the section and line levels. So some overhead functions only need to be done once for the entire STS-3.

The figure shows just which overhead bytes are used for transport overhead, which bytes are now undefined, and which bytes now take on other functions. The six overhead bytes marked as growth bytes are for future use. The six overhead positions marked as media dependent are used mainly when SONET frames are sent on wireless links (a capability undreamed of when SONET/SDH was first defined in 1988). Overhead positions marked for national use mean just that: in the United States, the responsible national regulatory body (ANSI or another) gets to specify the meaning and use of these four bytes. In another country, these bytes can take on other meanings.

Before considering higher-rate SONET frame structures, mainly with regard to transport overhead positions, one other point about SONET (and SDH) frames in general should be made. It has already been mentioned that SONET/SDH frames are *scrambled* before transmission at the optical level and unscrambled at the receiver before the SONET/SDH frame is processed. (As a result, scrambled SONET/SDH frames are invisible to almost all electrical diagnostic equipment.) Scrambling helps to prevent long strings of 0 bits from causing a receiver to lose synchronization with the sender. Scrambling provides the same function for SONET/SDH as the "1's density" solutions did and do for T-1.

Optical networks like SONET and SDH still need to provide some form of 1s density because most SONET/SDH equipment works with *intensity modulation*, as discussed earlier in this book. If the received light in a bit interval is above a certain threshold, it's a 1 bit. If below the threshold, it's a 0 bit.

But what if there are long strings of 0 bits inside the frame? This never happens when SONET/SDH is used for voice, because voice has no "all 0 codeword," meaning that no voice sample can ever be sent as 0000 0000. But data applications generate long strings of 0 bits all the time. Even if one source prevents truly long strings of 0's, it might happen that many sources are multiplexed together and so produce very long strings where the received light is below the 1 bit threshold. The problem is that these intervals on "zero transitions" cannot be used to recover timing and clock signals from the received bit stream (which is the way modern equipment always recovers clock) and so the receiver might not know if (for example) a given interval was 37 or 38 0-bits long.

Although this "too many 0's" issue mainly addressed SONET/SDH fiber transmission methods, it is just as important to realize that SONET/SDH frames can also be carried on radio links or on short electrical cables. So there is an issue with the "1's density" with regard to the bits sent on all types of links.

What can be done? SONET/SDH frames will scramble the bytes to prevent loss of synchronization between sender and receiver. This concern is shown in Figure 6-15.

The important point here is that the *first row* of any SONET/SDH frame is *never* scrambled. The A1/A2 framing bytes always have a fixed pattern of mixed 0's and 1's, so that is not a problem. But the J0 trace byte, as will be shown, is not scrambled, nor are the following growth bytes. The value of these bytes can be set through configuration of the equipment to many different values, some of which might be full of 0's.

All SONET/SDH frame structures therefore carry a warning to the effect that "care is needed when selecting the values" for bytes beyond the A1/A2 framing bytes in the first row. As SONET/SDH frames become larger and larger, this "unprotected" first frame row becomes more and more of a concern. This brings us to the STS-12 frame structure and overhead.

The STS-12 Frame Structure

The STS-12 frame has 9 rows and 1,080 columns (the same as an SDH STM-4). Figure 6-16 shows the structure of the now 36 columns of overhead at the beginning of the STS-12 frame.

Now there is only room for the overhead designations, not the names. Z0, Z1, and Z2 bytes are for growth. Blanks bytes are undefined because their functions are fulfilled by the overhead bytes in the first STS-1 column position in most cases (such as K1/K2). Media dependent bytes, now grown to 22 bytes per frame (1.408 Mbps) are shown as the Greek delta (Δ). National use bytes, shown as X, also occupy 22 bytes (1.408 Mbps).

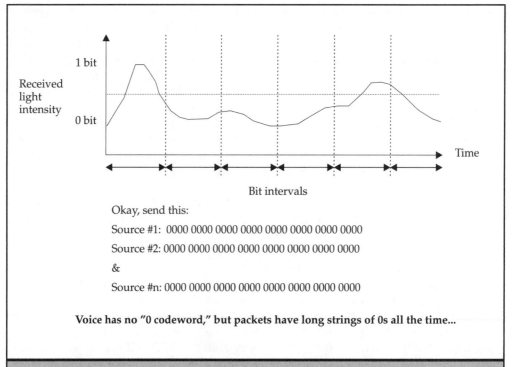

Figure 6-15. Scrambling need in SONET/SDH

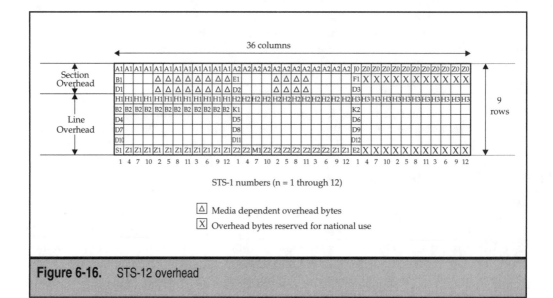

Figure 6-16. STS-12 overhead

The remaining point of interest is shown across the bottom of the overhead columns. Each of the columns representing the overhead from the STS-1s making up the STS-12 frame is labeled from 1 to 12. The familiar 1-4-7-10-2-5-8-11-3-6-9-12 interleave pattern from Figures 6-8 and 6-9 can be seen. The pattern repeats three times because there are three columns of overhead in the basic STS-1 frame. So the A1/A2/J0 first (unscrambled) overhead row for STS-1 number 7 appears in columns 3, 15, and 27 of the STS-12 frame.

ANSI's "wallpaper" representation of SONET frame structures makes it increasingly difficult to pinpoint overhead byte locations for the individual STS-1s. SDH documentation does not represent STM-N frames in this "wallpaper" fashion, as will be shown later on in this chapter.

It is important to realize that even if the entire STS-12 payload carries only a single stream of IP packets or ATM cells for a single customer, the frame overhead *must* look like it comes from 12 STS-1s byte-interleaved multiplexed together. This way, SONET equipment below the path level need not be concerned with the content of the frame. Only path terminating equipment needs to know the details of the frame content and format.

The STS-48 Frame Structure

Although STS-24 is still defined by ANSI, the specification details STS-48 after STS-12. The STS-48 frame has 9 rows and 4,320 columns (the same as an SDH STM-16). Figure 6-17 shows the structure of the 144 (3×48) columns of overhead at the beginning of the STS-48 frame.

Now the STS-1s at the bottom of the frame are numbered 1 to 48, again repeated three times for the three STS-1 overhead columns. Since the "wallpaper" format is stretched to the limit here, a kind of shorthand is used to condense many of the overhead columns into groups (such as 11-47, 10-46, and so on). These columns have exactly the same con-

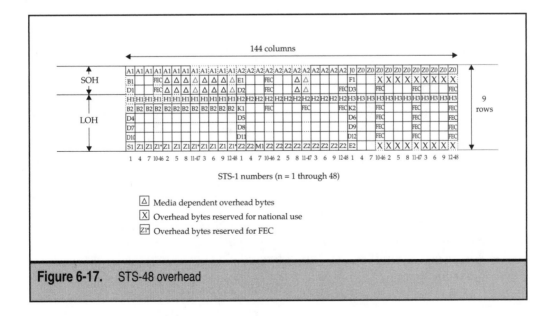

Figure 6-17. STS-48 overhead

tent, so this system works and is represented as dashed lines in the figure. Note that the media dependent and national use fields are larger than in the STS-12 frame.

Two new features make their appearance at the STS-48 level. Both concern the use of a forward error correction (FEC) codeword in the frame overhead. FECs are used to not only detect, but correct, certain patterns and numbers of errors in received frames.

A full discussion of the mathematically complex FECs used in SONET/SDH is far beyond the scope of this book. But as SONET/SDH frames grow larger and larger, the possibility of a frame arriving with an error (or errors) becomes higher and higher, even in the bit error rate remains constant. With a FEC, a lot of otherwise errored frames can have bit errors corrected and the frames processed as usual. Without a FEC, the entire frame must be discarded (network equipment should never, ever process a frame known to contain an error). Since resending is not possible at the SONET/SDH equipment level (buffer limits and time considerations prevent any thought of this solution), a standard FEC is added to the frame overhead at the STS-48/STM-16 level and above. This also makes more efficient use of the increasingly undefined overhead bytes as the frames grow and grow.

The second FEC-related point is the presence of certain Z1 growth bytes marked with an asterisk in the figure. Although these bytes are still defined by ANSI as SONET growth bytes, the positions marked are used in SDH for additional FEC bytes. So this is a compatibility feature.

The STS-192 Frame Structure

The STS-192 frame has 9 rows and 17,280 columns (the same as an SDH STM-64). Figure 6-18 shows the structure of the 576 (3 × 192) columns of overhead at the beginning of the STS-192 frame.

Now the STS-1s at the bottom are numbered 1 to 192, three times as usual. Although it is hard to tell from the figure, there are more bytes than ever for media dependent, national use, and FEC functions. For example, the number of media-dependent bytes is always 6N, where N is the SDH level or 1, 4, 16, 64, or 256. At STS-192, the SDH N value (STM-N) is 16, so the media-dependent bytes are 6 × 16 = 96 bytes. Each of these bytes is 64 Kbps, so the media-dependent bytes at this level are 6.144 Mbps. Only one feature is new in this frame structure, and it concerns SONET/SDH compatibility. Notice the arrow pointing to the Z2 growth byte in the second column of the second set of STS-1 column numbers at the bottom of the frame. This column "belongs" to STS-1 number 4, as shown in the figure.

SDH documentation defines this overhead byte position as the M0 byte for an error indicator known as the remote error indicator, line level (REI-L). The details of the M0 bytes and REI-L are discussed in later chapters. All that need be realized now is that the M0 (and M1) bytes are used as a kind of "error counter" to inform the sender of bit errors in the previously received frame. Because senders and receivers in SONET/SDH use separate fibers for transmit (outbound) and receive (inbound) frames, senders do not know about receiver errors or problems on the outbound fiber unless informed on the return fiber by the receiver.

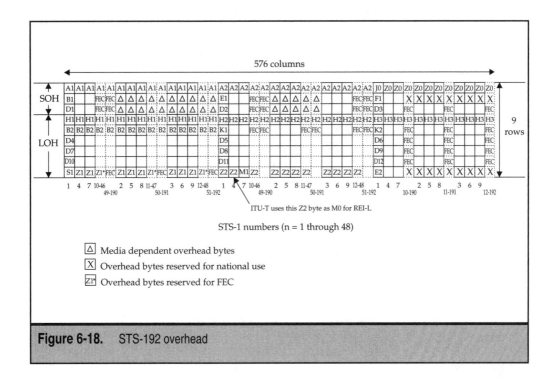

Figure 6-18.   STS-192 overhead

The STS-768 Frame Structure

Until recently, the SONET specification topped out at the STS-192/OC-192 speed of about 10 Gbps. But serial bit rates of about 40 Gbps are currently possible. It is a tribute to the extensibility of the SONET/SDH frame structures that SONET/SDH is able to accommodate this new speed with relative ease. The STS-768 frame has 9 rows and 69,120 columns (the same as an SDH STM-256). Figure 6-19 shows the structure of the 2,304 (3×768) columns of overhead at the beginning of the STS-768 frame.

Now the STS-1s at the bottom are numbered 1 to 768, three times. Many columns are the same and condensed as in the other frame figures. Although it is hard to tell from the figure, there are more bytes for media dependent, national use, and FEC functions.

Two things are different in the STS-768 frame structure from the frame structures that have gone before. One concerns the A1/A2 framing bytes and the other concerns the data communication channel available for network management of SONET systems.

Note that there are *not* 768 sets of A1/A2 bytes as might be expected. There are only 64 A1 bytes and 64 A2 bytes (the shorthand 12-48 does not mean 12-13-14…47-48, but 12-15-18-21…45-48, counting by threes in the usual multiplexing pattern).

The reason for this reduction in framing bytes is simply that modern SONET/SDH equipment seldom checked *all* of the A1/A2 bytes in higher-rate frames. Only the A1/A2 framing bytes around the A1/A2 "boundary" were really checked by these newer chipsets.

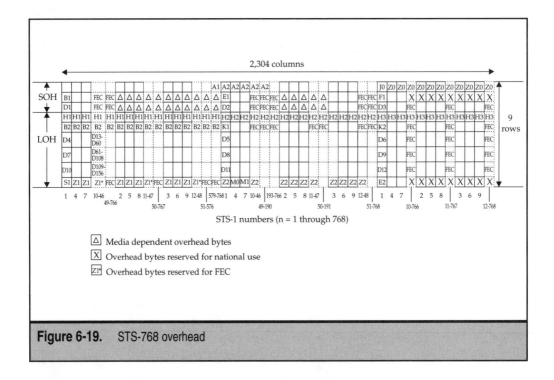

Figure 6-19. STS-768 overhead

So why not just define, create, and send the framing bytes actually needed? This is what was done in the STS-768 frame (and also in the SDH STM-256 frame).

The other addition is the presence of additional data communications bytes at the line level, shown as D13 through D156. These additional 144 bytes give a data communications channel at the line level not at 567 kbps (D4-D12 run at 9 × 64 Kbps) as at all other SONET/SDH speeds, but at (9 + 144) × 64 Kbps = 9.792 Mbps. This is fast enough to essentially run raw 10 Mbps Ethernet frames over this channel, although that is not the intention. Technically, these bytes are independent of the other D bits and form a channel of their own at 9.216 Mbps. The point is that more speed was needed for 40 Gbps links, and the additional D-overhead bytes provided this bandwidth.

This tour through SONET frame structures was undertaken for two reasons. First, to show how each frame structure is different, sometimes subtly. Second, to enable you to better comprehend the system used to define the SDH frame structure, which has essentially the same overhead but is shown very differently in SDH documentation. But first, a look at the overall SDH architecture is in order.

SDH ARCHITECTURAL LAYERS

There is no need to repeat the general points about the SONET architecture that also apply to SDH. The emphasis here will be on how SDH architecture and terminology differs from the SONET concepts and terms already introduced.

As in SONET, SDH has a PTE device known as a terminal multiplexer or TM. Beyond the TMs at each end of the SDH link there is no SDH at all, only non-SDH higher order or lower order tributaries (such as voice) or payloads (such as IP packets). In fact, the higher order path is essentially the SONET path level.

SDH also has an add-drop multiplexer (ADM) performing the same role as SONET LTE devices. That is, the ADM never examines or processes the path overhead, with the exception of the optional tandem connection byte in the path overhead, which also exists in SDH and has already been extensively discussed in this chapter. Regenerators are spaced along the usual maximum SDH span of 40 km (25 miles) between ADMs.

This more realistic look at the overall SDH architecture, contrasted with the SONET architecture shown earlier in Figure 6-5, is shown in Figure 6-20.

Both the new SDH terms and older SONET terms from Figure 6-5 are shown for ease of comparison. In the upper part of the figure, physical links, pairs of unidirectional fiber optical cables, connect the STM TMs, ADMs (still usually called LTEs in SDH), and regenerators (R). These five links form the photonic layer (not shown) and regenerator layers in SDH. Multiplex section layers are found between "major" SDH NEs, the TMs and ADMs in the figure, or everything but the regenerators. The ADMs usually multiplex other STM-Ns onto the fiber. In SDH, the SONET STS path layer between the TMs is now called the higher

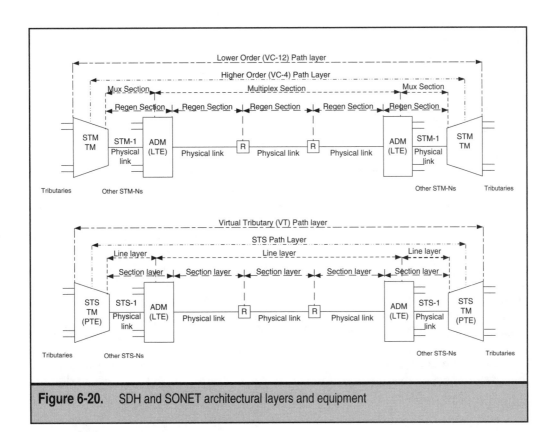

Figure 6-20.   SDH and SONET architectural layers and equipment

order path layer, and in this case carries a *virtual container type 4* (VC-4) inside the STM-1. The SONET virtual tributaries are now SDH lower order paths, in this case a VC-12 (read as "virtual container type one-two," not "VC twelve") when the STM-1 is carrying the 32 channels of E-1. All types of SDH frame contents will be detailed in Chapter 8.

SDH also has a payload, just like SONET, but the use of the terms higher order and lower order path, as well as related payload terms, makes the structure of the content of an SDH frame very difficult to grasp and understand. SONET/SDH frame contents will be discussed in full in Chapter 8. This chapter will compare and contrast the various SDH frame overhead structures with the SONET frame overhead structures presented already in this chapter.

SDH FRAME STRUCTURES

As with SONET, we'll start by exploring the STM-0 frame overhead structure in considerable detail, but only so that as the other, high-rate frame structures appear, they will seem less complicated than might otherwise be the case.

STM-0 Frame Structure

The basic building block of the SDH digital transmission hierarchy is *not* the STM-0 frame. The STM-0 frame was used for years as an educational tool, mainly because jumping right in to the STM-1 frame was quite confusing to people new to SDH. Now STM-0 is an officially defined SDH frame. The STM-0 frame looks much like a SONET STS-1 frame when it comes to overhead, but there are important differences, and not just in terminology.

Figure 6-21 shows the basic structure of the SDH STM-0 frame overhead. Like an STS-1 frame, the STM-0 frame is 9 rows of 90 columns.

Most of the overhead bytes have the same names and positions as the SONET equivalents. But the SONET section level overhead is now SDH regenerator section overhead (RSOH). Most of the SONET line level overhead is now multiplex section overhead (MSOH), but with one important difference.

While SONET defines the last six rows and first three columns of the STS-1 frame as line overhead (LOH), SDH essentially leaves out row 4 of the frame overhead. These are the H1/H2/H3 pointer bytes, used in both SONET and SDH to find the path overhead leading off the payload envelope. These pointer bytes are still present in SDH, and have exactly the same function and names, but are bundled into what SDH calls an administrative unit (AU). The AU is technically neither RSOH nor MSOH, but just the AU and that is that. AUs are supposed to be generated and processed separately from all other SDH overhead bytes (SDH now usually uses the term "byte" instead of "octet" just as in SONET). One thing that drove early implementers of interoperable SONET and SDH chipsets and equipment crazy was the need to process the H1/H2/H3 bytes one way in SONET (as part of the LOH) and another way in SDH (as separate from the MSOH). More powerful and agile chipsets perform this task with ease today, but in the early to mid-1990s it was a real issue.

The only other real difference between SONET and SDH overhead is in the path overhead. The path overhead does not change from frame size to frame size and so did

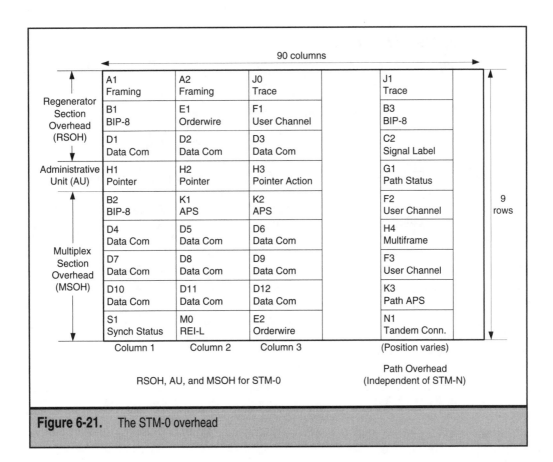

A1 Framing	A2 Framing	J0 Trace		J1 Trace
B1 BIP-8	E1 Orderwire	F1 User Channel		B3 BIP-8
D1 Data Com	D2 Data Com	D3 Data Com		C2 Signal Label
H1 Pointer	H2 Pointer	H3 Pointer Action		G1 Path Status
B2 BIP-8	K1 APS	K2 APS		F2 User Channel
D4 Data Com	D5 Data Com	D6 Data Com		H4 Multiframe
D7 Data Com	D8 Data Com	D9 Data Com		F3 User Channel
D10 Data Com	D11 Data Com	D12 Data Com		K3 Path APS
S1 Synch Status	M0 REI-L	E2 Orderwire		N1 Tandem Conn.

90 columns

Regenerator Section Overhead (RSOH)

Administrative Unit (AU)

Multiplex Section Overhead (MSOH)

9 rows

Column 1   Column 2   Column 3   (Position varies)

RSOH, AU, and MSOH for STM-0

Path Overhead
(Independent of STM-N)

Figure 6-21. The STM-0 overhead

not appear in the detailed SONET frame structures. SDH path overhead will not appear in the other frames either, so this is the time to point out the differences. Details will be discussed in Chapter 7.

SONET defines Z3 and Z4 growth bytes as rows 7 and 8 of the path overhead. SDH defines these bytes, technically in the higher order path overhead, as an F3 user channel and a K3 path automatic protection switching (APS) byte.

SDH FRAMES BEYOND STM-0

In addition to the STM-0 frame, SDH specifications currently define STM-1, STM-4, STM-16, STM-64, and STM-256 frame structures. Just as in SONET, SDH has a concept of "N STM-1s make an STM-N." Note that the basic SDH building block is *not* the STM-0 (STS-1 in SONET), but the STM-1 (STS-3 in SONET). As in SONET, however, the basic SDH frame format complicates this simple formula. Let's start with a look at the basic STM-1 frame.

The STM-1 Frame Structure

An STM-1 frame has 9 rows and 270 columns, the same as the SONET STS-3. The overhead is not really three times larger that the STM-0, since the STM-0 was more or less invented as a SONET STS-1 equivalent. The nine columns of overhead leading off the STM-1 frame technically stand on their own. The nine overhead columns still have the three overhead columns of three STS-1s in the familiar 1-2-3 byte-interleaved multiplexing pattern, which sounds odd for SDH. But since SDH was firmly based on SONET, this still makes sense.

The transport overhead of the STM-1 frame is shown in Figure 6-22. Because there are really no STM-0s (STS-1s) inside the STM-1 frame, no numbering appears at the bottom of the frame as in SONET.

However, just as in the SONET STS-3 frame, the three sets of A1/A2 framing bytes are there, just as if there were three STM-0s/STS-1s inside the STM-1. The J0 trace byte appears only once, in column 7. The other two columns, columns 8 and 9, are for national use. In SONET, these are Z0 growth bytes, a good example of national use of overhead bytes.

Many overhead positions are undefined or available for other functions. There are now only four overhead bytes marked as growth bytes for future use. The six overhead positions marked as media dependent are used mainly on wireless links, and there are four other national use bytes in the STM-1 frame besides those in the first row. The only difference is the M1 byte, which is called M2 in SONET, but serves the same purpose. The M0/M1/M2 bytes are discussed in full in Chapter 7.

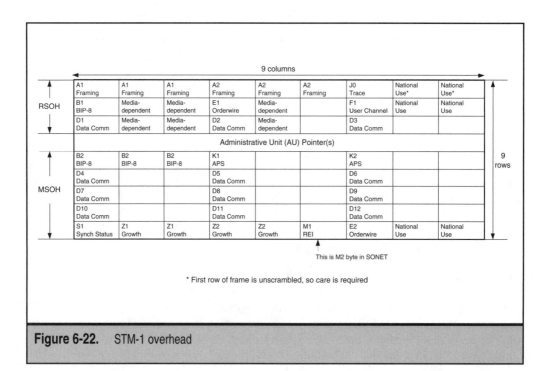

Figure 6-22.   STM-1 overhead

As in SONET, the first row of any SDH frame is never scrambled. The J0 trace is not scrambled, nor are the following national use bytes. So the SDH frame structures all carry a warning to the effect that "care is required when selecting the values" for bytes beyond the A1/A2 framing bytes in the first row.

Note the presence of the AU and lack of H1/H2/H3 pointer details. This aspect of SDH frames is universal. The real multiplexing in SDH begins with the STM-4 frame, so we will next take a look at the STM-4 frame structure and overhead.

The STM-4 Frame Structure

The STM-4 frame has 9 rows and 1,080 columns (the same as a SONET STS-12). Figure 6-23 shows the structure of the now 36 columns of overhead at the beginning of the STM-4 frame.

This frame looks almost identical to the STS-12 frame in Figure 6-16. Again there is only room for the overhead designations, not the names. Z0, Z1, and Z2 bytes are for growth. Blank bytes are undefined because their functions are fulfilled by the overhead bytes in the first STM column position in most cases (such as K1/K2). Media dependent bytes, now 22 bytes per frame (1.408 Mbps, as in SONET STS-12) are shown as the Greek delta (Δ). National use bytes, shown as X, now occupy 30 bytes (1.920 Mbps, in contrast to SONET STS-12's 22 bytes at 1.408 Mbps).

The remaining point of interest is shown across the bottom of the overhead columns. There are no longer any "STS" numbers to identify columns. This is because SDH does not consider an STM-4 or any higher number of STM to consist of a series of STM-0s. The basic unit is the STM-1 (similar to the STS-3). And STM-1 overhead is nine columns, not three.

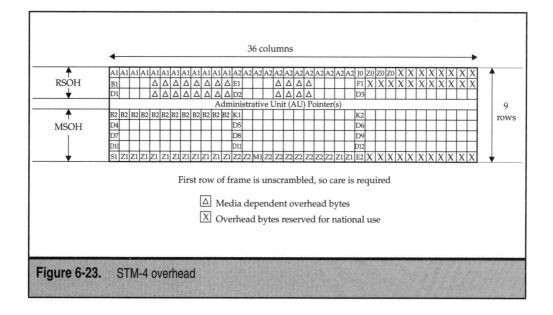

Figure 6-23.   STM-4 overhead

Because there are no STS numbers to provide guidance along the lines of ANSI's "wall-paper" representation of SONET frame structures, SDH documentation does not represent STM-N frames in this "wallpaper" fashion, as mentioned earlier in this chapter. Instead, SDH documentation adapts a "vector" notation to pinpoint overhead bytes in STM-N frames, where N is greater than or equal to 1 (in others words, N = 1,4, 16, 64, and 256).

The vector takes the form S(a,b,c). S (perhaps meaning "SDH frame location") relies on the three variables a, b, and c to locate any overhead position in the frame. The first variable, a, just represents the row number and so can take on values between 1 to 3 and 5 to 9 (row 4 is the special AU). The b variable represents the number of multiplexed STM-1 overhead columns in the STM-N frame, and so can take on the values from 1 to 9 as well. This is because there are nine basic overhead columns in the STM-1 that must be multiplexed into the STM-N frame, not just three columns as in SONET. SONET STS-1 numbers always repeated three times across the frame overhead columns, but SDH overhead positions cycle nine times. The b value is called the "multi-column" value in SDH. Finally, the c variable represents the column within the multi-column "b group" and always runs from 1 to N for a given level of STM-N, where N is greater than or equal to 1.

This S(a,b,c) "locator" system is shown in Figure 6-24 and applied to a generic STM-N frame structure.

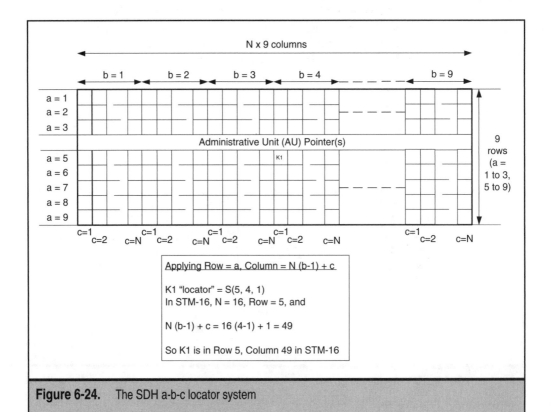

Figure 6-24.   The SDH a-b-c locator system

The S(a,b,c) value for any overhead byte position can be converted to a row and column position in any STM-N frame through two formulas:

Row = a

Column = N (b–1) + c

The figure shows the locator applied to find the row and column position of the K1 byte in any STM-N frame (as long as it is not an STM-0). The K1 locator is always S(5,4,1) in any STM-N frame. The find the row and column where the K1 byte is located in the STM-4 frame (N=4), apply the formula:

Row = 5

Column = N(b–1) + c = 4(3) + 1 = 12 + 1 = 13

So in an STM-4, the K1 byte is always at row 5, column 13. This can be verified by inspection of Figure 6-23. Figure 6-24 also applies the K1 "locator" formula to an STM-16 (N=16) frame. The result is row 5, column 49. Is this correct? Let's see.

The STM-16 Frame Structure

The STM-16 frame has 9 rows and 4,320 columns (the same as an SONET STS-48). Figure 6-25 shows the structure of the 144 columns of overhead at the beginning of the STM-16 frame.

This figure looks much different that the SONET "wallpaper" version of the same bit rate frame. In place of the STS-1 numbers, the figure shows the "b" multi-column numbers at the top of the frame. As a reminder of the relationship between the 9 b-groups and the STM-1 overhead columns, a note shows that there are 48 columns (1/3rd of the 144 total) in 3 b-groups for an STM-16, all with A1 bytes in their first row. There follow 48 A2 bytes, and then 48 J0/Z0/X bytes to finish up row 1.

The frame structure basically shows the "special" first column of the b-group in detail, then condenses the following 15 columns into one. This takes some getting used to, but it is much more elegant that the wider-is-better ANSI frame figures.

But does it work? The K1 formula applied to S(5,4,1) puts the K1 byte in row 5 column 49. It is impossible to tell from the official SDH frame pictures where this column really is (this figure adds the a,b,c numbers absent in the ITU-T Recommendation G.707). However, by comparing this figure to Figure 6-17, it can be seen that the K1 byte is right where it belongs: the first column following the 48 B2 bytes.

The figure shows where the M1 overhead byte at locator S(9,4,3) is found in the STM-16 frame. The presence of the M1 byte at row 9 column 51 can be verified again in the STS-48 figure (where it appears as STS-1 number 7, three columns beyond the 48 bytes of S1/Z1 and after 2 bytes of Z2).

The FEC feature of the STS-48 level also makes its appearance in the STM-16 frame. Note that the SONET Z1 growth bytes in the STS-48 are now SDH FEC bytes alone.

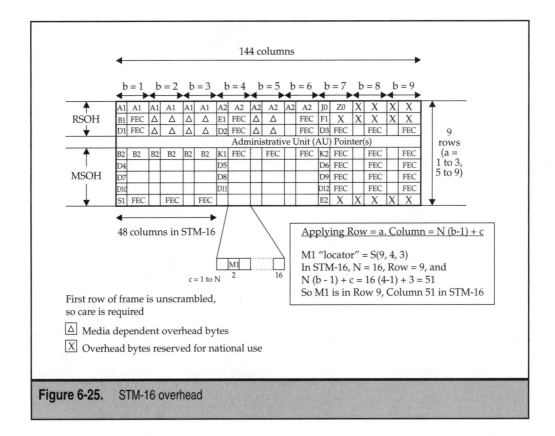

Figure 6-25.   STM-16 overhead

The STM-64 Frame Structure

The STM-64 frame has 9 rows and 17,280 columns (the same as an SONET STS-192). Figure 6-26 shows the structure of the 576 columns of overhead at the beginning of the STM-64. This can be compared to the SONET STS-192 frame shown in Figure 6-18.

Now there are 192 columns in the first three b-groups and the groups still total 9 for the 9 basic STM-1 overhead columns. Although it is almost impossible to tell from the figure, there are more bytes than ever for media dependent, national use, and FEC functions, just as in SONET.

Only one feature is new in this frame structure, and it concerns SONET/SDH compatibility. SDH documentation defines an M0 overhead byte position as an "error counter" in the STM-64 frame at S(9,4,2). The details of the M0/M1 bytes are discussed in Chapter 7.

But what STM-64 column is this M0 byte located in? The figure applies the formula and shows that the M0 byte is at row 9 column 194 in an STM-64. Looking at the STS-192 frame in Figure 6-18 shows that this is exactly the location referenced by the arrow below the frame. The Z2 location in SONET is in STS-1 number 4 and precedes the M1 byte, of course. That's the nice thing about the SDH overhead "locator" vector system: it always works, regardless of STM-N frame, as long as N is known.

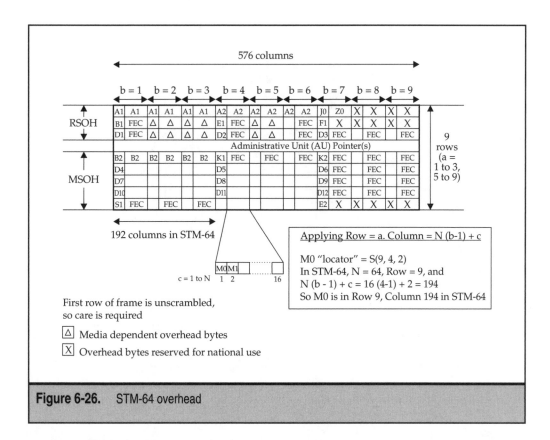

Figure 6-26. STM-64 overhead

The STM-256 Frame Structure

Until recently, SDH, like SONET, topped out at the STM-64/STS-192/OC-192 speed of about 10 Gbps. But now SDH has the STM-256 frame with 9 rows and 69,120 columns (the same as an SONET STS-768). Figure 6-27 shows the structure of the 2,304 columns of overhead at the beginning of the STM-256 frame.

Note that even as the SDH frames grow larger and larger, the frame representations remain as nine b-groups with a detail column and a section of repeated columns. The first three b-groups now stretch 768 columns on an STN-256.

In addition to the usual more bytes for media dependent, national use, and FEC functions, the STM-256 frame adds (or subtracts, depending on perspective) the 64 bytes of A1 and A2, and supports the extra D13 through D156 data communication bytes. These additional 144 bytes give a data communications channel at the line level not at 567 Kbps (D4-D12 run at 9 × 64 Kbps) as at all other SONET/SDH speeds, but at (9 + 144) × 64 Kbps = 9.792 Mbps.

There is one drawback to the highly condensed a,b,c locator system used in SDH. When the columns following a major (c =1) b column *do* have a structure rather that repeating the same pattern 15 or 63 or whatever number of times, the details need to be added outside of the frame.

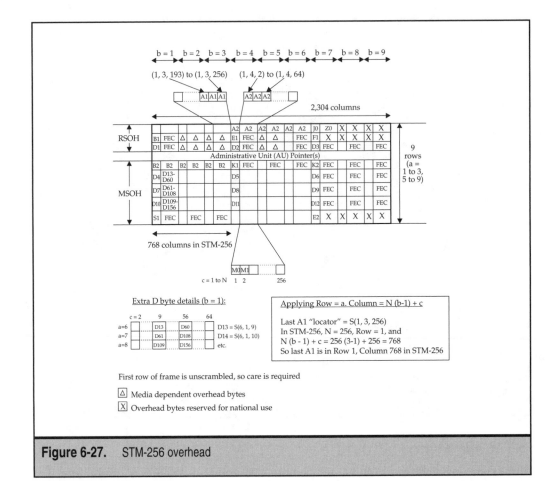

Figure 6-27.   STM-256 overhead

The M0/M1 bytes were a simple example of this idea. In the STM-256 frame, two more detailed additions are needed. The details show the locators S(a,b,c) for the 64 A1/A2 sets, and the extra D bytes. Consider the D bytes first. All have b =1, and a = 6, 7, or 8. The D13 through D60 bytes are in columns 9 through 56 inclusive, as shown in the figure. So D13 = S(6,1,9), D14 = S(6,1,10), D15 = S(6,1,11), all the way up to D59 = S(6,1,55) and D60 = S(6,1,56). D61 to D108 and D109 to D156 follow the same locator pattern for rows 7 and 8.

An STM-256 or STS-768 has 64 A1 bytes and 64 A2 bytes. The locators for the first and last A1 bytes (shown in the figure) are S(1,3,193) and S(1,3,256). Sure enough, this is 64 A1 bytes (256 − 193 inclusive = 64). This first A1 byte is in column 705 and the last A1 byte (calculated in the figure) is shown to be in row 1, column 768 in an STM-256. The figure shows the last A2 byte as S(1,4,64). Since S(1,3,256) is followed by S(1,4,1) in an STM-256, this gives 64 A2 bytes as well (64 − 1 inclusive = 64). These columns are 769 for S(1,4,1) and 832 for S(1,4,64) respectively.

This tour through SONET/SDH frame structures is now complete. Before moving on to a complete consideration of SONET/SDH overhead in the next chapter, one other frame-related topic should be discussed. This is the idea of concatenated payloads and, by extension, the tributaries that make up SONET/SDH payloads in many cases.

CONCATENATED SONET/SDH FRAMES

This chapter has stated several times that a SONET STS-N frame contains N STS-1s. This is just another way of saying that SONET is as channelized as T-carrier. Thus, a DS-1 contains 24 DS-0 operating at 64 Kbps each. However, because a single DS-1 operates at 1.5 Mbps, perhaps the user's needs would be better addressed with a single channel operating at the full 1.5 Mbps, instead of 24 channels operating at only 64 Kbps each. This is, of course, what an unchannelized T-1 does for a user. Technically, the term should be *nonchannelized*, but everyone just says unchannelized.

This need may be even more critical in SONET/SDH. The unprecedented speeds available in SONET/SDH are capable of solving a number of user problems. However, this might not be the case when an OC-12 operating at 622.08 Mbps is only useful as twelve STS-1s running at 51.84 Mbps. SDH has the same concern above STM-1. An unchannelized SONET/SDH link is needed. In SONET, this would give a user just a raw bit rate at 622.08 Mbps (minus the STS-12 overhead). Of course, such an unchannelized frame structure exists in SONET/SDH but is known as a *concatenated* SONET/SDH frame that gives a user a transfer higher than 51.84 Mbps (SONET) or 155.52 Mbps (SDH).

Consider a SONET example. Using railroad boxcars to represent STS-1 frames, a comparison can be drawn between channelized multiplexing of SONET frames to achieve higher transmission rates, and the unchannelized concatenation of SONET frames, which leads to the creation of larger payload envelopes that operate at the same higher rates. This is shown in Figure 6-28.

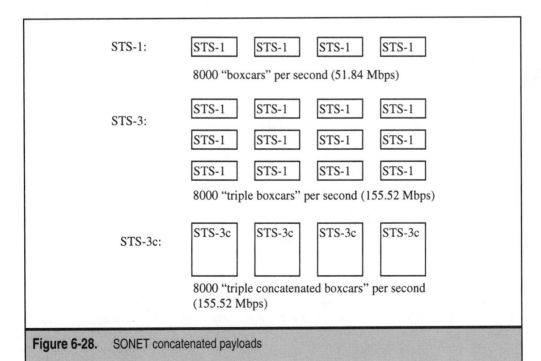

Figure 6-28.   SONET concatenated payloads

At the top of Figure 6-28, STS-1 frames are transmitted at a rate of 8,000 frames per second. This results in a data rate of 51.84 Mbps. In order to build up STS-N frames structures that operate faster, STS-1 frames can be multiplexed (by byte interleaving them) and these new larger frames are still transmitted at a rate of 8,000 per second. The middle of the figure shows three STS-1s multiplexed together to create an STS-3. Note that although the STS-1s are multiplexed in a single STS-3 sent on an OC-3, the STS-1s are still considered three independent frame streams and each has its own set of payload pointers. All of the user information must fit within an STS-1 frame SPE, and the frames are simply multiplexed prior to transmission so that the high speeds of fiber can be used more efficiently. This is just like multiplexing 28 DS-1s into a DS-3. The original user bit stream is still less than or equal to the DS- 1 rate. The DS-3 allows greater speeds on the transmission media, but that is all.

At the bottom of Figure 6-28, SONET concatenation is illustrated. Here, three STS-1 frames are "pasted together" (concatenated) to create a single, large frame structure which, in this case, is called an STS-3c. Note the presence of the little "c." This lets people know that they are dealing not with three STS-1 frames inside the STS-3, but a single STS frame payload structure.

Because the individual frames are concatenated, one SPE is present and this only has one set of payload pointers. Concatenation is used when data is too fast to fit within an STS-1 frame. Larger frames need to be created for this situation. This is similar to providing a customer with a DS-3 that is not channelized—it is one big, 45 Mbps pipe. Note that an unchannelized DS-3 looks just like the DS-3 containing the 28 DS-0s, but the internal structure of the frame is not quite the same.

Before looking at a concatenated SONET STS-3c frame in detail, it might be a good idea to look at more details of the way payloads are carried inside a channelized (default) STS-3.

STS-3 SPEs

Because channelized STS-3 are the rule rather than the exception, the STS-3 frame structure contains a number of features designed expressly for combining three STS-1s into a single payload. First, the overall frame structure is derived by simply multiplexing the three input STS-1s a single byte at a time into the outgoing STS-3 frame structure. Of course, this overall STS-3 frame is still 9 rows, but has 270 columns (90×3). The effect of this byte-wise multiplexing is to effectively interleave the *columns* of the input STS-1s.

Thus, the 1st, 4th, 7th, and so on up to the 268th column of the STS-3 frame is totally derived from the bytes of the first STS-1. Likewise, the 2nd, 5th, 8th, and so on up to the 269th column of the STS-3 frame is totally derived from the bytes of the second STS-1. Naturally, the 3rd, 6th, 9th, and so on up to the 270th column of the STS-3 frame is totally derived from the bytes of the third STS-1. Because the first 3 columns of the STS-1s are the SONET overhead bytes, the net result is that the first 9 columns of the STS-3 frame are SONET overhead, while the remaining 261 columns are information payloads derived from the three input STS-1s.

There are three sets of overhead bytes, and three column interleaved payloads in each STS-3 frame. Because the error checking and many other SONET overhead functions only have to be done once for the whole STS-3 frame, overhead bytes such as the

B1 byte and the K1/K2 APS bytes are undefined and ignored except in the "first" STS-1 of the STS-3 (columns 1, 4, and 7). Also, what should be done if two fields applying to the same frame did not agree, such as two B1 bytes? The undefined status of "extra" overhead bytes neatly solves the problem. The general structure of the STS-3 frame is shown in Figure 6-29.

Figure 6-29 shows a normal, completely channelized STS-3 with three STS-1s inside. Note that there is one frame (the STS-3 frame), but three sets of overheads and payloads (the three STS-1 payloads). Therefore, there are three sets of live payload pointers and three completely separate POHs. Of course, it is still possible to lock the three STS-1 payloads to the beginning of the STS-3 frame, as with the individual STS-1s; however, this introduces a great delay and then requires buffering.

Multiplexing occurs by columns, as usual, and the three sets of columns are "exploded out" of the STS-3 frame in Figure 6-29.

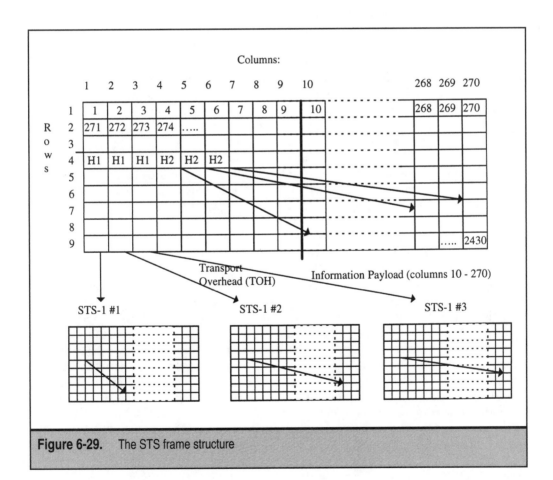

Figure 6-29. The STS frame structure

STS-3c (OC-3c) SPE

As nice as it is to combine three STS-1s into a single STS-3 for transmission over a single OC-3 fiber link (which is the whole point), this provides little advantage to the customer. Many customers do not need three STS-1s, but rather a single STS-3 operating at 155.52 Mbps (including overhead) to satisfy their bandwidth needs. Fortunately, SONET allows for the creation of *unchannelized* STS-3s, as well as other speeds. In SONET, this process is known as *concatenation*; thus, a concatenated STS-3 is an unchannelized STS-3 and contains not three, but one SPE. The designation for a concatenated STS-N is to add a lower case "c" to it. So a concatenated STS-3 becomes an STS-3c.

It is important to distinguish the lower case SONET "c" in terms of function from the upper case "C" used in T-carrier. Both mean "concatenation," but each uses the term in a different sense. An FT-3C, for instance, is two T-3s transmitting on fiber "pasted" together. But there are still two distinct T-3s, each operating at 45 Mbps. It makes no difference whether either is channelized into 28 T-1s or not. Two bit steams still operate at 45 Mbps. However, in SONET, an STS-3c is strictly unchannelized into a single bit stream operating at 155.52 Mbps. There are 9 columns of overhead (the overhead stays in SONET, no matter what), but only one SPE that spans the entire 261 columns of the information payload.

It is technically proper only to speak of an OC-3 as the physical transport for either an STS-3 or STS-3c. After all, optical senders and receivers care nothing about presence or lack of structure in the frames they transfer. However, just as T-carrier people came to speak of a "T-1" as an overall term for "DS-1" and much else, SONET people have come to speak of "OC-3c" to indicate that the STS-3 frames sent on the link are STS-3c frames; whether referred to as STS-3c or OC-3c, the structure is the same.

Figure 6-30 shows the structure of an STS-3c or OC-3c frame. The concatenation is indicated by putting special values in the H1/H2 pointers (called the *concatenation indicator*) where the pointers for the other STS-1s would ordinarily be positioned. The concatenation value is not a valid pointer value (i.e., not between the allowed values of 0 and 782); thus, no confusion at the receivers exists. Obviously, there is only one SPE, so one set of H1/H2 pointers will do. Because only one SPE exists, only one set of POH bytes is needed. But it is still the first column of the SPE. As with STS-1 SPEs, the H1/H2 pointers must point to the first byte of the POH.

The SPE may still be set and locked into column 4, but this is another matter. There is, however, an important point regarding the pointer behavior when the SPE is allowed to roam within the STS-3c frame. Many people quickly observe that the single STS-3c SPE can start anywhere within the 261 columns of the information payload; the SPE can never occupy the overhead column positions. This gives 9 rows × 261 columns = 2,349 or 2,349 potential starting positions; however, the H1/H2 pointer is a single 10-bit field with valid values only up to 782 (pointers will be explored further in Chapters 7 , 8 and 9). The other H1/H2 fields are lost to the concatenation indicator(s). How then can only the values 0 to 782 indicate starting positions from 0 to 2,348?

The answer is actually a pretty neat trick. SONET receiving equipment cannot distinguish between an incoming STS-3 frame or an STS-3c frame, except for the presence of concatenation indicators. When these indicators are detected, the concatenation indicator basically tells the receiver to increase the values of the H1/H2 pointers in the first over-

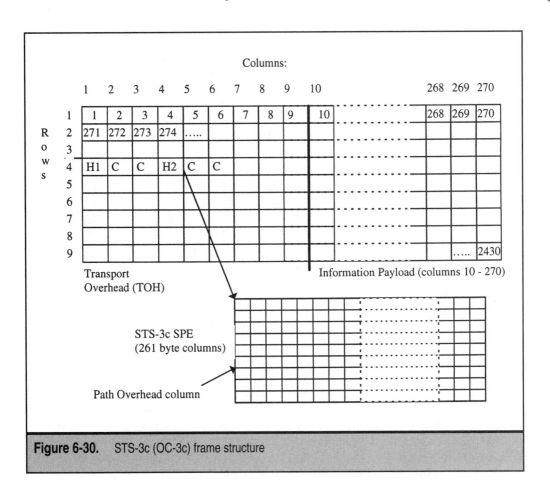

Figure 6-30.   STS-3c (OC-3c) frame structure

head positions (columns 1 and 4) by a factor of 3. Thus, an H1/H2 offset of 1 really means 3 and an offset of 782 really means 2,346. The problem is neatly solved. (It should be noted that SDH pointers in their special AUs can operate in a totally different fashion—and do in some cases. SDH AU pointers are discussed in Chapters 7 and 8 as well.)

Of course, the sending SONET equipment has a role to play in this scheme as well. The sender must buffer not 8 bits before moving the payload pointer value as before, but 24 bits (3 bytes) instead. Buffering also increases by a factor of three. When SONET STS-3c SPE pointers adjust themselves due to link timing jitter (discussed in Chapter 9), whether positive or negative, they do so in 3-byte units. For instance, a value of 2,346 points to the last 3-byte unit of the SONET frame. The whole idea is simple and effective.

As has been stated, an STS-3c has only a single SPE, which cannot be used for transporting existing DS-1s or DS-3s (that is what STS-1s are for). So what good is an STS-3c? The STS-3c was created expressly for the transport of ATM cells as part of a B-ISDN network. This corresponds to the STM-1 in the SDH hierarchy. In fact, the initial ITU B-ISDN specification for ATM cell transport had no provisions for the transport of ATM cells at speeds lower than 155.52 Mbps. Broadband needs speed above all.

What about SDH concatenation? SDH frames use exactly the same notation for concatenated frames. So an STM-4c is a concatenated STM-4 and does not contain four STM-1s, an STM-64c is a concatenated STM-64 and does not contain sixty-four STM-1s, and so on. In keeping with SDH's lack of H1/H2/H3 direct pointer bytes references, SDH refers to pointers to normal, non-concatenated traffic as administrative unit type 3 (AU-3) and pointers to concatenated payloads as administrative unit type 4 (AU-4). More details about SDH use of AU-3 and AU-4 pointers are discussed in Chapter 8.

Concatenation in SONET/SDH can be *contiguous* or *virtual*. Contiguous concatenation has been described here, meaning that inside an STS-12 (for example), the three STS-1s that are concatenated into an STS-3 must be contiguous, or located in physically adjacent columns in the STS-12 frame. Contiguous concatenation can be extremely wasteful of SONET/SDH bandwidth because SONET/SDH speeds do not line up especially well with common user LAN bit rates. The effective payload rate for an STS-1 (minus the transport and path overhead) is 49.536 Mbps. If a 100 Mbps Ethernet is to be transferred at Ethernet line rate across a concatenated SONET link, three, not two, STS-1s are needed. Almost 50 Mbps is wasted. The same is true of FDDI, which runs on fiber at 125 Mbps, not 100 Mbps (actually, both examples are more complex that this simple calculation, but the point is not changed by adding details).

But doesn't concatenation gain some bandwidth by eliminating some of the path overhead? There is now only one path (end equipment) between STS-3c endpoints, not three. Why use three sets of path overhead? In SONET, the answer is that concatenated payloads cannot use the "extra" path overhead columns for user data above STS-3c. Those columns must be used for *fixed stuff* bytes and set to all 0 bits. Oddly, the number of fixed stuff columns follows the SDH convention, not SONET. For example, an STS-12c has one column of path overhead, but three columns of fixed stuff bytes following, not twelve, as if this was an STM-4. This replaces the "normal" four columns of path overhead in an STM-4.

The formula $N/3 - 1$ is used in SONET to calculate the number of fixed stuff columns in an STS-Nc payload. The formula $(N \times 87) - N/3$ then gives the number of useful payload columns for user data. N must be greater than or equal to three. The N maps to STM values as 3X, where X = 1,4, and 16 (higher values are under study). This concept and an example using STS-12c is shown in Figure 6-31. SDH has similar fixed stuff columns, of course, but the math is easier.

Virtual concatenation does not require the concatenated STS or STM channels to be physically associated. Virtual concatenation creates *logical* associations of SONET/SDH frames. Because the concept of virtual concatenation is intimately tied up with SONET/SDH payloads, virtual concatenation is more fully discussed in Chapter 8.

CONTAINERS AND TRIBUTARIES: LOWER RATE PAYLOADS

One of the advantages of SONET/SDH is that it will transport higher bit rate signals in concatenated frames. An STS-12c or STM-4c can carry a stream of ATM cells, but what about the lower rate signals of the T-carrier or E-carrier hierarchy?

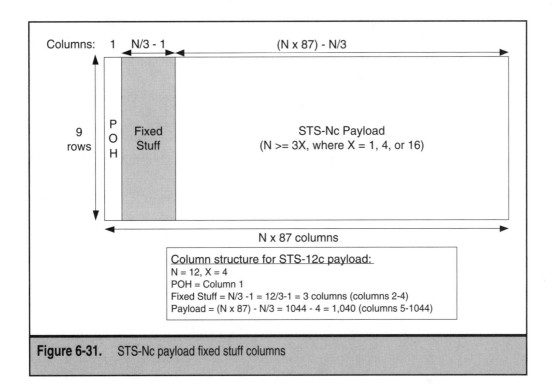

Figure 6-31.   STS-Nc payload fixed stuff columns

SONET/SDH handles this situation by mapping the "lower" into "sections" of an STS-1 or STM-1 frame. These sections are each called a *virtual tributary* in SONET and usually a *virtual container* in SDH (the terms in SDH are very complex). In SONET, each of the virtual tributary sections are independent of each other and can carry different types of *sub-rate payloads*. A sub-rate payload is just a payload that does not map directly to an STS-1, but requires virtual tributaries.

This concept can be visualized by extending the boxcar example introduced to illustrate concatenated payloads. For sub-rate payloads, each boxcar can be divided into sections, or virtual tributaries. Each virtual tributary can contain any type of lower rate data; therefore, they all do not need to contain the same type of information. DS-1s, E-1s, and DS-2s could all be in one "boxcar," which is the basic STS-1 frame.

In SONET, each STS-1 frame is divided into exactly seven virtual tributary groups (VTGs), and it is all or nothing. A single STS-1 frame cannot have, for example, only four VT groups and use the rest of the payload for something else (like ATM cell transport). Either the STS-1 frame is chopped up and sectioned off into exactly seven VT groups, or it is not.

Figure 6-32 shows this VTG process. For illustration purposes, the figure shows only three VTGs: a single boxcar can be divided into sections for the transport of different types of things.

Chapter 8 will examine SONET/SDH payloads and tributaries in more detail, especially for voice and IP packets. But first, a more detailed look at the SONET/SDH overhead bytes is presented in Chapter 7.

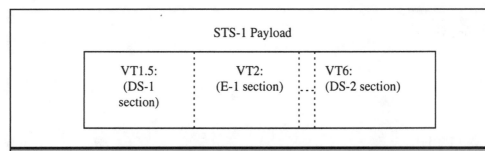

Figure 6-32.   Lower rate payloads

CHAPTER 7

SONET/SDH Overhead

It is always a challenge to try and figure out the best way to introduce key SONET/SDH terms and concepts. SONET/SDH standards tend to start off with frame tributary/container content, introducing overhead bytes as needed as they go along. Then the overhead bytes themselves are detailed, and finally the frame structures that contain both overhead and content are presented. The problem is that frame tributary/container content is one of the most complex of SONET/SDH topics, especially when it comes to SDH. As a result, most people barely make it through the start of the tributary/container content before quitting.

After a lot of thought, I decided to *reverse* the order of major topics as presented in the SONET/SDH standards. The previous chapter already detailed the SONET/SDH frame structures, introducing overhead bytes with little mention of their use or meaning. This chapter addresses that issue and details all of the overhead bytes in SONET/SDH, including the line/regenerator section, section/multiplex section, path/higher order path, and virtual tributary/lower order path, just to illustrate the proper SONET/SDH term for each level of overhead.

As a result, there will be terms and ideas introduced in this chapter that will not be very meaningful until payload contents are discussed in Chapter 8. But no matter which end of SONET/SDH is chosen as a starting point, high up at the frame or low down at the frame content, there are always topics that must be mentioned but detailed later. SONET/SDH is a whole, but must be learned as a sequence of parts. If nothing else, the parts of SONET/SDH as introduced will form a contrast to the sequence usually presented in the standards. Perhaps this sequence will make the standard sequence more understandable, or maybe the reverse will be true and the standards will be more enlightening. Either way, this group of chapters is much more than just a "commentary on the standards."

The previous chapter treated SONET in full before turning to SDH. This approach will not be followed in this chapter. There are several reasons for presenting a SONET overhead byte structure and function, and immediately following it with the SDH equivalent. The most important reason is that almost all of the overhead bytes discussed in this chapter are *exactly* the same in SONET and SDH. So there is little to no advantage in having all of the SONET bytes in one section, followed by a lot of repetition in an SDH section. Moreover, by delaying the re-introduction of an overhead byte, there is risk of forgetting key aspects regarding the use of that particular byte. The method followed here reduces repetition and reinforces understanding immediately.

A lot of detailed information about many of the overhead bytes is gathered into SONET and SDH "references" in the appendices and in later chapters in this book. This chapter explores the overhead bytes together, but at a high level, for efficiency purposes.

SONET/SDH OVERHEAD

The overhead in SONET/SDH is both a distinctive feature of SONET/SDH as well as a much needed change from T-carrier or E-carrier overhead methods. This reflects both the more ambitious uses for SONET/SDH and the byte-oriented nature of SONET/SDH itself, which is in stark contrast to T-carrier/E-carrier higher speed multiplexing. Because overhead bytes in SONET/SDH play such an important role in many of the chapters that follow, this is a good place to outline the functions of the overhead bytes in SONET/SDH.

Many of these functions overlap because there are separate section, line, path, and tributary components on a SONET link, and regenerator section, multiplex section, higher order path, and lower order path on an SDH link, that all need overhead benefits like error monitoring and failure protection. (Technically, SDH specifications speak of *higher order virtual container (VC) path overhead* and *lower order virtual container (VC) path overhead*, but this chapter just uses the more concise terms higher order and lower order path overhead.)

Mention is made during this discussion of *composite* SONET/SDH signals. When several SONET/SDH data streams are multiplexing into one higher rate data streams (such as from three STS-1s to a single STS-3), the resulting SONET/SDH frame structure is called a *composite*. Every frame except a basic STS-1 frame (or STM-1 frame in SDH) is a composite frame of one form or another. Composites still retain *all* of the overhead bytes from each signal source. Obviously, an STS-3 frame that has nine columns of transport overhead, instead of just three from each of the STS-1s, still forms a unit and not just three frames traveling together. Thus, many of the "repeated" overhead bytes are essentially ignored ("undefined" in SONET/SDH talk) in composite frames. The full set of 27 transport overhead bytes is retained only in the first STS-1 or STM-0 of any composite. In the other levels of SONET/SDH, all of the transport overhead bytes may be present, but many of them are neither examined nor processed.

Although the complete, basic sets of transport and path overhead for SONET and SDH were presented in Chapter 6, this is a good place to repeat this information for ease of reference. Figure 7-1 shows the SONET overhead bytes and Figure 7-2 shows the SDH equivalents.

There is even another level of overhead, called *virtual tributary overhead* in SONET and *lower order path overhead* in SDH. This overhead will be introduced and discussed later on in this chapter.

The discussion of virtual tributary overhead in SONET and lower order path overhead in SDH will unfortunately introduce some concepts about SONET and SDH payload content that will not be discussed until the next chapter. In fact, the details of operation regarding several of the overhead bytes outlined in this chapter (such as the H1/H2/H3, S1 and K1/K2 bytes) will be investigated in more depth in Chapters 9 and 16 of this book. The topics of SONET/SDH overhead is just too large and sprawling a subject to handle all at once.

Each section that follows discusses the SONET definition and use of the overhead byte first, then explores the SDH use of the same overhead byte location.

Section/Regenerator Section Overhead

The SONET section layer is called the regenerator section layer in SDH. This layer contains overhead information used by all SONET/SDH equipment along a network path, including signal regenerators. Now, all transport overhead in SDH is section overhead, but there is regenerator section overhead (RSOH) and multiplex section overhead (MSOH). When forward error correction (FEC) information appears in RSOH bytes, it is always MSOH and ignored and not processed or terminated by RSOH equipment in SDH.

This overhead is contained in the top three bytes of the first three columns in the basic STS-1 or STM-0 frame structure. The section/regenerator section's nine constituent bytes function as follows.

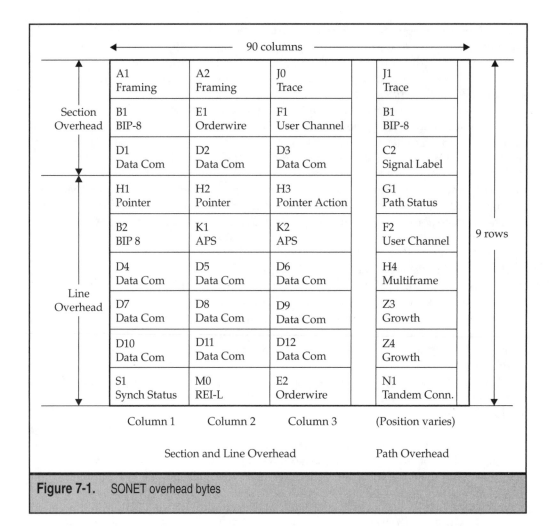

Figure 7-1.   SONET overhead bytes

Framing Bytes (A1, A2)   This two-byte code (1111-0110-0010-1000, or F628 in hexadecimal notation) is used for frame alignment. These bytes uniquely identify the start of each STS-1/STM-0 frame and are not scrambled during the transmission process. Lack of scrambling makes their detection that much easier, and there is little chance of these bytes containing long strings of 0's. When multiple STS-Ns are sent in a higher rate STS-N, framing bytes must still appear in every STS-1 of the composite signal.

For all STS-Ns except STS-768, there are A1 framing bytes in columns 1 through N followed by A2 framing bytes in columns N+1 through 2N of row 1. An STS-768 frame has 64 columns of A1 in columns 705 to 768 of the first row and 64 columns of A2 in columns 769 to 832. The other row 1 columns, 1 to 704 and 833 to 1536, are reserved for future standardization and their value must be chosen with care, since the first row of any SONET/SDH frame is always unscrambled.

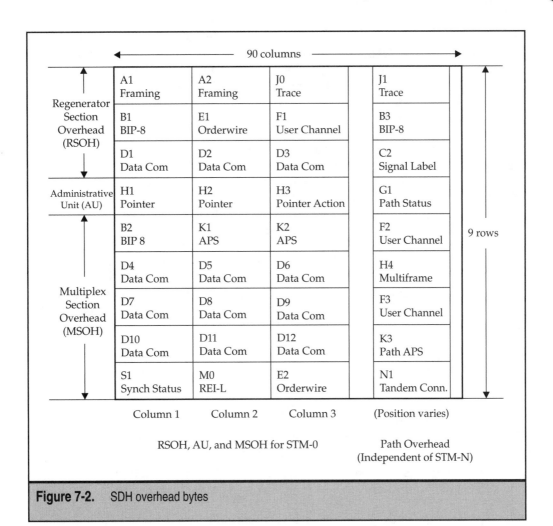

90 columns				
Regenerator Section Overhead (RSOH)	A1 Framing	A2 Framing	J0 Trace	J1 Trace
	B1 BIP-8	E1 Orderwire	F1 User Channel	B3 BIP-8
	D1 Data Com	D2 Data Com	D3 Data Com	C2 Signal Label
Administrative Unit (AU)	H1 Pointer	H2 Pointer	H3 Pointer Action	G1 Path Status
Multiplex Section Overhead (MSOH)	B2 BIP 8	K1 APS	K2 APS	F2 User Channel
	D4 Data Com	D5 Data Com	D6 Data Com	H4 Multiframe
	D7 Data Com	D8 Data Com	D9 Data Com	F3 User Channel
	D10 Data Com	D11 Data Com	D12 Data Com	K3 Path APS
	S1 Synch Status	M0 REI-L	E2 Orderwire	N1 Tandem Conn.
	Column 1	Column 2	Column 3	(Position varies)
	RSOH, AU, and MSOH for STM-0			Path Overhead (Independent of STM-N)

9 rows

Figure 7-2. SDH overhead bytes

In SDH, the A1/A2 bytes are the "frame alignment word." There are 3xN A1/A2 bytes in an STM-N where N = 1, 4, 16, and 64. There are 64 A1 and A2 bytes in an STM-256 in the same places as in the SONET STS-768 frame. In SDH, they are from locator S(1,3,193) [1,705] to S(1,3,256) [1,768] for A1 and S(1,4,1) [1,769] to S(1,4,64) [1,832]. (SDH uses the square bracket notation to give row, column equivalents for locator vectors in parentheses.) Locator positions S(1,1,1) [1,1] to S(1,3,192) [1,704] and S(1,4,65) [1,833] to S(1,9,256) [1,1536] are reserved for future international standardization (for media dependent, additional national use, or other purposes). As in SONET, the value of these "unused" row 1 bytes should be set so that the unscrambled row 1 should not be an issue when it comes to "1's density" (also called "DC balancing" when SONET/SDH is carried on radio links, short electrical cables, and so on).

Section Trace (J0) In SONET, this byte is used to trace the origin of an STS-1 frame as it travels across the SONET network. As SONET frames make their way across a series of links, the frames are often multiplexed to higher and higher levels, and then most likely back down again as they near their destination. Many senders process the SONET frames, and create new packages for the payloads inside. It is obviously a plus to be able to trace the origin of any particular STS-1 frame to a particular piece of equipment from any point on the network. In the case of multiple STS-1s in an STS-N, the J0 byte is defined only in the first STS-1 because all of the frames must come from the same device. In other STS-1s in an STS-N, the use of this byte is just for Growth (Z0).

This overhead byte used to be known as the STS-1 ID (C1) byte. This single byte was used to uniquely identify the STS in question. The intent was to keep track of which frame is being transmitted at any time. The STS-1 ID was a binary number that corresponds closely to the number of STS-1 frames being sent as a composite STS-N. For example, if the signal being sent was a single STS-1, then the octet would be 0000-0001 (01 in hex). If an STS-N were being sent, the first STS-1 inside would be numbered as 0000-0001, the second as 0000-0010 (02 in hex), the third as 0000-0011 (03 in hex), and so on, up to N. Naturally, the STS-1 ID appeared in each STS-1 of an STS-N frame. To make sure newer equipment interoperates with older SONET gear, all SONET equipment at OC-48 or below must be able to generate C1 STS-1 ID bytes if needed.

ANSI specifications allow the J0 byte in SONET to be a repeated one byte field, or use the SDH 16-byte frame format, discussed below. If the section trace is not supported (by older equipment) or if no value for this field has been configured, then SONET uses a 01. ANSI is also studying a 64-byte message format. This format uses an ASCII carriage return (CR) or carriage return line feed (CR LF) to delimit the string and does not use a checksum.

In SDH, this is the regenerator section trace byte. The SDH locator is S(1,7,1) or [1,6N+1]. The J0 byte sends what the ITU-T calls a section access point identifier (API). Within a country, like the U.S. with SONET, this can be single byte value 0-255. Internationally, the "clause 3/G.831" API message format is used. This message format, also supported in SONET, is shown in Figure 7-3.

The section API message consists of 16 bytes. The first byte is a header with a 7 bit cyclical redundancy check (CRC-7, an error detection method, also called the *checksum*) computed on the previous frame. Next are fifteen T.50 (ASCII) characters. Once the CRC-7 checksum has been calculated to match the sent value at the receivers, the receivers normally only check the value of the CRC-7 every 16 SONET/SDH frames. If it has not changed, the other 15 bytes should not have changed either.

Similar to SONET, this was formerly the C1 STM identifier and basically carried the "c" value of the S(a,b,c) locator system. (Technically, c is the "multi-column, interleave depth coordinate.") This might have one time helped receivers to find the right columns ("assist in frame alignment") but now SONET/SDH chipsets do not need to be informed of the value of c in every frame to function correctly.

For compatibility with this older equipment using the C1 single-byte system, J0 in SDH must be able to generate a 0000 0001 pattern.

Section Growth (Z0) In all frame structures beyond STS-1, except STS-192, SONET defines the row 1 bytes following J0 as Z0 section growth bytes. All frames except STS-192

Byte #	Value (bit 1, 2, ..., 8)							
1	1	C_1	C_2	C_3	C_4	C_5	C_6	C_7
2	0	X	X	X	X	X	X	X
3	0	X	X	X	X	X	X	X
:	:	X	X	X	X	X	X	X
16	0	X	X	X	X	X	X	X

Alternate Method:
64 byte "frame"
Delimited by CR or LF CR
No checksum

(Receivers only check X values
if checksums do not match)

| XXXXXXX | = ASCII (T.50) character (15 in all) |

| C_1 - C_7 | = CRC-7 on previous frame |

Figure 7-3.   The J0 16 byte message format

should be able to conform to the C1 byte older use mentioned above if necessary. STS-192 frames set this field to 0xCC (1100 1100) for the time being.

In SDH, this is the Z0 Spare byte. The locator is S(1,7,2) [1,6N+2] to S(1,7,N) [1,7N]. STM-0 does not have any Z0 spare bytes, just a single J0. STM-1 has bytes for "national use" here, such as SONET's Z0. An STM-4 frame has three Z0 bytes after the J0 byte, and the locator formula shows that the STM-16 has 15 Z0s after the J0, the STM-64 has 63, and the STM-256 has 255. The "tail-end" bytes of row 1 after this Z0 string in every case are national use bytes. In SONET, frames always end with an N-1 string of Z0 bytes. For example, STS-12 has 12 – 1 = 11 Z0 bytes at the end of row 1.

Officially, these spare bytes are reserved for future international standardization. As in SONET, some very old SDH equipment might still require 0x01 in these byte positions for C1 compatibility.

Section BIP-8 (B1)   A single, interleaved parity byte is used to provide STS-N error monitoring. BIP is an abbreviation for *bit-interleaved parity*. The BIP performs a routine even-parity check on the previous STS-1 frame, after scrambling. The parity is then inserted in the B1 BIP-8 field of the current frame before scrambling. SONET relies on even parity for its error-checking calculations. During parity checking, the first bit of the BIP-8 field is set so the total number of ones in the first positions of all octets in the previously scrambled STS-1 frame is always an even number. The second bit of the BIP-8 field is used in exactly the same way, except it performs a check on the second bits of each octet, and so on. This octet is only defined in the first STS-1 frame of a composite STS-N signal. The idea of the BIP check is shown in Figure 7-4.

The nice thing about the BIP check is that the number of BIP errors detected over a given time period can be directly translated into a bit error rate (BER) at that level of oper-

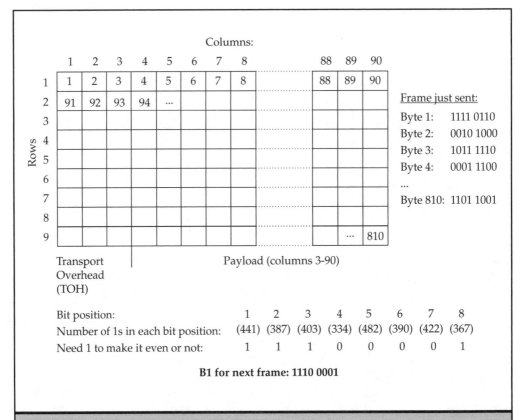

Bit position: 1   2   3   4   5   6   7   8
Number of 1s in each bit position: (441) (387) (403) (334) (482) (390) (422) (367)
Need 1 to make it even or not: 1   1   1   0   0   0   0   1

B1 for next frame: 1110 0001

Figure 7-4.   The idea behind the B1 BIP-8 check

ation (there are other BIP checks in SONET/SDH overhead that apply to other levels). That is, unlike T-carrier, BER testing is ongoing and non-intrusive. As example, consider a SONET link that sends 514,403 frames at the OC-3 rate of 155.52 Mbps. This takes about a minute and generates about 9,999,994,320 bits, just short of 10 billion or a thousand million. (What follows is just an example of BIP use, and not taken from any specification.)

If the count of BIP *violations* is 100, this translates to a BER of 10^{-8}, or about one in 100 million. If the BIP violation counter is 10, this is a BER of 10^{-9}, or about one in a billion. And if the BIP violation counter is 1, then the BER is 10^{-10}, or about one in 10 billion. In reality, SONET (and SDH) network equipment has standard tables built in to convert BIP violations to BERs for standard intervals. This whole system works well because fiber links are not as prone to burst errors as copper links like T-carrier and E-carrier. Bit errors on fiber tend to come one at a time, rather than in bursts as on electrical links. On today's fiber links, a BER of 10^{-10} is considered acceptable, a BER of 10^{-11} is expected, and a BER of 10^{-12} is what customers want.

The BIP is important because it is always important to isolate the "weakest link" when end-to-end BERs become high enough to cause concern. Consider an end-to-end SONET/SDH link between two users' TMs as shown in Figure 7-5. There are two ADMs and two regenerators on the link, each with their own BIP error counter. The figure shows the BER on each fiber span.

The BER as measured by the customer end-to-end is marginally worse than 10^{-8}. This is because the other spans do nothing to clean up the errors introduced between the regenerators, and add very few errors of their own to the weak link in the middle. With the BIPs, SONET/SDH users can watch the link BER independently of the carrier and check the observed (and recorded) BER against the tariff or contract. BIPs should reflect overall link performance, and sudden jumps in the BER at any level indicate damage to the fiber or signal tapping.

In SDH, the B1 BIP-8 is for "regenerator section error monitoring" and is "bit interleaved parity 8 code using even parity." As in SONET, this is done on all bits of the previous STM-N frame after scrambling, and this B1 value is placed in the current frame before scrambling.

Orderwire (E1)   This channel was a leftover from the days of copper wire carrier systems and historically was available to craft personnel for maintenance communications. In SONET, this channel is a 64 Kbps voice path used for communication between remote terminals and regenerators. A telephone can be "plugged in" at the repeater location into a more or less standard voice jack, allowing craft personnel instant voice access on the fiber span. This octet is only defined in the first STS-1 of an STS-N frame. Signaling use, such as for "telephone numbers" on this orderwire, is for further study. Implementation is totally optional. It is unlikely that anyone ever uses the orderwire overhead voice channel in a SONET network, except perhaps in some rural areas, since craft personnel all carry cell phones today, but it seemed like a good idea at the time.

In SDH, the E1 byte is also an optional channel for voice communication. There is no more real information.

User (F1)   The section user channel is reserved for use by the network services provider for its own application management activities. Its design is unspecified because it can be tailored by the network provider. F1 is only defined in the first STS-1 of an STS-N frame. Currently, many vendors of SONET equipment use the F1 byte, which operates at 64 Kbps, for

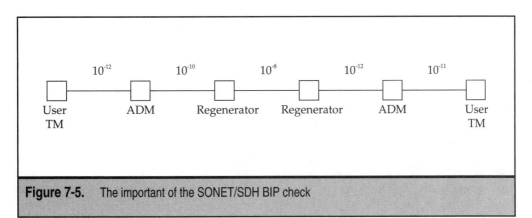

Figure 7-5.   The important of the SONET/SDH BIP check

configuration and maintenance communications. A standard implementation of the F1 field is undefined, giving vendors a lot of flexibility, but limiting interoperability. The F1 use is optional in both SONET and SDH.

In SDH, this byte is "reserved for user purposes (e.g. to provide temporary data/voice channel connections for special maintenance purposes)."

Data Communications Channel (DCC; D1-D3)   These three bytes are used as a 192 Kbps data channel for message-based operations functions, such as OAM&P. These capabilities constitute a (if not *the*) primary advantage offered by SONET technology. The DCCs ship such things as alarms, administration data, signal control information, and maintenance messages. D1 through D3 are only defined in the first STS-1 of an STS-N frame. The use of this channel should comply with Telecommunications Management Network (TMN) standards and is discussed more fully in Chapter 15. Several interim methods are also used. The line DCC is often called the embedded operations channel (EOC) in SONET and the embedded communications (or control) channel (ECC) in SDH.

SDH also calls this overhead the D1-D3 RS data communications channel (DCC_R).

All other byte positions in the first three rows of all SONET/SDH frame are undefined. They can be used by a network services provider or vendor for additional functions within a single "provider's domain." For interoperability, receivers can just ignore these undefined bytes. Of course, both SONET and SDH essentially reserved the right to assign formerly undefined bytes for standard use, as they have in some cases already with the bytes now used for media dependent and FEC purposes.

Line/Multiplex Section Overhead

In SONET, line overhead is contained in the bottom six bytes of the first three columns. This overhead is processed by all pieces of SONET network equipment, except for the regenerators. This SONET line overhead roughly corresponds to the SDH multiplex section overhead, with one exception. In SDH, row 4 of the section overhead is considered to stand as a unit all its own and is neither part of the regenerator section overhead nor the multiplex section overhead. The H1/H2/H3 bytes in row 4 of the section overhead in SDH form the *administrative unit* (AU). In an STM-1, the basic SDH multiplexing frame structure, the AU pointer bytes can point to three separate tributary units (TUs) as an AU-3 (similar to a SONET STS-3 with three STS-1s inside) or point to a single payload tributary unit as an AU-4 (similar to a SONET STS-3c). More details of the AU pointers in both SDH and SONET are discussed in Chapter 9.

What follows with regard to the H1/H2/H3 pointer bytes technically applies only to SONET. However, even though the AU is present in SDH as row 4, there is no doubt that the AU consists of H1/H2/H3 pointer bytes. So, all of the discussion also ultimately applies to the SDH AU row.

Pointer (H1, H2)   The H1/H2 byte form a pointer that is comprised of two bytes used to indicate the offset between the pointer bytes themselves and the beginning of the STS SPE. It allows for the dynamic alignment of the SPE within the allowable capacity of the envelope itself. In other words, because the SPE can begin anywhere within the information payload of a SONET frame (that is, anywhere but inside the overhead fields in the first columns of an STS-N), it would be nice to give the receiver a hint as to exactly where this may be. This is not to say that SPEs jump around haphazardly. After all, SONET is supposed to be synchronous. But once in a while, due to jitter and timing differences, the start of the SPE may move one byte forward or backward within the information payload. The pointer bytes provide a mechanism for the sender to inform the receiver of the correct position of the SPE at all times. The H1/H2 pointer always points to the J1 byte of the path overhead.

The actual inner workings of the H-bytes are straightforward. The first four bits of the H1 byte are called the new data flag (NDF). The NDF allows the pointer position to change when a major change occurs in the payload position in the payload. When an arbitrary new pointer value is introduced, the NDF is inverted (in this case 0110 changes to 1001), but the remaining fields stay the same. In this way, the H1 and H2 pointer bytes always point to the first byte of the SPE, which is the start of the POH bytes as well.

The next two bits in the H1 byte, which are zeros during normal SONET operation, have no significance in current SONET specifications: they are simply "place keepers." However, these two bytes once formed the *size* (s) bits, written as ss or often s1s2. The size bits in SONET were once used to communicate payload mapping information, but this information is carried in other fields now.

However, SDH still assigns meaning to these s1s2 bits. The ANSI SONET specification sadly notes that "G.707 will no longer use these ss bits in a manner that allows connectivity between SONET and SDH equipment." Of course, all international links are supposed to use SDH formats anyway. But the point is that SONET sets these bits to 00, but SDH normally uses the value 10 for both AU-3 and AU-4 (some older ADMs use 01.) The value 11 is reserved. This use of the s1s2 bits is summarized in Table 7-1.

The following ten bits, which overlap the H1 and H2 bytes, constitute the actual pointer to the payload J1 byte. In normal STS-1 operation, this pointer field can have any value between 0 and 782 in decimal (the STS-1 frame size minus the overhead bytes). Values greater than 782 are undefined. When the pointer is all zeros (00 0000 0000), this indicates that the SPE begins immediately to the right of the H3 byte, in row 4, column 4. As another example, a value of 87 indicates that the payload begins with the J1 byte immediately to the right of the K2 overhead byte, one row down. More details on pointer bytes will be examined in Chapter 9.

Binary value of s1s2 bits	Use
00	SONET (receivers ignore these bits)
01	Older ADMs might use this
10	AU-3 or AU-4 in SDH
11	Reserved

Table 7-1.   Use of the s1s2 bits

Pointer Action (H3)   The pointer action byte is allocated to compensate for the payload timing variations (ANSI calls them "frequency justifications") mentioned above. In the course of otherwise normal STS-1 operation, the arriving payload data rate might exceed the SONET frame capacity. This means that within a 125-microsecond period, more than 783 bytes may be ready to be sent out in the payload. If this excess is less than 8 bits, the extra bits would be just buffered sequentially and sent out as the first bits of the next frame; however, when a full byte (8 bits) has accumulated in the buffer, the pointer action byte is used to carry the "extra" 784th byte. This is called a *negative timing justification* and is governed by strict rules of SONET equipment operation. Naturally, a companion *positive timing justification* is used when 783 full bytes are not in the CPE buffer to fill the payload exactly. Details about positive and negative justifications are covered in more detail in Chapter 9.

The H3 byte must be provided in all STS-1 signals within an STS-N signal. The value contained in this byte, when it is not used to carry the payload data from a positive justification, is not defined and ignored. The H1/H2 pointer bytes tell the receiver when the H3 byte is used for useful information. This byte is necessary due to the variations in timing across different service provider's networks or when imprecise end equipment clocking feeds a SONET link.

The whole mechanism of "floating SPEs" with the H1, H2, and H3 bytes is quite complex and some service providers used to try to minimize its use when SONET chipsets were not as powerful as they are today. In those cases, the value of the H1/H2 byte pointer is locked to 522 decimal (0x20A in hex), which points to row 1, column 4 of the next SONET frame. This method of "locking" the payload to a fixed position in the STS-1 frame minimizes pointer justifications, but increases buffer requirements and management in the face of continued timing jitter.

With the end of this introduction of SONET H1/H2/H3 pointers and the SDH AU pointers, the rest of this section describes overhead bytes in the SDH multiplex section overhead.

Line/Multiplex section BIP-8 (B2)   Similar to the B1 byte, this BIP byte is used for bit error monitoring. It calculates the parity check based on all bits of the line overhead and the STS-1 frame capacity of the previous frame prior to scrambling. Note that the section overhead is specifically excluded, reflecting the hierarchical nature of the SONET overhead bytes. Then, the line overhead processing places the BIP-8 into the B2 field of the current frame before the current frame is scrambled. This field must be present and used in all STS-1s of an STS-N signal.

The line/multiplex section level aspect of the B2 byte explains something that at first seems odd in SONET (and SDH) frame structures. Higher order STS-N and STM-N frames have N B2 bytes, but only one B1 byte. This is because there is really only one frame (B1 scope) but as many lines (or multiplex sections) as there are STS-Ns. However, there is more to it than that. SONET specifications still expect that N B2 bytes will provide a single error measurement, even if there are N BIP-8 checks all operating independently. So the parity errors measured by N BIP-8s or a single BIP-8xN (a BIP-24 in an STS-3, for example) must give the same result.

In SDH, the B2 BIP is a BIP-Nx24, where N is the STM-N level. This is expected given the basic STM-1 = STS-3 correspondence. The B2 byte is the "multiplex section error monitoring" Nx24 code with even parity on previous frame except rows 1-3. the B2 byte is placed in the B2 bytes of the current frame. STM-1 uses a BIP-24 (3 bytes), STM-4 uses a BIP-96 (12 bytes), and so on. STM-0 uses N = 1/3, giving a BIP-8, the same as in the SONET STS-1.

Automatic Protection Switching (APS) (K1/ K2)   These two bytes communicate automatic protection switching (APS) commands and error conditions between pieces of line equipment. They are specifically used for link recovery following a network failure. These two bytes are only used in the first STS-1 of an STS-N signal. The full purpose and use of the K1/K2 bytes are somewhat flexible, but are used in many pieces of SONET equipment to switch lines on SONET *rings* (more on SONET/SDH rings is covered in Chapter 16). The K2 byte is also used to detect other types of alarms on SONET/SDH links.

In SDH, the format and protection protocol of the K1/K2 bytes is exactly the same as in SONET, except that line level alarms become multiplex section alarms. For example, bits 6-7-8 in the K2 byte (written as K2 (b6-b8)) is for the SDH MS-RDI (remote defect indicator) alarm. This means that the far end network element is receiving an SDH section defect or MS-AIS (alarm indication signal). The MS-RDI is sent by setting the K2 b6-b8 bits to 110 before scrambling. Chapter 16 details the SONET equivalents of these error signals.

The complete K1/K2 byte protocol and alarms are also investigated in detail in Chapter 16.

Line DCC (D4–D12)   These bytes represent a 576 Kbps message-based channel used for shipping OAM&P messages between SONET line level network equipment. This channel uses the same protocols as those used in the D1–D3 components of the section overhead. Typical messages may be for maintenance, administration, or alarms, and can be either internally generated or obtained from an outside source. In some cases, they may be manufacturer-specific. The DCC bytes are only defined in the first STS-1 of an STS-N signal. As with section overhead DCC bytes, TMN is supposed to be used. However, interim methods such as transaction language 1 (TL1) are often still used today.

In a SONET STS-768 frame, there are additional D13-D156 bytes defined. This is the *extended line DCC* running at 9.216 Mbps. The D13 to D60 bytes are located in row 6 columns 2 to 49, directly after the D4 byte. The D61 to D108 bytes are located in row 7 columns 2 to 49, directly after the D7 byte. The D109 to D156 bytes are located in row 8 columns 2 to 49, directly after the D10 byte.

In SDH, the D4-D12 bytes form the multiplex section (MS) data communications channel (DCC$_M$). More details of the operation of the SONET/SDH DCCs are discussed in chapters 14 and 15 on network management.

In an SDH STM-256 only, as in a SONET STS-768 only, the D13-D156 bytes form the extended MS data communications channel (DCC$_{Mx}$). The SDH locators are:

S(6,1,9) to S(6,1,56) for D13-D60

S(7,1,9) to S(7,1,56) for D61-D108

S(8,1,9) to S(8,1,56) for D109-D156

Synchronization Messaging (S1)   This byte is used to carry the synchronization status message (SSM) between SONET network elements. The use of this byte is defined for the first STS-1 in an STS-N. Only bits 5–8 are currently defined, leaving bits 1–4 for future functions. The purpose is to allow SONET equipment to actually choose the best clocking source from among several potential timing sources. This helps to avoid the creation of disastrous timing loops within the network. The S1 byte cannot prevent timing loops all but itself, however. In other STS-1s present in an STS-N, this byte is defined as a Growth (Z1) byte. Growth bytes are used for future functions as yet undefined by SONET standards.

In SDH, the S1 byte is also the SSM (b5-b8). The locator is S(9,1,1) or [9,1]. There are code values for the S1 byte for an "unknown" clock level and "do not use" this link for clocking purposes, four code values for the four standard ITU-T clock levels, and the rest of the S1 values are "reserved for quality levels defined by individual administrations." Details of the SSM protocol in the S1 byte are discussed in Chapter 16.

STS-1 line REI-L (M0) or STS-N line REI-L (M1)   The use of this byte is complex, and requires some review of SONET frame basics. SONET frames may be a basic STS-1, or have a channelized STS-N structure with N STS-1s inside. (Concatenated STS-Nc structures do not affect the present discussion.) Because the two major frame structure types exist, two purposes for this overhead byte are necessary. It all depends on whether the STS-1 stands alone or sits inside an STS-N.

When the SONET link is an OC-1, which can only contain an STS-1 or an electrical STS-1, then this byte is the M0 byte. The M0 byte is used for a line level remote error function (REI-L), formerly called the line far end block error (FEBE). This conveys the line BIP (B2) error count back to the source. Bits 5–8 of the M0 byte are used for this purpose, and bits 1–4 are undefined. Only counts 0 to 8 makes sense, of course, and all other values carried in bits 5-8 of the M0 byte are interpreted as 0 errors.

For an STS-N frame structure, this becomes the M1 byte. Not all STS-1s in an STS-N have the M1 byte. Only the "third" (as defined by the SONET frame structures shown in Chapter 6) STS-1 in the STS-N will carry the M1 byte. Obviously, the value of N must be three or greater. The M1 byte has the same purpose and function as the M0 byte, with a minor difference above the STS-48 level. The number of valid values is always equal to 1+ 8N, with everything else indicating 0 errors. So an STS-12 has 1+ 96 = 97 values, for the 0 to 96 possible B2 errors in the 12 byte units of the basic STS-12 frame structure. At STS-48 and above, the single M1 byte can only report a maximum of 255 errors back to the source (STS-48 should have 1 + 384 = 385 legal values).

In SDH, the use and format of the M0/M1 bytes are different and do not impose a limit on the number of B2 errors reported to the source. Since SDH defines and uses these bytes differently, a somewhat lengthy explanation is required.

In SDH, there are two M0/M1 bytes possible for the multiplex section remote error indicator (MS-REI). Only M1 is used in STM-0, 1, 4, and 16. STM-64 and 256 use an M0/M1 combination. There is no "automatic" interoperability with equipment not using M0/M1 system, such as SONET; manual configuration is needed. The STM-64 frame formerly used only the M1 byte. Newer STM-64 equipment must be able to be configured to support only M1.

The count in these fields is of "interleaved bit blocks that have been detected in error by the BIP-24xN B2." Each BIP bit position is a "bit block." Bit 1 of M1 is always ignored below the STM-16 level. Bits 2 thru 8 are the counter itself (max = 127) in STM-0, 1, and 4.

STM-0 uses a BIP-8, naturally. The valid value range depends on N. For STM-0, the range is 0 to 8 (maximum of 8 bits positions in the B2 BIP can be in error). The STM-1 range is 0 to 24. For an STM-1, the range is 0 to 24, STM-4 has a valid range 0-96, and an STM-16 has 0 to 255 (the specification notes this is "truncated" from an expected maximum of 256). This makes complete sense, because the B2s are 24 bits for an STM-1, 96 bits for an STM-4, and 256 bits for an STM-16. Table 7-2 shows the interpretation of the M1 bits for these four SDH levels.

M1 bits (bit 1 ignored by all but STM-16)	STM-0 interpretation	STM-1 interpretation	STM-4 interpretation	STM-16 interpretation
0000 0000	0 BIP violation	0 BIP violation	0 BIP violation	0 BIP violation
0000 0001	1 BIP violation	1 BIP violation	1 BIP violation	1 BIP violation
0000 0010	2 BIP violations	2 BIP violations	2 BIP violations	2 BIP violations
0000 0011	3 BIP violations	3 BIP violations	3 BIP violations	3 BIP violations
:	:	:	:	:
0000 1000	8 BIP violations	8 BIP violations	8 BIP violations	8 BIP violations
0000 1001	0 BIP violation	9 BIP violations	9 BIP violations	9 BIP violations
:	:	:	:	:
0001 1000	0 BIP violation	24 BIP violations	24 BIP violations	24 BIP violations
0000 1001	0 BIP violation	0 BIP violation	25 BIP violations	25 BIP violations
:	:	:	:	:
0110 0000	0 BIP violation	0 BIP violation	96 BIP violations	96 BIP violations
0110 0001	0 BIP violation	0 BIP violation	0 BIP violation	97 BIP violations
:	:	:	:	:
1111 1110	0 BIP violation	0 BIP violation	0 BIP violation	254 BIP violations
1111 1111	0 BIP violation	0 BIP violation	0 BIP violation	255 BIP violations

Table 7-2. M1 for STM-0, STM-1, STM-4, and STM-16

Note how the valid STM-0 count ends at 8, the STM-1 count ends at 24, and so on. But what about the STM-64 and STM-256 counts? The STM-64 uses a BIP-1536 (24×64). So SDH simply adds the M0 byte in front of M1! (Older equipment still uses 255 maximum for STM-64, and this 255 maximum should be configurable.) The STM-256 frame uses BIP-6144 (24×256). The M0 and M1 byte counts are shown in Table 7-3 for the STM-64 and STM-256 frames.

Growth (Z1, Z2)   When the first two bytes of the last row of the SONET frame are not used for the S1 and M0 or M1 functions, these overhead bytes are defined as growth bytes. Growth bytes are reserved for future functions as yet undefined by SONET standards. The Z1 and Z2 byte positions are shown in the structure figures in Chapter 6 for all SONET frame structures. SDH reserves these bytes for future international standardization.

Orderwire (E2)   Similar in function to the E1 byte, this is a 64 Kbps voice channel at the SONET line level. It is only defined in the first STS-1 of an STS-N signal. SDH uses this byte for the same purpose at the multiplex section level.

In addition to the overhead bytes defined above, both SONET and SDH define media dependent and forward error correction bytes.

M0	M1	STM-64 interpretation	STM-256 interpretation
0000 0000	0000 0000	0 BIP violation	0 BIP violation
0000 0000	0000 0001	1 BIP violation	1 BIP violation
0000 0000	0000 0010	2 BIP violations	2 BIP violations
0000 0000	0000 0011	3 BIP violations	3 BIP violations
⋮	⋮	⋮	⋮
0000 0110	0000 0000	1536 BIP violations	1536 BIP violations
0000 0110	0000 0001	0 BIP violation	1537 BIP violations
⋮	⋮	⋮	⋮
0001 1000	0000 0000	0 BIP violation	6144 BIP violations
0000 0110	0000 0001	0 BIP violation	0 BIP violation
⋮	⋮	⋮	⋮
1111 1111	1111 1111	0 BIP violation	0 BIP violation

Table 7-3.   M0/M1 for STM-64 and STM-256

In SDH, the media dependent bytes are composed of 6N bytes at S(2,2,X) [2,N+X], S(2,3,X) [2,2N+X], S(2,5,X) [2,4N+X], S(3,2,X) [3,N+X], S(3,3,X) [3,2N+X], or S(3,5,X) [3,4N+X], where X = 1…N.

This notation is confusing at first. Consider an STM-1 example. For the STM-1, the three bytes in row 2 are located at S(2,2,1) [2,2], S(2,3,1) [2,3], and S(2,5,X) [2,5]. The other three bytes are right below in row 3. However, in an STM-4, there are 24 bytes (6 × 4). In row 2, these 12 bytes are at:

S(2,2,1) [2,5], S(2,2,2) [2,6], S(2,2,3) [2,7], S(2,2,4) [2,8] (as X cycles through 1, 2, 3, and 4)

S(2,3,1) [2,9], S(2,3,2) [2,10], S(2,3,3) [2,11], S(2,3,4) [2,12] (X = 1, 2, 3, and 4)

S(2,5,1) [2,17], S(2,5,2) [2,18], S(2,5,3) [2,19], S(2,5,4) [2,20] (X = 1, 2, 3, and 4)

The other 12 bytes are right below in row 3. These positions are the same in SONET, as can be verified by examining the figures of STS-768 and STM-256 in Chapter 6.

Radio use for these media dependent bytes is defined in ITU-R Rec. F.750.

The use of FEC in higher SONET/SDH bit rate levels is optional and not discussed further in this book.

Path/Higher Order Path Overhead

In addition to user data, the payload envelope contains path overhead bytes in SONET and higher order path overhead bytes in SDH (also called the VC-4-Xc/VC-4/VC-3 overhead bytes). These are processed in SONET by the STS-1 terminating equipment (usually CPE, but not always) because they travel as part of the payload envelope and are processed everywhere it is processed. The payload contains nine bytes of path overhead. The SONET payload is a 9 row by 87 column structure, and consists of nine bytes called J1, B3, C2, G1, F2, H4, Z3 (F3), Z4 (K3), and Z4.

These bytes form a "column" in the payload portion of the basic SONET/SDH payload, meaning that the payload bytes always line up in a column of the payload. However, because the position of the payload envelope can "float" within the frame payload area, the position of the path overhead bytes can float as well. The first byte of the path overhead is always pointed to by the H1/H2 pointer bits in SONET and the AU pointer in SDH.

In SONET, path overhead is required for each STS-1 in an STS-N. Only one set of path overhead bytes is required for an STS-Nc, in the first STS-1 position of the payload. The path overhead performs four major types of functions, called *classes*:

▼ **Class A** Payload-independent functions with standard format and coding. All path terminating equipment must be able to read, interpret, and modify these bytes.

■ **Class B** Mapping-dependent functions with standard format and coding specific to the payload type. These bytes are usually needed for more than one type of payload mapping, but not all types. Path level equipment that generate and receive these payloads must be able to read, interpret, and modify these bytes.

■ **Class C**   Application-specific functions. The formats and codings for these functions are not covered in full in SONET specifications. Path level equipment that generate and receive these application payloads must be able to read, interpret, and modify these bytes.

▲ **Class D**   Undefined functions for future use. These bytes will be defined in the future for support of the functions of class A, B, or C. When assigning overheads functions, class A functions have priority over classes B and C.

In SDH, the higher order VC POH should be called the VC-4-Xc/VC-4/VC-3 overhead, where VC-4-Xc indicates contiguous concatenation of payloads. VC-4s, oddly enough, can in a sense be *both* concatenated and tributary in an STM-1, in spite of the presence of the single set of AU-4 H1/H2 pointer bytes. This is discussed more fully in the next chapter, but deserves a few more comments here.

In a VC-4 (which carries a concatenated payload, the same as a VC-4-Xc in many ways) the path overhead is in the first column of the 261 column VC-4 structure (the 270 STM-1 columns minus the 9 section overhead columns).

In a VC-3 (called a *tributary payload*), the path overhead is in the first column of an 85 column VC-3 structure, similar to the STS-1. But an STS-1 payload has 87 columns. In SDH, 85 columns × 3 VC-3s in STM-1 = 255 + 9 columns of section overhead = 264 columns. But the STM-1 frame has 270 columns. What are the other six columns used for? Each VC-3 has two columns of "fixed stuff" bytes. So each VC-3 "structure" is really 87 columns wide, just as in SONET. Now 87 columns × 3 = 261 + 9 section overhead columns = 270 columns as it should. Much more about VC-4-Xc/VC-4/VC-3 SDH payloads will be presented in the next chapter.

In SDH, the higher order path overhead (HO POH) consists of nine bytes: J1, B3, C2, G1, F2, H4, F3, K3, and N1. Note that SONET considers the F3 byte Z3 and the K3 byte Z4, although there is no doubt that SONET will follow the SDH definitions of the functions for the Z3/Z4 bytes.

These nine SDH bytes fall into four general categories, similar to the SONET classes:

▼ Bytes or bits for end-to-end communications, independent of payload type: J1, B3, C2, G1, K3 (b1-b4).

■ Bytes specific to a certain payload type: H4, F2, F3.

■ Bits reserved for future international standardization: K3 (b5-b8)

▲ Byte that can be used inside an "operator domain": N1

STS Path Trace (J1)   This Class A field transmits a repeated, 16-byte string that enables the receiving path terminating equipment to verify continued connection to the device sending the payload. This is a user programmable field of the same format as the J1 byte. SONET defines this as a 64-byte field, but most implementations use the SDH 16-byte format shown in Figure 7-3. If no message has been loaded by the user, then a string of 64 null characters is sent. This field can be something as simple as the IP address or E.164 address (i.e., telephone number) of the CPE device. Because few customers have SONET CPE-retaining T-carrier equipment that just feeds SONET links, this field is typically used by the service provider.

When used by the service provider, in most cases the local telephone company, the J1 byte often contains the CLLI™ code (pronounced "silly code") which stands for common language location identifier. The CLLI code is 8 to 11 bytes long and identifies a particular end office (central office) within the national telephone system. Because the CLLI code is much less than 64 bytes long, the remaining J1 bytes are padded with nulls and terminated with ASCII carriage return (CR) and line feed (LF).

In SDH, this is the J1 Path trace. This is always the first byte of VC, pointed to by AU-3 or AU-4 or TU-3 pointer. (The TU-3 pointer is how a "concatenated" VC-4 gets split up into three tributary unit groups, or TUG-3s.) J1 repeats the path access point identifier. This is a 16-byte frame with the same format as the J0 byte. At national or network boundaries, this format is always used unless an alternative is implemented by mutual agreement. Inside country or network a 64 byte frame can be used, as often with SONET.

The use of the J1 byte is simple, yet effective. Figure 7-6 shows an example of the use of J1 on an SDH link between Paris and Marseilles. The link involves four different service providers, which is perhaps a stretch, but this is just an example. Each of the service providers has their own cross-connections, so there is always a chance that a mistake can be made in configuration and suddenly the bits inside the SDH frames are coming from the wrong place.

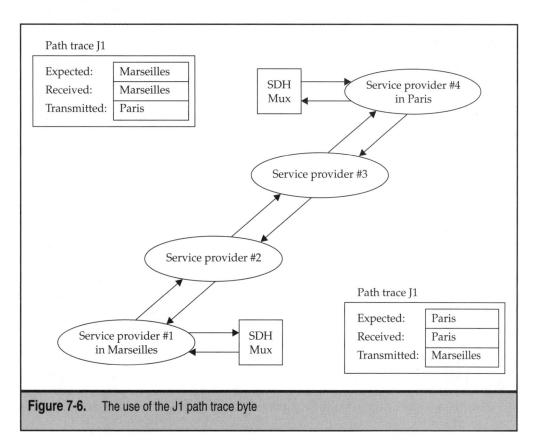

Figure 7-6.   The use of the J1 path trace byte

The SDH multiplexer in Paris is configured to transmit the string "Paris" in the J1 path trace byte. The multiplexer is also configured to expect the string "Marseilles" in the received J1 path trace bytes, although of course this has nothing to do with the sending process. The figure shows that the expected string ("Marseilles") matches what is actually received in Paris. At the Marseilles end of the link, the reserve configuration is shown.

What happens when the J1 path trace content does *not* match the configured value? This is an error condition called a trace identifier mismatch (TIM). Any mismatch results in the outgoing bits any the payload to the end equipment attached to the SDH multiplexer being filled with an "all 1's" bits pattern called the alarm indication signal (AIS). The sender is notified of this action through a remote defect indication (RDI) on the return path. (More on SONET/SDH error conditions is discussed in Chapter 14.)

The same error conditions and actions apply to J0 and J2 (there is also a J2 byte, described below) mismatches. In practice, the J0 trace is not as useful as the J1 byte because the J1 byte flows end-to-end. The J0 RDI is sent to the remote end in the K2 bits mentioned above, and the J1 RDI is found in the G1 byte discussed below.

Path BIP-8 (B3)   This Class A byte's function is analogous to that of line and section BIP-8 fields. It uses even parity, and is derived from the parity of the previous payload envelope prior to scrambling. Note that the line and section overhead bytes are specifically excluded, again reflecting the hierarchical nature of SONET overhead.

In SDH, this is the path BIP-8, a BIP-8 code with even parity over all bits of previous VC-4-Xc, VC-4, or VC-3. However, SDH does not include the fixed stuff bytes if present, while SONET always includes all columns of the payload envelope.

In summary, all three BIP checks have essentially the same function, but differ in the *scope* of operation in terms of the parts of the STS or STM they cover. The difference in the scope of the B1, B2, and bytes are shown in Figure 7-7.

STS Path Signal Label (C2)   The Class A signal label tells network equipment what is contained in the payload envelope (that is, the format of the bits inside the payload and how it is constructed). The intent is to allow the transport of multiple services simultaneously, especially for the projected B-ISDN services. This means that a single SONET/SDH device can actually *interleave* frames (technically, the payloads of the frames) containing a DS-3 with frames containing ATM cells, or fiber distributed data interface (FDDI) data frames, and so on.

The C2 signal label byte has some interesting history and carries a kind of warning to SONET/SDH vendors and implementers. SONET/SDH is essentially a point-to-point, trunking technology. That is, SONET/SDH does no switching based on payload content: the payloads can be cross-connected, but that is all. So it came as a surprise to equipment vendors when a label was included in the path overhead to identify payload content. After all, end equipment is usually configured to send and receive one type of traffic and no other. Otherwise, it just won't work. If one end is set up to send IP packets, and the other end expects ATM cells, that is easy enough to find and correct.

So why bother labeling the payload contents? The end equipment still has to be configured independently of the label. The answer is that given the huge bandwidths SONET/SDH can offer, it was seen as entirely possible that, on one SONET/SDH link, equipment could carry both IP packets and ATM cells and much else at the same time.

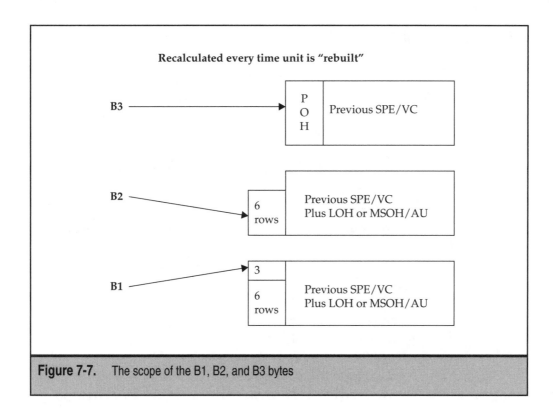

Figure 7-7.   The scope of the B1, B2, and B3 bytes

But then the sender had to tell the receiver what was inside this particular frame's payload. That is what the signal label was invented for.

But none of the early implementers used SONET/SDH for much other than traditional voice, which was carried inside payloads as *virtual tributaries* (VTs) in SONET or *virtual containers* (VCs) in SDH. The signal label value for the most common way of carrying VTs or VCs was 02 (0000 0010), so this became effectively the default value for the C2 byte in a lot of SONET/SDH gear. It could not even be changed in many cases, but the equipment hummed merrily along because it was configured to handle IP or ATM cells and did not bother to check the C2 byte. However, once SONET (and SDH) specifications matured to the point where mismatching the signal label to the traffic type generated an alarm, new SONET/SDH NEs went into alarm condition (or even took itself out of service) if the C2 byte value did not match the traffic type in the path overhead! Many vendors had to upgrade or patch their software and firmware quickly.

The moral of the story is that ignoring a specification just because it is not currently enforced can be a risky proposition. This is especially true of international standards.

SONET uses the C2 byte for more than just the payload format. Many values are used for an error count called the path-level payload defect indicator (PDI-P). This counter reflects what are called *payload defects* at the VTx level (there are several types of virtual tributaries), usually carrying a number of voice channels. The exact mechanisms used to find these VT payload defects is beyond the scope of this book. But the pattern of C1 values assigned to the PDI-P is similar to the idea behind the use of the M0/M1 bytes. The whole point is to allow the sender to figure out how many errors there are in the bits being sent in the (voice) VTs to the receiver.

Currently defined values of the C2 byte in SONET are included in Table 7-4. Some of these will be discussed more fully in the next chapter.

Hex Code	Interpretation of VC-4-Xc/VC-4/VC-3 Content Mapping
00	Unequipped or supervisory unequipped: contains no payload on "open" connection
01	Equipped-nonspecific payload. Payload format determined by other means
02	Floating VT mode: mainly for "voice" tributaries (discussed in next chapter)
03	Locked VT-mode: for locked byte synch mode (no longer supported – see next chapter)
04	Asynchronous mapping for DS-3 (44.736 Mbps) – see next chapter
05	Mapping under development: an experimental category
12	Asynchronous E-4 (139.264 Mbps) mapping – see next chapter
13	ATM Cell mapping – see next chapter
14	Metropolitan Area Network (MAN) Dist. Queue Dual Bus (DQDB) mapping
15	Fiber Distributed Data Interface (FDDI) mapping
16	Mapping for HDLC over SONET: used for IP packets – see next chapter
17	Simple Data Link (SDL) for scrambler mapping (under study)
18	HDLC/LAPS (link access procedure – SDH) frame mapping (under study)
19	Simple Data Link (SDL) for scrambler mapping (under study)
1A	10 Gbps Ethernet frame mapping
1B	Flexible Topology Data Link mapping (called GFP in SDH)
1C	Fibre Channel at 10 Gbps
CF	Reserved: former value used HDLC/PPP frame mapping (obsolete)

Table 7-4. SONET C2 byte values

Hex Code	Interpretation of VC-4-Xc/VC-4/VC-3 Content Mapping
E1	STS-1 payload with 1 VT-x payload defect
D0-DF	Reserved for proprietary use
E2	STS-1 payload with 2 VT-x payload defects
:	:
FB	STS-1 payload with 27 VT-x payload defects
FC	STS-1 payload with 28 VT-x payload defects or a non-VT payload defect
FE	Test signal (also used, like 05, for experimental traffic)
FF	AIS (generated by source if there is no valid incoming signal)

Table 7-4. SONET C2 byte values *(continued)*

Some of these payloads will be detailed in the next chapter. For example, the high level data link control (HDLC) or point-to-point protocol (PPP) mapping is used for carrying IP packets inside SONET frames. The AIS "all 1's" error condition was already discussed.

In SDH, this is the C2 signal label byte. This determines the "composition of VC-4-Xc/VC-4/VC-3." There are 209 codes for future use. The currently defined values are shown in Table 7-5.

Some are these discussed in more detail in the next chapter. Note the difference from SONET's definition of the 01 value, the use of Generic Packet Framing (GFP) for 1B, and SONET's use of the values E1-FC.

Hex Code	Interpretation of VC-4-Xc/VC-4/VC-3 Content Mapping
00	Unequipped or supervisory unequipped: contains no payload of "open" connection
01	Reserved: Formerly "Equipped-non-specific." New receivers ignore 01 from old NEs
02	TUG structure: mainly for "voice" tributaries (discussed in next chapter)
03	Locked TU-n: for locked byte synch mode (no longer supported – see next chapter)

Table 7-5. SDH C2 byte values

Hex Code	Interpretation of VC-4-Xc/VC-4/VC-3 Content Mapping
04	Asynchronous mapping of E-3 (34.368 Mbps) or DS-3 (44.736 Mbps) – see next chapter
05	Mapping under development: an experimental category
12	Asynchronous E-4 (139.264 Mbps) mapping – see next chapter
13	ATM Cell mapping – see next chapter
14	Metropolitan Area Network (MAN) Dist. Queue Dual Bus (DQDB) mapping
15	Fiber Distributed Data Interface (FDDI) mapping
16	HDLC/PPP frame mapping: used for IP packets – see next chapter
17	Simple Data Link (SDL) for scrambler mapping (under study)
18	HDLC/LAPS (link access procedure – SDH) frame mapping (under study)
19	Simple Data Link (SDL) for scrambler mapping (under study)
1A	10 Gbps Ethernet frame mapping (under study)
1B	Generic Packet Framing (GFP) mapping – see next chapter
1C	Fibre Channel at 10 Gbps
CF	Reserved: former value used HDLC/PPP frame mapping (obsolete)
D0-DF	Reserved for proprietary use
E1-FC	Reserved for national use
FE	Test signal (also used, like 05, for experimental traffic)
FF	VC-AIS (generated by tandem connection source if there is no valid incoming signal)

Table 7-5. SDH C2 byte values *(continued)*

Path Status (G1)   This Class A byte notifies the originating end of the path about performance and status of the entire path. It carries two *maintenance signals*, the B3 error count known as the path remote error indicator (REI-P), formerly called Path FEBE, in bits 1–4, and a path remote defect indicator (RDI-P) in bits 5–7, in both old and new formats. This allows the monitoring of the entire path from either end. The remaining bit is undefined.

SDH says much more about the G1 byte. In SDH, this is the G1 path status byte. It provides path status and performance information back to the "trail" (what SONET calls the "drop side") termination source. Figure 7-8 shows the structure of the G1 byte.

```
            G1 REI              RDI      Reserved    Spare
      ┌─────┬─────┬─────┬─────┬─────┬─────┬─────┬─────┐
      │  1  │  2  │  3  │  4  │  5  │  6  │  7  │  8  │
      └─────┴─────┴─────┴─────┴─────┴─────┴─────┴─────┘

      Bits 1-2-3-4: Remote Error Indicator count from B3 (0-8 valid)
      Bit 5:        Remote Defect Indicator (details of REI)
      Bits 6-7:     Reserved
      Bit 8:        Spare (receiver required to ignore)
```

Figure 7-8. The G1 path status byte

The 4-bit REI field is for the B3 BIP-8 error count. Values 0-8 are valid, with all other values interpreted as "no error." The RDI bit is for connectivity and signal failure conditions at the receiver. (In 1993, this bit used to mean "remote loss of cell delineation (LCD).") Bits 6 and 7 are reserved for E-RDI (enhanced RDI) for VC-4-Xc/VC-4/VC-3, which adds details to the RDI bit values (bit 5 is used as well). The enhanced use of these two bits are optional, and set to 00 or 11 if the E-RDI is not supported. Bit 8 is for future use and is ignored by receivers.

Path User Channel (F2)   This class C byte (the first so far), like the F1 byte, can be used by the network service provider for internal network communications. The F2 byte is also used in one odd case to carry layer management information generated by distributed queue dual bus (DQDB) networks.

In SDH, there are two bytes, F2/F3, called the Path User channels. These are payload dependent bytes for "user" (really vendor) path terminating equipment communications. With aVC-4 DQDB payload (C2 = 0x14), the F2/F3 bytes carry the DQDB layer management information bytes, a role played by F2 alone in SONET.

Multiframe Indicator (H4)   This Class B byte (the first Class B byte so far) is used when a frame is organized into certain types of mappings. Mappings determine the actual structure of the portion of the payload carrying user data. One of the most common mappings is for virtual tributaries used for voice channels. Several types of virtual tributaries are defined, but all exist to provide some degree of backward compatibility for SONET links. With virtual tributaries, it is possible for users to have a T-1 at each end of a private line, but SONET in the middle. SONET transports virtual tributaries as easily as ATM cells or anything else, using different mappings. This byte is also used in SONET for the *virtual concatenation* of payloads, as explained below with regard to the SDH use of the H4 byte.

In ATM networks, the indicator byte was used to denote cell boundaries, but this use is now officially "not recommended" (ANSI and ITU talk for "totally obsolete"). This feature allowed SONET to transport such cell relay services as B-ISDN (ATM).

In DQDB networks, the indicator byte is used to carry link status information.

In SDH, this is the H4 position and sequence indicator. It is still a multiframe indicator, but also a sequence indicator for *virtual* VC-3/4 concatenation (more of this in the next chapter). If payload is VC-2/1, this is a multiframe indicator. This byte has special meaning for VC-4 DQDB payload, as in SONET.

When used as a multiframe indicator, the last two buts of the H4 byte form a counter that cycles 00, 01, 10, 11 (0, 1, 2, 3) over and over.

Growth (Z3–Z4)   These bytes in SONET are reserved for future use. However, in DQDB the Z3 byte is used to carry more DQDB layer management information. It is likely that SONET will use these bytes for the same purpose as SDH: F3 and K3.

The SDH use of the F3 byte has already been discussed with the F2 byte. The K3 byte is of more interest, especially bits 1-4. These bits form the K3 automatic protection switching (APS) channel. The K3 byte provides APS protection at the VC-4/3 path levels. (No other details about the use of the K3 byte are included in G.707.)

Bits 5-6 of the K3 byte are spare and for future use. The receiver is required to ignore these bits. Bits 7-8 form the higher order APS data link. The structure of the K3 byte is shown in Figure 7-9.

Tandem Connection (N1)   This Class C byte is used as a tandem connection maintenance channel and a tandem connection data link. In some documentation, especially older SONET documents, this is still the Z5 byte, but defined as tandem connection anyway. As discussed in the previous chapter, a tandem is a switching office that switches between trunks and, therefore, has no CPE devices on the end of the links. This can be a problem with payload handling because no direct communication occurs with the originator of the payload. So the N1 byte forms a kind of "logical path termination" for tandem bundles of STS-1s. Bits 1-4 form the tandem connection incoming error count (IEC) between the tandem end points. Bits 5-8 can be used for a tandem connection data link (Option 1) or just as 4 bits for tandem connection level error conditions such as RDI or REI (Option 2).

In SDH, this is the N1 network operator byte used for tandem connection monitoring (TCM). This has essentially the same use as in SONET and is not discussed further.

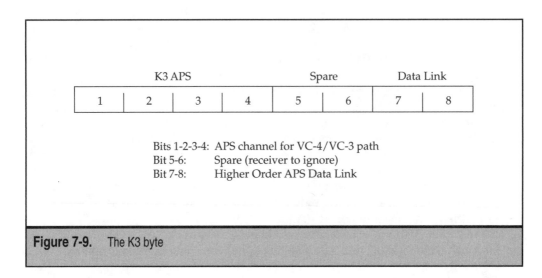

Figure 7-9.   The K3 byte

Bits 1-4 form the IEC (incoming error count) based on B3 errors across the tandem connection terminating equipment (TCTE). Bits 5-8 are unassigned in SDH. Figure 7-10 shows the structure of the N1 byte and the IEC values.

Notice that the error count in the IEC field is quite similar to the format of the M0/M1 bytes. Undefined values are interpreted as no B3 BIP violation errors.

This tour of SONET/SDH overhead can be quite overwhelming at first. When a SONET frame is built, different parts of the overhead are placed on the payload at different stages of the process. To try and sort out the overhead processing, Figure 7-11 shows

IEC   Data Link

| 1 | 2 | 3 | 4 | 5 | 6 | 7 | 8 |

Bits 1-2-3-4: Incoming Error Counter (IEC)
Bits 5-6-7-8: TC Data Link

IEC Value	Interpretation
0000	0 BIP violation
0001	1 BIP violation
0010	2 BIP violations
0011	3 BIP violations
0100	4 BIP violations
0101	5 BIP violations
0110	6 BIP violations
0111	7 BIP violations
1000	8 BIP violations
1001	0 BIP violation
1010	0 BIP violation
1011	0 BIP violation
1100	0 BIP violation
1101	0 BIP violation
1110	0 BIP violation, Incoming AIS
1111	0 BIP violation

Figure 7-10.   The N1 byte and the IEC

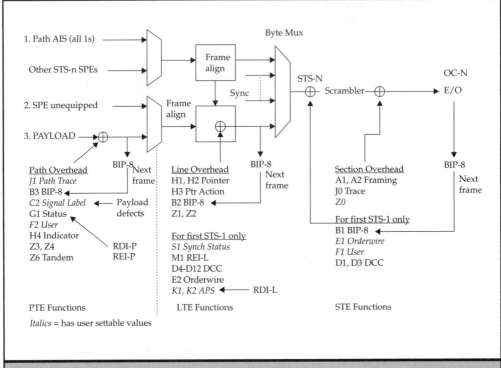

Figure 7-11. Building STS-1 overhead into a SONET frame

how an STS-1 frame is built in terms of overhead. SDH frames are more complex, so SONET is used for this example.

Basically, path overhead is added to a basic STS-1 first, even before any multiplexing has taken place. Other STS-1s can have alarms (AIS "all 1's"), be unequipped, or carry other payloads. Line and section overhead is added as shown, along with pointer additions (frame alignment). Overhead bytes shown in *italics* have user settable values determined through configuration of the SONET link. The other values are determined more or less automatically by the SONET equipment itself. The tandem connection byte is shown as Z5, just to illustrate this variation. Arrows show where major error indications (if there are any) are placed for the other end of the link. Other arrows show the BIP calculation points. Note the unscrambled row 1 overhead.

Figure 7-12 shows how a basic SDH frame is constructed, but not in as much detail as the SONET figure.

This is only one way to build the STM-1 frame, and the emphasis here is not on the details of how an E-4 source signal running at about 140 Mbps is placed inside an STM-1. The emphasis is on where in the process the path overhead, AU pointer, and regenerator section/multiplex section overhead is added. More details on the steps involved in making an STM-1 frame will be investigated in the next chapter.

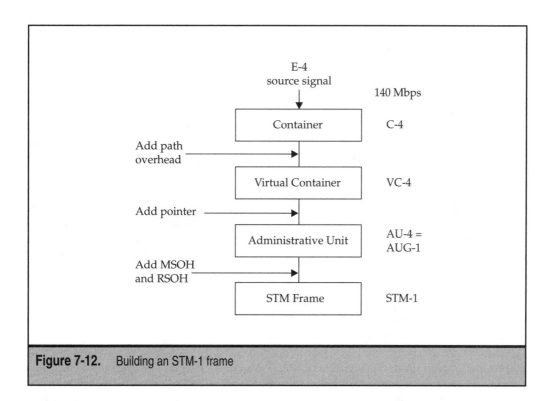

Figure 7-12. Building an STM-1 frame

Virtual Tributary and Virtual Container Overhead

There is another level of overhead included in the SONET and SDH specifications. This is the virtual tributary overhead (VTOH) in SONET and the lower order (or low order) path overhead in SDH. SDH specifications also commonly call this the VC-2/VC-1 path overhead. This SDH overhead is only defined for these VC-2/VC-1 structures, which carry the lower levels of the E-carrier and T-carrier voice channels multiplexing hierarchy.

SONET uses the virtual tributary overhead for the same purpose as SHD: to carry information about the (mainly) voice channels inside the payload. This voice channel information is very similar to the management and performance information transported at the SONET/SDH proper. Both SONET and SDH rely on the multiframe counter in the H4 path overhead byte to identify these lower order tributary overhead bytes (in the rest of this chapter, these bytes will just be called the "tributary overhead" bytes.

This is a good place to end this chapter on SONET/SDH overhead, because the next chapter will explore tributaries and the tributary overhead in much more detail. In this chapter, the bytes will be identified and their functions defined, mostly to keep the main overhead information all in one chapter.

The tributary overhead bytes consist of V1 through V5, and three other bytes: J2, N2, and K4. These last two bytes are still defined as Z6 and Z7 in a lot of SONET documentation, but these bytes are still reserved for the SDH N2 and K4 functions and will be described as such in this section. Technically, only V5, J2, N2, and K4 form the tributary overhead, so these are the bytes discussed in full here.

The role of the V1 through V4 bytes will be discussed more fully in the next chapter. The V1/V2 bytes play a role similar to the H1/H2 pointers in the sense that the V1/V2 overhead bytes always allow a receiver to find the V5 byte, which is the first byte of the tributary overhead proper. So, tributary units can be offset from the beginning of the payload envelope in the same way that the payload envelope can be offset from the start of the SONET/SDH frame. The goals are the same—to minimize the need for buffering and adjust for timing differences, this time at the tributary level.

In SDH, the V5 byte is found using by the TU-2/TU-1 pointer, but these are just other names for the V1/V2 bytes, depending on the exact structure of the tributary payload.

Once down to the tributary level in SONET/SDH, the bit rates are on the order of 64 Kbps to a few megabits per second. At these channel speeds, there is no longer the overhead bandwidth available to send all four tributary overhead bytes with every frame, nor is there a real need to. Things do not have to happen 8,000 times a second at the tributary level, and few E-1 or T-1 signals, status bits, and alarms are generated in each and every E-1 or T-1 frame. So the four tributary overhead bytes occupy the same position in four successive payloads. First this is the V5 byte, then the J2 bytes, then the N2 bytes, and finally the K4 byte.

But how are the receivers to know which frame is which? That is where the H4 multiframe indicator comes in. The last two bits of this path overhead byte, as mentioned above, form a counter that cycles the values 00 (0), 01 (1), 10 (2), and 11 (3) over and over. So identifying the role of the byte pointed to by the V1/V2 pointers (which appear in every tributary payload overhead) is simple. When the H4 bits are 00, the byte indicated is V5. When the H4 bits are 01, the byte indicated is J2. When the H4 bits are 10, the byte indicated is N2, and when the H4 bits are 11, the byte indicated is K4. This relationship between the H4 bits and the tributary overhead bytes is shown in Figure 7-13.

The figure shows the role of the path overhead bytes, mostly for review purposes. The names and general functions of the tributary overhead bytes are shown in the figure also. These are detailed below for SONET and SDH.

Indicator (V5)   This single byte performs the same roles for the tributary as the B3, C2, and G1 bytes provide for the overall payload envelope. There is a structure to the V5 byte that allows for this, as shown in Figure 7-14.

The figure shows the position and structure of the V5 byte. The V5 byte always leads off the tributary overhead in the same way that the J1 byte leads off the path overhead. The fields defined for error checking, signal label, and status are shown in the figure. The fields are:

Bits 1 and 2   These bits are used for error performance monitoring. It is a BIP-2 even parity check. Bit 1 is even parity on all odd bits in the previous tributary unit (VT in SONET, VC-2/VC-1 in SDH). The even parity on the even bits is carried in bit 2. This check excludes the V1 through V4 bytes (except when V3 has data, as can happen in the same way as the H3 path overhead byte).

Bit 3   This is the VT or VC-2/VC-1 path remote error indicator (REI-V). A 1 bit here means one or more errors in the received BIP-2, otherwise the value of this bit is 0.

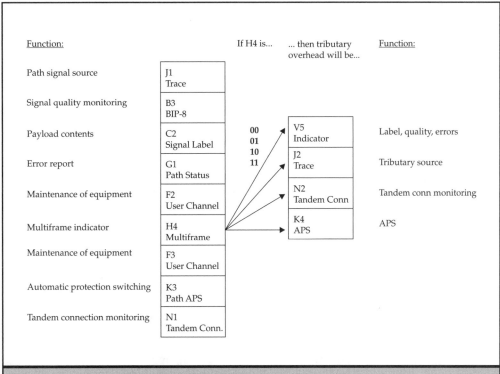

Figure 7-13.   Tributary overhead and the H4 indicator bits

Bit 4   This is defined only for the VT1.5 (SONET) or VC-11 (SDH) byte synchronous mapping path remote failure indication (RFI-V). This bit is 1 for RFI-V for VT1.5/VC-11. This bit is undefined for related mappings, such as for VC-2 and VC-12. A failure is defined by the ITU as essentially a persistent defect (the time frame to "persist" is configurable).

Bits 5-6-7   These bits form the signal label for the tributary (VT in SONET, VC-2/VC-1 in SDH). The use of the signal label values is different in SONET and SDH, but related, as discussed shortly.

Bit 8   This is the VT or VC-2/VC-1 remote defect indicator (RDI-V). A value of 1 means there is a path RDI-V present indicating a connectivity and signal failure at this level.

It is much simpler to present the SDH version of the V5 signal label values. There are eight values possible, as shown in Table 7-6.

Only the lower order virtual container AIS is new, and this is just another "all 1's" error indicator. The mappings are for SDH-specific tributary content (VC-2/VC-1 and so on), but the SONET version essentially just substitutes the SONET equivalent mappings. But SONET also repeats the table four times over, for the four defined SONET VT types: VT1.5, VT2, VT3, and VT6.

VT Path Trace J2   This byte is yet another repeated access point identifier similar to J0 and J1. In SDH, this is the low order path access point identifier. In both cases they are

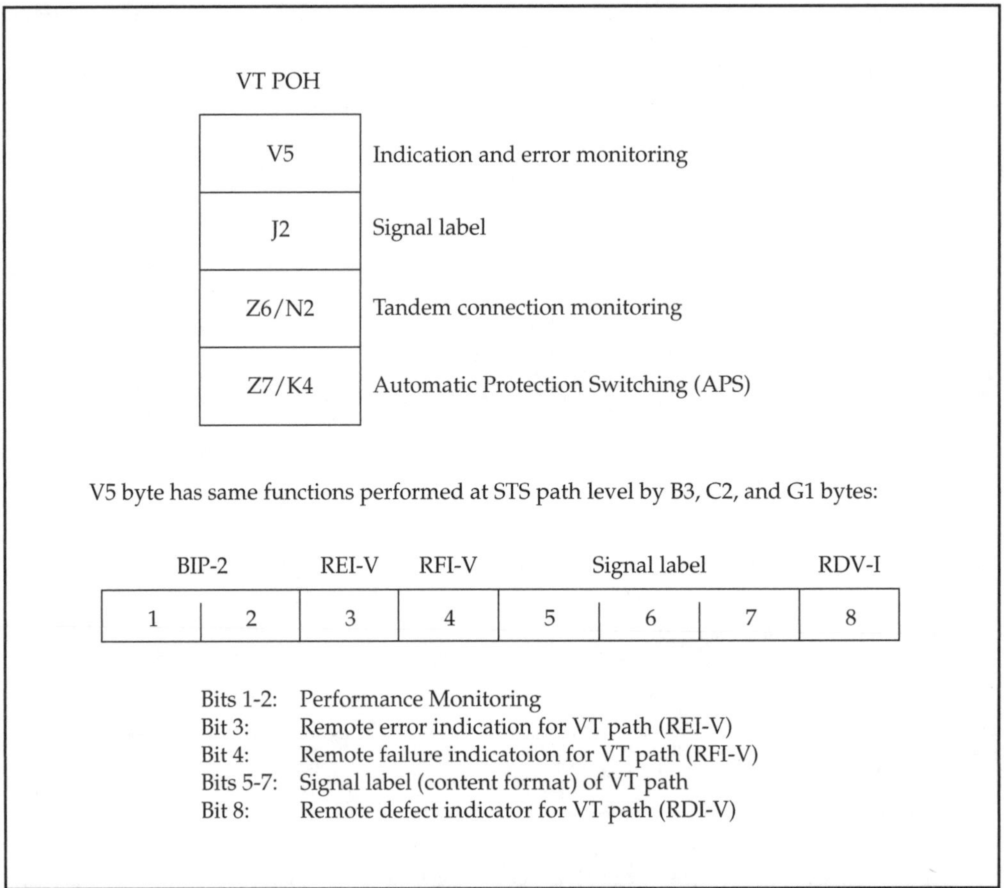

VT POH

V5	Indication and error monitoring
J2	Signal label
Z6/N2	Tandem connection monitoring
Z7/K4	Automatic Protection Switching (APS)

V5 byte has same functions performed at STS path level by B3, C2, and G1 bytes:

BIP-2		REI-V	RFI-V	Signal label			RDV-I
1	2	3	4	5	6	7	8

Bits 1-2:   Performance Monitoring
Bit 3:   Remote error indication for VT path (REI-V)
Bit 4:   Remote failure indicatoion for VT path (RFI-V)
Bits 5-7:   Signal label (content format) of VT path
Bit 8:   Remote defect indicator for VT path (RDI-V)

Figure 7-14.   The V5 byte position and structure

used so a receiver can verify continued connection to expected source. This byte has the same 16-byte structure as J0 and J1 bytes.

Each of the three trace bytes has a different scope along a SONET/SDH link. Figure 7-15 shows the scope of the J0, J1, and J2 trace bytes in an SDH context. The SONET equivalent would be similar except for the terminology used.

Note the use of the J1 byte at the lower level for a VC-3 payload and the role of the J2 byte when the payload is a VC-12.

Network Operator N2   This byte provides low order tandem connection functions essentially the same as the higher order N1 byte. The scope is now at the tributary level, however. Bits 1 and 2 form a BIP-2, bit 3 must be a 1 bit, bit 4 when set to 1 indicates incoming AIS, bit 5 is yet another form of REI, bit 6 is used as an outgoing error indication, and bits 7 and 8 are used for many error and performance purposes.

b5-b6-b7	Interpretation of VC-2/VC-1 Content Meaning
0 0 0	Unequipped or supervisory-unequipped
0 0 1	Reserved: Formerly "Equipped-non-specific." New receivers ignore 001 from old NEs
0 1 0	Asynchronous mapping – see next chapter
0 1 1	Bit synchronous E-1 (2.048 Mbps) VC-12 mapping (no longer supported – see next chapter)
1 0 0	Byte synchronous mappings – see next chapter
1 0 1	Extended signal label: Used with K4 byte to provide mapping details at LO level
1 1 0	Test signal: for any non-virtual concatenated payload
1 1 1	VC-AIS (generated by tandem connection source if there is no valid incoming signal)

Table 7-6. The SDH V5 signal label (b5-7)

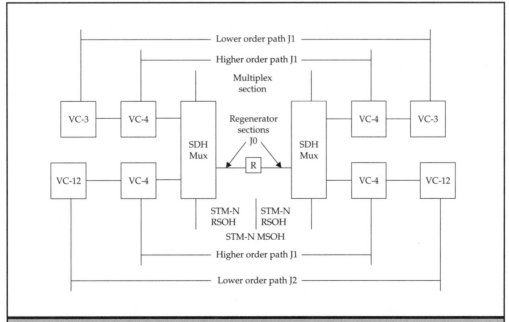

Figure 7-15. The scopes of the J0/J1/J2 trace bytes

Low Order Path APS K4   It is somewhat misleading to call this an APS byte at all. But to be consistent with the K1/K2/K3 byte definitions, that is what this byte is called. Like most tributary overhead functions, this byte is loaded with meaning, and few have anything to do with APS. The main role of the K4 byte is as an extended signal label used along with the V5 byte. Extended signal labels, used when the V5 signal label field (bits 5-6-7) is 101, are used with *virtual concatenation* in SONET/SDH. This requires some explanation.

Bit 1 of the K4 byte is the extended signal label. When the V5 bits 5-7 are set to 101, this bit is used as part of a 32-frame multiframe used in virtual concatenation of tributaries (more details on virtual concatenation are presented in the next chapter). Figure 7-16 shows the structure of this 32-frame sequence.

This single bit can do a lot of work when spread across 32 frames. The structure starts out with the multiframe alignment signal (MFAS) pattern, which is 0111 1111 110 in frames 1-11. This just keeps everything in the K4 bit1 multiframe synchronized between sender and receiver. Bit 20 must be 0, and bits 21-32 are reserved for future standardization (these bits should be all 0, and must be ignored by the receiver).

Bits 12 to 19 have the extended signal label itself. There are 242 codes that are currently spare. The most important thing to understand is that when SONET/SDH virtual concatenation is used, the V5 signal label must be 101 and the K4 byte *must* use this multiframe structure. This is how virtual concatenation works in SONET/SDH, as will be explained in the next chapter.

The extended signal label tells the receiver what's inside the VT or VC-2/VC-1 when virtual concatenation is used. The currently defined values for the K4 bit 1 multiframe extended signal label are shown in Table 7-7.

The rest of the K4 byte is much easier to describe. Bit 2 is the low order virtual concatenation bit. This is used with bit 1 (the extended signal label) 32-frame multiframe for the sequence numbers of the channels to be "virtually concatenated." So there are really 32 bits available, used as shown in Figure 7-17.

There is a frame counter in bits 1-5 (counting 0-31 over and over) and a sequence indicator number in bits 6-11 defining the sequence or order in which the lower bit rate channels are combined to form the virtual concatenation itself. How this works will be shown in the next chapter.

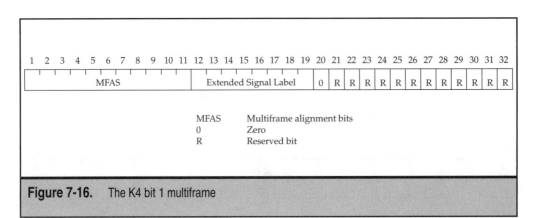

Figure 7-16.   The K4 bit 1 multiframe

Hex Code	Interpretation of K4 bit 1 multiframe Extended Signal Label (bits 12-19)
00-07	Reserved so not to overlap with values 0-7 of V5 signal label
08	Mapping under development: content not defined in this table
09	ATM mapping
0A	HDLC/PPP mapping
0B	HDLC/LAPS
0C	Virtually concatenated test signal: for virtual concatenation content not defined here
0D	Flexible topology data link mapping (under study)
FF	Reserved

Table 7-7. The interpretation of K4 bit 1 multiframe extended signal label (bits 12-19)

Bits 3 and 4 are for the APS of the K4. This use is for further study.

Bits 5-7 are reserved for something called the enhanced remote defect indicator (E-RDI) for the VT or VC-2/VC-1, which adds details to the RDI at this level. This function is optional, and these bits are set to 000 or 111 if E-RDI not supported.

Bit 8 is reserved for a lower order path data link. This is not discussed further in this book.

This last section on tributary overhead has introduced many terms and bit structures not treated in full here. This was the price to be paid for presenting an overview of the SONET/SDH overhead in one chapter. Many comments about "details next chapter" were the unavoidable result of this approach. But the promises of further details are about to be fulfilled. The time has come to turn to one of the most complex and yet crucial topics in SONET/SDH: the actual formats of the user information bits inside the SONET/SDH payloads. SONET/SDH does not exist to shuttle SONET/SDH frames back and forth across the link. The whole purpose is to carry user/customer bits. The next chapter shows how this is done.

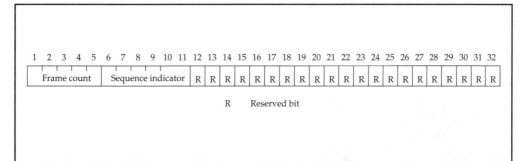

Figure 7-17. The K4 bit 2 multiframe

CHAPTER 8

SONET/SDH Payloads

Probably the most challenging area for newcomers to SONET/SDH trying to form an understanding and grasp of SONET and SDH is the concept of payloads, especially in SDH if people have learned SONET first. This prior SONET knowledge is often the case, and the hardship is not only due to the differences in terminology between SONET and SDH, but due to the insistence in SDH of giving a special name to most of the stages in taking a simple collection of voice channels and turning them into an SDH frame.

Not that the corresponding SONET procedure is simple, far from it. But if SONET/SDH was a twist-off bottle cap in need of opening, the simple SONET directive of "twist to open" would be replaced in SDH by "(1) Grasp bottle firmly (2) With other hand, twist to remove (3) Dispose of cap properly." This may be a gross exaggeration to prove a point, but it is a fact that where SDH frame-formation consists of mini-steps to map (add overhead), align (add pointers), and multilplex (usually add both overhead and pointers), SONET just conceptually adds the necessary overhead and pointers in one operation without feeling the need to name each mini-step's outcome separately (although in fairness to SDH, all of the same mini-steps must be taken by SONET, whether named or not).

What makes matters even worse is that the standard SONET and SDH specifications usually start out with virtual tributary (voice) payloads as the first topics explored in depth. Even those somewhat familiar with T-carrier and E-carrier voice find themselves bombarded with terms, procedures, and rules that are usually presented so abstractly that it is hard to believe that point-to-point voice trunking can be so complicated. The basic ITU-T SDH specification starts out with about 75 pages of voice-related payloads before even showing the STM-1 overhead bytes. ANSI is better, but mainly because the complex figures and tables are separated from the text (although this practice poses challenges of its own).

This chapter therefore can only take an introductory look at all there is to say about SONET/SDH payloads. But it is a very complete introduction. Enough about virtual tributaries (SONET) and tributary units and virtual containers (SDH) will be explored to at least make the next visit to the specifications less painful than a trip to dentist's office. At the end of the chapter, the latest advances in SONET/SDH payloads will also be investigated. These are the key areas of virtual concatenation and the generic framing procedure (GFP). Along the way between tributaries and virtual concatenation/GFP, stops will be made at mappings for ATM and IP packets in SONET/SDH.

This chapter will handle SONET tributaries before SDH. But the rest of the chapter treating ATM and IP mappings, virtual concatenation, and GFP will only discern SONET and SDH when such differences are important. Of necessity, this will be a very long chapter. The organization into distinct sections should help to keep things organized.

This chapter provides some details on three other popular payload contents in addition to voice. The four payload types considered in this chapter are:

▼ Virtual tributaries and tributary units (VTs and TUs; channelized and digitized 64 Kbps DS-0s)

■ Asynchronous DS-3 (unstructured DS-3 frames)

■ ATM cells (SONET/SDH in its designed role as a B-ISDN transport)

▲ Packet over SONET/SDH (POS; any layer 3 packet, but almost always IP)

It is safe to say that the vast majority of SONET/SDH links around the world carry one of these four types of traffic in the payload. As the values of the signal label (C2) byte listed in the previous chapter show, there are more than just these four possibilities for SPE content. Most service providers have never seen anything else inside a SONET/SDH payload but one of the four types of traffic listed here. (Some SONET/SDH service providers have seen nothing but 64 Kbps voice inside their payloads.)

All of these payload types apply to the basic level of the SONET/SDH hierarchy. In addition, the ATM cell and packet over SONET/SDH payloads can be carried inside the larger payloads available on STS-3c/STM-1 and STS-12c/STM-4 frames and links and even higher in the case of POS.

VOICE AND SONET/SDH PAYLOADS

By now it is apparent that SONET/SDH is much more complex than either T-carrier or E-carrier when it comes to information-carrying capabilities. In the simpler T-carrier world, the structure and information content of an individual DS-0 or even an unchannelized DS-1 was totally up to the user or customer. Anything that fit into a 64 Kbps DS-0 channel (in some cases, the DS-0 was limited to 56 Kbps) popped out at the other end of the T-1 link. T-carrier expected voice in the DS-0, but anything else that "looked like voice" went right through the link. However, care was required in setting up T-1 links that carried computer data and not voice. A service provider could cross-connect voice, but it made no sense at all to try to reach inside a DS-0 carrying IP packets 8 bits at a time and "cross-connect" a piece of an IP packet.

However, since T-carrier and E-carrier expected to find voice inside their individual channels, there was nothing at all in the frame formats of either T-carrier or E-carrier to specifically identify the content of a T-1 or E-1 frame. Things are very different in SONET/SDH.

One crucial point needs to be made about "voice" and SONET/SDH. SONET/SDH is optimized for carrying voice in the form of 64 Kbps digital channels. Virtual tributaries and tributary units, which lead off this chapter, are in many ways the expected content of SONET/SDH payloads. But as has been pointed out earlier in this book: what is loosely called "voice" by the telephone companies is really four things:

1. Two people talking. This always comes to mind first when voice is mentioned.
2. Touch-tone signaling. Technically this is dual tone multi-frequency (DTMF) signaling and is done not only at the start of a conversation, but even during ("Press 1 for…").
3. Analog modem data. From Internet access to computer-to-computer communications, analog modems are still the workhorses of dial-up applications.
4. Fax. It is easy to forget that a large number of "voice" calls are made to and from fax machines. Fax is really just a special case of regular analog modems.

As shown in Chapter 6, equipment vendors and service providers alike got used to the "default" signal label of the path overhead indicating voice channels inside the SONET/SDH payload. But at least SONET/SDH allow for the possibility that payloads might not be digitized voice channels. And that makes all the difference.

This chapter begins by considering the structure of the SONET payload when used for voice channels. These "voice" channels might not actually be carrying speech, as pointed out already. The point is that this 64 Kbps bit stream *looks exactly like voice* to the service provider and can be cross-connected or otherwise processed without regard to the details of the channel bit stream structure.

SONET Virtual Tributaries

Important as virtual tributaries (VTs) are as SONET payload contents, there is a tendency to over-emphasize the details of VT operation in SONET, so discussions of SONET VTs tend to take over a chapter or even a whole book. This section will therefore necessarily be somewhat limited in the amount of detail offered (and still be very lengthy). All the essentials are here, but the operational details of things like VT alarms are not given exhaustive treatment. Most SONET equipment vendors are more than willing to supply details regarding how their particular SONET NE handles these issues. There are also slight differences from vendor to vendor. The interested reader is referred to the individual vendors in such cases.

The figures in this chapter will offer slightly different representations of the SONET frame seen in previous chapters. Since the topic in this chapter is the form and function of the SONET payload, the figures in most cases will always represent only the payload itself. Keep in mind that the payload envelope would still be offset or locked in position in the overall SONET frame structure and found by the value of the H1 and H2 pointer bytes. But this chapter will deal with a 9 row, 87 column synchronous payload envelope (SPE) structure, not the full frame. The first column will always be the path overhead (POH) column. So the total size of the SPE is 783 bytes (9 × 87), but the area available for information is 9 bytes less (the POH column), or 774 bytes. This is shown in Figure 8-1.

When the signal label path overhead byte (C2) is set to a value of 02x (02 in hex, or 0000 0010), then the SONET receiver knows that the arriving SPE must be carrying VTs. The first thing to realize is that the number of VTs present is not arbitrary. Each SPE must

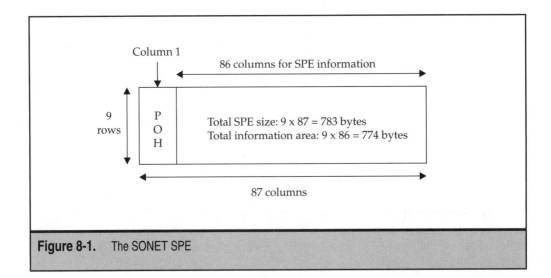

Figure 8-1. The SONET SPE

be divided into exactly seven *virtual tributary groups* (VTGs). This is the first major division of the SPE for VTs.

The columns assigned to the seven VTGs in the SPE mimic the pattern established for multiplexing three STS-1s into a single STS-3 frame. That is, multiplexing proceeds byte by byte from each source until an entire row of the frame is filled. The result is that the multiplexing process creates a column-by-column structure where each column belongs to a separate multiplexing source. This frame creation process was detailed in Chapter 6. An SPE consisting of seven multiplexed VTGs follows the same pattern. The intent is the same: to minimize the impact of delay on voice by interleaving sources instead of employing buffers.

The only complication is that 7 does not divide 86 columns equally: there are two columns left over. So 86 columns ÷ 7 VTGs = 12 columns for each VTG, with a remainder of two columns. The solution here is to "skip over" two columns when multiplexing the seven VTGs into an SPE. The two columns "reserved" (really, "ignored"—their content is defined as fixed stuff bytes that must be ignored by the receiver) are columns 30 and 59 of the SPE, counting the POH column as column 1. Once this interleaving process is defined, the structure of an STS-1 SPE containing seven VTGs can be represented as in Figure 8-2.

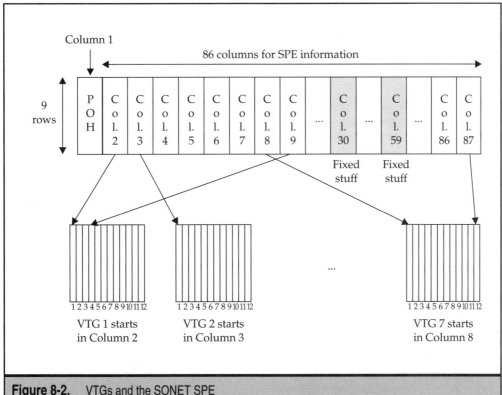

Figure 8-2.   VTGs and the SONET SPE

Now that this structure for the VTGs has been defined, it is a simple matter to generate a table showing the columns occupied by each of the VTGs, which are always numbered VTG 1 through VTG 7. This is shown in Table 8-1.

Several features of the table are noteworthy. First, the table starts with column 2 of the SPE. This just reflects the position of the POH column always as column 1 of the SPE. Next, the VTG column assignments skip over columns 30 and 59, columns that would normally be between VTG 7 and VTG 1.

It is important to realize that when an SPE is divided into seven VTGs, the SPE cannot carry information organized into anything *but* VTGs. Once the VTG columns are established inside the SPE, that is all that the SPE can be used for. This does *not* mean that the signal label cannot change with the arrival of the *next* SPE: signal labels are actually allowed to change, SPE by SPE. However, this feature of SONET was controversial when introduced and has never been used except experimentally. Certainly it makes no sense to flip-flop signal labels on an STS-1 link when the intention is to carry DS-0 voice. There is just no room for anything else on that link. The 8000 SONET frames (SPEs) generated each second will carry 8-bit voice samples in each column and row byte, yielding 64 Kbps, but no more.

Although this discussion of VTGs is limited to SONET in this section, it is worth mentioning that the SONET VTG structure also exists in SDH. Of course, the name and acronym must be different. So in SDH, a VTG is called a *tributary unit group-level 2*, or TUG-2.

So far, the process of dividing the SPE into VTGs has produced seven identical structures consisting of 12 columns each, spread across 84 columns of the SPE. So each VTG has exactly 108 bytes (9 bytes in a column, and 12 columns). The next step is to look inside each of the VTGs and see the structure of this basic 108-byte unit.

Inside a VTG

Just because all VTGs form identical 108-byte units at different places in the SPE does not mean that the content of all VTGs is the same. There are four different forms that individual VTGs can take on. Although VTGs are optimized for the transport of 64 Kbps voice

Virtual Tributary Group	SPE Columns Used by the VTG
1	2 9 16 23 31 38 45 52 60 67 74 81
2	3 10 17 24 32 39 46 52 61 68 75 82
3	4 11 18 25 33 40 47 54 62 69 76 83
4	5 12 19 26 34 41 48 55 63 70 77 84
5	6 13 20 27 35 42 49 56 64 71 78 85
6	7 14 21 28 36 43 50 57 65 72 79 86
7	8 15 22 29 37 44 51 58 66 73 80 87

Table 8-1.   SPE columns occupied by the VTGs

channels (DS-0s), not all DS-0s arrive at a SONET NE inside the same level of the T-carrier hierarchy. So the four possible VTG formats are:

▼ VT1.5, used for DS-1 transport (1.544 Mbps)

■ VT2, used for E-1 transport (2.048 Mbps)

■ VT3, used for DS-1C transport (3.152 Mbps)

▲ VT6, used for DS-2 transport (6.312 Mbps)

The only real surprise on the list is the definition of the VT2, used for E-1 transport. E-1 support is provided in SONET because whenever links cross a border beyond the United States, international standards apply. With a VT2, a service provider in another country can easily hand off an E-1 to a SONET NE and all the DS-0 channels inside will be carried to the end-user equipment. As an aside, a real problem before SONET was that 30 voice channels on an E-1 arriving in the United States had to be handed off to a DS-1 with only 24 voice channels! Either six channels could not be used (but were paid for) or a second DS-1 had to be purchased (but only six channels were used). SONET VT2s are a much better solution. (The preceding does not consider that the byte coding for U.S. voice channels (called μ-law) is different than in E-carrier (called A-law). Supposedly, μ-law is optimized for deeper tones of American English but A-law is better suited for an environment with many languages. So conversion must still be done somewhere.)

Of the four possible VTG structures, the most important is the VT1.5. Since there are many more DS-1s in service in North America than DS-1Cs or DS-2s, SONET links will most likely encounter DS-0s inside a DS-1 than in any other form. Even when T-carrier links arrive at a SONET NE in the form of a DS-3, chances are that the DS-3 is carrying exactly 28 DS-1s, each with 24 DS-0s. As it turns out, when an STS-1 is divided into seven VTGs and each VTG is structured as a VT1-5, the carrying capacity of the STS-1 is exactly the same as the capacity of a DS-3: 672 DS-0 channels running at 64 Kbps.

So a SONET SPE consisting of VT1.5s can accept and carry and cross-connect 28 separate DS-1s or a single DS-3 containing 28 DS-1s. This is no accident. As has been pointed out, the whole reason that the STS-1 level exists in SONET is for backward compatibility with existing DS-3 links (and the DS-1s most likely inside them). This allows an easy, gradual migration to an all-SONET network with flash cutovers and the like.

VT Structures

Each VTG is 108 bytes long (9 rows, 12 columns). However, the structure and carrying capacity of the VTG varies with VT type. In other words, a VTG containing DS-1s is fundamentally different from a VTG containing E-1s, and so on, for all four VT types. The carrying capacity of each VT type is determined by the frame structure of the T-carrier or E-carrier link feeding the SONET NE. It is probably easier to look at each VT type in turn by example rather than attempting to describe the structure in the abstract.

Consider VT1.5. It is designed to carry DS-1s. DS-1s consist of frames generated 8,000 times per second with a structure of 193 bits. The 24 DS-0s, each 8 bits, make 192 bits. The 193rd bit is the framing bit. The framing bit is important, but look at the 24 DS-0s by themselves. The 24 DS-0s do not divide the 108-byte VTG evenly. There would be 12 bytes left over ($108 \div 24 = 4$, with 12 bytes remainder). So four DS-1s would fit nicely inside a single VTG. But what about the leftover 12 bytes? Well, there are four DS-1s. The 12 bytes are distributed between the four DS-1s inside the VT1.5 so that each DS-1 gets 3 bytes of *VT overhead* (VTOH). What is this overhead for? The same types of things as SONET overhead: alarms, performance information, and other OAM&P functions.

So each DS-1 now requires 24 bytes of the VT1.5 for the DS-0 and gets another three bytes for overhead. That makes 27 bytes total. And since SPEs (and SONET frames) are nine rows, the DS-1 fits nicely in three columns of the 12-column VTG! Further, four DS-1s of three columns each precisely fill the 12-column VTG. So the structure of a VTG according to VT1.5 looks like Figure 8-3.

In the figure, each column of the VTG is labeled A, B, C, and D. These represent the four DS-1s inside the VTG. The multiplexing of the DS-1s inside the VTG is done column by column, as expected and in keeping with the general SONET/SDH column-wise multiplexing approach.

Above the VTG, the columns that comprise each VT1.5 are given under their respective representations. The first VT1.5 (labeled VT1.5 #1 [A]), has arrows pointing to the A columns of the VTG. The other three VT1.5s do not have arrows, but only to keep the clutter in the figure to a minimum.

The key point is that each VTG structured according to VT1.5 carries four interleaved DS-1s. If each VTG is structured according to VT1.5, then the total DS-1 capacity of the SONET STS-1 SPE is 28 DS-1s (4×7). It is no accident that this is also the number of DS-1s inside a DS-3. Notice that the DS-3 *frame structure* is not carried directly inside the SONET SPE. (Recall that DS-3 uses bitwise multiplexing, not byte-multiplexing, as does SONET/SDH.)

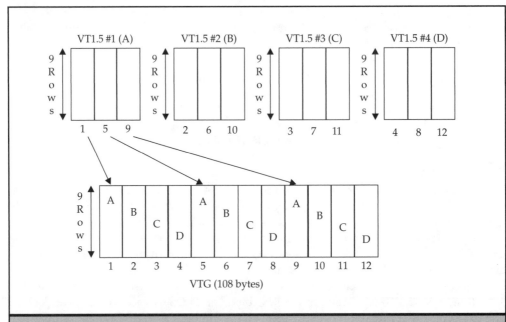

Figure 8-3. A VTG structured according to VT1.5

Rather, the 28 DS-1s inside the DS-3 frame are extracted by the SONET NE and then repackaged as 28 DS-1s inside seven VTGs.

The other three VT types' carrying capacities can be determined in the same way by considering the size of the frame each VTG must carry. For VT2, the E-1 frame has 30 DS-0s, plus one byte for synchronization and one byte for signaling. These 32 bytes do not divide the 108-byte VTG evenly. There can be three E-1s inside the VTG and, once again, 12 bytes left over ($3 \times 32 = 96$). These 12 bytes form VT overhead as before, but in the case of VT2s, the 12 VTOH bytes are distributed among three E-1s, not four DS-1s. So the number of overhead bytes allotted to each E-1 is four, not three. When all seven VTGs are carrying VT2s, the carrying capacity of the SONET SPE is 21 E-1s (3×7).

A VT3 is designed to carry a DS-1C. A DS-1C is basically two DS-1s pasted together with some additional overhead bits. So the DS-1C frame structure is twice as large as the DS-1 frame and carries 48 DS-0s. So a VTG can carry only two DS-1Cs, again with 12 bytes left over for VTOH ($48 \times 2 = 96$). The number of overhead bytes allotted to each DS-1C is now six. When all seven VTGs are carrying VT3s, the carrying capacity of the SONET SPE is 14 DS-1Cs (2×7). Note that this is half the DS-1 capacity, but each DS-1C carries twice as many DS-0s, so the overall channel capacity is still the same (672 DS-0 channels).

A VT6 is for DS-2. The frame structure of a DS-2 is rather complex, but the bit rate of a DS-2 is 6.312 Mbps and a DS-2 frame carries 96 DS-0s. These 96 bytes fit inside the 108-byte VTG nicely, with the same 12 bytes left over for VTOH. This time, all the VTOH goes for the single DS-2 inside the VTG. When all seven VTGs are carrying VT6s, the carrying capacity of the SONET SPE is 7 DS-2s (1×7).

All of this is most easily expressed in a table. Table 8-2 lists all of the important parameters of the four VT types.

In the table, the numbers in the "Columns per VT" and "Number in VTG" must multiply out to 12, which is the number of columns in each VTG. The "Maximum Number in SPE" column applies when all seven VTGs carry that type of VT, as does the "Maximum DS-0s" entry.

The VT structures of VT2, VT3, and VT6 are shown in Figure 8-4. This figure follows the same conventions used in Figure 8-3.

Before looking at each VT type in a little more detail, there are a few rules that apply to VTGs and the VT types they contain that should be examined first. The first rule has al-

VT Type	Tributary	Columns per VT	Number in VTG	Maximum number in SPE	Maximum DS-0s
VT1.5	DS-1	3	4	28	672
VT2	E-1	4	3	21	630
VT3	DS-1C	6	2	14	672
VT6	DS-2	12	1	7	672

Table 8-2. SONET virtual tributaries

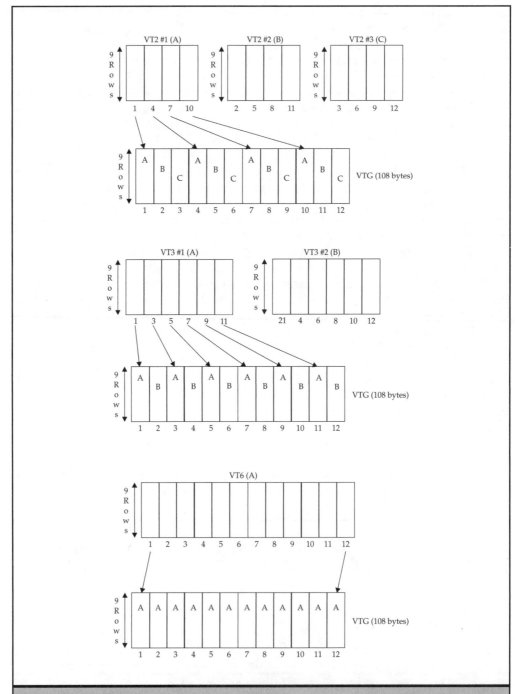

Figure 8-4. VT2, VT3, and VT6

ready been stated: when a SONET SPE is divided into VTGs, there are exactly seven VTGs, as detailed in Table 8-1. The two rules that apply to VTs themselves are given as follows.

Each VTG can have a different VT type. That is, VTG 1 can be structured according to VT1.5, VTG 2 structured according to VT6, and so on through all seven VTGs. Although this mixing of VT types is allowed, in practice this is seldom done.

Each individual VTG can contain only one VT type (VT1.5, VT2, VT3, or VT6). That is, if a given VTG is structured according to VT1.5, then no other type of VT can be mixed into that particular VTG, either in that frame or in subsequent SPEs. VT types are an all-or-nothing proposition. VT types are configured at each end of the SONET link to be one type. There is no real justification for this limitation, besides a wish not to make life too hard for SONET NEs. There *is* a VT signal label in each VTG, but this signal label is not used for the same purpose as the POH signal label (C2). A further problem is that the basic unit of the VTG is *not* a single SPE. The basic unit of the VTG is a sequence of four SONET SPEs, called a VT superframe. Allowing changes of VT types would require careful handling of VT superframes. So each individual VTG must contain only one VT type, as identified by the configuration of the SONET NEs.

But each VTG can carry a *different* VT type, if that is desired. Figure 8-5 is a summary figure that shows all of the ideas presented already. The STS-1 frame payload at the bottom has 87 columns. At the top, each of the first four VTGs is of a different type, and the columns are now shown as A-B-C-D for the VT1.5, X-Y-Z for the VT2, M-N for the VT3, and Q for the VT6. This makes it easier to show just where each column loads into the STS-1 SPE. Arrows point to some of the columns.

It is tempting to look at the 108-byte VT1.5 structure, for instance, and decide that each and every SPE should carry a single DS-1 frame locked into position so that each DS-0 always appears at the same place in the VT1.5. This would make cross-connecting the DS-0s very easy—as easy as finding a DS-0 inside a DS-1. This is what *payload visibility* in SONET is all about.

This *locked mode* method of operation was allowed in SONET, but is no longer supported. VTs must be allowed to float within the SPE, in the same way that SPEs float inside their SONET frames and are found with the H1/H2 pointer bytes. The issue with locked VT mode was that the arriving DS-1 clocks varied so much that processing, buffering, and loading locked VTs proved to be very troublesome and more effort than could be easily justified. So *floating mode* is the rule today. (The POH signal label [C2] value of 03x [03 in hex, or 0000 0011 in binary] is still defined, but only for backward-compatibility to prevent future mappings from using the 03x value and thus conflicting with locked mode legacy equipment.)

However, if VTs are allowed to float inside their respective VTGs, a whole system of pointers and other forms of VTOH had to be specified. This is exactly what was done, especially for the key VT1.5 type. Note that locked mode offered little advantage over floating mode and required a lot of extra work, since the floating mode VTOH had to be supported in SONET NEs handling VTs anyway.

So each VT type was given one or more special *mappings* to determine the form and function of the VTOH in each case. The key component of the VT mappings is the unit in SONET SPEs known as the *VT superframe*.

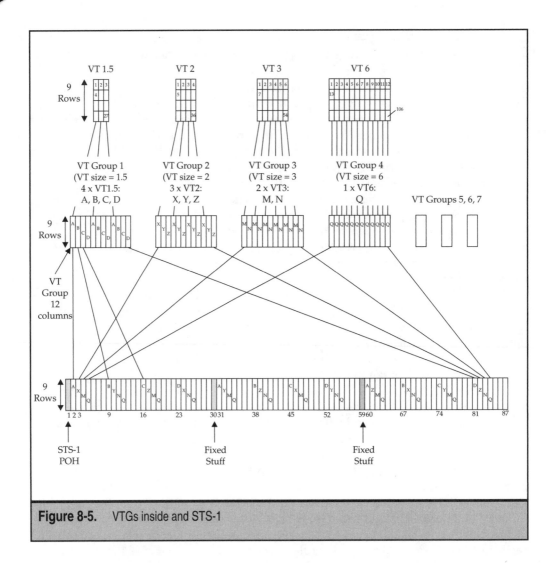

Figure 8-5. VTGs inside and STS-1

VT1.5 Superframes

In this section, the basic unit under discussion is the VT itself. In other words, each of the VT1.5s (for example) in Figure 8-3 will be presented as if it were a unit in and of itself. There is nothing wrong with thinking of the VT1.5 like this, but it should be kept in mind that things are really a lot more complicated. For instance, consider the VT1.5 in Figure 8-3 labeled "VT1.5 #3 (C)." Note that the three columns of this VT1.5 actually occupy columns 3, 7, and 11 of the VTG that the VT1.5 is configured in. And suppose that this is the 5th VTG of the SPE. A look a Table 8-1 will show that this VTG occupies columns 6, 13, 20, and so on, of the SPE itself. Not only are the VT1.5 columns spread out, but so are the VTG columns, as shown in Figure 8-5. Where in the overall 87 column SPE is VT1.5 #3 of VTG #5? Table 8-2 shows this.

As it turns out, once all of this reverse mapping is done, VT1.5 #3 of VTG #5 occupies columns 20, 49, and 78 of the overall SPE. Table 8-3 shows the location of all 28 DS-1s inside the SPE. Note that this table is completely valid only when all seven VTGs are VT1.5s. The

DS-1 No.	VTG no.	VT No. (letter)	VTG Columns	SPE Columns
1	1	1(A)	1,5,9	2,31,60
2	2	1(A)	1,5,9	3,32,61
3	3	1(A)	1,5,9	4,33,62
4	4	1(A)	1,5,9	5,34,63
5	5	1(A)	1,5,9	6,35,64
6	6	1(A)	1,5,9	7,36,65
7	7	1(A)	1,5,9	8,37,66
8	1	2(B)	2,6,10	9,38,67
9	2	2(B)	2,6,10	10,39,68
10	3	2(B)	2,6,10	11,40,69
11	4	2(B)	2,6,10	12,41,70
12	5	2(B)	2,6,10	13,42,71
13	6	2(B)	2,6,10	14,43,72
14	7	2(B)	2,6,10	15,44,73
15	1	3(C)	3,7,11	16,45,74
16	2	3(C)	3,7,11	17,46,75
17	3	3(C)	3,7,11	18,47,76
18	4	3(C)	3,7,11	19,48,77
19	5	3(C)	3,7,11	20,49,78
20	6	3(C)	3,7,11	21,50,79
21	7	3(C)	3,7,11	22,51,80
22	1	4(D)	4,8,12	23,52,81
23	2	4(D)	4,8,12	24,53,82
24	3	4(D)	4,8,12	25,54,83
25	4	4(D)	4,8,12	26,55,84
26	5	4(D)	4,8,12	27,56,85
27	6	4(D)	4,8,12	28,57,86
28	7	4(D)	4,8,12	29,58,87

Table 8-3. SONET VT1-5 mapping into an STS-1

only complexity in constructing such a table is that the DS-1s are organized and numbered by VTG order, not VT order. So all of the first VT1.5s (the A's) in all seven VTGs load first into the SPE. The VT1.5 #2s (B) follow, and so forth. In other words, the DS-1 numbering is consecutive by SPE position, not VT position. To find the 15th DS-1 inside a VT1.5-structured SPE (perhaps in order to cross-connect the whole DS-1 or a DS-0 inside), all a SONET NE need do is extract the 16th, 45th, and 74th columns of the SPE. Naturally, other VTG structures locate other VTs in the SPE using other rules.

The whole point of the exercise is to help you realize that even when a VT1.5 superframe is treated as a unit in this section, 12 columns of the SPE are actually involved.

The VT superframe is defined as four consecutive SPEs. Since each SPE arrives in 125 μseconds, the entire VT superframe repeats every 500 μseconds. The SPEs are numbered 00, 01, 10, and 11 in binary to help receivers determine the VT superframe boundaries. All VT superframes begin with SPE 00 and end with SPE 11. The two bits involved are in the H4 byte of the SPE POH and are called the *multiframe pointer* (some SONET documentation calls the VT superframe the *VT multiframe*).

When structured according to VT1.5, each of the four VT1.5s in the VTG occupies three columns and is therefore composed of 27 bytes. This is just another way of saying that there are four DS-1s in the VT1.5, and each VT1.5 gets three VTOH bytes per SPE. The first VTOH byte in each of the four SPEs making up the VT superframe is special. These form the V1, V2, V3, and V4 VT OH bytes. They are found by the value of the H4 multiframe pointer in the POH of the *previous* SPE. So a 00 value in the H4 POH byte of an SPE says, essentially, "get ready for the V1 VTOH byte in the *next* SPE."

Subtracting the V1, V2, V3, or V4 VTOH bytes from the 27-byte VT1.5 leaves 26 bytes per SPE. Now, 24 of these bytes are for the DS-0 channels inside the VT1.5, so this leaves two other bytes per SPE for other functions. Since there are four SPEs in a VT superframe, a total of 8 bytes in the VT1.5 form additional overhead. One of these is the V5 VTOH byte. The position of the V5 byte can be anywhere in the VT1.5 superframe, except for the initial byte position reserved for the V1, V2, V3, or V4 bytes in each SPE.

What are the functions these five V overhead bytes introduced so far? The V1 and V2 bytes perform many of the same tasks for floating VTs as the H1 and H2 bytes in the LOH do for floating SPEs. For example, some of the bits in the V1 and V2 bytes form a pointer to the V5 byte, which is the VT path overhead byte. The V5 byte always determines the beginning of a sequence of four DS-1 frames spread over the VT1.5 superframe. The V3 byte helps the receiver move the location of the V5 VT path overhead byte, most likely because of timing variations between the arriving DS-1 clock and the SONET clocks. The V3 byte works much like the H3 byte in this regard. More details about these timing considerations will be explored in the next chapter. The V4 byte is reserved and sometimes called the PTR-X byte. For now, it is enough to realize that the V1/V2 bytes point to the V5 byte in the VT superframe.

The relationship between the SPE POH H4 multiframe pointer and the VT1.5 superframe V1 through V5 VTOH bytes is shown in Figure 8-6. The position of the V5 byte in the figure is totally arbitrary and for illustration only. The byte immediately following the V3 byte, marked with an *, is the *positive stuff VT byte* and performs in ways similar to the location following the H3 byte in the SONET line overhead, discussed in the next chapter.

So the 108-byte VT1.5 superframe spans four SPEs. It is important to realize that the size of the VTG, also 108 bytes, is just a coincidence. A 108-byte VTG is inside a single SPE.

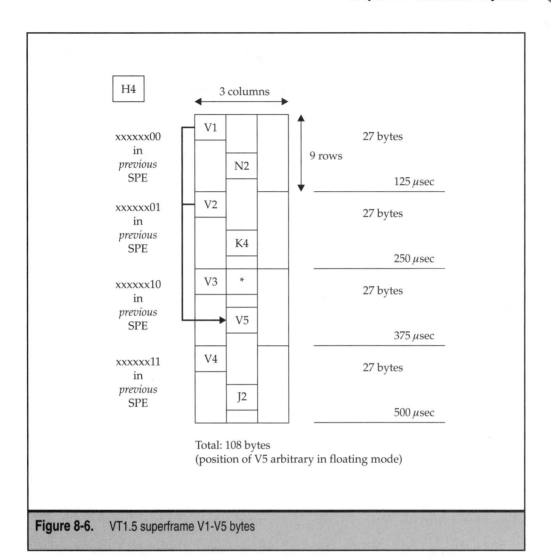

Figure 8-6.   VT1.5 superframe V1-V5 bytes

VT1.5 superframes span four SPEs. Once the four V bytes are subtracted, the VT1.5 has 104 bytes to carry four frames of the DS-1. In four frames the 24 DS-0s inside a DS-1 generate 96 bytes. This leaves eight bytes for additional overhead.

Once the V5 VTOH byte is located by the V1/V2 pointer field, three other bytes of VTOH can be located. These are the J2, N2(Z6), and K4(Z7) VTOH bytes. These bytes were discussed at the end of the previous chapter. By definition, once the V5 byte is located, the positions of the J2, N2(Z6), and K4(Z7) bytes are exactly 125 μseconds away. In other words, if the V5 byte is found as the 17th byte of a 27-byte VT1.5 frame within the VT1.5 superframe in an SPE, then the J2 byte is the 17th byte in the following SPE. The N2(Z6) and K4(Z7) byte positions are determined the same way. This is also shown in Figure 8-6.

It is important to realize that this overall VT superframe structure, consisting of four SPEs and the V1 through V5, J2, N2(Z6), and K4(Z7) VTOH bytes, is exactly the same for each of the four VT types. However, only the VT1.5 superframe has a 108-byte structure. Once the eight VTOH bytes are subtracted, there are only 25 bytes per SPE to carry the 24 DS-0s for each VT1.5 and one additional byte that is used mainly for carrying the DS-1 framing bit.

A VT2, which has three E-1s in a VTG, needs four columns, or 36 bytes, for each E-1. So VT2 superframes are $4 \times 36 = 144$ bytes long. Subtracting the eight VTOH bytes leaves 136 bytes split among the four—SPE VT superframe sequence, giving 34 bytes per SPE for the 32 DS-0s. This is called the *payload capacity* of the VT superframe. For a VT3, the numbers are 6 columns and 54 bytes for each DS-1C. The VT3 superframe is 216 bytes, with 208 bytes of payload, or 52 bytes per SPE. For a VT6, the numbers are 12 columns and 108 bytes for each DS-2. The VT6 superframe is 432 bytes, with 424 bytes of payload, or 106 bytes per SPE. This comparison in VT superframe size is shown in Table 8-4.

To finish the discussion of VT1.5 superframes, all that is necessary is to map the DS-0 around the 8 VTOH bytes inside the VT superframe. However, there are two ways to accomplish this for VT1.5s. The two ways are *byte-synchronous mapping* and *asynchronous mapping*. Either method may be used and both are equally standardized. It all depends on what needs to be done to the DS-0s inside the DS-1s riding the VT1.5. If the DS-0 channels need to be found in the SONET NE, probably for cross-connection purposes, then byte-synchronous mapping is indicated. In byte-synchronous mapping, the DS-1 framing bit is carried in the VT1.5, as are the signaling bits giving status information on each DS-0. In byte-synchronous mapping, each and every DS-0 inside each and every VT1.5 in each and every VTG can be found and extracted directly from the SONET frame. This is DS-0 payload visibility. Today, asynchronous mapping is more common.

With asynchronous mapping, the channel structure of the DS-1 is neither visible nor preserved. The DS-1 framing bit is not transported through the SONET portion of the DS-1 link, and DS-1s are transported as just 192-bit units that might span DS-1 frame boundaries (or not). When asynchronous mapping is used, SONET NEs can extract and cross-connect DS-1s, but not the individual DS-0 channels inside them. Of course, the extracted DS-1 could

SPE Sequence	VT1.5 Payload	VT2 Payload	VT3 Payload	VT6 Payload
1 (00 in previous H4)	25	34	52	106
2 (01 in previous H4)	25	34	52	106
3 (10 in previous H4)	25	34	52	106
4 (11 in previous H4)	25	34	52	106
Total payload capacity, less 8 VTOH bytes:	100	136	208	424

Table 8-4. VT payload capacity

always be sent to a separate DS-1 and DS-0 cross-connect for further processing. In fact, this was one reason that asynchronous mappings were invented in the first place: why force SONET NEs to perform a function that legacy equipment still does quite well?

The V5 byte starts off the *VT1.5 SPE* (distinct from the STS-1 SPE, of course). The VT1.5 SPE is 104 bytes long (108 bytes for the VT1.5 superframe, less the V1-V4 VTOH bytes). The structure of the 104-byte VT1.5 SPE for both byte-synchronous and asynchronous mapping is shown in Figure 8-7.

In Figure 8-7, the VT1.5 itself is represented as a *linear* array 104 bytes long instead of as three columns, but this should cause no confusion. The V1–V4 bytes are shown, but only to emphasize the fact that the VT1.5 unit is now 104 bytes long, not 108 bytes. The J2, N2, and K4 bytes (the Z6/Z7 ANSI legacy designation for N2/K4 is now dropped) are shown in relation to the V5 byte, not the V1–V4 bytes. The notes identify the function of all the overhead bits in both mappings. The asynchronous mapping does not carry the DS-1 framing bit in the overhead. The framing bit is included in the VT1.5 bit stream, of course, but the SONET NE cannot find the frame boundaries directly.

Although byte-synchronous mapping offers DS-0 payload visibility directly to SONET NEs, asynchronous mapping is much more common. This is because cross-connecting at the DS-0 level is usually done by a separate piece of equipment, so cross-connecting is not really all that essential a function of SONET NEs. Also, unchannelized DS-0s, which consist not of 24 channels running at 64 Kbps but of a single bit stream running at 1.544 Mbps, cannot be cross-connected anyway. The more unchannelized DS-0s there are, the less sense byte-synchronous mappings make. Byte-synchronous mapping renders the DS-0 channels visible, while asynchronous mapping renders the DS-1 frame visible (but only by examining the VT1.5 bit stream), and this is all that is needed to add and drop DS-1s from the SONET stream. (On a more technical level, byte-synchronous mapping severely limits the ability of SONET clocks to interface properly with DS-1 clocks, due to the requirement to align the DS-0s into position.)

Some older SONET documentation also lists the *bit-synchronous mapping* for VT1.5 and other VT types. Bit-synchronous mapping is just a special case of the asynchronous mapping and uses the same VT1.5 structure as the asynchronous mapping. The only difference is how the stuffing bits (S bits) are used. The mapping type used is determined by the value of the VT signal label field in the V5 byte of the VT superframe. The VT signal label field is a three-bit field.

A couple of other points should be made about DS-1s and VT1.5. All of the VT1.5 mappings will work with either D4 superframe or ESF DS- 1s. All of them will work with either bipolar AMI or the B8ZS T-1 line codes as well. The form of the DS-1 transported inside the VT1.5 does not matter. The transport of the DS-1 signaling bits is not considered here, but is discussed at length in the specifications.

So in summary: byte-synchronous mapping lets SONET NEs see the DS-0s, asynchronous mapping lets SONET NEs see a stream of bits, and bit-synchronous mapping lets the SONET NEs see a slightly different stream of bits. Any of the three mappings can be used in VT1.5. It all depends on what the SONET NE needs to do with the DS-1s.

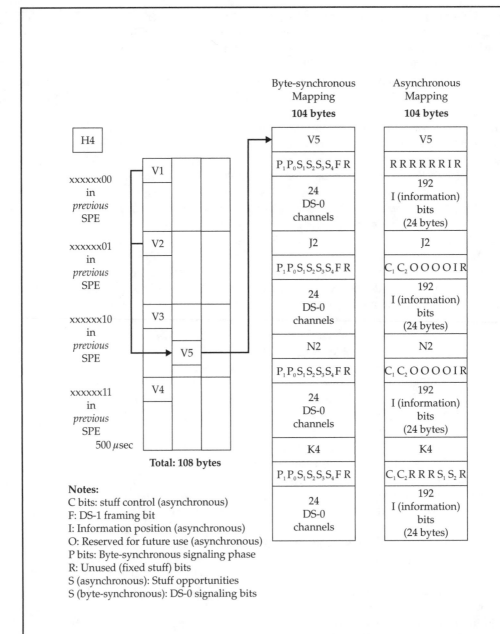

Figure 8-7.   VT1.5 byte-synchronous and asynchronous mappings

Mappings for Other VT Types

Just as VT1.5 has a distinctive format for the VT superframe, the other VT types also have distinctive VT superframe formats. There are also mappings established for the other three VT types (VT2, VT3, and VT6). It is not necessary here to detail all three to the degree afforded VT1.5, if only because this chapter would then be very, very long. A few words will suffice for VT3 and VT6. More details will be explored for VT2s, but only because E-1 transport is the way in which international circuits must be supported.

VT3s and VT6s have only asynchronous mappings. That is, only the overall DS-1C (VT3) and DS-2 (VT6) frames can be found and delimited by the SONET NE. VT2s, like VT1.5s, define three types of mappings: byte-synchronous, asynchronous, and bit-synchronous (no longer supported). The type of mapping employed in the VT is indicated by the value of the three bit signal label field of the V5 byte. The defined values of the signal label field by VT type are shown in Table 8-5. The bit synchronous mappings are still there, but for backward compatibility only.

Only the VT2 mappings, intended for E-1 links, running at 2.048 Mbps and carrying 30 DS-0 channels, require some additional comment. First of all, a SONET SPE carrying all VT2s in the seven VTGs transports 21 E-1s (3 VT2s × 7 VTGs) and thus 630 DS-0 channels. The positions of the E-1 columns in the SPE columns is shown in Table 8-6, which might be compared to Table 8-3 for the DS-1 equivalents. Note the absence of SPE columns 30 and 59 in the table. These are always fixed stuff columns and cannot be used in any VT type.

The VT payload capacity of the VT2 is not 108 bytes (27 bytes in 4 SPE frames) as are VT1.5s. Because a VTG structured according to VT2 contains only three E-1s instead of four DS-1s, the columns unit is four columns, not three as in VT1.5. So the basic unit for an E-1 is 36 bytes (nine rows, four columns) and the VT2 superframe is 144 bytes long. Naturally, there are the usual V1–V4 bytes, and it is still the position of the V5 bytes that is pointed to by the V1/V2 field. The structure of the VT2 superframe is shown in Figure 8-8, which can be compared to Figure 8-5 for VT1.5. As in Figure 8-6, the position of the V5 byte is determined by

Value	VT1.5	VT2	VT3	VT6
000	Unequipped	(not used for live traffic)		
001	Equipped for nonspecific payload			
010	Asynchronous DS-1	Asynchronous E-1	Asynchronous DS-1C	Asynchronous DS-2
011	Bit-synchronous DS-1	Bit-synchronous E-1		
100	Byte synchronous DS-1	Byte-synchronous E-1		

Table 8-5. VT signal labels

DS-1 No.	VTG no.	VT No. (letter)	VTG Columns	SPE Columns
1	1	1(A)	1,4,7,10	2,23,45,67
2	2	1(A)	1,4,7,10	3,24,46,68
3	3	1(A)	1,4,7,10	4,25,47,69
4	4	1(A)	1,4,7,10	5,26,48,70
5	5	1(A)	1,4,7,10	6,27,49,71
6	6	1(A)	1,4,7,10	7,28,50,72
7	7	1(A)	1,4,7,10	8,29,51,73
8	1	2(B)	2,5,8,11	9,30,52,74
9	2	2(B)	2,5,8,11	10,31,53,75
10	3	2(B)	2,5,8,11	11,32,54,76
11	4	2(B)	2,5,8,11	12,33,55,77
12	5	2(B)	2,5,8,11	13,34,56,78
13	6	2(B)	2,5,8,11	14,35,57,79
14	7	2(B)	2,5,8,11	15,36,58,80
15	1	3(C)	3,6,9,12	16,37,59,81
16	2	3(C)	3,6,9,12	17,38,60,82
17	3	3(C)	3,6,9,12	18,39,61,83
18	4	3(C)	3,6,9,12	19,40,62,84
19	5	3(C)	3,6,9,12	20,41,63,85
20	6	3(C)	3,6,9,12	21,42,64,86
21	7	3(C)	3,6,9,12	22,43,65,87

Table 8-6. E-1s inside a VT2 SPE

the value of the V1/V2 field pointer. The position of the J2, N2, and K4 bytes is determined by the position of the V5 byte in the VT2 superframe: each follows 125 µseconds after the other.

In order to complete this brief look at VT2s, all that is needed is to present the byte-synchronous and asynchronous mappings for VT2s. These mappings are shown in Figure 8-9, which can be compared to Figure 8-7 for VT1.5.

Figure 8-9 assumes that the byte-synchronous mapping employ channel-associated signaling (CAS), in which the E-1 frame byte between DS-0 channels 15 and 16 is used for signaling the status of each channel. There is also a special case of the asynchronous VT2 mapping for bit-synchronous operation.

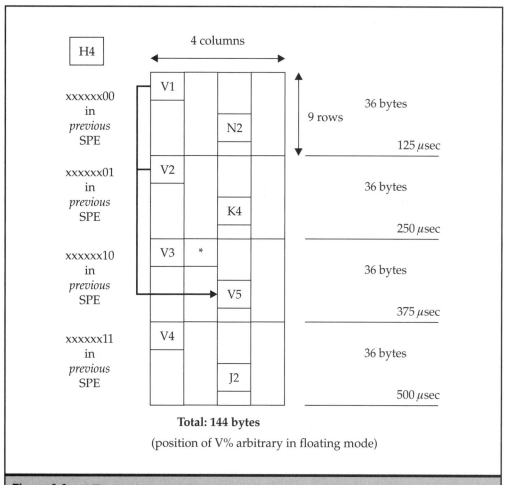

H4

4 columns

xxxxxx00
in
previous
SPE

V1

N2

36 bytes

9 rows

125 μsec

xxxxxx01
in
previous
SPE

V2

K4

36 bytes

250 μsec

xxxxxx10
in
previous
SPE

V3   *

V5

36 bytes

375 μsec

xxxxxx11
in
previous
SPE

V4

J2

36 bytes

500 μsec

Total: 144 bytes

(position of V% arbitrary in floating mode)

Figure 8-8.   VT2 superframe V1-V5 bytes

SDH Tributary Units

There is no place where SDH terms and functions differ so radically from SONET as in the area of SDH tributary units. SDH payloads use radically different terms and SDH builds payloads and frames differently than SONET. SDH distinguishes every phase of the operation with a unique term to describe the operation, and considers mapping (adding overhead), aligning (adding pointers) and multiplexing (combining units) as different things. This has already been pointed out, but the SDH "road map" of STM-n content should be seen in full to be appreciated.

Figure 8-10 shows the official ways that SDH STM-n content can be structured. Note the different arrow types for mapping, aligning, and multiplexing. All of the SDH terms

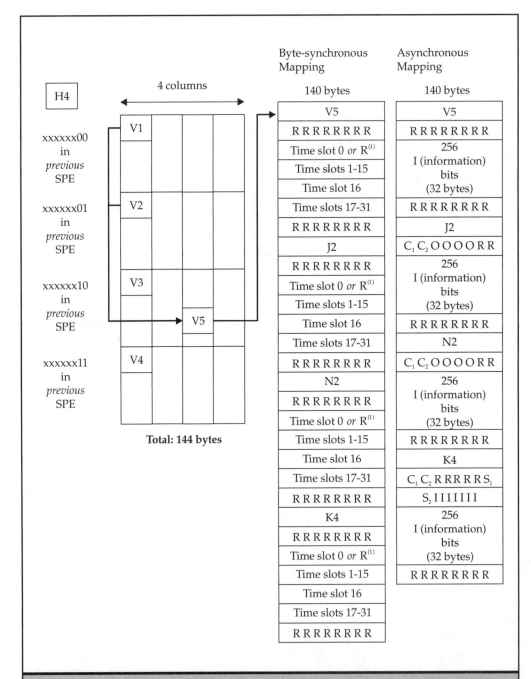

Figure 8-9.   VT2 byte-synchronous and asynchronous mappings

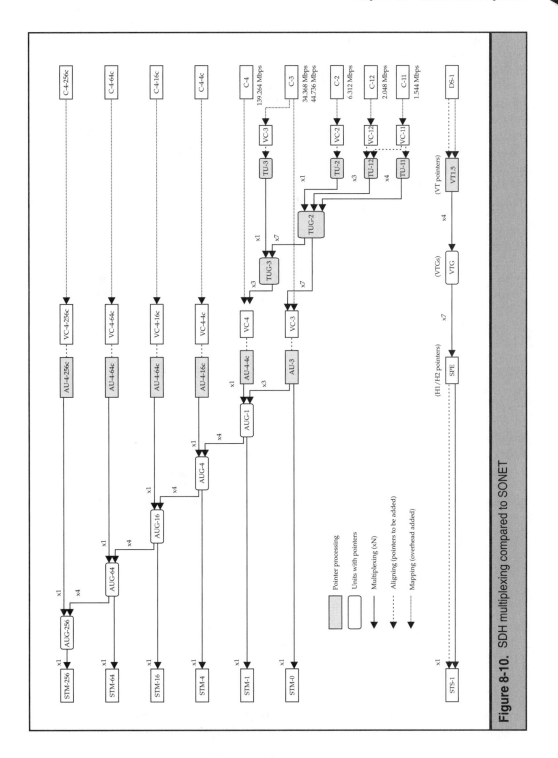

Figure 8-10. SDH multiplexing compared to SONET

will be defined in this section, but the figure also shows (completely un-officially) how a SONET frame is built, but by using some of the terms and conventions of SDH. Note that SONET uses fewer steps, and combines many arrow types.

SDH uses the term *containers* for payload content types, such as a SONET DS-1. Higher rate concatenated containers fill the STM-*n* completely. When multiplexed, everything of interest in this section is packed into the basic STM-1 package. This happens to the C-11 (read as C-one-one, not C-eleven, at the DS-1 rate of 1.544 Mbps), C-12 (read as C-one-two, not C-twelve, at the E-1 rate of 2.048 Mbps), C-2 (6.312 Mbps DS-2), and C-3 (34.368 Mbps E-3 or 44.736 Mbps DS-3). Overhead added to these form the *virtual container* (VC). Adding SONET-type VT pointers creates the *tributary unit* (TU, also read as TU-one-one and TU-one-two where appropriate). In SONET, this would be the VT1.5.

SDH next forms the *tributary unit group* (TUG). In the SONET example, this would be the VTG with four VT1.5s. TUGs become other types of virtual containers (also VCs) by adding path overhead, which in SONET would be the SPE. When the pointers (H1/H2 this time) are added, this forms the *administrative unit* (AU), gathered when necessary into *administrative unit groups* (AUGs). SONET puts the SPE into the frame. Note how the SONET "path" through the SDH process combines units and arrows into consolidated steps.

Table 8-7 shows how some of these SDH terms more or less equate to SONET terms.

Some terms are exact. For example, a C-11 *is* a DS-1. But the important point is that the TUG-2 is closest to the SONET VTG and the TUG-3 is pretty much an incomplete STS-1.

Why is SDH seemingly so complex? SONET was made specifically for the U.S. environment. SDH must apply globally, and SDH TUs must be able to carry all of the tributary types from the plesiochronous digital hierarchy (PDH) shown in Table 8-8 as SDH containers.

For European systems, 32 digitized voice signals (or equivalent signals) at 64 Kbps are multiplexed into a 2.048 Mbps digital signal, known as a CEPT-1 or E-1 signal. Four E-1 signals are multiplexed into an E-2 signal operating at 8.448 Mbps. Four 8.448 Mbps signals form a 34.368 Mbps E-3 signal. Finally, an E-4 operating at 139.264 Mbps consists of four E-3s.

For Japan, the DS-1 and DS-2 signals are the same as those in North America. However, five DS-2 signals are multiplexed into a 32.064 Mbps signal. Three of these 32.064 Mbps signals form a 97.728 Mbps signal. It should be noted that for SDH purposes, Japanese digital signals may carry E-1 or E-4 signals for ease of implementation.

SDH term	Closest thing in SONET
AU	SPE
TUG-2	VTG
TUG-3	STS-1
VT1.5	C-11
VT2	C-12
VT6	C-2

Table 8-7.   SDH and SONET terms

Digital Level	North America	Japan	Europe	ISDN	SDH Signal
0	64 Kbps	64 Kbps	64 Kbps	—	—
1	1.544 Mbps	1.544 Mbps	—	H-11	C-11
	—	—	2.048 Mbps	H12	C-12
2	6.312 Mbps	6.312 Mbps	—	—	C-2
	—	—	8.448 Mbps	—	—
3	—	32.064 Mbps	—	—	—
	—	—	34.368 Mbps	H31	C-3
	44.736 Mbps	—	—	H32	C-3
4	—	95.728 Mbps	—	—	—
	—	—	139.264 Mbps	H4	C-4
	(274.176 Mbps)	—	—	—	—

Table 8-8.   SDH containers

SONET was really intended only for the North American PDH (T-carrier) hierarchy. SDH, however, must accommodate all three regional standards, including all ISDN signals. Otherwise, international use may become a nightmare (as it is now with only PDH). For this express purpose, SDH has defined several containers of various types. These are designated as "C–n", where n = 11, 12, 2, 3, or 4, as shown in Table 8-8. These are intended for all the various PDH signals. These containers are described in detail later in this section. It should also be noted that ISDN H11 signal is best known as the *primary rate ISDN signal*. This has a format of 23B + D (B for information *bearer* channel and D for *data* communication channel.) Both types of channels have the same rate of 64 Kbps in North America. The counterpart of H11 in ITU-T digital networks is known as N12.

The C-11 is designed to carry a 1.544 Mbps DS-1 signal or an H11 ISDN signal. A C-12 is designed to carry a 2.048 Mbps E-1 signal or an H12 ISDN signal. There were initially two types of second-level container C-2 signals (C-21 for transporting a DS-2 signal and C-22 for transporting an E-2 signal). However, the lack of common E-2 deployments has diminished the need for C-2 containers. Thus, current SDH standards define only one type of second-level container, the C-2 for transporting a DS-2 signal. Confusingly, it is now merely called a C-2, and not a C-21. This leads to an interesting problem: What is an E-2 on SDH to be called? The designation "P-22" can be used to indicate a PDH level 2 and type 2 signal and used to transport an E-2 signal, if ever the need arises. But this is not standardized. These containers described, C-11, C-12 and C-2 are known in SDH as the *lower order tributaries*.

At the third level of the PDH hierarchy, another interesting discussion arose with regard to SDH. Obviously, an STM-1 operating at 155.52 Mbps can easily handle four E-3s operating at 34.368 Mbps each; however, the STM-1 line rate is the same as the STS-3 in SONET. An STS-3 also only carried three DS-3s, the third level of the T-carrier digital hierarchy. For the sake of consistency, it was decided to carry only three E-3s in an STM-1, just as three DS-3 were carried in an STS-3. Therefore, where two types of third-level containers were defined—C-31 for an E-3 signal and C-32 for a DS-3 signal—there is now just C-3 for both. An SDH STM-1 carrying three containers (now just C-3s) can be used to transport either three E-3 signals or three DS-3 signals.

As far as the fourth level of the PDH hierarchy is concerned, the "DS-4" signal that was proposed for the North American digital hierarchy was never widely implemented. This was mainly due to the rapid deployment of higher-speed fiber optical communications systems. Thus DS-4 never got the chance to be deployed in North American digital networks except for a very few isolated locations. For these reasons, in SDH standards, it was decided to have only one type of fourth-level container, C4. This is used to transport an E-4 or an H4-ISDN signal. Note that there is no provision for any "DS-4" container. These containers, C-3 and C-4, are known in SDH as the *higher order tributaries.*

It is important to realize that STM-1 SDH frames can carry multiple DS-1s just like an STS-1 in SONET, just more of them. These are sometimes called "DS-1As" to reflect the fact that A-law digital coding is used to digitize the voice, not μ-law digital coding as is used in T-carrier. When an STM-1 is structured in this fashion to carry virtual tributaries, called virtual containers (VCs) in SDH, the STM-1 pointers have a particular structure that conforms to what is called an administrative unit-type 3, or AU-3. When an STM-1 frame does not carry VCs and is used for traffic like ATM cells or IP packets (like an STS-3c), the pointers are structured according to the AU-4 specification.

This AU-3 or AU-4 (AU-*n* is the generic term) takes some getting used to in SDH, but is no more complex than the difference between an STS-3 (channelized for VTs) and an STS-3c.

What's in the SDH Containers?

Currently, in the SDH standards, there are five sizes of containers (the others are concatenated payloads, such as C-4-4c). These are C-11, C-12, C-2, C-3, and C4, as shown earlier in Figure 8-10. The five containers are classified into two groups. The higher-order containers are C-4 and C-3; while the lower-order containers are C-2, C-12 and C-11. In some documentation, C-11 and C-12 are simply referred to together as C-1.

Either one DS-1 signal of 1.544 Mbps or one ISDN H11 signal can be placed in one C-11. The 2.048 Mbps E-1 (CEPT-1) signal or an ISDN H12 signal can be placed in one C-12. One DS-2 signal of 6.312 Mbps is mapped into one C-2, while a DS-3 signal of 44.736 Mbps or one ISDN H32 is carried inside one C-3. Note that the container for a 34.368 Mbps E-3 (CEPT-3) signal or an ISDN H31 signal also is inside a C-3. However, the mappings of these two C-3s are different. This is why the C-3 container in Figure 8-10 has two arrows. An E-3 (CEPT-3) takes the upper path (that is, VC-3, TU-3, TUG-3, VC-4, AU-4, AUG-1, and then STM-1). Conversely, a 44.736 Mbps DS-3 signal takes the lower path (that is, VC-3, AU-3, AUG-1, and then STM-1). This can cause a lot of confusion when reading the SDH specifications. Note that the E-3 path requires an "extra" set of pointers (TU-3) while the DS-3 path does not. This also explains the odd concatenated/not-concatenated use of the AU-4 pointers with a VC-4 mentioned in Chapter 6.

The highest rate ITU-T PDH signal of 139.264 Mbps, an E-4 (or CEPT-4), or an ISDN H4 signal is placed in one C4.

It should be mentioned that the DS-1C, which is carried by the T-carrier digital system, is not considered an SDH container at present. This is also true for the 8.448 Mbps E-2 (CEPT-2) signal. Neither the 32.064 Mbps signal nor the 95.728 Mbps signal used in Japan are mapped into any of the currently assigned containers.

"Paths" Through SDH

There is no need to repeat the discussion of loading containers into TUs or TUGs in SDH. The process is essentially the same as in SONET, but more complex. Details are beyond the scope of this chapter. Instead, it is more productive to consider the various "paths" through Figure 8-10 and the ways in which SDH frames are built. This section starts by considering the differences of the AU-3 and AU-4 types of pointers.

It is easiest to start in the middle of Figure 8-10. Here, a C-4 (read as container-level 4) for transporting a 139.264 Mbps signal, such as a E-4 (CEPT-4) signal is directly mapped, with all needed VC-4 POH bytes, into a VC-4 (read as virtual container-level 4). The simple arrow indicates there is no multiplexing involved, just the proper overhead added to go from a C-4 to VC-4. Thus, a VC-4 is basically a C-4 with the relevant overhead. Next, the AU-4 pointers—these are AU-4 pointers for a VC-4, naturally—are added to the VC-4 to form an AU-4 (read as administrative unit-level 4). This AU-4 is essentially the same as an AUG-1 (AU group-type 1) because there is only one AU-4 inside, but this is not always true. Finally, the MSOH octets and the regenerator SOH octets are added to this AUG-1 to form an STM-1 frame. A number n of STM-1 frames may now be multiplexed into an STM-n frame. All of the STM-1s must be frame-aligned before multiplexing if this is the case. Figure 8-11 shows this process. A more simplified view of this same procedure is shown later on in this section.

The C-3 is a little more complex. A C-3 (container-level 3) can be used for 34.368 Mbps signal mapping and is directly mapped, with the proper VC-3 POH bytes, into a VC-3. This is indicated by the upper arrow from the C-3 box in Figure 8-10. Three octets for TU-3 pointers are now added to the VC-3 to form a TU-3 (tributary unit-level 3). This TU-3 is the same as the TUG-3 (TU Group-level 3) in this case, because no multiplexing of TU-3s into a TUG-3 is involved with C-3s. This distinction is discussed further a little later. Next, three TUG-3s are multiplexed and mapped (indicated by the "×3" next to the arrow in the figure) into a VC-4 which, just as in the C-4 case above, is mapped into one AU-4, one AUG-1, and one STM-1.

However, a C-3 is more complicated than the C-4. A C-3 can also be used for 44.736 Mbps signal mapping (a DS-3). The DS-3 is directly mapped into a VC-3, as before, but this time following the *lower* arrow out of the C-3 box in Figure 8-10. This time, it is AU-3 pointers that are added to this VC-3, not AU-4 as before. Generally, AU-3s are used for T-carrier signals and AU-4s are used for ITU-T E-carrier or CEPT signals. This now forms an AU-3 signal frame. In this case, three AU-3s are multiplexed and mapped into an AUG-1, which is finally mapped into an STM-1. Note that in this case the use of TUs and TUGs is only for C-3s that follow the ITU-T digital (E-carrier) hierarchy.

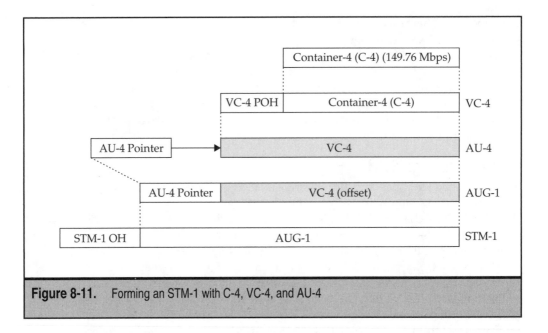

Figure 8-11. Forming an STM-1 with C-4, VC-4, and AU-4

This use of AU-3 pointers with a C-3 is shown in Figure 8-12. Generally, AU-4s are single H1/H2 pointer because the TUG pointers, if needed, can take care of finding the tributaries. AU-3s are three sets of H1/H2 pointers and work the same way as SONET STS-1s inside and STS-3.

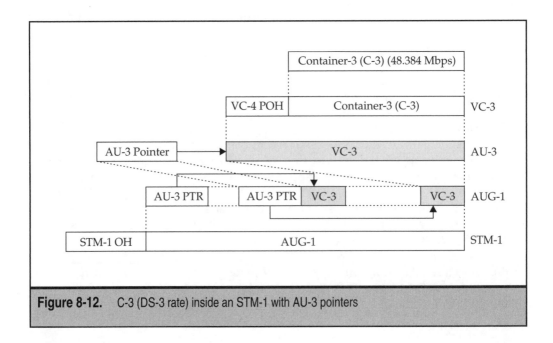

Figure 8-12. C-3 (DS-3 rate) inside an STM-1 with AU-3 pointers

Continuing the excursion through Figure 8-10, the next box down is the C-2. A C-2 (container-level 2) is intended for 6.312 Mbps signal mapping (a DS-2). The DS-2 is directly mapped, along with the VC-2 POH octets, into a VC-2. The TU-2 pointers are added to this VC-2 to form a TU-2 (tributary unit-level 2). Because there is no further multiplexing, a TU-2 is the same as the TUG-2 (TU group-level 2); however, seven TUG-2s are multiplexed and mapped into a VC-3 (TUG-2s are like VTGs, remember). Then the AU-3 pointers are added to this VC-3 to form an AU-3 signal. Finally, three AU-3 signals are multiplexed and mapped into an AUG-1 that, exactly as before, is mapped into an STM-1.

Next in the figure, a C-12 (container-level 1, type 2) is intended for 2.048 Mbps signal mapping (E-1). The E-1 is directly mapped, with VC-12 POH bytes, into a VC-12. The TU-12 pointers are added to this VC-12 signal to form a TU-12. In this case, however, three TU-12s are multiplexed and mapped into a TUG- 2, but a TUG- 2 with E- 1s inside cannot form a VC-3 or AU-3. These are basically for T-carrier. Thus, in the case of C-12, the arrow must next point to the TUG-3 box. Seven TUG-2s are now essentially the TUG-3 (think of a TUG-3 as a form STS-1 without the transport overhead). From TUG-3 to an STM-1 for this C-12, the process is exactly the same as in the C-2 case.

The C-11 (container-level 1, type 1) is intended for 1.544 Mbps signal mapping (a DS-1). The DS-1 is directly mapped, with VC-11 POH octets, into a VC-11. The TU-11 pointers are added to this VC-11 to form a TU-11. Next, 4 TU-11s are multiplexed and mapped into a TUG-2 (VTG). Alternatively, and somewhat confusingly, the VC-11 may be mapped into a TU-12 and the three TU-12s are then mapped into the TUG-2. In either case, the path from TUG-2 to an STM-1 for the C-12 is exactly the same as in the C-2 case.

Figure 8-10 is quite complex, and a fuller discussion of SDH tributaries would include notice of the distinctive letter system used (similar to the SDH locator vector for overhead). But it is more important to emphasize the relationships among all the formats and signals that have been discussed. These tributary relationships are more fully explored in subsequent sections.

The path from a generalized C-a to VC-a can be summarized as:
C-a + VC-a POH = VC-a; where a = 11, 12, 2, 3 or 4

The path from VC-b to TU-b can be summarized as:
VC-b + TU-b pointers = TU-b; where b = 11, 12, 2, or 3

And the path from VC-c to AU-c can be summarized as:
VC-c + AU-c pointers = AU-c; where c = 3 or 4

Finally, the general path from an AUG-1 to STM-1 is:
AUG-1 + RSOH + MSOH = STM-1

It is important to note that in the four relationships given above, it does not follow that the left-hand signal frame is identical to the right-hand signal frame. For instance, all that a relationship like the third one implies is that in order to form an AU-c signal, the AU-c pointer octets need to be added to the VC-c signal.

C-4, VC-4, AU-4, AUG-1, and STM-1 "Path"

This brief (compared the full treatment in the specifications) tour of SDH tributaries concludes with a little more detail about the "paths" in SDH from container to STM- 1 frame.

A C-4 occupies 9 rows by 260 columns in a 125 microsecond frame interval. Thus, a C-4 has a rate of 149.76 Mbps (which is 9 × 260 octets × 64 Kbps). By adding one column of VC-4 POH to a C-4, the result is a VC-4. This is a 9 × 261 frame structure.

A nine octet AU-4 pointer is then added to a VC-4 to form an AU-4. Next, 27 octets of RSOH and 54 octets of MSOH are then added to the AUG-1 (which is the same as the AU-4) signal to form the SDH STM-1 signal. This whole process is shown in Figure 8-13.

In this case, the AUG-1 is the same as the AU-4. Then what is the purpose of even inventing an AUG-1? This is because an AUG (administration unit group) can be used to carry either an AU-4 or an AU-3s. In typically T-carrier-based, AU-3 applications, a group of (actually 3) AU-3 signals will form the AUG-1. The case of three AU-3 applications is discussed later in this section. For now, it is enough to point out that when an AU-4 is involved, the AU-4 and the AUG-1 are one and the same.

C-3, VC-3, TU-3, TUG-3, VC-4, AU-4, AUG-1, and STM-1 "Path"

It has already been mentioned that either a 34.368 Mbps E-3 (CEPT-3) signal or a 44.736 Mbps DS-3 signal can be mapped into a C-3 container. First, the relationship between a C-3 container, which carries a 34.368 Mbps E-3 (CEPT-3) signal or an ISDN H31 signal, and an STM-1 is discussed.

A C-3 occupies 9 rows by 84 columns in a 125 microsecond frame interval. Thus, a C-3 has a rate of 48.384 Mbps (which is 9 × 84 × 64 Kbps). Nine octets of VC-3 POH is added to C-3 to form a VC-3; therefore, a VC-3 has an organizational structure of 9 rows

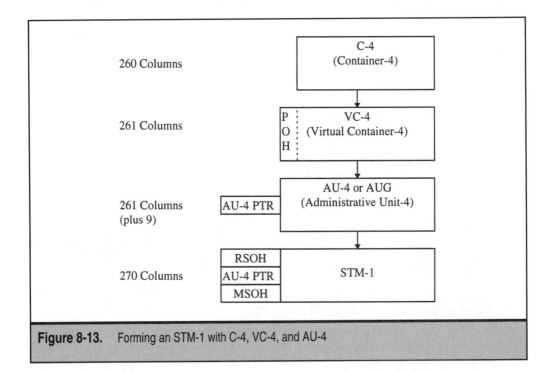

Figure 8-13. Forming an STM-1 with C-4, VC-4, and AU-4

by 85 columns and a 48.96 Mbps rate. In order to form the next level of signal, the TU-3, three octets of TU-3 pointer must be inserted into the eighty-sixth column. This column, nine octets long, has six octets of fixed stuffing octets occupying the remaining locations. These are shown in the bottom left (and other locations) in Figure 8-14, which illustrates the process.

This completes the 9 × 86 TUG-3. Note that a TU-3 signal has 85 columns plus only the 3-octet TU-3 pointers, while the full TUG-3 has 86 full columns. So a TU-3 with a rate of 49.152 Mbps is not quite identical to a TUG-3, which has a rate of 49.536 Mbps. Why is a separate name needed? Well, because one TUG-3 can be formed from one TU-3 (as in this case) or from a whole *group* of TUG-2 signals. More precisely, a TUG-3 contains 7 TUG-2 signals. Therefore, the TUG refers to the *group* signal.

Three TUG-3 signals are then multiplexed and mapped into an VC-4 which, besides these three TUG-3s, contains two fixed-stuffing columns, and one column of VC-4 POH. Next, nine AU-4 pointer octets are added to the one VC-4 signal to form an AUG signal. Then 27 octets of RSOH and 54 octets of MSOH will be added to an AUG-1 signal to form the SDH STM-1 signal This is also shown in Figure 8-14.

C-3, VC-3, AU-3, AUG-1, and STM-1 "Path"

There is another path from C-3 to STM-1 as well. A C-3 may also carry a 44.736 Mbps DS-3 signal. The C-3 and VC-3 for the DS-3 has the same structure as the previous case. After forming virtual container VC-3, which has 85 columns, two columns of fixed stuffing and

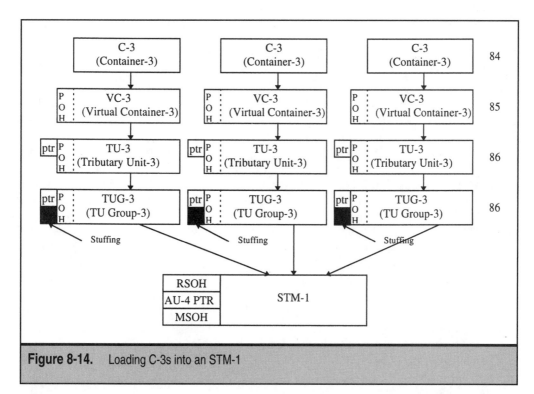

Figure 8-14. Loading C-3s into an STM-1

three octets of pointer are added to construct an AU-3. The major difference between a VC-3 and an AU-3 is the three added pointer octets.

Next, after having constructed three AU-3s, three octets of AU-3 pointer from each of these three AU-3s are multiplexed into the nine AU-4 pointer locations. The payload of these three AU-3s also are multiplexed into the AUG-1 payload location. Finally, just as all the previous cases, RSOH and MSOH are then added to the AUG-1 to form an STM-1.

C-2, VC-2, TU-2, TUG-2, VC-3, AU-3, AUG-1, and STM-1 "Path"

A C-2, carrying a 6.312 Mbps DS-2 or equivalent signal, has a payload capacity of 106 octets in a 125 microsecond interval. One octet of VC-2 POH (either V5, J2, N2, or K4) is added to form virtual container VC-2. Another octet, acting as the TU-2 pointer (either V1, V2, V3, or V4), is then added to form the 108 octet capacity of the TU-2. It is worth noting that there are four octets in total, one per each 125 microsecond interval, serving as the TU-2 pointer. For this application, the TU-2 is the same as TU Group-2 (TUG-2). In fact, a TUG-2 can be equivalent to one TU-2, three TU-12s, or four TU-11. More details are given at the end of this SDH investigation. Seven TUG-2s will then be multiplexed into the payload location of the VC-3.

The VC-3 has a structure of 9 rows by 85 columns. One column is VC-3 POH, and 84 columns are evenly distributed among the 7 TUG-2s. That is, because $84 \div 7 = 12$, the capacity for each TU-2 is 12 columns. In other words, a TU-2 requires 108 octets of capacity (12 columns $\times$ 9 octets/column = 108 octet). All of these are discussed later. For now, it is enough to note the similarity between a TUG-2 and the SONET VT Groups (seven in an STS-1). From VC-3 to AU-3, then to AUG-1 and STM-I, the process is the same as in the previous case.

C-12, VC-12, TU-12, TUG-2, TUG-3, VC-4, AU-4, AUG-1, and STM-1 "Path"

The C-12 to STM-1 path is quite complex. A C-12, carrying a 2.048 Mbps E-1 (CEPT-1) or equivalent signal, such as an ISDN H12 signal, has a payload capacity of 34 octets in a 125 microsecond interval. One octet of VC-12 POH (either V5, J2, N2, or K4) is added to C-12 to form a VC-12. Another octet (either V1, V2, V3, or V4), acting as the TU-12 pointer (sometimes known as TU-1 pointer for simplicity), is then added to the TU-12 signal to form the TU-12 itself with a 36-octet capacity. As in the case of C-2, it should be mentioned that there are four octets in total, one per each 125 microsecond interval, serving as the TU-12 pointer.

It is worth noting that this 36-octet unit of the TU-12 signal is equivalent to four columns in an STM-1 frame. Three TU-12s are multiplexed into a TUG-2, which now has a capacity of 12 columns. Seven TUG-2s will then be multiplexed into the payload location of a TUG-3. This seems odd at first, but note that the TUG-3 has a structure of 9 rows by 86 columns. There are 84 columns of the seven evenly distributed TUG-2s (12 columns each); one column of something called NPI (null pointer indication; not discussed further) plus fixed-stuffing octets, one column of fixed stuffing (which may be used for payload transport if needed in future standards), for a total of 86 columns. Because $84 \div 7 = 12$, the capacity for each TUG-2 is 12 columns. Each TUG-2 carries three TU-12s, each of which occupy four (which is $12 \div 3$) columns. Thus, a TU-12 requires 36 octets of capacity (4 columns $\times$ 9 octets/column = 36 octets). All of these TUs are discussed later.

From TUG-3, to VC-4, to AU-4, then to AUG-1 and STM-1, the process is the same as the previous case.

C-11, VC-11, TU-11, TUG-2, VC-3, AU-3, AUG-1, and STM-1 "Path"

The last case that needs to be discussed is from C-11 to STM-1. A C-11, carrying a 1.544 Mbps DS-1 or equivalent signal, has a payload capacity of 25 bytes in a 125 microsecond interval. One octet of VC-11 POH (either V5, J2, N2, or K4) is added to form VC-11. Another octet (either V1, V2, V3, or V4), acting as the TU-1 pointer, is then added to VC-11 to form a TU-11 signal with a 27-octet capacity. Note that there are four octets in total, one per each 125 microsecond interval, serving as the TU-11 pointer. This 27-octet capacity is equivalent to three columns in an STM-1 frame.

Next, four TU-11s are multiplexed into a TUG-2, which has a capacity of 12 columns. Seven TUG-2s will then be multiplexed into the payload location of VC-3. Note that the VC-3 has a structure of 9 rows by 85 columns. One column is VC-3 POH, and 84 columns are evenly distributed among the 7 TUG-2s. Because $84 \div 7 = 12$, the capacity for each TUG-2 is 12 columns. Each TUG-2 carries four TU-11s, each occupying three (which is $12 \div 4$) columns. Therefore, a TU-11 requires 27 bytes of capacity (3 columns × 9 octets/column = 27 octets).

The path from VC-3 to AU-3, then to AUG-1 and STM-1 is the same as the previous cases.

The SDH Tributary Unit

The previous section dealt with the container to STM-1 possibilities. But more details on the several types of intermediate signals formed that are important for the applications should be examined. For example, to map a C-11 into an STM-1, a TU-11 and TU-2 must first be formed. Therefore, a thorough understanding of these TUs would be in order. The counterpart of SDH TU-n in SONET is the VTn. That is, a VT1.5, VT 2, and VT6 are identical to the SDH TU-11, TU-12, and TU-2, respectively. The purpose of the TU is to transport and switch a container, which can be defined as a sub-STM-1 signal of various forms.

With the TUG-2, VC-3 and AU-3, either with 84 TU-11s or 21 TU-2s, can be mapped and multiplexed into one AUG-1. These applications are designed for North American digital systems using T-carrier.

With the TUG-2, TUG-3, VC-4, and AU-4, 63 TU-12s can be mapped and multiplexed into one AUG. Also, three TU-3s can be multiplexed and mapped into an AUG-1 signal through a TUG-3, VC-4, and AU-4 sequence. This set of applications is designed for ITU-T digital systems.

The last SDH topics to be discussed are the purpose and capacity of each TU-n: the relationship between a TU-n signal and a TUG-2 or a TUG- 3 signal; and the relationship between a TU-n signal and an STM-1 signal.

Purpose and Capacity of TU

The purpose of each individual TU-n, except TU- 3, can be seen in Table 8-9.

The TU-1 is designed to transport a DS-1 with a signal rate of 1.544 Mbps. Each column of an STM-n (n = 1, 4, 16, 64, or 256) or AUG-n has a rate of 0.576 Mbps (64 Kbps/octet × 9 octets/column). Therefore, it requires three columns of AUG-n or 27 octets per frame (or 125 microseconds) for a TU- 11 to accommodate a DS-1.

TU Type	Columns	Octets/frame	Octets/4 frames	Rate	Use
TU-2	12	108	432	6.912 Mbps	DS-2 (6.321 Mbps)
TU-12	4	36	144	2.304 Mbps	E-1 (2.048 Mbps)
TU-11	1	27	108	1.728 Mbps	DS-1 (1.544 Mbps)

Table 8-9.   TU types and their uses

The TU-12 is designed to transport an E-1 (CEPT-1) that has a rate of 2.048 Mbps. Therefore, it takes four columns (or 36 octets per frame) of AUG-1 for a TU-12 to accommodate an E-1.

The TU-2 is designed to transport a DS-2 signal with a rate of 6.312 Mbps. It takes 12 columns (108 octets per frame) of AUG-1 for a TU-2 to accommodate a DS-2.

With all TUs, just as in SONET, four 125 microsecond frame intervals are defined as a 500 microsecond *multiframe* or *superframe*. That means that in one superframe there are four 125 microsecond frames. Each of these four frames is started with one overhead octet (either V1, V2, V3, or V4), again as in SONET. A superframe starts with the V1 byte. The multiframe indicator octet, H4, of the VC-n (where n = 3 or 4) POH is used to indicate the *phase* of this 500 microsecond multiframe. Thus, if H4 = (XXXXXX00), where "X" is a "don't care" bit (this is not strictly true: the s1s2 "size" bits are also present and usually set to 10, as mentioned in Chapter 7), this 125 microsecond frame is the first frame of a 500 microsecond multiframe. The first octet in this 500 microsecond frame, therefore, is V1, the first TU-n pointer octet. In the same way, if H4 = (XXXXXX01), this 125 microsecond frame is the second frame of a 500 microsecond frame. The first octet in this 125 microsecond frame is V2, the second TU-n pointer octet. The same is true for H4 = (XXXXXX10) and H4 = (XXXXXX11) and works the same as in SONET.

The multiframe capacity of each TU type also is of interest and can be compared to SONET. For example, the TU-2 has a capacity of 12 columns or 108 octets for every 125 microsecond frame. Because there are four frames per superframe, the TU-2 can be said to have a capacity of 432 (which is 4×108) octets per 500 microsecond multiframe. Similarly, the TU-12 has a capacity of 144 (which is 4×36) octets per multiframe and the TU-11 has a capacity of 108 (which is 4×27) octets per superframe.

Relationship Between TU-n (n = 11, 12, or 2) and TUG-2

In addition to the TU-n (where n = 11, 12, or 2) signals, there is another intermediate signal that is required for the mapping and multiplexing of a C-11, C-12, or C-2 into an STM-1. This is known as the TUG-2 (tributary unit group-level 2).

It has already been shown that a TU-11 requires three columns of an AUG-1 frame to accommodate a DS-1 operating at 1.544 Mbps. Likewise, a TU-12 needs four columns to accommodate an E-1 signal operating at 2.048 Mbps and a TU-2 needs 12 columns to accommodate a DS-2 signal operating at 6.312 Mbps. This means that all three types (i.e., TU-11, TU-12, or TU-2) can all be carried as groups of 12 columns because the least common multiplier of 3, 4, and 12 is 12.

That means that if an AUG-1 frame (or an STM-1 frame) has 12, 24, 36, or any $n \times 12$ (where n = any positive integer) columns, this frame can be used to transport any TU-n (where n = 11, 12 or 2). The TUG-2, which is a special 12-column group of an AUG frame, was introduced for this very purpose. Therefore, a TUG-2 signal is equivalent to a VT group (VTG) signal in SONET.

A TUG-2 has 12 columns and can transport either (1) one TU-2 signal (one DS-2), (2) three TU-12 signals (three E-1s), or (3) four TU-11 signals (four DS-1).

TU and STM-1

Just to emphasize the essential identity of SONET and SDH, this section closes with a look at an SDH STM-1 carrying DS-1s. The same principle can be extended to TU-12s and TU-2s, but using SDH for DS-1s is still startling to some.

Each STM-1 can carry 84 TU-11s. This is the same as an STS-3 carrying three DS-3s in SONET, of course. Both have 84 DS-1s (3×28 in the case of the STS-3). In the STM-1, the first column after the POH column is allocated for the first TU-11; the second column for the second TU-11, and so on. This is done because SONET/SDH standards must apply byte-interleave multiplexing; therefore, frame alignment is critical. The AU-4 pointer will indicate the starting point of the STM-1 payload in the frame that is started out by the VC-4 path overhead. Once this POH is found, the three columns for the first TU-11 are uniquely located.

Just as in SONET, the VC-4 POH has a floating phase with respect to the STM-1 frame, but the phase between the three columns of the first TU- 11 signal has a fixed-phase relationship with respect to the VC-4 POH.

The overhead octets (or bytes because these are DS-1s) associated with each TU-11 within every 125 microsecond payload frame is either a V1, V2, V3, or V4. Each of these overhead bytes begins the 27-byte frame of the TU-11. The V1 and V2 bytes form a pointer that will point to the starting location of a VC-11 within the TU-11. This starting byte is called V5. The 500 microsecond interval of a multiframe headed by the V5 byte is called a VC-11 for this C-11 container. Thus, 84 DS-1 signals are carried by 84 C-11s, 84 VC-11s, 84 TU-11s, 21 TUG-2s, and 3 VG-3s (AU-3s), all inside 1 STM-1 signal.

This just points out the identity of SDH and SONET, but with all the names changed.

OTHER SONET/SDH MAPPINGS

Today, SONET/SDH is used for much more than for voice channel transport. This section discusses three non-channelized-voice structures for the SONET/SDH payloads. These are asynchronous DS-3 mapping, ATM cell mapping, and packet over SONET/SDH (POS). Although the first mapping applies only to SONET, there are asynchronous mappings for E-carrier equivalent speeds. ATM mapping and POS is essentially the same for SONET and SDH.

Asynchronous DS-3

Compared to the complexities of VT structures, superframes, and mappings, the rest of the SONET/SDH payload types explored in this chapter are quite simple. The next SONET/SDH SPE content to be investigated in this chapter is *asynchronous DS-3*. This section applies only to SONET.

In contrast to virtual tributaries, this type of SONET payload occupies all of the SPE. Just like virtual tributaries, asynchronous DS-3 transport is defined only in the basic STS-1 SPE. When the signal label POH byte (C2) is set to a value of 0x04 (0000 0100), then the SONET receiver knows that the arriving SPE must be carrying an asynchronous DS-3 (in SDH, the same value means the SPE carries an asynchronous E-3 at 34.368 Mbps).

As might be expected, asynchronous DS-3 transport allows the SONET NE no direct access to the DS-1s or DS-0s that might be contained inside the DS-3. This is to be expected, since DS-3 multiplexing is bit-oriented anyway. DS-3s can also be unchannelized as a single bit stream at 45 Mbps. With asynchronous DS-3, any type of DS-3 can be carried inside a SONET STS-1 SPE without regard to channelization issues. Cross-connections can always be done by non-SONET auxiliary equipment, exactly as is done when SONET is not in the equation.

In spite of this apparent DS-1 and DS-0 visibility limitation, the asynchronous DS-3 transport is popular and quite useful. If a service provider has a large national network, for example, a lot of legacy DS-3 and T-3 equipment might still be in place at the regional level, entirely operational, and nowhere near its retirement age. Beyond the cost aspect of DS-3 equipment replacement, there could also be regulatory issues involved with any large-scale equipment replacement scheme.

Many regulators limit the amount of physical equipment that can be replaced in any given time period. If this service provider has a large national SONET backbone, then the DS-3s can be transported from region to region over the SONET backbone without concern about the lack of payload visibility. This is really what asynchronous DS-3 transport is for.

When used for asynchronous DS-3 transport, the 87 columns of the SONET SPE are structured according to Figure 8-15. The information (I) bytes are what carry the DS-3 bits themselves. These I bits are spread out row by row into 25-byte units interspersed with control byte columns and reserved columns. It is a little misleading to label these information bits as "bytes" since DS-3 bytes are just bit-multiplexed streams. However, since the basic unit in the SONET SPE is the byte, it makes most sense to represent the I bits as SONET bytes, which is how the columns and rows always appear in SONET. Also, the control bytes in Figure 8-15 (for example, C1–C3) are SONET control bytes, not DS-3 frame bytes. It makes no difference if the I bits are DS-3 framing bits, or C-parity bits, or any other part of the overall DS-3 frame. All DS-3 framing bits are just other bits to SONET. Finally, bit stuffing is necessary, as is usual when the PDH meets the SONET/SDH world.

That is really about all there is to asynchronous DS-3 transport. The main point is that the bits in the I fields are just a meaningless jumble of bits to the SONET NE.

If a service provider actually wanted to carry the 28 DS-1s inside a DS-3 as organized bytes instead of the bit-multiplexed DS-3 stream, it would be necessary to de-multiplex the DS-3 and then load the 28 DS-1s as asynchronous VT1.5s or perhaps even as 673 DS-0s

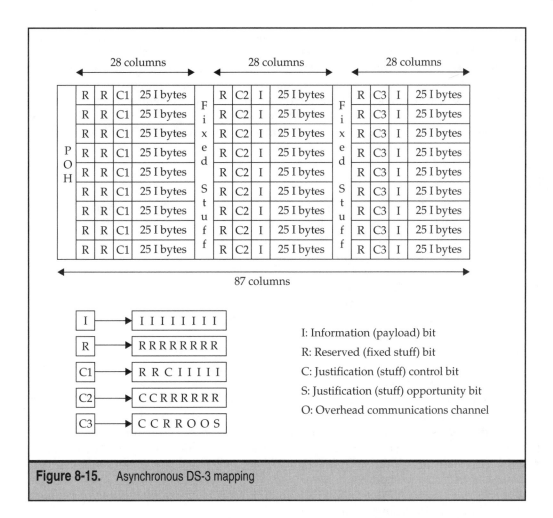

Figure 8-15. Asynchronous DS-3 mapping

with byte-synchronous VT1.5. But for raw DS-3 transport from here to there, asynchronous DS-3 is the quickest and easiest way to accomplish this task.

There is one other point of interest regarding asynchronous DS-3 transport that will be important when considering ATM cell and POS transport. Although it is not easy to tell by the structure of the SPE in Figure 8-15, the DS-3 bits in the SPE are really organized by *row*, not by *column*.

Multiplexing many sources in the virtual tributary world organizes the SPE into columns automatically. But an asynchronous DS-3 is a *single* source of bits, not many multiplexed sources presented to the SONET NE. Any multiplexing of the DS-3 has been done before the DS-3 hits the SONET NE.

Yet Figure 8-15 undeniably seems to organize the SPE into columns. This is only the effect of inserting additional overhead bytes such as C1–C3 systematically through the

SPE, which is always sent row by row across the fiber link. This overhead is required for the proper handling of the SPE bit stream in order to recover the DS-3 frame.

However, what if a single source of bits could fill an SPE and have its own framing? Then no column organization of the SPE would be necessary, and indeed would make no sense. And since the single source framed its own information content independently of the SONET SPE, no additional overhead bytes need be inserted into the SPE. This is exactly the philosophy followed by ATM cell transport inside the SONET SPE: organization by rows and no additional columns of overhead.

ATM Cells

The third major type of traffic carried by SONET/SDH SPEs is a stream of ATM cells. When the POH signal label byte (C2) is set to a value of 0x13 (0001 0011), then the SONET/SDH receiver knows that the arriving SPE must be carrying ATM cells.

ATM cells are fixed-length information blocks that are set at 53 bytes in length. ATM cells have a 5-byte header and a 48-byte payload structure, but that is not important to SONET/SDH. SONET/SDH cannot, will not, must not ever look at an ATM cell header (that is a switch function, and SONET/SDH NEs are not switches). ATM is the international standard for *cell relay* technology. Cell relay attempts to speed up networks, not by switching packets (OSI-RM layer 3 protocol data units, or PDUs), nor by relaying frames (frame relay uses OSI-RM layer 2 PDUs), but by essentially inventing a new, fixed, small PDU called a *cell.* Cells can carry frames or packets alike through a very fast switch.

ATM performs its magic by heavy use of statistical multiplexing, in which idle connections (ATM is connection-oriented) do not occupy "all the bandwidth, all the time," as do traditional time division multiplexing systems. Ownership of cell contents among the active connections is determined by a field in the ATM header. If a transmitter does not have enough live ATM cells to send on a link, then special idle cells are inserted in a process known as *cell rate decoupling.* ATM cells are often positioned as "very small and fixed-length PDUs" compared to frames and packets. It is just as correct, and often more accurate, to position ATM cells as "very big time slots" with variable ownership due to the statistical multiplexing involved.

This is not the place to even attempt an introduction to ATM technology. It is enough to note that ATM cells are fixed in size, and so can be packed directly, head to tail, on a communications link. Once the boundary of one cell tail and head is determined, the rest can easily be found without additional framing, overhead, or control bytes. This process of *cell delineation* is performed by the ATM switch on the receiving side of the link.

In reality then, there is really no need for a fiber or any other type of communications link to carry anything other than a stream of ATM cells. This is sometimes called *raw cell transport*, but it poses a problem for the service provider of the communications link. The ATM gear at each end of the link is presumably customer premises equipment (CPE). How is the link provider to perform OAM&P and detect alarms and failures, if the ATM cells on the link are "invisible" to the service provider? Service providers are not generally allowed to eavesdrop on user traffic, especially in regulated environments.

SONET/SDH offers a perfect solution. The SONET/SDH frame offers plenty of OAM&P overhead for the service provider, and the SPE provides plenty of bandwidth for the senders and receivers of ATM cells (SDH is capable of the same feat). The process of putting ATM cells directly inside a SONET/SDH SPE without the need for additional SPE overhead besides the required POH is called *direct cell mapping*. The difference between the use of raw cell transport and direct cell mapping is the use of a transmission frame, mainly for service provider OAM&P considerations, in direct cell mapping.

SONET/SDH specifications allow for the transport of ATM cells inside an STS-1 or STM-1 SPE using direct cell mapping. STS-3c/STS-12c and STM-4 (technically, a C-4-4c) speeds are also supported, but there is just no equipment to make ATM cells above the STS-12c/STM-4 data rate. When this mapping is used, the only alignment required in the SONET SPE is that the bytes of the ATM cell coincide with the bytes of the SPE. That is, the bit boundaries of the bytes must be preserved, but this is not a big deal.

The only other important thing to realize is that the SONET SPE is not an even number of ATM cells long. The 9 rows and 84 information-bearing columns (columns 30 and 59 are usually fixed stuff columns) of an SPE contain 756 bytes. Dividing this number by the 53-byte size of an ATM cell gives about 14.2 cells in each SPE. So a SONET SPE, assuming an ATM cell starts right after the POH J1 byte that defines the start of the SPE, will contain 14 full ATM cells and only a few bytes of the 15th ATM cell. The "tail" bytes of the 15th ATM cell will begin directly following the POH J1 byte of the next SONET SPE. This next SPE will have only the last few bytes of the last ATM cell in the SPE. The position of the ATM cell header keeps shifting from SPE to SPE. So any attempt to fix the beginning of an ATM cell to a particular place in the SPE is doomed to failure, since ATM cells are always packed head to tail on a link. This is one reason that cell delineation (the determination of ATM cell head-tail boundaries) is a task best left up to the ATM switch. (As an aside, on links other than SONET/SDH, ATM cell mappings that attempt to fix the relationship between ATM cell boundaries and transmission framing require additional overhead known as a *convergence protocol*. The advantage of direct cell mapping is that no convergence protocol is needed.)

So ATM cells always span the SPEs. This is of no concern to SONET/SDH, since cell delineation is an ATM switch function. SONET/SDH links deliver the bytes in the SPE to an ATM switch. It is up to the ATM switch to make sense out of the bytes.

All that is left is to look at a SONET/SDH SPE carrying ATM cells. A schematic of ATM cells inside a series of SONET STS-1 SPEs is shown in Figure 8-16. ATM cell transport is also supported on unchannelized ("concatenated") STS-3c (STM-1) and STS-12c (STM-4) links.

Not too much should be read into Figure 8-16, but it is informative. Not all 90 columns are shown, of course, and the exact sizes of the ATM cells are not to scale. Figure 8-16 points out that the full structure of the SONET frame is preserved, right down to the H1/H2 pointer bytes and fixed stuff columns 30 and 59. It is the structure of the SPE that changes. Figure 8-16 does show the ATM header creep from row to row, and emphasizes the fact that even if an ATM cell header happens to start right after the JI POH byte in a given SPE, this will not be true of the next SPE. Cells also jump over the POH column. The mixture of idle and data (information-bearing) cells is shown as well.

But by far the most important point of Figure 8-16 is the row orientation of the SPE when carrying ATM cells. This is due to the fact that when SONET/SDH is used as a link to or from an ATM switch, there is no longer any need for SONET/SDH to perform time

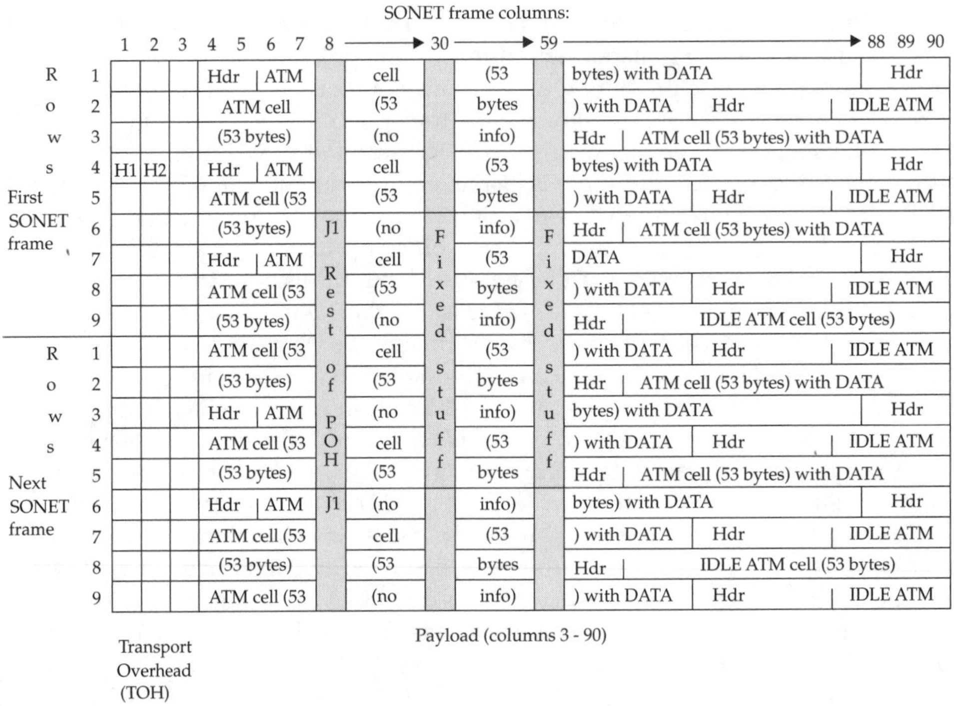

SONET frame columns:

		1	2	3	4 5 6 7	8 → 30 →	59 →		88 89 90
R	1				Hdr |ATM	cell	(53	bytes) with DATA	Hdr
o	2				ATM cell	(53	bytes	) with DATA | Hdr |	IDLE ATM
w	3				(53 bytes)	(no	info)	Hdr | ATM cell (53 bytes) with DATA	
s	4	H1	H2		Hdr |ATM	cell	(53	bytes) with DATA	Hdr
First	5				ATM cell (53	(53	bytes	) with DATA | Hdr |	IDLE ATM
SONET	6				(53 bytes) J1	(no	info)	Hdr | ATM cell (53 bytes) with DATA	
frame	7			Rest	Hdr |ATM	cell Fixed	(53 Fixed	DATA	Hdr
	8			of	ATM cell (53	(53	bytes	) with DATA | Hdr |	IDLE ATM
	9			POH	(53 bytes)	(no	info)	Hdr | IDLE ATM cell (53 bytes)	
R	1				ATM cell (53	cell	(53	) with DATA | Hdr |	IDLE ATM
o	2				(53 bytes)	(53	bytes	Hdr | ATM cell (53 bytes) with DATA	
w	3				Hdr |ATM	(no	info)	bytes) with DATA	Hdr
s	4				ATM cell (53	cell	(53	) with DATA | Hdr |	IDLE ATM
Next	5				(53 bytes)	(53	bytes	Hdr | ATM cell (53 bytes) with DATA	
SONET	6				Hdr |ATM J1	(no	info)	bytes) with DATA	Hdr
frame	7				ATM cell (53	cell	(53	) with DATA | Hdr |	IDLE ATM
	8				(53 bytes)	(53	bytes	Hdr | IDLE ATM cell (53 bytes)	
	9				ATM cell (53	(no	info)	) with DATA | Hdr |	IDLE ATM

Transport Overhead (TOH)

Payload (columns 3 - 90)

Figure 8-16. ATM cells inside the STS-1 SPE

division multiplexing, which gives the SPE its distinctive column-oriented look. Any multiplexing of the ATM cells is performed by the ATM switch, not by SONET/SDH.

The fixed stuff columns 30 and 59 in SONET are for SDH payload compatibility. For higher forms of SONET frames, the fixed stuff columns move, as in SDH, to the columns immediately following the path overhead column of the concatenated SPE. The number of columns depends on the value of N in the STS-Nc, as pointed out in Chapter 6. This is shown in Figure 6-17, along with the formula the figure out the number of fixed stuff and payload columns.

As an example of applying the formula, consider an STS-12c (the figure shows an STS-48c example). The number of fixed stuff columns after the path overhead column is given by $(N \div 3) - 1 = (12 \div 3) - 1 = 3$ columns. The number of columns available for ATM cells is given by $(87 \times N) - N \div 3 = (87 \times 12) - 12 \div 3 = 1044 - 4 = 1040$ columns. This makes the ATM payload structure for an STS-12c the same as for an STM-4 (actually, a C-4-4c), and the structure for the STS-48c the same as the STM-16 (C-16-16c).

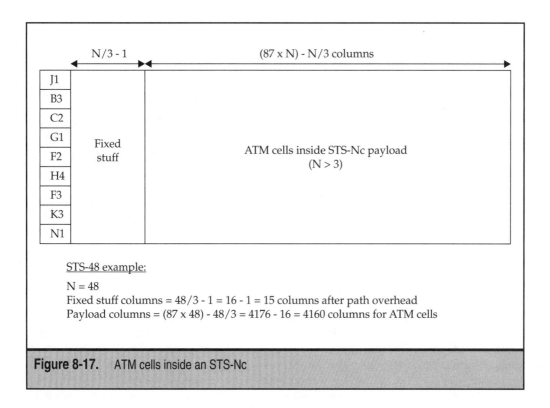

Figure 8-17. ATM cells inside an STS-Nc

The point is that whenever a SONET/SDH SPE carries traffic from a single source, it is more efficient to organize the SPE into rows, not columns. This is especially important when it comes to the last SONET/SDH payload type: packet over SONET/SDH.

PACKET OVER SONET/SDH (POS)

When packet over SONET/SDH (POS) was originally proposed, some observers re- marked that packets had been sent over SONET/SDH links since the inception of SONET/SDH, so what was the big deal? Take any stream of IP packets, for example, place them inside a DS-0 (64 Kbps) or VT1.5 (1.544 Mbps), and away they go. But the key to understanding POS is realizing that POS organizes the SPE into rows, like the ATM mapping, not columns, like virtual tributaries. As long as the packets emerge from one source, like a router port, row-wise SPEs are much more efficient than column-wise SPEs. POS is a way of sending packets between routers at much higher speeds than virtual trib- utaries allow, all the way up to one Gbps or beyond.

Of course, ATM cells can also carry IP packets. And since ATM cells can fill the SONET/SDH SPE, this is another way that IP packets can be sent on a SONET/SDH link. However, the issue regarding the use of ATM cells to carry IP packets boiled down to a question of overhead. A comparison of packet over SONET/SDH and ATM with IP inside over SONET/SDH overhead is given at the end of this section, after the basics of Packet over SONET/SDH have been discussed.

The oddest thing about packet over SONET/SDH (POS) is that, although it is a common enough acronym and term, there is really no such thing as POS. It is proper to talk about "IP-over-SONET" or "PPP over SONET/SDH" (several equipment vendors even call it just that) or even "HDLC-over-SONET," but technically there is no "packets over SONET/SDH." The term POS is just a shorthanded way of saying "allow the SPE to be structured in rows, as it is for ATM cells, but allow for the transport of any OSI-RM layer 3 packet structure, such as IP packets, inside an appropriate frame." The PPP and HDLC protocols, which *are* supported over SONET/SDH, are nothing more than the frame types used to transport the packets. PPP stands for the Internet Point-to-Point Protocol and HDLC stands for OSI-RM High-level Data Link Control protocol. Although intentionally closely related, PPP and HDLC are not exactly the same.

This section therefore makes heavy use of the Open Systems Interconnection Reference Model (OSI-RM), especially the bottom three layers, numbered layer 1 through layer 3. The OSI-RM, specified by the International Organization for Standards (ISO), has become the accepted way to refer to data network protocols and architectures. At the bottom of the OSI-RM, layer 1, or the physical layer, defines a serial stream of bits as the basic transport for data. The bit stream has no structure, so layer 2, the data link layer, creates a structure defined as the *frame*, a variable-length protocol data unit (PDU) from some minimum length up to some maximum length, which varies from implementation to implementation. Both PPP and HDLC define a frame structure and minimum/maximum length. Inside the frames are, by definition, packets. Packets are layer 3 (network layer) PDUs that also have a structure, variable size, and maximum length. Today, the primary example of a packet as layer 3 PDU is the IP packet without which the Internet and Web cannot be used.

The main difference between frames and packets in the OSI-RM is that frames have headers and trailers, while packets have only headers. The frame trailer is used mostly for error checking, which means that the packet header need not perform error checking at all, or at least not to the extent that the frame trailer provides error checking. Error detection, in OSI-RM talk, is primarily a layer 2 frame function. Also, frames are defined to flow on simple point-to-point links from NE to NE. Although frames might look exactly the same entering and leaving an IP router, for example, technically they are two different frames. Frames flow "hop by hop" between adjacent systems (those directly connected by serial link).

Packets, on the other hand, flow end to end from source to destination, whether there are 10 or 100 NEs, such as IP routers, in between. This is possible because each packet header has enough information to allow NEs to make routing decisions on a hop-by-hop basis. In IP, for example, this information is the source and destination IP network address. An important part of ISO-RM layer 3 is the concept of a network address, which exists at no other layer of the model. Generally, OSI-RM NEs (called *intermediate systems*) just extract packets from frames arriving on an input serial link, examine the destination network address, determine the proper output port, and then rewrap the packet inside a frame to queue for sending on the correct output serial link. (If this explanation makes data networking sound very simple, the answer is that data networking *is* very simple! It is the *networks* that are complex.)

So the first and foremost reason that it is not a good idea to attempt to put raw packets on a physical layer transport such as SONET is that the frame header and trailer provides error detection functions for itself and also for the packet inside. It would be hard for most packets to provide elaborate error checking for themselves: many packets have no error fields in the header at all, or if they are present, they provide only rudimentary functionality much weaker than the frame error detection (IP falls into this latter category).

There are two other reasons for placing a packet inside a frame before transporting the frame inside a SONET/SDH SPE. The first reason concerns the delineation of the packet and the second reason concerns the existence of multiple packet types.

Packets in general rely on the frame structure surrounding them to properly delimit the beginning and ending of the packet itself. Unlike ATM cells, packets and the frames that surround them are almost always of variable length between some minimum and maximum. So frame delineation is an important receiver function (transmitters, presumably, know exactly where a packet and frame begin and end). In HDLC and PPP, frame delineation is provided by means of a special *interframe fill* bit pattern, which is almost universally referred to as an *idle pattern*. When the idle pattern disappears at the receiver, the frame has begun. When the idle pattern reappears, the frame has ended. Everything in between is the frame. Both HDLC and PPP employ as interframe fill 7E in hex, or 0111 1110 in bits. (There are special rules for preventing the 7E pattern and thus *false framing* from occurring within the transmitted frame and packet, but these are not of concern here.)

And not all layer 3 packet structures are the same. Even IP, the most popular layer 3 protocol in use today, thanks to the Internet and the Web, technically employs different packet types at layer 3. Frequently, the same serial link, employing a single frame type, will have to carry multiple packet formats more or less at the same time. Both PPP and HDLC frame structures provide a method for receivers to determine the precise type of packet inside the arriving frame. This is necessary because the format and processing of each packet type can vary based on the exact type of packet involved. For PPP, the presence of a "normal" data-carrying IP packet inside the PPP frame is indicated by a protocol field in the PPP frame having the value 0021 in hex, or the 16-bit pattern 00000000 00100001.

So the three reasons *not* to place packets directly inside a SONET/SDH payload are:

▼   The frame trailer contains more robust error checking.

■   The frame provides delineation for the packet inside.

▲   The frame can carry multiple protocols (packet types).

HDLC or PPP?

So far, equal mention has been made of HDLC and PPP. Which one is used when packets are sent across a SONET/SDH link? The issue is important because PPP, like all Internet Protocol Suite (TCP/IP) protocols, is not an official international standard in the sense that ATM or HDLC are official international standards. So specifications on SONET/SDH, for example, only list "HDLC-over-SONET mapping" as an allowable SPE content. The POH C2 signal label assigned to this mapping is 16 in hex or 0001 0110.

But PPP, the serial link protocol most often used with IP packets, is not quite HDLC. HDLC has sometimes been called "the mother of all data link layer protocols," and this is not a bad analogy. HDLC defines a basic frame structure, but many implementations of HDLC vary one or more fields or the header or trailer in size and format. This is not to say that there are no rules; for instance, the *order* of the fields cannot be changed.

PPP was designed as an efficient serial link transport for IP and related packet structures. While PPP simplifies and modifies many of the functions of the entire HDLC protocol, PPP frames were intentionally designed to look as much like a basic HDLC frame as possible. But just to show how different the basic HDLC frame is from a PPP frame, Figure 8-18 shows both.

Keep in mind that many HDLC frames can also have 16-bit (2-byte) address, control, or cyclical redundancy check (CRC) checksum, also called the *frame check sequence*(FCS) fields. The PPP frame fixes some HDLC options, such as defining the address filed as 1 byte, and using the all 1's "anycast" address on the PPP link. The PPP control field is also 1 byte and is set to indicate that the content of the frame is an *unnumbered* (i.e. there is no sequence numbering used on the frame) *information frame*. The most unique aspect of the PPP frame is the presence of the 1- or 2-byte *protocol identifier* field. This is used to indicate the format of the *information* field that follows. This protocol ID field is 0021 in hex when an IP packet is inside the frame. The IP packet itself rides in the information field. Some protocols (not IP) might require some bytes of padding (usually just all 0's or null fields) to end the packet on the proper byte boundary. Finally, the PPP FCS field might be 2 bytes (CRC-16) or 4 bytes (CRC-32), depending mostly on the speed of the link (the higher the speed, the more likely a 4-byte FCS is used). SONET/SDH supports HDLC at all rates.

Interframe fill	Address	Control	Data	Checksum (FCS)	Interframe fill
01111110	1 or 2 bytes	1 or 2 bytes	0 to X bytes	1 or 2 bytes	01111110

Basic HDLC frame format

Flag	Address	Control	Protocol ID	Information	FCS	Flag
01111110	11111111	00000011	1 or 2 bytes	Variable/Pad	2 or 4 bytes	01111110

Basic PPP frame format

Figure 8-18.   Basic HDLC and PPP frame formats

The PPP protocol, when used with SONET/SDH, is defined in a series of requests for comment (RFCs) issued by the Internet Engineering Task Force (IETF), which acts as much as a central authority for Internet protocols as the loose structure of the Internet allows.

Three RFCs are especially important for POS:

▼ RFC2615, PPP over SONET/SDH

■ RFC1661, the Point-to-Point Protocol (PPP)

▲ RFC1662, PPP in HDLC-like framing

RFC1661 currently defines PPP procedures, but the PPP frame structure is defined in RFC1662. RFC2615 has the rules for using PPP inside SONET or SDH. Also of interest is RFC2153, which adds "vendor extension" fields to the basic PPP in RFC1661.

The frames inside the SONET SPE are organized into rows, as expected, and cross SPE boundaries. Interframe fill is used between frames. According to RFC2615, the POH C2 signal label should use a value of 0x16 (0001 0000) to indicate the presence of PPP frames when scrambling (added in RFC2615, which replaced RFC1619) is used on the frame. The scrambling is the same as that used in ATM, and if scrambling is not used, then the C2 signal label value is set to 0xCF (1010 11110). Scrambling was added because some payloads could cause synchronization problems. The default behavior is to scramble the payload. A 32-bit CRC is always recommended, but a CRC-16 is allowed for STS-3c/STM-1. The multiframe pointer byte (H4) is not used and must be set to all zeros regardless of C2 value. RFC2615, as an Internet standard, is silent on the status of payload columns 30 and 59, which are fixed stuff columns in STS-1 ATM mappings.

The POS SPE

So either HDLC or PPP can be used to frame packets sent on a SONET/SDH link. As long as both ends of the link understand the frame formats used and process the frames without generating alarms, the signal label is not all that important.

Figure 8-19 shows a series of PPP frames inside a SONET STS-1 SPE. This figure may be compared with Figure 8-16, which shows the ATM equivalent. The major differences are the variable lengths of the frames (packets) and the presence of the interframe fill pattern, or flag pattern of 7Es. Figure 8-19 is not very realistic, however, in the sense that IP packets are routinely 1,500 bytes long and so would easily fill an entire SPE at the STS-1 rate. Even the default maximum IP packet size of 576 bytes is quite large compared to the SPE. There are fixed stuff columns after the path overhead columns in all STS-Nc frames where N is greater than 3, exactly the same as in ATM. Since STS-1s are not seen much, this is the more common structure.

The only matter that should be discussed before closing this chapter is a brief discussion of using ATM cells inside a SONET/SDH SPE as opposed to using POS. IP packets can ride SONET/SDH inside ATM cells quite nicely. But the overhead needed to place IP packets inside ATM cells was considered much too large by the IETF, which then developed POS to cut down on this overhead burden. Consider an IP packet to be placed inside an ATM cell and then inside an STS-3c SPE. Assume that the IP packet is 576 bytes long. A spe-

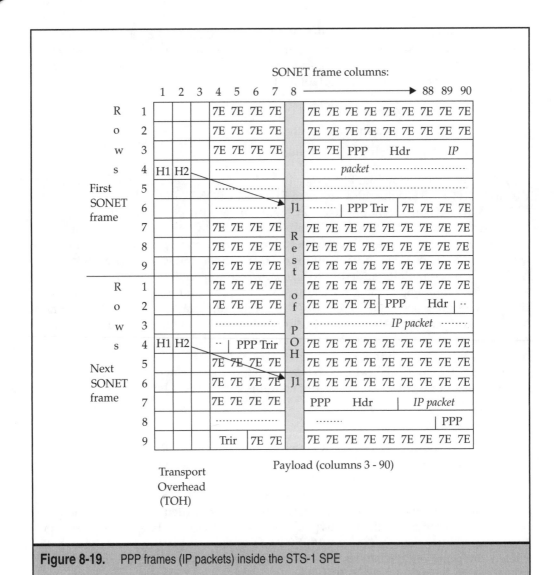

Figure 8-19.   PPP frames (IP packets) inside the STS-1 SPE

cial 8-byte field called the *logical link control/sub-network access protocol* (LLC/SNAP) header is first placed on the IP packet. Then a trailer called the *ATM adaptation layer type 5* (AAL5) trailer, which is at least 8 bytes long, and sometimes a lot more, is added to the end of the IP packet. Finally, this whole 592-byte (at least) unit is chopped up into 48-byte cell payloads, each with a 5-byte ATM cell header. The packet will occupy 13 cells (592 ÷ 48 = 12.333...) and require 13 × 5 = 65 more bytes of overhead, for a total of at least 81 extra bytes to send the original 576-byte IP packet. This translates to only about 80% of the STS-3c line rate being useful for IP packet transfer.

Now compare sending an IP packet on an STS-3c using PPP frames inside the SPE to carry the packets. Only the modest 6 bytes of overhead from the address (1 byte), control (1 byte), protocol identifier (2 bytes), and FCS fields (4 bytes) are needed, compared to 81 when using ATM, to transfer the same size packet. An approximation translating PPP and ATM overhead into line rates is shown in Table 8-10.

Even a quick look at Table 8-10 shows why Internet service providers (ISPs) and other service providers would rather send IP packets with POS than with ATM on SONET/SDH. Each line shows the line rate available to each layer, and then subtracts the overhead needed (about 45 cells fit inside an STS-3c/STM-1 SPE). The result is the line rate available to the next layer, and so on. These figures are approximations, but they are representative.

The bottom line is that an ISP "gains" about another 20 Mbps (125.92 Mbps with ATM to 147.15 Mbps) when using POS instead of ATM on the SONET/SDH links.

VIRTUAL CONCATENATION IN SONET/SDH

A lot of the payloads the SONET/SDH frames must carry are not a particularly good fit for the defined SONET/SDH line rates and frames sizes. For example, when SONET/SDH uses the payload mapping for the transport of fiber distributed data interface (FDDI) frames at 125 Mbps, the closest fit is the 155.52 Mbps SONET STS-3c or the SDH STM-1 with a C-4-4c inside. (Note that the FDDI data transfer rate is 100 Mbps, but the FDDI line rate is 125 Mbps due to "4 in 5" byte encoding.) This leaves a full 30.52 Mbps wasted on the link, since the signal label is always used as a single field in the path overhead and cannot be changed from payload to payload. It could be pointed out that FDDI is not particularly common on SONET/SDH networks, but the same is true for Fast Ethernet at 100 Mbps.

Layer	Line Rate	Less Overhead...	Layer	Line Rate	Less Overhead...
SONET/SDH	155.52 Mbps	90 bytes/frame	SONET	155.52 Mbps	90 bytes/frame
ATM	149.46 Mbps	5 bytes/cell	PPP	149.46 Mbps	8 bytes/PPP frame
AAL	135.36 Mbps	8+ bytes/packet	IP	147.15 Mbps	None
LLC/SNAP	126.94 Mbps	8 bytes packet			
IP	125.92 Mbps	None			

Table 8-10.   Effective line rates for PPP and ATM STS-3c SPEs

This is another very poor SONET/SDH match, and so are Gigibit Ethernet and 10 Gbps Ethernet, which are becoming more and more popular every day. There are other, less wasteful examples, but the point is that if SONET/SDH is to thrive in a networking world filled with more than just voice channels or a handful of alternative mappings for common transports, SONET/SDH must become more flexible in terms of allocating bandwidth for user data rates.

This section explores *virtual* concatenation in SONET/SDH, an alternative method to *contiguous* concatenation that allows the creation of payloads at low incremental rates in a non-disruptive fashion. Generally, virtual concatenation can break up a SONET/SDH payload into individual payloads at the higher order path or lower order tributary level, giving granularity for bandwidth allocation as low as 1600 bps. These virtually concatenated payloads do not have to be physically associated, as is the case with contiguous concatenation, so the network can transport each payload separately across the links and recombine them at the end point. There are complex mechanisms in virtual concatenation methods for adjusting for the differential delays that result from using separate physical paths across the SONET/SDH networks (these mechanisms are not discussed in any detail here). Virtual concatenation requires this enhanced capability only in the path terminating equipment.

Why wait until the payload chapter to explore virtual concatenation? Contiguous concatenation was introduced along with the basic frame formats in Chapter 6. There are two reasons for the delay. First, virtual concatenation is intimately tied up with considerations of payloads mappings, not discussed until this chapter. Second, virtual concatenation is relatively new, somewhat complex, and uses overhead bytes and structures not introduced until Chapter 7 and here in this chapter. This is no way downplays the importance of virtual concatenation, which is a fundamental change in the way the SONET/SDH operates at the payload level.

Virtual concatenation introduces new terminology for SONET/SDH frame structures. Although only SONET/SDH payloads need to be aware of virtual concatenation, this does not mean that the human operators of SONET/SDH networks should not be aware when virtual concatenation is in use. This avoids right from the start a repeat of the whole STS-3C/OC-3c controversy.

Virtually concatenated channels are administered together, as one, for network management purposes, but only at the end points. Channels are not limited to using the same path between end equipment, so "end point delay equalization" is required. The current specifications say the delay differential that can be allowed and still let virtual concatenation work must be at least one frame-time (125 microseconds) but larger values are not ruled out.

Channels do not have to be in the same STS-N, of course, and (for instance) two STS-12s can each have some STS-1s used for a single virtual concatenation channel. There are even defined ways to convert between contiguous and virtual concatenation for interoperability purposes, but these methods are not considered further.

When "high order" virtual concatenation is used for STS-1 or STS-3c payloads, the variable X indicates the number of payloads that have been logically associated to form the virtual path. Values of X from 1 to 256 are possible with an STS-1 or STS-3c. The new terms are:

▼ **STS-1-Xv**   Virtual concatenation of X number of STS-1 payloads

▲ **STS-3c-Xv**   Virtual concatenation of X STS-3c payloads

In SDH, which lacks an STS-1 (STM-0s are not used directly for virtual concatenation), the new term is STM-1-Xv, with possible values of X from 1 to 256. When STS-1s are used for virtual concatenation, columns 30 and 59 carry fixed stuff bytes for ease of interoperability with SDH VC-3 payloads.

The nicest thing about the values of X in virtual concatenation is that virtual concatenation provides a way to create "logical" SONET (and SDH) levels where none have existed. For example, a type of logical STS-9 could be created as an STS-1-9v, concatenating nine STS-1s for a given customer site. This "STS-9" could be provisioned on three separate STS-3 or STS-3cs (as an STS-3c-3v) running on different links and on separate paths. Naturally, the available bandwidth on the STS-1s or STS-3cs must meet or exceed the bandwidth created through virtual concatenation. SONET/SDH is not able to create bandwidth out of nothing.

When "low order" virtual concatenation is used for SONET virtual tributary types (VTn, where n = 1.5, 2, 3 or 6), the new terms are:

VTn-Xv   Virtual concatenation of X VTn payloads

In SDH's more complex lower order path formatting, the new terms are in the form VC-3-Xv, VC-4-Xv, VC-2-Xv, VC-12-Xv, and VC-11-Xv, but the principles are the same. For this reason, the basics of virtual concatenation explored in this section will mainly use SONET examples and terminology. Also, the SONET italics for X will be dropped in the rest of this section.

Two key acronyms make their appearance with virtual concatenation, SQ and LCAS. The sequence indicator (SQ) identifies the order in which the individual payloads of a virtual concatenation are combined to produce the "super-container." In the higher order forms of virtual concatenation, the SQ is carried in the H4 multiframe path overhead bytes as the VC sequence field. In lower order forms, bit 2 of the K4 (Z7) tributary overhead byte must be used with the extended signal label field as described at the end of Chapter 7 to provide the 32-bit multiframe structure necessary to support low order virtual concatenation. It is important to note that without this extended signal label (a value of 101 in bits 5-6-7 of the V5 tributary overhead byte) and the K4 multiframe in bit 2 support, virtual concatenation at the VT level is impossible.

The link capacity adjustment scheme (LCAS) is the mechanism used to "hitlessly" increase or decrease the link capacity to match the bandwidth needed by the user or application. The LCAS also provides a way to remove failed links from the virtual concatenation. LCAS is a quite complex protocol, and requires some 12 pages in the specifications to detail. Only the very basics of LCAS will be covered here.

This promised ability to adjust customer bandwidth up or down quickly and painlessly should not be underestimated. Provisioning link bandwidth has always been a problem to serviced providers offering bandwidth-based solutions, and has often taken an agonizingly long time to perform. Many applications are cyclic, with monthly, quarterly, or even yearly peaks. Paying for peak bandwidth all the time is wasteful, but there is nothing worse than not having adequate bandwidth when it is essential. Other public network methods to provide this type of flexible bandwidth allocation, such as ATM or frame relay, require "overlay" equipment and suffer from other issues beyond the scope of this discussion. If flexible bandwidth comes directly to SONET/SDH, this would be a huge benefit to customers and carriers alike. However, virtual concatenation is new enough so that it is not considered to be a routine benefit of SONET/SDH when these benefits are investigated and detailed in later chapters in this book.

Higher Order Virtual Concatenation with H4

Unlike the contiguous concatenation of an STS-3c or STS-12c and so on, each STS-1/STS-3c-Xv (the standard SONET notation for this practice) still has its own path overhead. So the STS-1-9v created above would still have nine J1 trace bytes and all other path overhead bytes intact, including, most importantly for virtual concatenation, the nine H4 multiframe indicator bytes. Usually, this H4 byte is of limited use, cycling bits 7 and 8 endlessly through the decimal values 0, 1, 2, 3 so that the V1/V2/V3/V4 tributary overhead bytes (and through them the V5/J2/N2/K4 bytes) can be found.

Virtual concatenation essentially redefines the role of the H4 byte. This redefinition does not interfere with the operation of this 0-3 multiframe sequence, since the tributary overhead bytes will still be present—or not—depending on the value of the C2 signal label (virtual concatenation at this level is totally transparent to the payload format indicated in the C2 signal label).

In a virtual concatenation role, the H4 byte now performs an intricate ballet through the interplay of two separate fields called the first multiframe indicator (MFI1) and the second multiframe indicator (MFI2). The MFI1 is four bits and carried in bits 5-6-7-8 of the H4 byte. This count of 0-15 also forms the basis for the LCAS delay compensation mechanism: 16 frames are needed to provide 125 microsecond delay differential.

Confusingly, MFI2 is an 8-bit field, carried in bits 1-4 of the H4 bytes of the first two multiframes (0 and 1) as indicated by MFI1. This is complex enough already, so Figure 8-20 shows the interplay of the two MFI fields and the new structure of the H4 byte.

Even this representation is complex. Basically, the MFI1 value cycles from 0 to 15 every 16 frames. That is simple enough. The 8-bit MFI2 field, split between MFI1 frame 0 (the most significant bits of the MFI2, bits 1-4) and MFI1 frame 1 (the least significant bits of the MFI2, bits 5-8), will increment every one cycle of MFI1 values. So in frame n, the value of the MFI2 bits will be 0000 0000 (as an example). Once the MFI1 pattern starts to repeat in frame $n+1$, the MFI2 value increments to 0000 0001, again for 16 frames. From the table, if the value of MFI2 is 0000 1111 (15) in frame $n-1$, the MFI2 field will be 0001 0000 (16) in frame n, and 0001 0001 (17) in frame $n+1$, and so on. These values are shown in the figure. So the total possible bit combinations for MFI1 and MFI2 are $16 \times 256 = 4096$. This covers 4096 frames, which take 4096×25 microseconds = 512 milliseconds.

What's all this housekeeping for? The main goal of MFI1/MFI2 is to find the virtual concatenation sequence indicator (SQ). The SQ is carried in the 14th and 15th frames (last two) of the MFI1 sequence. This is also shown in the table. The SQ is itself an 8-bit field and the most significant bits of the SQ, bits 1-4, appear in MFI1 frame 14, and the least significant bits of the SQ, bits 5-8, appear in MFI1 frame 15. The SQ range of values allows up to 256 individual STS-1s to be concatenated. The SQ also identifies the sequence or order in which the individual STS-1s are combined in the specific virtual concatenation. Sequence numbering starts with 0 and ends with X-1. The STS-1-9v example would contain SQ numbers from 0 to 8 in the 256 possible MFI2 value fields.

H4 byte								1st multi-frame number	2nd multi-frame number
Bit 1	Bit 2	Bit 3	Bit 4	Bit 5	Bit 6	Bit 7	Bit 8		
				1st multiframe indicator MFI1 (bits 1-4)					
Sequence indicator (SQ) MSBs (b1-4)				1	1	1	0	14	n-1 (ex.15)
Sequence indicator (SQ) LSBs (b5-8)				1	1	1	1	15	
2nd multiframe indicator MFI2 (b1-4)				0	0	0	0	0	
2nd multiframe indicator MFI2 (b5-8)				0	0	0	1	1	
CTRL				0	0	1	0	2	
GID ("000x")				0	0	1	1	3	
Reserved ("0000")				0	1	0	0	4	
Reserved ("0000")				0	1	0	1	5	n (ex.16)
CRC-8				0	1	1	0	6	
CRC-8				0	1	1	1	7	
Member status				1	0	0	0	8	
Member status				1	0	0	1	9	
RS-ACK				1	0	1	0	10	
Reserved ("0000")				1	0	1	1	11	
Reserved ("0000")				1	1	0	0	12	
Reserved ("0000")				1	1	0	1	13	
Sequence indicator (SQ) MSBs (b1-4)				1	1	1	0	14	
Sequence indicator (SQ) LSBs (b5-8)				1	1	1	1	15	
2nd multiframe indicator MFI2 (b1-4)				0	0	0	0	0	
2nd multiframe indicator MFI2 (b5-8)				0	0	0	1	1	
CTRL				0	0	1	0	2	
GID ("000x")				0	0	1	1	3	
Reserved ("0000")				0	1	0	0	4	n+1 (ex.17)
Reserved ("0000")				0	1	0	1	5	
CRC-8				0	1	1	0	6	
CRC-8				0	1	1	1	7	
Member status				1	0	0	0	8	

Figure 8-20. The H4 multiframe used with virtual concatenation

As a simple example of how this all works, consider two sites with two STS-1s between them. The customer wants to combine these two STS-1s into an STS-1-2v to provide 103.68 Mbps between the two locations (perhaps for Fast Ethernet speed connectivity for a particularly busy period). So the customer sets the SQ values on one payload to 0 and the SQ values on the other payload to 1. A single, virtually concatenated payload can now be generated and sent (presumably, the return path works the same way). That's all there is to it.

Between the sites, the STS-1s can take different paths, even on different rings, and so have different delays to the other site. At the receiving end, the payload order is recovered from the SQ numbers and the arriving bits inside the payloads pasted together in the order specified. This sending process is shown in Figure 8-21.

But how do the endpoints know that virtual concatenation is in effect or not? What if there is an STS-3 between the sites? How is the number of STS-1s to be changed to add or subtract bandwidth for the virtual concatenation? This is where LCAS comes in.

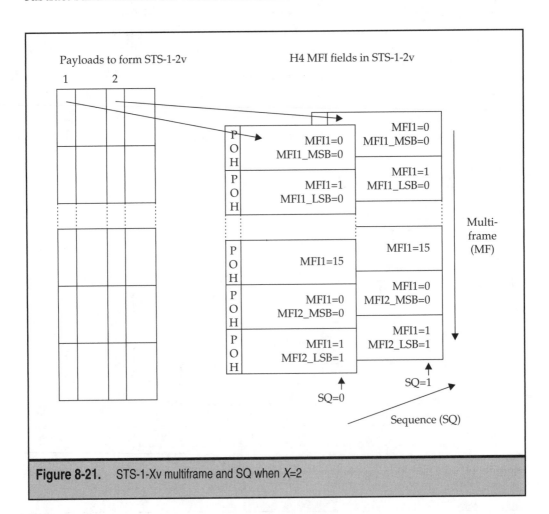

Figure 8-21. STS-1-Xv multiframe and SQ when *X*=2

The link capacity adjustment scheme (LCAS) is a complete protocol addressing all aspects of virtual concatenation. For example, the SQ numbers are assigned automatically when the concatenation is set up and are not individually configurable. LCAS performs this task. Figure 8-21 shows more than just MFI and SQ bits in the H2 multiframe. These other bits are used with LCAS and not discussed in further detail, with the exception of the GID field. The group identifier (GID) allows LCAS to distinguish multiple virtual concatenation groups on a single STS-N. The GID is generated by the path equipment "pseudo-randomly" and is not configurable. Regarding the full LCAS protocol, it is enough to note that in addition to adjusting frame differential delays, LCAS allows a SONET channel to be in one of five states:

▼ **IDLE** This channel is not provisioned to participate in the concatenated group.

■ **NORM** This channel is provisioned to participate in the group (identified by GID) and has a good path to the endpoint. (As an aside, SQ = 11111111 initially, since the exact sequence of channels has not yet been determined.)

■ **DNU (DO NOT USE)** This channel is provisioned for the group but has failed. Failures show up as reduced bit rates, as long as one path remains to the destination.

■ **ADD** This channel is being added to the group to increase the bandwidth.

▲ **REMOVE** This channel is being deleted from the group to reduce bandwidth.

As an aside, SQ = 11111111 initially when a channel joins a group, since the exact sequence of channels has not yet been determined. A new sequence number is accepted if the received SQ has the same number in m consecutive H4 multiframes (m must be greater than or equal to 3 or less than or equal to 10).

In SDH, virtual concatenation closely follows the same mechanisms as SONET, but with the STM-1-Xv terminology and STM-1 units. The definition of the H4 byte is the same in SONET and SDH.

Lower Order Virtual Concatenation with K4(Z7) bit 2

Virtual concatenation at the STS and STM level is not the only form of virtual concatenation allowed in SONET/SDH. SONET allows virtual concatenation of virtual tributaries of the form VTn where n = 1.5, 2, 3, and 6. These form VTn-Xv payloads. SDH allows virtual containers of type VC-3-Xv, VC-4-Xv, VC-2-Xv, VC-12-Xv, and VC-11-Xv, but the principles are the same as in SONET. As before, to avoid repetition and to keep explanations as simple as possible, the basics of virtual concatenation explored in this section will mainly use SONET examples and terminology.

This section on lower order virtual concatenation will be briefer than the introduction in the previous section. Now that the basic mechanisms have been explained in principle if not in detail, the main issues to be addressed in this section are where the extended signal label, MFI, SQ, and LCAS fields are placed in the tributary overhead. No H4 byte is available at the tributary level, of course.

Simply put, lower order virtual concatenation uses the K4(Z7) path overhead byte bit 1 multiframe in place of the extended signal label in the C1 byte, and bit 2 in place of the H4 multiframe. The SQ function is embedded in the bit 2 multiframe, as might be expected. The LCAS functions are embedded in the same type of multiframe structure in what were presented as the reserved fields in Chapter 7 (LCAS support is not strictly required for virtual concatenation, but it certainly makes configuration at both sites easier). The structure of the K4 (Z7) bits 1 and 2 when used for lower order virtual concatenation is shown in Figure 8-22. As before, the details of LCAS operation are not considered further.

The general idea of tributary virtual concatenation in SONET is shown in Figure 8-23. The idea is the same as at the STS (or STM) level: associate and order various payloads *of the same type* as a unit. All of the other rules and functions of higher order concatenation also apply to the lower order form, so there are fewer details. The number of columns that make up the VTn-Xv payload vary according the VT size, from 25 column units in a VT1.5 to 106 column units in a VT6. The value of X determines the number of VT payloads needed to carry the concatenated information, and the value of n determines the column structure. The value of X has a maximum value determined by the VTn type.

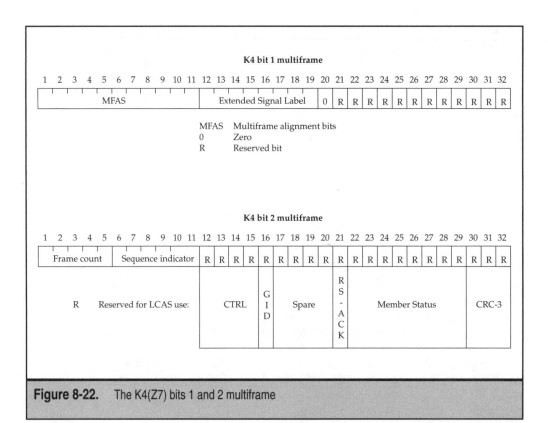

Figure 8-22.   The K4(Z7) bits 1 and 2 multiframe

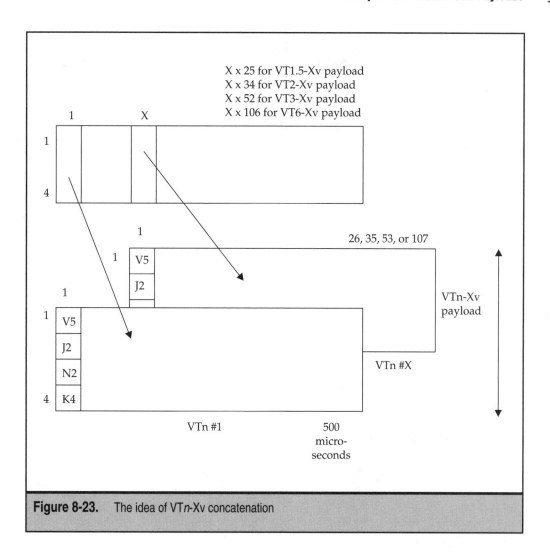

Figure 8-23.   The idea of VT*n*-Xv concatenation

Lower order virtual concatenation can be very granular, added and dropping bandwidth in units as small as 1.6 Mbps. The exact amount of each bandwidth step varies with the size of the tributaries being concatenated. Table 8-11 shows the defined capacities of virtually concatenated VT*n* payloads and the incremental steps.

It should be noted that because the SQ field at the VT level is six bits and not eight bits as in H4 level concatenation, the value of X is limited to 64 values. Also, as SDH specifications note, it would be very inefficient to concatenate more than 63 1.5 Mbps channels. There are just too many other more effective ways to obtain about 100 Mbps and above.

Naturally, in an STS-1 (for example), more VT channels than present cannot be concatenated. So X for a VT1.5 is limited to a maximum of 28, and so on. Note also the presence of SDH container type options for some levels above STS-1.

VT*n*-Xv	Carried in	X values	Capacity (Mbps)	Steps (Mbps)
VT1.5-Xv	STS-1	1 to 28	1.6 to 44.8	1.6
VT2-Xv	STS-1	1 to 21	2.176 to 45.969	2.176
VT3-Xv	STS-1	1 to 14	3.328 to 46.696	3.328
VT6-Xv	STS-1	1 to 7	6.784 to 47.448	6.784
VT1.5-Xv/VC-11-Xv	STS-3c	1 to 64	1.6 to 102.4	1.6
VT2/VC-12-Xv	STS-3c	1 to 63	2.176 to 137.088	2.176
VT3-Xv	STS-3c	1 to 42	3.328 to 139.776	3.328
VT6	STS-3c	1 to 21	6.784 to 142.464	6.784
VT1.5-Xv/VC-11-Xv	Unspec'd	1 to 64	1.6 to 102.4	1.6
VT2/VC-12-Xv	Unspec'd	1 to 64	2.176 to 139.264	2.176
VT3-Xv	Unspec'd	1 to 64	3.328 to 212.992	3.328
VT6	Unspec'd	1 to 64	6.784 to 434.176	6.784

Table 8-11. Capacities of virtually concatenated SONET VT*n* payloads

Although SDH lower order virtual concatenation will not be discussed in any detail, Table 8-12 shows the SDH equivalent information for the defined levels. Note the obvious overlaps for similar structures.

VC-*n*-Xv	Carried in	X values	Capacity (Mbps)	Steps (Mbps)
VC-11-Xv	VC-3	1 to 28	1.6 to 44.8	1.6
VC-11-Xv	VC-4	1 to 64	1.6 to 102.4	1.6
VC-11-Xv	Unspec'd	1 to 64	1.6 to 102.4	1.6
VC-12-Xv	VC-3	1 to 21	2.176 to 45.696	2.176
VC-12-Xv	VC-4	1 to 63	2.176 to 137.088	2.176
VC-12-Xv	Unspec'd	1 to 64	2.176 to 139.264	2.176
VC-2-Xv	VC-3	1 to 7	6.784 to 47.448	6.784
VC-2-Xv	VC-4	1 to 21	6.784 to 142.464	6.784
VC-2-Xv	Unspec'd	1 to 64	6.784 to 434.176`	6.784

Table 8-12. Capacities of virtually concatenated SDH VC-*n* payloads

The VT11-Xv concatenation inside the VC-3 is possible only when an STM-0 is used. That is, when VT1.5s are multiplexed as C-11 –> VC-11 –> TU-11 –> TUG-2 –> VC-3 –> AU-3 –> STM-0 and so looks just like an STS-1 structure.

All in all, both higher order and lower order virtual concatenation make SONET/SHD more flexible and useful than ever before.

GENERIC FRAMING PROCEDURE (GFP)

Virtual concatenation is all about compensating for the initial optimization of SONET/SDH for carrying channelized voice. Expanded payload mappings not only for ATM cells, but mappings that can be used for frame relay and IP packet streams were only a first step. Flexibility in the form of higher and lower order virtual concatenation was a good second step because many serial bit streams are highly variable over time and not as easy to handle my just adding more channels as in voice. But the most common layer 2 frame structure in use by the customers of SONET/SDH is not frame relay or the HDLC/PPP frames commonly used with IP. The most common framing of all is the ordinary Ethernet frame, which stays the same across all speeds of Ethernet (jumbo frames are not considered here).

The Generic Framing Procedure (GFP) started as an initiative to define a way to map Ethernet frames to a SONET/SDH payload. This is not a trivial task, but it is not especially difficult either. The only reason it is *not* trivial is that the idea is not to use a new value of the C2 signal label in the path overhead for every possible data frame structure of payload content. The value of 1B is now used in the C2 signal label for GFP payload mapping. Without a unique value for every possible payload content, when the C2 signal label is set to 1B, the path equipment on each end must use other means to determine the precise format of the arriving payload. This is really not a big issue. As pointed out in Chapter 7, almost all networks before SONET/SDH relied not on network overhead but user configuration to set the format of the arriving and departing payload content.

The protocol stack established for GFP in SONET/SDH is shown in Figure 8-24. Although Ethernet, HDLC, and Token Ring are mentioned by name, anything that corre-

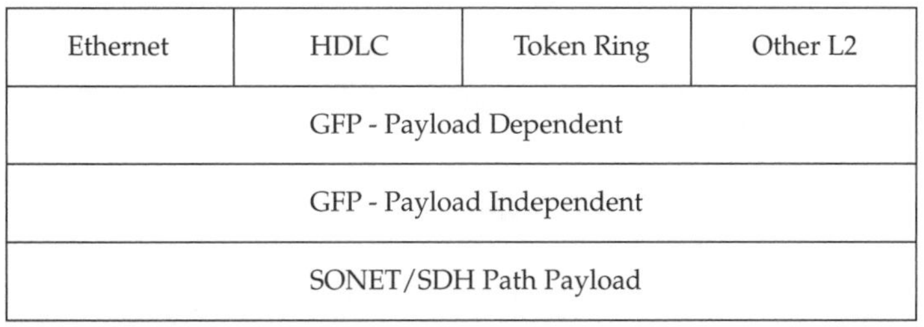

Ethernet	HDLC	Token Ring	Other L2
GFP - Payload Dependent			
GFP - Payload Independent			
SONET/SDH Path Payload			

Figure 8-24.   The GFP protocol stack

sponds to a standard layer 2 protocol frame structure should work. For example, switched digital video streams, which usually have distinct framing, would be no problem at all for GFP.

Note that the fields that GFP itself adds to the layer 2 frame are divided into payload dependent and payload independent functions. Usually this adds up to a series of headers with inner headers representing payload dependent functions and outer headers that are always the same. There is not much more to it.

At its most basic, GFP uses the SONET STS-1 87 column payload, with one column used for path overhead, and columns 30 and 59 used for fixed stuff bytes for SDH compatibility. This leaves 756 bytes for the GFP information itself. Most GFP frames will be larger than that (and there is no maximum defined), so GFP information will routinely span payload frames. GFP frames arrive as a continuous stream because special idle frames are always inserted for delineation and there is no bit-level fill pattern. The GFP frames are scrambled when the GFP headers are added, but not again when they are mapped to SONET or SDH.

With GFP, SONET/SDH takes a big step toward becoming as useful for data applications as it has always been for channelized voice.

CHAPTER 9

SONET/SDH Synchronization and Timing

A key feature, if not *the* key feature, of SONET/SDH is its synchronous operation. This feature is so important to SONET that it appears first in the name Synchronous Optical Network, even before the fiber optic aspect of SONET. After all, there are many forms of fiber optic network links and communications networks; however, it is the synchronous operation of SONET that makes it truly distinctive. And SDH is the *synchronous* digital hierarchy.

This chapter explores the synchronous aspects and features of SONET/SDH in more detail than has been discussed to this point. This chapter begins with a look at network synchronization in general, and points out that distribution of timing information is not an exclusive feature of SONET/SDH, but is indeed necessary in all networks that cross-connect voice. What SONET/SDH brings to network synchronization and timing is not so much invention, but rather innovation. SONET/SDH is synchronized to a much higher degree than is the older T-carrier or E-carrier networks, which are now considered to be part of the plesiochronous digital hierarchy (PDH). SONET/SDH, of course, is part of the newer synchronous digital hierarchy.

The chapter also considers exactly how this stringent timing information is distributed in a SONET/SDH network, and some of the requirements imposed on SONET/SDH equipment to enforce this timing requirement. Processor clocking has improved tremendously over the years, as have the processors themselves; SONET/SDH takes full advantage of this fact.

The chapter concludes with the admission that SONET/SDH is not perfectly synchronous, and details why. This being the case, SONET/SDH's lack of perfect synchronization leads to a reconsideration of the use of SONET/SDH overhead pointers, and gives a real-world example of precisely how these pointers are used.

NETWORK SYNCHRONIZATION

The presence of pointers in SONET/SDH overhead for payload envelopes and even tributaries was briefly introduced in Chapter 6. Some mention was made of clock phase differences and timing jitter, but the chapter included no in-depth exploration of SONET/SDH synchronization and timing. This chapter is the proper place to investigate this most distinguishing characteristic of SONET. The need for payload pointers in SONET is related to the differences in timing (or, interchangeably, clocking information) in modern digital networks.

It may come as a surprise to many that network synchronization is important in private-line networks, which is what a SONET/SDH network is at heart, of course. Many organizations deploy and/or lease private lines with T-carrier or E-carrier and have never considered the issue of network synchronization or the distribution of clocking information (or just "clock" to many) among the many pieces of PDH equipment. Yet, the network seems to hum along just fine day in and day out. Other organizations have added a few pieces of T-carrier or E-carrier equipment to an already functioning network and suddenly experienced problems that the organization has never encountered before. Usually after much head-scratching and shoulder-shrugging terms like "timing loops" and "incorrect stratum" begin to make their way into the expert's reports on the situation. However, right down the block, at another organization with just as many PDH links, no such problems are encountered—ever. What's going on here?

Clearly, there is more to the situation than meets the eye. What is it about one private-line network that makes timing so critical, while another has no need to worry about timing? It all depends on what the network is trying to accomplish.

A truly thorough understanding of network timing requirements on the part of many telecommunications personnel is about as common as a truly thorough understanding of grounding on the part of many electricians installing telecommunications wiring in a building. Most electricians know that unless a number 6 gauge copper wire is installed from the communications rack to a water pipe or other suitable ground, the building distribution wiring will not work properly. However, a full and insightful explanation of why this should be done is seldom available from the electricians involved. This is not to downplay the expertise of cable installers, which is considerable and more than adequate for the task. Rather, the whole point is that network timing, like proper grounding, seldom becomes an issue until it is critical. (By the way, grounding a communications rack provides a return path, or ground, for normal direct current signals, which will otherwise quickly build up a net charge on the wire and prevent the modest voltages used from pushing signals through the wire.)

Because network synchronization and the distribution of clock generally plays such a large role in private-line networks, and SONET/SDH in particular, this chapter will first attempt in simple terms to explain exactly what a network "clock" is doing. Along the way, the configurations in which clock distribution becomes critical will be examined in detail.

Consider two T-carrier, private-line networks. The example is simple, but has all the ingredients needed to understand network timing requirements. The example also applies to E-carrier or any other PDH link. The first network is shown in Figure 9-1.

The T-carrier network in Figure 9-1 consists of three T-1 point-to-point links connecting three sites in an organization's network. The customer premises equipment (CPE) at the end of the T-1 link in each case is a T-1 multiplexer (or just "multiplex") that usually combines the inputs from 24 DS-0 ports and multiplexes them onto the outgoing T-1. Of course,

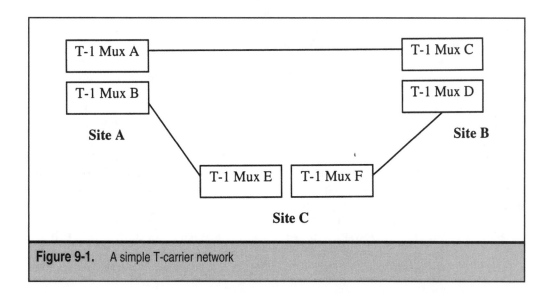

Figure 9-1. A simple T-carrier network

demultiplexing takes place at the other end of the link. For the sake of simplicity, the 64 Kbps inputs and outputs are not shown in the figure.

As long as a network consists of a series of point-to-point links, there is really no need to "distribute clock" among the network devices in order to "synchronize" the network. Point-to-point links in this configuration can easily use "loop" timing and recover clock from the received data stream. This loop timing process needs an explanation.

First, it is important to realize that all digital links, including SONET/SDH fiber links, consist of two paths: a transmit and receive path. This is due to the fact that digital signals are not like analog signals. It is easier to send digital signals in a simplex, unidirectional fashion than it is to try to send digital signals full duplex, in both directions, as with analog signals (such as a simple PC, full duplex analog modem arrangement). In a T-1, these two paths are provided by two twisted pairs of copper wire, one for transmit (typically abbreviated TX) and one pair for receive (RX).

When a T-1 multiplexer sends digital signals from a buffer, which is usually a full frame or more long (193 bits in the case of T-1), timing the output signal is not much of a problem. This means that the sender obviously knows where one bit stops and another begins because the bits are fed serially from the buffer to the transmitter. In a T-1, the 193 bits in a full T-1 frame must be sent 8,000 times a second, or 193 bits in 125 microseconds (1/8,000th of a second). Thus, each bit lasts about 0.65 microseconds. Obviously, the clock in the T-1 multiplexer must be capable of "ticking" faster than 1 million times per second (1 million microseconds = 1 full second) for this to be done properly.

This is not a problem for T-1 equipment, thankfully; the T-1 transmitter merely reads the next bit in the output buffer. When it is a "1" bit, some voltage is placed on the output pair for half of the 0.65 microsecond bit time interval (called a "50% duty cycle"). When the output buffer contains a "0" bit, no voltage is placed on the output pair for the entire 0.65 microsecond bit interval. The point is that the sender always knows when its own bit interval starts and ends.

However, the situation is different on the receiving pair. There is no explicit bit timing and interval information sent along with the data from the transmitter at the other end of the receive pair. It is up to the T-1 receiver electronics to determine when it should check the input pair to see what the voltage is. Presence of voltage indicates a "1" arriving in that bit interval and absence of voltage indicates a "0" arriving in that bit interval. Of course, because even a "1" bit only has voltage on the receive pair for ½ of the full 0.65 microsecond bit interval, if the bit interval is not "looked at" by the receiver electronics at the end on the pair at just the right instant, errors will result when "1" bits are mistakenly interpreted as "0" bits. In practice, the process is a little more complex, but not enough to make the discussion invalid.

In modern digital networks, it is sometimes possible to lose sight of just how short a time interval a microsecond really is. A microsecond is below the threshold of human perception. Those who play golf can sometimes hear the brief "tick" of the club hitting the ball amid the whoosh of the driver. Studies by golf club manufacturers have revealed that the golf ball is on the club head for about 450 microseconds. The sound expands in the time it takes to reach the ear, so the golf ball "tick" is 500 microseconds in duration, half a millisecond, a full four T-carrier, E-carrier, or SONET/SDH frames long.

T-1 bits are short, about ½ a microsecond in duration, but this is nothing compared to 10 Gbps or 40 Gbps serial links. Consider 10 Gbps as an example. Sending 10 billion bits per second means each bit lasts only $1/10^{th}$ of 1 billionth on a second. A microsecond is one millionth of second, and a billionth of a second is a nanosecond. A nanosecond is sometimes called a "light-foot" in the United States because at the speed of light, a signal can only travel about a foot in one nanosecond. So, on a 10 Gbps link, each bit is only about an inch long.

Keeping these short time intervals in mind, a hazard exists in this lack of an explicit timing circuit in PDH network links. Because no explicit timing information is sent with the data, how is the receiver on the T-1 multiplexer supposed to determine exactly when the bit interval chosen by the sender T-1 multiplexer starts and ends? Each clock in the respective T-1 multiplexers at each end of the link may be equally accurate, but out of phase. The same may happen when two identical wristwatches show a two-minute time difference, although they are both equally accurate on their own. Clearly, three o'clock occurs at different times for each wristwatch wearer. If something vital were to happen at exactly three o'clock and would require the presence of both wristwatch wearers, a need would exist to "synchronize watches." Humans can do this by exchanging explicit timing information through direct communication ("I have five minutes to three"). But how are T-1 multiplexers to synchronize bit intervals when there is no contact between them for exchanging explicit timing information? In technical terms, T-1 multiplexers cannot "recover clock" from the raw, received bit data stream.

Fortunately, there is a simple trick employed by the T-1 multiplexers in the example network to solve this limitation. The six T-1 multiplexers can simply employ "loop" timing to provide timing information to the receiver pair. With loop timing, the sending clock "ticks" are used to provide clocking information for the receiver pair electronics. Because both T-1 multiplexers do the same thing on their receiver pairs, this whole system forms what is called a "phase locked loop" because each sender clock prevents the receiver clock from getting too far out of phase with their counterparts. In practice, there is a "double-phase locked loop" employed, but the principle is the same. Figure 9-2 shows such loop timing in principle, although the whole process is performed internally and electronically in the T-1 multiplexers.

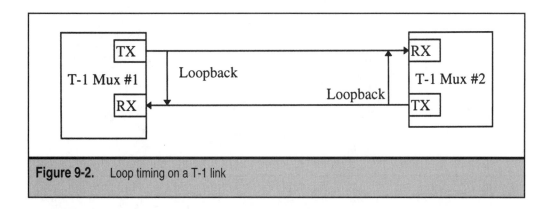

Figure 9-2. Loop timing on a T-1 link

Loop timing, locked or not, still has limitations. The clocks can still wander from "ticking" at the proper point in a bit interval, and if this timing jitter were to persist in one direction or another ("faster" or "slower") for long enough, a timing slip would invariably result. In a slip, a bit is either lost or duplicated, because the incoming bit stream is not sampled and interpreted electronically at the correct time. Clocks running slower will invariably lose a bit now and then, and clocks running faster will invariably duplicate a bit sooner or later. This factor is shown in Figure 9-3.

In order to prevent persistent jitter faster or slower from inducing slips, T-1 equipment clocks exhibit a characteristic *pull-in range* over which the clocks can adjust their bit intervals and compensate for the effects of jitter. All clocks, therefore, have a characteristic accuracy (which limits slips caused by jitter to a specific number per time interval) and pull-in range (which is how they ensure accuracy).

On a point-to-point T-1 link, the bits originate from a send buffer and finish in a receive buffer. The buffers are filled and emptied by the digital end equipment at each end of the T-1 link. This limits the effects of jitter and their resulting slips. Simple loop timing, therefore, is adequate for point-to-point T-1 link operation. In the example above, the six T-1 multiplex and three link T-carrier networks operate correctly with simple loop timing employed on each of the three links.

As a result, the organization with this type of T-1 network, with just a series of point-to-point T-1 links, no matter how many, will employ loop timing everywhere and

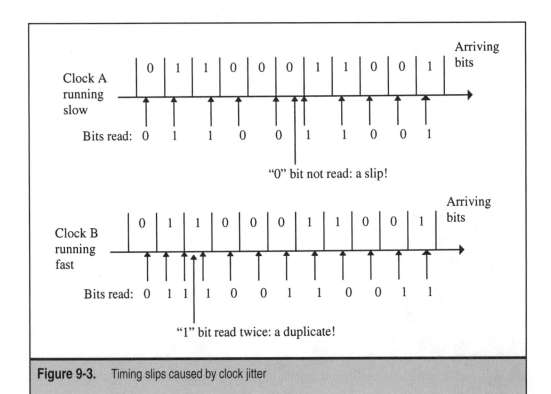

Figure 9-3.   Timing slips caused by clock jitter

wonder exactly what the debate about network timing is all about. As long as the bits originate at a user end device, and are delivered to another user end device, there is no problem with loop timing, except for occasional slips caused by jitter.

T-1 multiplexers are not the only devices that may be deployed in a T-1 network. Mention has been made in an earlier chapter of add/drop multiplexing (ADM) in T-carrier and SONET networks. When applied to T-carrier equipment, adds and drops may be done by equipment normally known as a digital cross-connect system (DCS), but the net result is the same. That is, specific DS-0 channels can be terminated at a site, while others are added to the outgoing data stream, while still other DS-0 channels continue essentially untouched through the DSC or ADM.

In SDH, the ideas are the same, but the names of the network elements (NEs) in the specifications vary somewhat from those in SONET. However, the generally simpler SONET terminology of ADM, terminal multiplexer (TM), and DCS is becoming more common even in SDH documentation, especially from the equipment vendors. Because the next few chapters use SONET NE terms almost exclusively, even in an SDH context, this is a good time to distinguish SDH equipment terms from SONET terms.

In SDH, the following NEs are typical path terminating devices, using their proper SDH names:

▼ Low-order (or low-speed) multiplexer

■ Wideband cross-connect system

▲ Subscriber loop access system

An SDH multiplex section is defined as the transmission medium, together with the associated equipment, required to provide the means of transporting information between two consecutive NEs. One of the NEs originates the line signal and the other NE terminates the line signal in each direction. An SDH Multiplex Section Terminating Equipment (MSTE) is a network element that originates and/or terminates STM-n signals. The MSTE can originate access, modify, or terminate the multiplex section overhead, or can perform any combination of these actions on the STM-n signal. The definition of an SDH multiplex section is the same as the "line" definition used in SONET. SDH specifications offer examples of SDH MSTEs called:

▼ Optical line terminal

■ Radio terminal

■ High-order multiplexer

▲ Broadband cross-connect system

Regardless of the name, timing plays an important role in all SONET/SDH NEs. Now, to return to the PDH, consider the second network shown in Figure 9-4.

Three sites are still connected by three links, as before; however, the end devices are no longer simple multiplexers, but rather more complex DCS devices. These have considerable add/drop capability, and it is possible for any number of DS-0s on any T-1 to be dropped, added, or passed through unchanged at any site, depending on how the DCS is

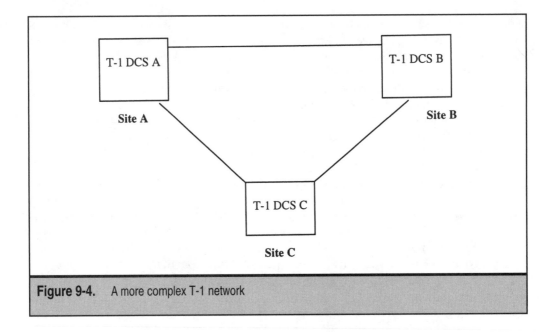

Figure 9-4.   A more complex T-1 network

configured by the users. The DS-0s that start out from or end up in a buffer destined for an end user device are not the problem. The problem is with the bits representing the DS-0s that are passed directly from an incoming T-1 frame to an outgoing T-1 frame. These "pass through" DS-0 bits are the reason that other timing arrangements must be made beyond simple loop timing.

Here is why. As soon as one or more of these devices involve the taking of bits off of one link and putting them directly on another link, such as occurs in an add/drop multiplexer (ADM) or digital cross-connect system (DCS), then the relationship between frames starts to become critical, especially in a synchronous multiplexing scheme like SONET. A DS-0 that is dropped and a DS-0 that is added must occupy exactly the proper bit positions in the overall T-1 frame structure. In this case, the receive clock on an input port must agree precisely with the transmit clock on the output port, and not by just a simple loop mechanism.

Therefore, networks that consist of more than just simple point-to-point links, links where bits are terminated in endpoint devices at each end, need to distribute some kind of network timing signal (i.e., distribute clock) among the separate devices on the network.

In order to see this more clearly, extend the wristwatch analogy used earlier. A quick agreement to adjust one wristwatch or the other suffices to allow both humans to engage in some mutual activity at precisely three o'clock. Consider this an example of loop timing. But when DCS devices are used, the analogy is more like a conference room without a clock on the wall. When twenty people need a quick break (perhaps they have been reading too much about SONET/SDH for one session) and all must return at the same time, they cannot all easily coordinate their wristwatches so they all agree on an time to return. One person's 3:01 may be another's 2:59, and so forth. If everyone had to check with everyone else, it is easy to see that this process could easily consume most of the time allotted for the break. Jitter effects would require the process to be repeated quite often to ensure adequate accuracy for split-second break timing.

However, if there were a master clock in the room (perhaps the instructor's own wristwatch) with which all watches could be synchronized, the process would become much simpler and efficient. This is naturally the whole idea behind hierarchical *clock distribution* in digital networks. Just to make the analogy more realistic, it could be added that all of the wristwatches have been assigned a certain degree of accuracy, dictating how often they should be synchronized against the *primary reference clock* on the instructor's wrist and how long they could be expected to operate alone without error. It is to be hoped that the instructor's watch would be the most accurate of all, and that some system could be established for ensuring this.

In the example DCS network, each of the DCS devices establishes a special link (normally just another DS-0) to receive timing signals on that keeps their individual clocks within proper operational ranges. This allows each device to accurately take bits directly off of an incoming link and put them on an outgoing link directly, no matter how complex the DCS network. Organizations with many interconnected T-1 links, as in this second example, quickly gain a healthy respect for network timing synchronization and clock distribution techniques.

One issue that always comes up at this point is how NE clocks that must tick off microseconds or less accurately can be synchronized by a DS-0 "ticking" at 64 Kbps. At the risk of offending electrical engineers, the simple answer is that internal clock timing signals are not all that bad to begin with. They are usually very stable, and the jitter is limited over short intervals. So even checking things 64,000 times a second is enough to adjust for any small phase changes with reference to the equipment clock.

The need for master clocks and a timing distribution scheme is totally lost on people with experience with IP networks and the Internet. There could be many DS-1s in an ISP's network, but never a thought of timing concerns. There is a simple reason for the IP world's lack of concern about network synchronization: no one ever worries about "reaching inside" an IP packet and pulling out a particular byte. IP packets can also be switched with something like multiprotocol label switching (MPLS), or simply routed, but always as a whole unit, even when voice is inside the IP packet. In other words, no one cross-connects IP, so there is no need to synchronize IP packet byte streams. But voice bytes inside DS-1s and SONET/SDH links are cross-connected all the time. SONET/SDH equipment routinely requires cross-connecting. This is true whether the SONET/SDH links ever carry voice or not between devices like central offices or routers. SONET and SDH require stringent timing mechanisms *just in case* someone wants to use the link for voice (yet another demonstration that SONET/SDH is "optimized for voice").

Where do the DCSs in the private network find the services of a master clock? Organizations that are not service providers (or carriers) with their own ADMs or DCSs can get timing signals from the service provider. Where, then, does the service provider get it? In most modern networks, clock ultimately comes from such standard systems as an atomic clock or LORAN (the U.S. coastal long-range radio navigation system, less common today) or the global positioning system (GPS), the master clock for which is in Boulder, Colorado. (There used to be U.S. military and civilian versions of GPS, but both are now one today.) Use of GPS is more common today. Both LORAN and GPS are navigational aids that fix positions of the earth's surface by radio signal triangulation. LORAN is less common today because it was initially deployed only in coastal areas for ocean navigation, and deployment inland was slow. GPS signals are extremely stable timing pulses and the pulses are very well-shaped. Because there are multiple GPS signals available anywhere on earth (they are satellites), usually at least three at a time, the signals can be cross-checked among each other constantly. As a result, very little wander or jitter occurs in the GPS signal; if you try to position a tank using this signal, jitter potentially can lead the tank to trample your own troops, so great efforts were made to eliminate jitter.

Usually in SONET/SDH systems, the GPS signal is used to set a primary reference clock (PRC in SONET) for the entire network. To use SONET terminology, this primary reference clock is known as the Stratum 1 clock for the whole service provider's network. A Stratum 1 clock is the most accurate in the entire network. When this clock says it's noon, it's noon and that's that. The trick is to let all the other devices in the network know that it's noon. Not long ago, it was common to have only one Stratum 1 clock in the whole network. But with reliable external timing references, such as GPS, it is possible to have many Stratum 1 clocks without running the risk of any of them contradicting the others. AT&T, for instance, had 17 Stratum 1 PRCs in 2000, and will deploy more in the future.

The network Stratum 1 clock (or clocks) uses regular leased lines (usually of modest speed) to distribute these clock pulses directly to other devices in the network. These are Stratum 2 clocks, and they are directly connected to the Stratum 1 clock. It would be too expensive to hook everything up directly to the Stratum 1 clock, so yet another set of network devices gets clock not from the Stratum 1 clock, but from a Stratum 2 clock. These are, not surprisingly, Stratum 3 clocks. Yet other devices, the Stratum 4 clocks, get their timing from the Stratum 3 clocks.

The whole structure of the network timing hierarchy, which is strictly spelled out in major networking standard documents, is shown in Figure 9-5. This architecture is usually called the *hierarchical source-receiver method* of clock distribution.

In Figure 9-5, networks are shown getting their PRC absolute timing information from LORAN or GPS. There may actually be more than one PRC in a network, used for backup or other purposes. The two networks in the figure may be carrier networks, two private corporate networks, one of each, or even other more exotic combinations.

In the real T-carrier world, most Stratum 2 clocks are in large toll offices for carrier networks or major network nodes in the case of private networks. Stratum 3 clocks are located in smaller end offices, such as a local central office. Finally, the Stratum 4 clocks are usually in CPE devices, such as corporate PBXs or T-carrier multiplexers. The whole scheme is quite effective, as long as a lower level clock receives timing pulses from a higher level clock. It is possible, when proper care is not exercised, to create situations like "timing loops" where the ultimate source of a clock signal may be a lower level clock. These situations must be avoided at all costs. A simple example of a timing loop occurs when someone is unsure the time on their watch is correct. They ask someone else the time, who asks someone else, who now looks at the *first* person's watch to verify the information! This is not helpful, so the rule is to *never* rely on a lower-level clock for timing signals. This prevents timing loops. Timing loops are considered in Chapter 16 on SONET/SDH rings, along with the operation of the S1 overhead byte.

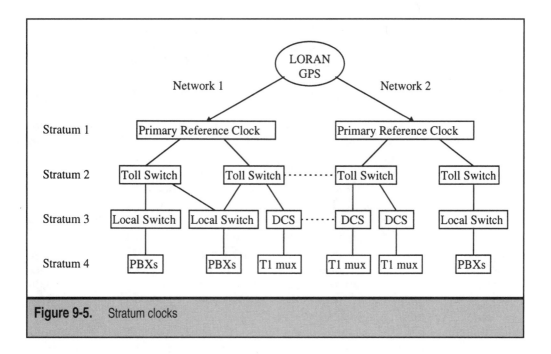

Figure 9-5.   Stratum clocks

Note that a local switching office may maintain more than one link to a higher stratum clock, which is okay, as long as only one link (the *primary*) is used for timing signals while the other(s) are maintained solely as a backup (the *secondary* or *secondaries*). Note also that the digital network from another service provider would have its own clock distribution scheme. That is also okay. The only time to worry is when a link goes from one service provider to another (i.e., the mid-span meet situation). This is what makes the T-carrier network truly plesiochronous because one network cannot impose its timing on another. Unless, of course, it is a SONET/SDH network. Then the more stringent SONET/SDH timing makes such a mid-span meet at least feasible.

What happens when a link to a reference clock is lost? In this case, each stratum clock has guaranteed performance characteristics that have different slip and pull-in parameters, depending on stratum. Naturally, the higher the Stratum level, the more accurate the clock must be. The idea is to prevent slips from occurring until the timing links are restored to service. Many service providers maintain physically separate facilities just for clock distribution, but nothing prevents any organization from distributing clock through the T-carrier network itself.

The ITU-T characterizes receiver clock performance based on whether the clock is operating in one of three modes:

▼ **Ideal operation**   the short-term behavior of the clock

■ **Stressed operation**   the typically operational mode where timing is received from elsewhere

▲ **Holdover operation**   the "rare" cases when all timing references are lost

In ideal operation mode, the receiving clock never loses the input timing reference. This is certainly true in the short term, but is not typical of normal day-to-day operation. But how the clock operates in the short term is a good baseline for overall timing performance. For example, ideal operation displays the short term "noise" of the clock, which will manifest itself as DS-3 level payload pointer adjustments in SONET and similar actions in SDH.

Stressed operation always expects that there will be short interruptions of the timing reference signal, which is certainly true of normal network operations. There can be from 1 to 100 interruptions per day according to the ITU-T. Naturally, during the interruption, the timing reference cannot be used. If the interruption is short, the receiver will restore the timing signal when it reappears. Otherwise, the receiver clock will have to switch reference if the outage is protracted. In either case, there will be some error between the time kept locally and the newly restored or secondary reference. This difference should be less than one microsecond. A timing error of even one microsecond can cause up to seven pointer adjustments at the DS-3 level in a short period of time, but SONET can handle this without error. And a 1-microsecond timing difference will not affect the DS-1 level payloads in SONET (or E-1 in SDH). However, timing errors will accumulate and eventually cause a pointer adjustment at lower SONET/SDH levels.

Holdover operation occurs when a SONET/SDH NE loses all timing references, primary or secondary, for an extended period of time. In SONET/SDH, if an NE is on holdover operation or receiving clock from a source in holdover operation, the pointer adjustments come often and at steady intervals.

If all is working as planned, there should be few DS-1 or E-1 pointer adjustments each day. Even in holdover operation, a SONET/SDH NE should have no more than one DS-1 pointer adjustment in nine seconds, and one per seven seconds for E-1.

STRATUM CLOCKS

When a link is lost to a higher-level Stratum clock in the timing hierarchy, what happens depends on at which level of hierarchy the link loss occurs.

Stratum 1 clocks, common even when GPS is used, are typically Cesium or Rubidium atomic clocks that derive their signals from the vibrations of atomic nuclei. These clocks, which used to cost $100,000 not too long ago but now go for about $10,000, are accurate to better than 0.00001 parts per million (i.e., a hundredth of a nanosecond—about the same relationship as one second is to 100 billion seconds, or some 3,000 years). Hence, these atomic clocks may lose one second every 3,000 years. If the link to GPS were lost, there obviously would be little to worry about. Newer atomic clocks have been prototyped at about 100 times more accuracy, or down to the picosecond levels; thus, these may lose one second every 300,000 years.

The Stratum 2 clocks are "only" accurate to 0.016 parts per million (less than a hundredth of a microsecond). The important thing about Stratum 2 is that if the reference link were lost to the Stratum 1 clock, the Stratum 2 clock would not wander off of the mark far before the link would be restored. Otherwise, the frame sending time would be out of synchronization with the other devices on the network. This wandering would eventually cause a timing slip. The number and interval of such resulting slips are strictly controlled by the stratum network timing standards. Therefore, all Stratum 2 clocks must have less than 255 slips in the first 86 days after a loss of reference. In the case of timing signals, a slip is just a lost or repeated timing pulse. Moreover, the first slip cannot occur within seven days of the reference loss.

A common enhancement is to add GPS capabilities to Stratum 2 or 3E (Stratum 3 "enhanced" clocks, known as "2 GPS" and "3E GPS," which effectively makes them into Stratum 1 clocks as long as the GPS link is functional). Stratum 3E was invented by Telcordia (Bellcore) and is not acknowledged by standards bodies (it is one of those "national use" reserved items in SDH).

Stratum 3 clocks are accurate to 4.6 parts per million (almost five microseconds, pretty sloppy in comparison to Stratum 2 clocks). These clocks must have fewer than 255 slips in the first 24 hours after a loss of reference. The first slip cannot occur less than six minutes after the reference loss.

Stratum 4 clocks have no guarantees along these lines. They can slip all over the place and no one can complain. Most CPE devices at end user sites, PBXs or T-1 muxes, are Stratum 4 clocks, as are all PCs and even many wristwatches.

Here is the main point regarding Stratum clocks. Even central office switches with plenty of T-carrier links to customer sites and other switching offices usually have only Stratum 3 accuracy. The end-user devices where the T-carrier bits originate and terminate are even worse. Therefore, it is not particularly unusual to have only 192 bits in a buffer instead of 193, or even 194. This is why T-carrier employs "slip buffers" in equipment and must use "bit stuffing" in T-3s. For this reason, SONET uses only Stratum 3 clocks or better to achieve "synchronous" multiplexing.

The Larus Company, a common vendor of carrier timing equipment, has extended the Stratum hierarchy with its own widely used enhancements to Stratum 3 and 4 clocks. These are known as Larus 3E and 4E, but they are not strictly part of the timing hierarchy.

The major accuracy, pull-in ranges, stability, and time to first slip for all stratum clocks are shown in Table 9-1. This applies to SONET only, and even then, ANSI does not recognize the enhanced 3E and 4E clock levels.

Stratum	Accuracy	Pull-in Range	Stability	Time to First Slip
1	1×10^{-11}/day	Not applicable	Not applicable	72 days
2	1×10^{-8}/day	Synchronizes to clock with accuracy of +/-1.6×10^{-8}/day	1×10^{-10}/day	7 days
3E	1×10^{-6}/day	Synchronizes to clock with accuracy of +/-4.6×10^{-6}/day	1×10^{-8}/day	3.5 hours

Table 9-1.   Stratum 1, 2, and 3E characteristics

Stratum	Accuracy	Pull-in Range	Stability	Time to First Slip
3	4.6×10^{-4}/day	Synchronizes to clock with accuracy of +/-4.6×10^{-6}/day	3.7×10^{-7}/day	6 minutes (255 in 24 hours)
4	32×10^{-6}/day	Synchronizes to clock with accuracy of +/-32×10^{-6}/day	Same as accuracy	Not yet specified
4E	32×10^{-6}/day	Synchronizes to clock with accuracy of +/-32×10^{-6}/day	Same as accuracy	Not applicable

Table 9-1.   Stratum 1, 2, and 3E characteristics *(continued)*

PLESIOCHRONOUS CLOCKING

In spite of highly accurate clocks in the T-carrier network, synchronous operation of multiplexers and other pieces of equipment has never been possible without the need for bit stuffing to make up for these timing differences. There are two reasons for this.

First, the T-carrier links to be multiplexed may come from two different networks. For example, a DS-4 may need to be created from three input DS-3s. One comes from the service provider's own network, the second comes from another service provider's network, and the third comes from an "asynchronous island," which is another way of saying a private network with its own internal clocking scheme.

Loss of reference also allows the clocks on isolated devices to wander out of phase. They do not necessarily slip, but they may be off just enough to deliver the "wrong" number of bits per frame time. Of course, these devices still need to send and receive data.

Because each of the clocks may not be in phase with each other, and probably will not be, it is necessary to buffer, usually a whole frame, and insert stuffing bits and control bits. The stuffed bits make up for a shortfall in the number of bits in a frame and the control bits enable the receiver to detect the stuffed bits. In the case of a T-1, the slip buffers will repeat an entire frame if the input clock constantly runs slower than the output clock in a DCS. An entire frame will be dropped if the input clock constantly runs faster than the output clock in a DCS. This is just another way of saying that clock distribution and timing accuracy is critical in networks where T-carrier equipment is hooked up back to back, instead of consisting of a small (or even large) number of point-to-point links.

This whole process of adjusting for timing variations translates into much effort and overhead. For example, a DS-4 uses more than 5 Mbps (more than three T-1s worth of bandwidth!) just for overhead to package the plesiochronous DS-3s. The bit-stuffing bandwidth would be much better used delivering users' bits (and adding revenue) to the carrier's network. This bit-stuffing and control process for DS-3 to DS-4 NA is illustrated in Figure 9-6.

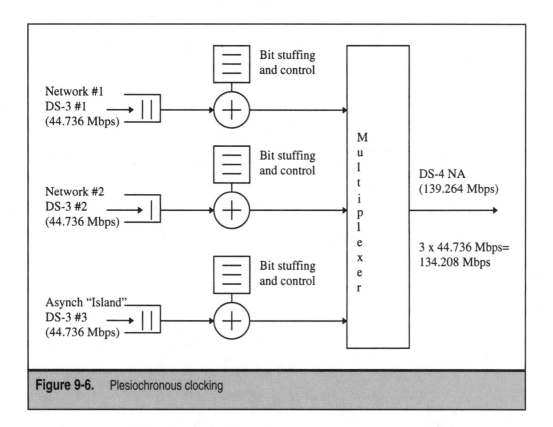

Figure 9-6. Plesiochronous clocking

SONET/SDH Synchronous Multiplexing

More accurate network clocking allows SONET/SDH to use more straightforward, byte-interleaved, synchronous multiplexing than the awkward bit-stuffing methods used with the higher levels of the T-carrier or E-carrier digital hierarchy. Thus, four STS-3/STM-1s can easily be combined into one STS-12/STM-4 without the use of large buffers, bit stuffing, or much control overhead.

This is not achieved without a price, however. There can be no Stratum 4 clocks in SONET/SDH. All clocking in all SONET/SDH devices must be Stratum 3 or better for SONET, or the SDH equivalent, ITU-T G.813 Option 1. This used to be a major issue and network expense, but the affordability of accurate clocks makes stringent SONET/SDH timing that much easier to accomplish.

The major benefit is shown in Figure 9-7, using a simple SONET example. In the figure, synchronized STS-1 inputs can be directly byte-interleaved onto the synchronized STS-3 output stream. The STS-3 is then converted to optical signals and sent on the SONET fiber. No complex bit-stuffing and control procedures are needed. As a result, SONET speeds are simple multiples of the basic rate of 51.84 Mbps. No additional overhead is ever needed for bit stuffing and control.

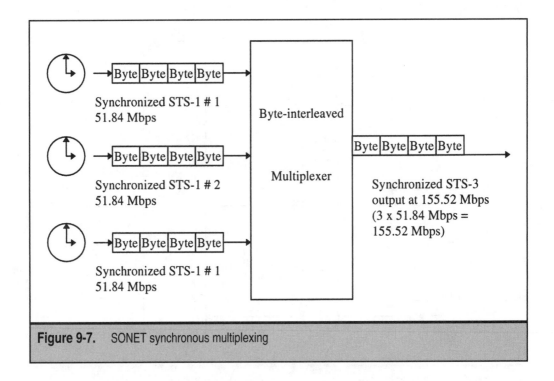

Figure 9-7.   SONET synchronous multiplexing

Synchronizing a SONET/SDH Network

It is important to realize that it is the potential need for DS-1 (and E-1) cross-connecting (pretty much a given in the voice circuit world) that SONET and SDH need to synchronize network clocks. Most telecommunications authorities use a technique called the *hierarchical source-receiver method* to accomplish this.

In the hierarchical source-receiver method, the master clock used in the network is one (or even more) primary reference source (PSR), which SDH calls the ITU-T G.811 clock. The timing reference is distributed to the rest of the network through a series of receiver clocks. Receiver clocks pass signals along to other receivers in the form of a pyramidal hierarchy.

In practice, SONET/SDH NEs can use one of five types of timing, presented here in a SONET context:

▼   **External timing**   There is a direct Stratum 1 clock reference available.

■   **Line timing**   The NE derives its own clock from the scrambled SONET signal input.

■   **Through timing**   The clock in this case is also derived from input, but then the NE sends the signal in a different outbound direction.

■   **Loop timing**   This is the same as line timing, with one major difference. Line timing can be used in an add/drop multiplexer (ADM) used as customer premises equipment (CPE) and there can be other SONET NEs beyond the ADM. Loop timing is used for straight CPE and the SONET ends at this NE. In terms of SONET equipment discussed more fully later, the ADM is configured for terminal multiplexer (TM) mode only.

▲   **Free running**   The SONET NE has access to an internal clock only.

It would be useful to give an example of how SONET NEs actually use these timing modes. The same argument applies to SDH, but this example uses SONET equipment names and clock terminology.

It is all well and good to describe Stratum clocks and how SONET timing allows for synchronous multiplexing. However, this does not explain exactly how SONET networks distribute clock among the various SONET network NEs.

Figure 9-8 shows how a SONET network is usually synchronized. In the figure, timing for outgoing OC-*n*s is shown as a dotted arrow. The internal clock of a piece of SONET equipment in a switching node (for example, a central office) may derive its timing signal from a building integrated timing supply (BITS). Of course, the BITS system

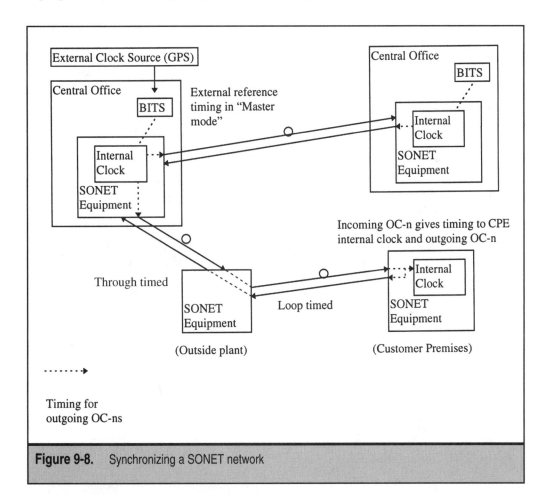

Figure 9-8.   Synchronizing a SONET network

draws its own timing from a higher layer of Stratum clocking system. The office may even draw timing directly from the GPS signals. Such building timing systems can be used to provide clock to all of the SONET devices in the switching office.

These BITS supply external reference timing in what is known as "master node" for each of the internal clocks in the service provider nodes' SONET equipment. This equipment serves as a master for other SONET network nodes, providing timing signals on its outgoing OC-*n* link at literally any level of the SONET hierarchy. Other switching offices use their own BITS to drive internal clocks and time the outgoing SONET link.

Of course, not all SONET equipment is located in switching nodes where good external timing signals or BITS signals are easily available. For instance, there may be SONET equipment deployed as *outside plant* equipment away from the central office. These other SONET nodes derive their timing from the incoming SONET bit stream in what is called a "loop timed" arrangement, also shown in Figure 9-8. Intermediate SONET equipment, such as ADMs (add/drop multiplexers), on SONET links use loop timing.

What about the SONET equipment on the customer's premises? This CPE operates in "slave mode" (the term is unfortunate, but universal), which employs "loop timing" for the internal clock. Note that this arrangement times both the incoming and outgoing OC-*n* signal from the CPE, as well as provides a reference clock for the internal clock in the CPE.

Of course, all clocks in Figure 9-8 should be Stratum 3 or better for SONET because of the lack of adequate performance and accuracy standards for Stratum 4 clocks.

Figure 9-9 shows some details of a BITS operation. For reliability purposes, except in more modest private networks, the timing network has links to both a primary and secondary clocking source. Both should derive from a Stratum 1 source but, of course, other arrangements are possible, including direct GPS synchronization. The secondary link acts as a standby or backup. A selector unit chooses the proper timing signal. If a short period of

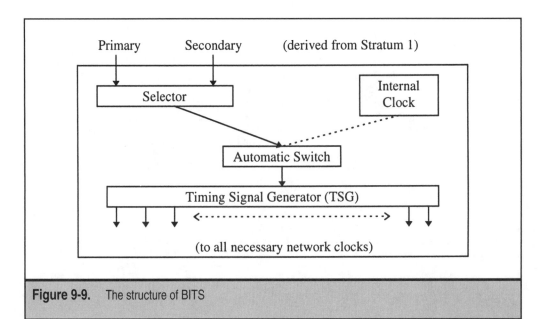

Figure 9-9. The structure of BITS

signal disruption were to occur from both the primary and secondary sources, the building's internal clock could be used. The internal clock must be accurate to +/− 20 parts per million if used for this purpose.

Within a single building, one of these clock sources is designated as the BITS, which has the highest accuracy. The timing signal generator (TSG) makes the decision. For example, if the primary source were determined to be performing poorly and not meeting Stratum 1 quality, the TSG could tell the selector to try the secondary source. If this were found to be unacceptable as well, the TSG then could be switched automatically to the internal clock as a timing source.

BITS provides all the timing for the digital links in the serving office. These are shown at the bottom of Figure 9-9.

This is a good place to show the timing hierarchy used explicitly for SONET and their SDH equivalents in a more systematic fashion. Figure 9-10 shows a simple picture of clock distribution in a SONET network, called the SONET *synchronization chain*.

Stratum 1 clocks (ST1) feed Stratum 2, usually the BITS itself. These clocks, capable of recovery, feed "chains" of SONET NEs through BITS, with the NEs themselves operating at Stratum 3 (ST3). The figure shows the accuracy of the clocks, and places Stratum 4 PDH clocks where they belong, completely outside of SONET.

How does SHD fit in? Table 9-2 shows SONET clocking on the left, and the SDH equivalent terms on the right. Both SONET and SDH have separate specifications for

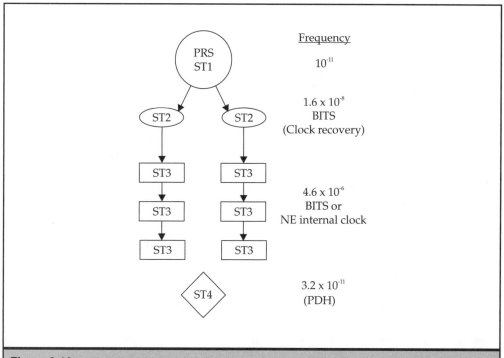

Figure 9-10.   The SONET synchronization chain

SONET Definitions	ANSI/Telcordia	ITU-T	ETSI	SDH Definitions
Definitions	T1.101/GR-253	G.810	ETS 300 462-1	Definitions
Network	T1.101/GR-253	G.825	ETS 300 462-3	Network
PRS (ST1)	T1.101	G.811	ETS 300 462-6	PRC
ST2	T1.101	G.812	ETS 300 462-4	SSU
ST3	T1.101/GR-253	G.813 (G.81s)	ETS 300 462-5	SEC

Table 9-2. SONET and SDH timing standards

timing definitions and networks, as well as the three strata. These are shown in the table, ANSI/Telcordia for SONET, and ITU-T/ETSI for SDH.

SDH calls the PRS the primary reference clock (PRC). There are actually two levels of synchronization supply unit (SSU), SSU-A and SSU-B (A is better), essentially at the BITS level. The synchronization equipment clock (SEC) is found in the NEs.

Benefits of SONET/SDH Timing

The benefits of requiring stricter clocking in SONET/SDH are numerous; some are listed as follows:

▼ This is the mechanism that allows SONET/SDH to byte multiplex into the gigabit ranges. Without the extremely accurate timing needed for byte multiplexing, SONET/SDH would be forced to revert to bit-stuffing and control at very high bit rates.

■ Multiplexing equipment is much more compact and simple. The same timing that makes byte multiplexing possible also results in much simpler designs because microprocessors work well with groups of 8 bits (the byte).

■ The circuitry in SONET/SDH can easily be placed on a simple circuit board, not in a monstrous cabinet; as a result, most SONET/SDH equipment is quite modest in size.

▲ Compared to T-carrier or E-carrier devices, SONET/SDH equipment is acquired at a lower cost and with less complexity. Smaller components draw less power and can be mass-produced much more effectively.

There is no longer any need for complete demultiplexing to find individual voice channels (which are nothing more than groups of 8-bit bytes) inside a higher bit-rate transport. Everything in SONET/SDH is byte multiplexed; therefore, it is easy in SONET to find a specific STS-1 inside an STS-N, and just as easy to find a DS-0 inside a given STS-1 when the payload is carrying DS-0s, that is.

Why SONET/SDH Needs Pointers

Much of this chapter has emphasized that SONET/SDH is tightly synchronized and employs accurate clocks to dispense with the T-carrier or E-carrier need for bit-stuffing and control above the T-1 or E-1 level. However, mention has been made that SONET/SDH employs pointers in the transport overhead to locate the synchronous payload envelope (SPE) within a SONET/SDH frame. However, if SONET/SDH were truly synchronous, why couldn't the position of the SPE just be "locked" into position and then rely on the SONET/SDH clocking to keep the SPE there?

The fact is that SONET/SDH still may need pointers, even when all of the clocks on the SONET/SDH network are derived from exactly the same primary reference source. This is mainly because the speed of light is finite, but accumulated timing jitter is also a factor. This section examines SONET pointer use and structure first, then considers the differences and similarities in SDH.

The reason for this need of pointers is shown in Figure 9-11. The figure illustrates what happens when a SONET ADM must take payloads from a "West" SONET terminal multiplexer (TM) and send the payload directly an "East" SONET TM device. This is not an unusual activity for SONET equipment.

First, note that if the fiber link were 40 km long, it would take about 200 microseconds for a frame to reach the SONET ADM. The transmission time is given by the formula $t = d \div (c \div n)$, where d is the distance in kilometers, c is the speed of light, and n is the fiber's

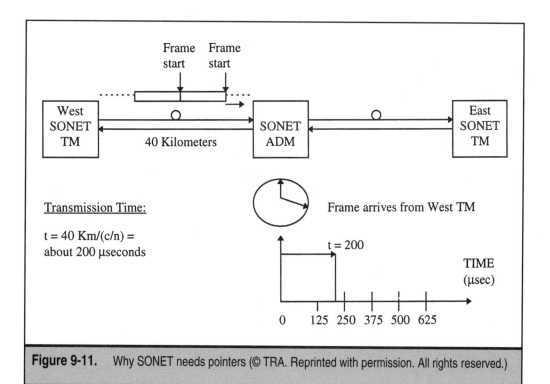

Figure 9-11. Why SONET needs pointers (© TRA. Reprinted with permission. All rights reserved.)

core index of refraction. It is easy to see that if d = 40 km (as given) and n = 1.5 (typical for SONET/SDH fiber), then t = 40 km ÷ (300,000 km/sec ÷ 1.5) = about 200 microseconds.

Now, here is the problem: The frame arrives at the ADM after 200 microseconds, but the ADM is sending one frame eastbound every 125 microseconds, as usual, based on the internal clock. Therefore, the ADM may potentially need to wait another 50 microseconds to send the payload on to the east if the payload must always be aligned with the start of a SONET frame. The trouble is that at the OC-48 rate (about 2.488 Gbps), there will be about 20 Mbps arriving at the ADM every microsecond (2.488 gigabits/125 microseconds)! This is about 1 gigabit in 50 microseconds and would require a buffer of at least 125 megabytes. These buffers also would be required in each piece of SONET equipment along the way on the SONET network.

Nevertheless, some service providers do indeed buffer the payload in this situation, all in the interest of preserving the simple fourth column position of the SPE inside the SONET frame. (The same is probably true for SDH, but definitely true for SONET.) However, the SONET pointers (i.e., H1/H2/H3) provide a way to offset the start of the payload eastbound from the start of the SONET frame itself.

Furthermore, the pointers themselves are capable of "positive" and "negative" pointer adjustments (or justifications) due to timing variations (jitter) in the Stratum 3 SONET network clocks. They do this not one bit at a time, but one byte at a time. Pointer justification is necessary because the H1/H2 pointers cannot merely point anywhere in the SPE, but must always point to the position of the first byte of the Path Overhead (POH). If this moves due to byte buffer overrun or underrun, so must the H1/H2 pointer to the POH. Thus the use of pointers in SONET reduces the need for buffer space and delays on the SONET links.

SONET Pointer Justification

In spite of the improvements in accuracy of SONET clocking, timing errors do occur on SONET networks and there is still a need for adjusting the positions of the H1 and H2 payload pointers on an active SONET link. Standards control the frequency of these pointer adjustments, known as *pointer justifications*, but they are still necessary periodically to avoid the bit-stuffing pitfalls of the older PDH systems, such as T-carrier.

One of the nice things about pointer justification in SONET is that everything is done by the byte, not by the bit. This preserves the ability to do byte multiplexing and easy DS-0 drops and adds while, at the same time, allows high-speed frame alignment without large buffers in each piece of SONET equipment. This section takes a closer look at this process.

Mention has been made about network timing synchronization being a problem in networks where bits make their way directly from an input signal through a buffer to an output signal. The device's internal clock is used to write bits to the buffer and read bits from the buffer. A need for bit stuffing occurs when the buffer is empty at a time when a bit must be read. If the buffer ever fills, bits will be dropped. This process is shown in Figure 9-12.

SONET buffers are organized in groups of eight bits, which are commonly called bytes. When multiplexing occurs, bytes from a variety of buffers must be combined onto a single output signal. However, because of timing variations (jitter), there may not always be exactly eight bits in the buffer when they are needed for the output signal.

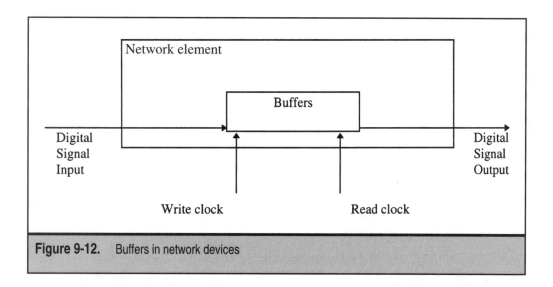

Figure 9-12. Buffers in network devices

Consider a clock that is accurate to 60 parts per million (60 ppm) on a link operating at a modest bit rate of one Mbps (a T-1 sends about 1.5 Mbps). Thus, in a minute, this equipment should send and receive 60,000,000 bits. But at 60 ppm accuracy, the minimum received could be 59,999,940 (60 less) and the maximum received could be 60,000,060 (60 more). This is the limit of timing accuracy. Of course, the goal is to still send 60,000,000 bits per minute, and perhaps the jitter cancels out over time. Nonetheless, when a group of bits or bytes needs to be sent every 8,000th of a second, there is a need to buffer the bits to a greater or lesser extent, depending on the accuracy of the clocks and the speed of the link.

With SONET, there are three situations that must be accounted for in a discussion of pointer justifications. All three involve considering the number of full bytes and "leftover" bits in the buffer. First, when a byte is needed for building the SPE within the frame on the output link of a multiplexer, there may be between 8 bits and 16 bits in the buffer (one full byte and some bits left over). Next, there may be more than 16 bits, but less than 24 bits in the buffer (two full bytes and some bits left over). Finally, there may be less than eight bits in the buffer (less than one full byte). This example assumes that no more than 24 bits will ever be in the buffer, perhaps because the buffer is only 24 bits deep, or the timing is accurate enough to rule this out.

These possibilities are shown in Figure 9-13, along with the desired result. Normally, it would not be a problem to have a few bits left after a byte is sent. These bits are just the first bits of the next byte, which have arrived a little early. But when there are more than 16 bits to send, it would be nice if a mechanism existed to allow more than one byte to be sent during this particular frame time. Naturally, if there were less than one byte to send, it would be nice if there were some way to allow no byte to be sent during this frame time. These bits are arriving a little late, and the input clock may need the time to "catch up" with the output clock.

In Figure 9-13, a SONET device is shown taking bytes directly from a input link and placing them on an output link. In the first case, there are ten bits in the buffer. In the second

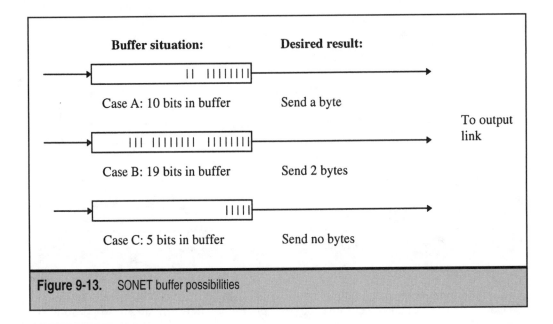

Figure 9-13. SONET buffer possibilities

case, there are 19 bits in the buffer. Finally, the last case shows only 5 bits in the buffer. Jitter errors will only accumulate when there is a mechanism in SONET for allowing not only one byte to be output (Case A), but also 2 bytes (Case B), or even no byte (Case C).

A SONET frame is exactly a fixed number of bytes long. An STS-1 frame has 810 bytes, or 9 rows of 90 columns. There are exactly 27 bytes of overhead on the first three columns (9×3) and 783 bytes of payload in the SPE area (9×87). Where could an extra byte be added or omitted when needed? Fortunately, SONET overhead handles this relatively easily. The H1 and H2 pointer bytes always point to the start of the SPE, which is the position on the first POH byte, by definition. Immediately following the H1 and H2 pointer bytes is the H3 overhead byte. This is the pointer action byte, and as may be expected, the pointer action byte plays a role in this process of adjusting the value of the H1 and H2 pointer when more or fewer bytes need to be sent in the SPE.

When more than 783 bytes need to be sent, the H3 pointer action byte location can hold an extra byte. When fewer than 783 bytes need to be sent, the SPE immediately following the H3 byte can be used to hold a special "stuff byte" which is ignored by the receiver. The values of the H1/H2/H3 determine which is the case. Thus, SONET overhead has a mechanism for sending and receiving SPEs that contain 783 bytes (normal), 784 bytes (when input clocks run fast), or 782 bytes (when input clocks run slow). How often these adjustments are made depends on exactly how frequently send buffers fill and empty and how stable the clocks are in the long run. Naturally, SONET standards cover all of this in considerable detail.

The interplay between the H1/H2/H3 pointer bytes and the input buffers solves the jitter problem in the SONET network. Of course, the trade-off is in processing power at the sender and buffer management techniques. But the prize is the synchronous operation of SONET. Purists may point out that instead of bit-stuffing, SONET merely uses byte-stuffing;

however, until all network clocks are equally accurate, some allowance for jitter effects in any network is needed. It is undeniably true that SONET is not "100%" synchronous in the sense that all network clocks guarantee flawless sender and receiver synchronization. This is not the goal of SONET anyway. SONET is a better way of dealing with the reality of phase differences and jitter on the network.

The relationship between the three cases outlined above and these four bytes in the SONET frame (i.e., H1/H2/H3 bytes and following payload byte) are shown in Figure 9-14. One of these will handle any buffer situation as the sender.

In Figure 9-14, the same three cases are addressed. In normal operation, there is always a byte to send in the outbound SPE, and a few bits left over (exactly eight bits would be a special case). However, when the input clock runs slightly faster than the output clock, there will be more than 16 bits to send. Finally, when the input clock lags behind the output clock, there will be fewer than eight bits to send (no bits would be a special case in this scenario).

All of these conditions are handled by the interplay of the H1/H2/H3 pointer bytes and the payload byte immediately following. The H1/H2/H3 bytes are located in the fourth row and first three columns of the SONET frame and, thus, form the initial line overhead bytes.

The payload byte in question, therefore, is the byte located in row 4, column 4 of the SONET frame.

Figure 9-14 shows that, in normal operation, the H1 and H2 bytes taken together form a 10-bit pointer field that indicates the offset of the first byte of the SPE, which is also always the first byte of the POH. In normal operation, the H3 byte is always ignored by the receiver, whatever its value or content, and the SPE byte in row 4, column 4 is valid user payload. The SPE is 783 bytes long. However, in the case where two bytes should be sent with the SPE in order to adjust for jitter, the H3 byte now contains user payload as well. The SPE pointed to by the H1 and H2 bytes will now contain 784 bytes. The receiver knows this is the case by the value of the H1 and H2 pointers, which also indicate to the

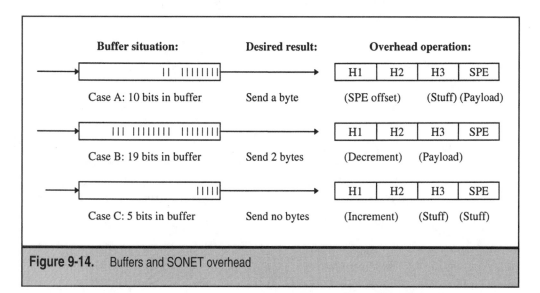

Figure 9-14.   Buffers and SONET overhead

receiver that the position of the SPE in the next SONET frame will move back one position in the SONET frame; thus, the H1 and H2 pointer value must be *decremented* to reflect this change. This is also shown in the figure. Case B is known as a "negative stuff opportunity" or "negative frequency justification" in SONET documentation.

Finally, the figure shows the case where there are less than eight bits in the buffer. The SPE pointed to by the H1 and H2 bytes will now contain 782 bytes. Not only is the H3 byte ignored as a stuff byte, but so is the SPE position in row 4, column 4, which now also contains a byte "stuffed" in by the sender. As before, the receiver knows this is the case by the value of the H1 and H2 pointers. They also now indicate to the receiver that the position of the SPE in the next SONET frame will move forward one position in the SONET frame. So the H1 and H2 pointer value must be *incremented* to reflect this change. This is also shown in the figure. Case C is known as a "positive stuff opportunity" or "positive frequency justification" in SONET documentation.

The fact that SONET contains a mechanism for jitter compensation involving the sending of one byte more or less is almost intuitive to grasp. Less intuitive is the need to adjust the value of the H1 and H2 pointers at the same time. After all, why should there be a need to change the pointer value just because one byte more or less has been sent? The reason is that the H1 and H2 pointers cannot point just anywhere in the SPE, but must always point to the first POH byte.

To see why there is a need for pointer justification, consider a very truncated SONET frame. Although truncated, this frame has all of the pieces needed to show why the pointer values must change with positive or negative byte stuffing. The special frame is shown in Figure 9-15. It is only three rows by five columns for a total frame size of 15 bytes. The first two columns are overhead, so the SPE is only nine bytes, while the overhead is six bytes. The pointer byte—this frame size needs only a single byte to act as a pointer—is located in row 2, column 1. The "stuff" overhead byte occupies the next position.

Just as in a full SONET frame, this special SONET frame contains an SPE. Or rather, the frame holds the beginning of an SPE that, in this case, is only nine bytes long. In normal operation, this special SONET frame contains the end of one 9-byte SPE and the beginning of the next. The pointer indicates the offset of this first SPE byte. This case is shown in Figure 9-16.

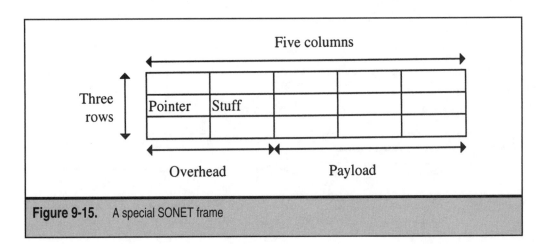

Figure 9-15. A special SONET frame

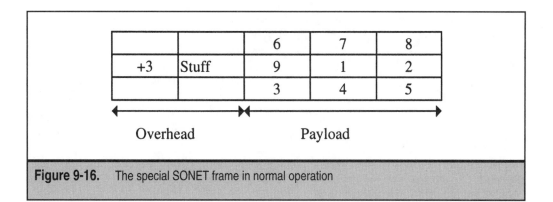

Figure 9-16.   The special SONET frame in normal operation

In Figure 9-16, the nine bytes of SPE occupy the payload area of the SONET frame. The pointer byte contains the value "+3" (the plus indicating that the value is a pointer), which points to the first byte of the SPE (the first POH byte). This +3 offset points to row 2, column 4 of the truncated SONET frame, which is correct. The stuff byte is never examined by the receiver.

Now consider another example: There are two bytes in the buffer that should be sent in the next outgoing SONET frame. In this case, the stuff byte position is now used to send this "extra" payload byte. This solves the jitter problem and results in a frame that contains not nine, but ten payload bytes. The next frame then would revert to nine bytes again. This situation would look like the two frames in Figure 9-17.

In Figure 9-17, the effect on the "extra" byte of payload on the pointer byte is shown. Not only has the stuff byte been used to transport the extra byte from the buffer, but this

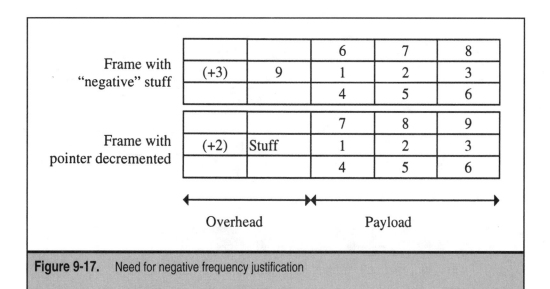

Figure 9-17.   Need for negative frequency justification

action has affected the position of the first byte of the SPE where the POH begins. Note that in order to continue to point to the first byte of POH, it is necessary to *decrement* the pointer value (from +3 to +2). It is called a "negative stuff" because one less byte of "stuff" is inserted into the SONET frame, which now has one more byte of payload.

The "extra" payload byte throws off the start of the SPE in the next SONET frame, making it appear to the receiver that the start of the SPE has moved one byte to the left in the SONET frame. The decrementing of the pointer compensates for this problem. All of this is necessary because the pointer must point to the first byte of POH in the SPE, not just anywhere.

The last case is easy to visualize as well. The case where there are only eight bytes of payload in the buffer and the subsequent effect on the start of the SPE and corresponding pointer value is shown in Figure 9-18.

In Figure 9-18, the effect on the "missing" (i.e., late-arriving) byte of payload on the pointer byte is shown. Not only has the following payload byte been used to transport another stuff byte, but this action has affected the position of the first byte of the SPE where the POH begins. Note that in order to continue to point to the first byte of POH, it is necessary to *increment* the pointer value (from +3 to +4) in the next frame. It is called a "positive stuff" because one more byte of "stuff" than usual is inserted into the SONET frame, which now has one less byte of payload.

The "missing" payload byte throws off the start of the SPE in the next SONET frame, making it appear to the receiver that the start of the SPE has moved one byte to the right in the SONET frame. The incrementing of the pointer compensates for this factor. All of this is necessary because the pointer must point to the first byte of POH in the SPE, not just anywhere.

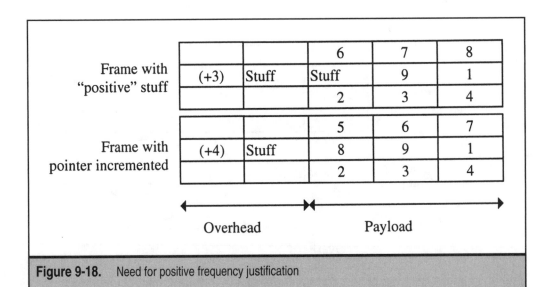

Figure 9-18.   Need for positive frequency justification

SONET Pointer Values

In a full SONET frame structure, the values of the H1/H2/H3 pointer bytes tell the receiver when a positive or negative stuff has occurred, and when the pointer values, therefore, need to be incremented or decremented, respectively. This section details the actual values used in SONET operation.

The structure of H1 and H2 pointer bytes is a little complicated, and their use fairly awkward, mostly due to the related needs for both sophistication and yet compactness of operation. The pointer itself is a 10-bit field and, thus, must span the H1 and H2 bytes. Ten bits are needed because 2^{10} is 1,024, which gives more than enough offset value to point anywhere within the 783 byte SONET SPE. Valid values of the H1/H2 pointer are in the range of 0 to 782. Other values indicate error conditions. Offsets do not include the SONET overhead columns, but point to payload areas in the SONET frame exclusively.

The structure of the H1 and H2 pointer fields is shown in Figure 9-19. In the figure, the first four bits form the new data flag (NDF) bits. In normal operation, these bits are set to 0110, which means that the 10-bit pointer field is to be interpreted by the receiver as a valid offset to the SPE start. The NDF value of 1001 (which is the exact inverse of 0110) indicates that the previous pointer value is incorrect (due to a startup or return to normal operation after a failure) and that the correct value is now indicated by the content of the pointer field. All other values are undefined. Any other bit configurations besides 0110 and 1001 are interpreted as one or the other by a simple "3 of 4" rule. Therefore, 1110 is interpreted as 0110, 1101 is interpreted as 1001, and so on.

The next two bits of the H1 byte are reserved and set to 00 during normal SONET operation. This is not strictly required, but is strongly "recommended." The last 10 bits of the H1 and H2 bytes are the pointer bits, which form a simple binary offset to the start of the SPE. A value of zero means that the SPE starts in row 4, column 4, immediately after the H3 byte. A value of 87 means that the SPE starts in row 5, column 4, immediately after the K2 overhead byte, and so on. A value of 89 would be carried by the H1 and H2 bytes as 0110 00 0001011001.

Note that the pointer bits are alternately labeled "I" and "D", which stand for "increment" and "decrement," respectively. The compactness of the H1 and H2 bytes is reflected in the use of these bits. When the need for a positive- or negative-frequency

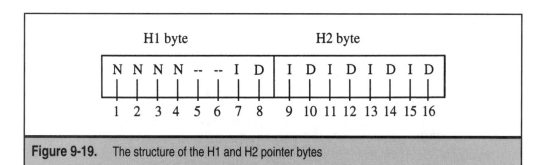

Figure 9-19.   The structure of the H1 and H2 pointer bytes

justification arises, requiring a new pointer value to be conveyed to the receiver, the I and D bits are used to indicate whether this is to be a positive (I bits used) or negative (D bits used). A positive-pointer adjustment inverts, or flips, the values of the I bits and a negative-pointer adjustment inverts, or flips, the values of the D bits. These changes are easy for electronic components to detect and avoids the need for another field in the H1 and H2 bytes for this sole purpose.

The value of the I or D bits also indicates to the receiver whether there is a positive stuff and, therefore, the payload byte immediately after the H3 byte is to be ignored or whether there is a negative stuff. A negative stuff indicates that the H3 byte itself is now valid payload. Of course, the pointer value of the new SPE offset must be sent as well. This is done by repeating the new pointer value at least three times with the NDF bit pattern (1001). Only then can the pointer change again. Therefore, SONET SPEs can only change position every four frames, and most SPE positions are quite stable because of the accuracy of the network clocks used in SONET.

Conflicts between the values of I bits and D bits are handled by an "8 of 10" rule. This means that if eight of the bits seem to indicate to a receiver that the I bits are inverted, then that is exactly the assumption that is made. Ambiguous bit configurations will generate "loss of pointer" error conditions when they persist for several frames.

This whole pointer operation with I bits and D bits may sound confusing, but it really is not. Figure 9-20 shows the whole sequence of H1-and H2-pointer values used when moving the SPE offset from a value of 89 (in decimal) to 90 and back to 89 again. Such movement is not unusual because the long-term stability of SONET clocks typically cancels out persistent movement of the SPE in any direction.

In Figure 9-20, the values of the I and D bits are shown, along with the value of the NDF field, although a normal shift due to positive or negative justification does not involve the NDF. No bit errors are included, but the simple 3 of 4 and 8 of 8 rules would resolve any conflicts. Note that in frames with I or D bits set, there is no valid pointer value. The next frame is to be used for the new value, which must be repeated for three frames to be valid.

Table 9-3 shows the actions performed by transmitters and receivers during positive- and negative-pointer justifications. The table is relatively straightforward but helps to put all of the operations required for frequency justification in perspective.

SONET STS-3c Pointer Justification

Only one more topic needs to be explored in this discussion of SONET pointer justifications: it concerns the use of pointers in higher rate payloads, such as STS-3 or STS-12, or any other SONET speed. When the frame structure is "channelized" and the STS-N contains N STS-1s, then the use of pointers is straightforward. Each STS-1 has a pair of H1- and H2- pointer bytes to locate its own SPE. Everything will be multiplexed by columns, of course, making it easy for each STS-1 pointer pair and SPE to be broken out correctly. However, an STS-N with N STS-1s is not the only structure allowed in SONET.

Mention has been made of the contiguous concatenated STS-3c frame structure used to transport payloads, such as ATM cells on B-ISDN networks. These frames are carried on OC-3 fiber links (typically called OC-3c) and have 9 rows and 270 columns in the frame

Action	NDF Value	I D I D I D I D I D	Pointer Value	Frame #
Normal operation	0 1 1 0	0 0 0 1 0 1 1 0 0 1	89	X
Positive stuff	0 1 1 0	1 0 1 1 1 1 0 0 0 0	undefined	X + 1
New pointer value	0 1 1 0	0 0 0 1 0 1 1 0 1 0	90	X + 2
	0 1 1 0	0 0 0 1 0 1 1 0 1 0	90	X + 3
	0 1 1 0	0 0 0 1 0 1 1 0 1 0	90	X + 4
		•		
		•		
		•		
Normal operation	0 1 1 0	0 0 0 1 0 1 1 0 1 0	90	Y
Negative stuff	0 1 1 0	0 1 0 0 0 0 1 1 1 1	undefined	Y + 1
New pointer value	0 1 1 0	0 0 0 1 0 1 1 0 0 1	89	Y + 2
	0 1 1 0	0 0 0 1 0 1 1 0 0 1	89	Y + 3
	0 1 1 0	0 0 0 1 0 1 1 0 0 1	89	Y + 4

Figure 9-20.   Examples of positive and negative frequency justification

structure. There are 27 columns of overhead in an STS-3c and 243 columns of payload, but there is only one SPE in the STS-3c.

Because there is only one SPE, only one pointer and pair of H1 and H2 bytes are needed to locate the SPE. These are in the first and fourth columns of the STS-3c frame, where one would expect to find the first STS-1 in a channelized STS-3. The other H1 and H2 bytes contain a special *concatenation* bit configuration to let the receivers know that this is an STS-3c with only one SPE. This concatenation indicator is an NDF value of 1001 and has a pointer field value of all 1's, which is an otherwise invalid pointer value; therefore, a value of 1001 00 1111111111 would be placed in the second and third pair of H1 and H2 bytes on an STS- 3c. Note the use of the NDF 1001 value in this instance.

This concatenation pattern applies to contiguous concatenation, such as in an STS-3c. What about *virtual* concatenation, as discussed at the end of Chapter 8? This is not a problem at all for the H1/H2/H3 bytes. Recall that there is still the same number of path over-

Input running slow (positive justification)	Transmitter	1) Invert 5 I bits 2) Insert stuff byte after H3 3) Increase pointer value by one
	Receiver	1) Detect inverted I bitsg 2) Remove stuffed byte after H3
Input running fast (negative justification)	Transmitter	1) Invert 5 D bits 2) Transmit "extra" payload in H3 byte 3) Decrease pointer value by one
	Receiver	1) Detect inverted D bits 2) Gets and inserts H3 byte into payload

Table 9-3. Pointer justification operations.

head columns, and thus H1/H2/H3 pointer bytes as there are STS-1s. So the STS-1-2v shown in Figure 8-21 still had two payloads with path overhead and H1/H2/H3 bytes to handle justifications. This is one reason that it is *virtual* concatenation—the concatenation takes place as the H4 level (or below, in the case of lower order virtual concatenation) and is invisible at the SONET (and SDH) frame level.

Now, contiguous concatenation with one set of H1/H2 pointers seems simple and elegant. There is only one problem, which astute readers may have picked up already. This pointer use for STS-3c cannot possibly work as described. Why not?

Think about the valid pointer values. The H1 and H2 pointer field for an STS-3c is still only 10 bits. This is from 0 to 1023 only. Yet, a full STS-3c frame is 9 × 270 = 2430 bytes. Granted, the payload area is only 9 × 261 = 2349 bytes, but this is still much larger than 1023. How can the H1 and H2 pointers work with an STS-3c?

The answer is to realize that everything about an STS-3c is three times "bigger" than an STS-1 is every respect. An STS-3c operates three times faster, and the frame is three times larger. It turns out that *the values of the pointers in an STS-3c are also three times bigger than an STS-1.*

This is why the pointers still work in an STS-3c. The single SPE cannot start anywhere in the payload area, as the SPE can in an STS-1. An offset value of "0" still points to the first payload byte after the H3 byte (in this case, row 4, column 10); however, an offset value of "1" in the H1 and H2 pointer field points *three bytes* beyond that. An offset value of "1" points to row 4, column 13 in an STS-3c. It is impossible for the pointers to point to Columns 11 or 12 and, therefore, also impossible for the SPE to start in these columns.

Naturally, this means that the bytes are also buffered in "threes" in an STS-3c piece of SONET equipment: This only makes sense. Instead of worrying about two bytes accumulating in a buffer as in an STS-1, an STS-3c only performs a negative stuff when more than three bytes have accumulated in the buffer. Of course, the three bytes are sent using all three available H3 bytes, but only one set of H1 and H2 pointers (the first) is ever used. Positive stuffs occur in an analogous fashion, using the three payload locations after the last H3 byte.

Justifications on even high-speed SONET equipment are performed in the same fashion, with N bytes being stuffed for an STS-Nc and the pointer values multiplied by N for normal operation.

Both SONET and SDH use the V1/V2/V3 tributary overhead in the same fashion as H1/H2/H3 to allow tributaries to adjust their positions inside tributary groups and units. Discussion of these V-pointer justifications is beyond the scope of this book.

SDH POINTERS AND JITTER

This chapter has dealt primarily with SONET timing and pointers. However, many of the introduced concepts apply equally to SDH. Of course, there are differences in how SDH handles the H1/H2/H3 pointer bytes, due mainly to the fact that the STM-1 frame (technically, the VC-4-1c) is like the STS-3c frame in SONET. So in SDH there are essentially always three sets of H1/H2/H3 pointers to deal with, at least with the AU-4. The AU-3, which is basically used for T-carrier payloads, does have three sets of H1/H2 pointers, but that's what the AU-3 is for. So AU-3 pointers act like STS-1 pointers and AU-4 pointers act like STS-3c pointer. However, SDH also defines other structures for the AU-3 pointer, used for SDH Tributary Units (TUs), which must be treated differently than in SONET. The actions of these TU-3, TU-2, and TU-1 pointers mean that vendors of SONET equipment cannot always employ the same chipsets developed to process SONET H1/H2/H3 pointers to process SDH AU-3 pointers in all cases.

There is no sense repeating the full discussion of pointers all over again in SDH terms. The NDF is handled exactly the same. Justification occurs for the same reasons and for exactly the same rules. The only real difference is in bits 5 and 6 of the H1 pointer, where the s1s2 bits reside. In SONET, these bits are usually 00, and in SDH they are set be to 10 (other values are allowed, but AU-4s and AU-3s always use 10).

The differences beyond operation are really only in terminology. But it is important to emphasize that in the fourth row of the STM-1 frame, where in SONET the H1/H2/H3 pointer bytes are found, SDH places the administrative unit (AU).

One of the biggest differences between SONET and SDH is that SONET tends to treat the H1/H2/H3 pointer bytes as just another line overhead function, while SDH treats the same pointer fields as a type of special overhead row (the administrative unit). So differences that seem trivial tend to become magnified when moving from SONET and SDH and vice versa.

This chapter closes with a closer look at all of the currently defined types of jitter in SONET/SDH. Pointer justifications must handle them all, and they do. SONET/SDH links (and rings) suffer from a greater or lesser extent from all of these types of jitter.

▼ **Pointer jitter**   This is just the proper term for the pointer justifications described in this chapter.

■ **Mapping jitter**   Bit stuffing is always required when mapping asynchronous payload signals into SONET/SDH because there is no clock information to match up bit rate information. Mapping jitter is the result, properly applied when the payload is "demapped" at the end of the SONET/SDH link.

■ **Intrinsic jitter**   Any timing jitter that appears at the output of a device when the input is defined as "jitter-free" is intrinsic jitter. This is jitter that is just there and must be dealt with.

■ **Stuffing and wait-time jitter**   This one is rather subtle and usually worries only engineers. When stuff bits (or bytes) are used in SONET/SDH, they must be removed at the far end. But this leaves "gaps" in the arriving bit stream that must be "equalized" out by a "smoothed" clock signal. This smoothing is never perfect, and results in stuffing (sender) and wait-time (receiver) jitter.

■ **Pattern jitter**   This is a consequence of what engineers call "inter-symbol interference" (also an analog term) or "time-domain impulse crosstalk." All it means is that consecutive pulses (usually 1's) in a digital signal can interfere, and the extent of this timing variation depends on the bit pattern (therefore, *pattern* jitter).

▲ **Wander**   SONET engineers tend to use the terms jitter and wander interchangeably. In SDH, wander is a type of persistent jitter in one direction, such as causing repeated positive justifications at one time, repeated negatives at another. Jitter should smooth out over time, but wander does not. Wander is a "slow drift" from a digital signal's ideal "significant instants" positions in time. These are typically caused by daily temperature changes in fiber optic cables.

More detailed information about timing and network synchronization for SONET and SDH can be found on the Web. There are excellent papers in PDF format from Hewlett-Packard (now at the Agilent web site). They are very helpful real-world guides to understanding the entire issue. The three form a sort of series:

Synchronizing Telecommunications Networks: Basic Concepts
literature.agilent.com/litweb/pdf/5963-6867E.html

Synchronizing Telecommunications Networks: Fundamentals of Synchronization Planning
literature.agilent.com/litweb/pdf /5963-6978E.html

Synchronizing Telecommunications Networks: Synchronizing SDH/SONET
literature.agilent.com/litweb/pdf /5963-9798E.html

As with all URLs, these references are subject to change.

PART III

SONET/SDH
Network Equipment

P art II of this book provided a lot of detailed information about how SONET and SDH function on a byte and bit level. This type of information is essential, naturally. However, SONET/SDH networks consist primarily of SONET/SDH equipment and components that embody the byte and bit functions as much as an automobile consists of components that embody the functions of a carburetor or radiator. Although it is necessary for mechanics to understand the role of carburetors and radiators in automobiles, most people who encounter a car are unconcerned with the detailed operation of these components. Their main concern is what the entire car can do when all parts are in place and running properly. This part of the book emphasizes the "as a whole" aspect of SONET/SDH networks.

This section is intended to be much more "real-world" oriented than earlier portions of this book. Although this book previously described both SONET/SDH and non-SONET/SDH equipment, the components were examined more or less piecemeal, in isolation from not only each other, but also from the networks they comprise. This section of the book is meant to extend this analysis to the SONET/SDH network as a whole.

This section of the book explores the implementation of SONET/SDH standards in the various pieces of equipment that make up a network. An attempt is made to classify the diverse SONET/SDH components into a small number of categories as a means to better understand the relationship of one type of SONET/SDH equipment to another. Admittedly, this scheme is not standard in a sense, but it is instructive nonetheless. Finally, SONET/SDH equipment is manufactured by any number of vendors and sold for profit to those who build the SONET/SDH networks. This section of the book takes a look at the vendors involved in the manufacture of SONET/SDH equipment.

This part of the book is organized into three chapters. The first chapter (Chapter 10) explores how the standards for SONET/SDH operation are embodied in the various pieces of equipment needed to build a SONET/SDH network. There are different pieces of SONET/SDH equipment designed and built to accomplish different networking tasks, and these are all detailed in this chapter. A high-level profile of a typical SONET/SDH carrier network is also contained in the chapter, and an explanation on the use of SONET/SDH links to connect routers on the Internet. This chapter includes a detailed look at one of the most overlooked aspects of SONET/SDH networks: the calculation of the span power budget. The chapter concludes with a high-level first look at SONET/SDH rings, their purpose, and operation.

The next chapter (Chapter 11) divides all of the various pieces of SONET/SDH equipment introduced into six major categories. The chapter then examines the functioning of each major piece when the device is used in a SONET/SDH network. The whole idea is to provide a framework for classifying SONET/SDH equipment types according to function.

The final chapter of this part (Chapter 12) deals with SONET/SDH equipment providers. The chapter details the vendors that build and supply SONET/SDH network components and provides some information about their products. The chapter concludes with a look at several example SONET/SDH network configurations that use some of the equipment described in the chapter.

CHAPTER 10

SONET/SDH Networks

There is nothing wrong with learning as much as one can about the inner workings of SONET and SDH on a byte level and related topics, such as timing distribution; however, SONET/SDH networks consist of devices that are real boxes full of hardware. In addition to an intimate working knowledge of SONET bytes and frame structures, it is also desirable to have a intimate working knowledge of what SONET networks look like from an equipment standpoint.

Nothing is more frustrating to network engineers and managers than to have to deal with a person whose knowledge about any technology is limited to "book learning" on an entirely abstract and conceptual level. In the early days of LANs, it was not unusual to come across a supposed expert in Ethernet or Token Ring who had never seen, let alone worked with, an Ethernet hub or Token Ring MAU. They knew all the rules of Ethernet, for instance, from the 64-byte minimal frame size to the strange structure of the IEEE 802.3 version's frame header. But asking them to reset a hub in a telecommunications closet was like asking them to disarm a nuclear weapon. Chances were, nothing good was going to come from the effort.

This chapter is an attempt to address some of these issues. Saying SONET/SDH equipment can interface with T-carrier equipment does not say how this can be accomplished in a device. The most detailed study of SONET/SDH technology does not begin to scratch the surface as to how all of the nifty features of SONET/SDH can be brought to the real world of network hardware and software.

This chapter begins by looking at some of the details of pre-SONET carrier systems like T-carrier. Although SONET is the main topic in this section, the arguments are easily extended to SDH and E-carrier systems. The perspective in this chapter focuses on the equipment used.

PRE-SONET CARRIER SYSTEMS

Older fiber optic transmission systems in the United States were equipment-intense, proprietary, and an operational nightmare. A visit to a central office employing fiber optic trunks revealed a snake pit of coaxial cables leading to and from the fiber optic transmission equipment, a Medusa-like corner with patch panel and physical cross-connections that threatened to entangle an unwary technician, and an utter lack of up-to-date paperwork that was absolutely essential to make sense of it all. Proprietary connectors did not match any other piece of equipment in the building, and it was a truism that any delivered equipment or device would not be able to be hooked up to anything else—in some cases, even from the same vendor—without extensive connector conversion or customized interface components.

Still, deployment of SONET transmission *overlays* in these systems has been cost-effective for two major reasons: An overlay indicates that a particular link or trunk is converted to SONET using as much of the existing physical plant as possible. When the fiber conforms to SONET standards, it can be used. Naturally, as much of the existing input and output multiplexers and cross-connects are retained to minimize expenses. Thus, a SONET overlay replaces non-SONET fiber links and trunks one by one, with no concerted attempt to integrate the SONET equipment into a complete SONET network. The overlaying is common practice in SONET carrier networks.

The two major reasons for the continued use of SONET overlays are easy to describe. First, the capability of SONET multiplexers to accept multiple data rates (i.e., DS-1s and DS-3s) has reduced the equipment necessary for the creation of a high-rate optical link; therefore, the expense of the SONET equipment is easily offset by the reduced amount of equipment needed when conversion to SONET is implemented. The second area of cost reduction is in the operations, administration, maintenance, and provisioning (OAM&P) of the SONET links as compared to the cost associated with the proprietary links. SONET equipment complies with standard operations systems and allows a single point of entry into the network for OS functions.

Figure 10-1 shows pre-SONET carrier systems equipment deployment. The equipment shown is a full complement and is able to *groom* DS-0 channels in a high-speed (i.e., greater than the DS-3 rate of 45 Mbps) fiber optic system. *Grooming* is the process of combining and distributing feeds from many sources operating at various speeds onto one higher speed trunk. Grooming is one giant step up from simple muxing, which is usually point-to-point. (Sometimes the term *hubbing* is used to describe the same process. The *hub* is where grooming takes place.)

Grooming is common practice in carrier systems and is done for reasons of efficiency. Suppose a central office has 10 T-1s coming from one customer location and 7 from another customer at another location. Because the T-1s are coming from two locations, it may seem that two separate carrier systems are needed to send the T-1s onto the trunking network and on to their destination. A single T-3, however, can carry 28 T-1s. It makes

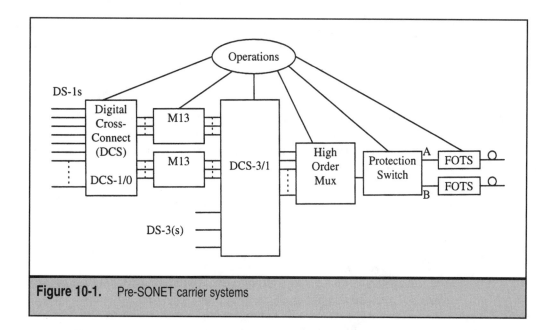

Figure 10-1.   Pre-SONET carrier systems

much more sense to *groom* the user traffic and install digital cross-connect system (DCS) equipment to combine the total of 17 T-1s onto a single T-3. Therefore, only one T-3 is needed to aggregate the traffic from both locations onto the carrier backbone. Often, when grooming combines traffic streams, it is called *concentration* or *traffic consolidation*, and when grooming is used to split up traffic streams (as may be done at the opposite central office), it is called *distribution* or *traffic segregation*. Whatever the name, the intention is the same: to make more efficient use of otherwise under-utilized high-capacity trunks and links.

In Figure 10-1, there are many pieces of equipment. In order to better understand these devices and the role they play, and not coincidentally how they relate to the equivalent SONET devices, some words of explanation are necessary. In the figure, the equipment and its functions include the following:

▼ **DCS 1/0** DCS-terminating DS-1 trunks and cross-connecting DS-0 channels. The "1/0" designation is used to indicate that the DCS handles both DS-0s at 64 Kbps and the entire DS-1 operating at 1.5 Mbps. Some DS-0s from some customer sites may terminate at other sites served by the same switching office. The DCS-1/0 equipment handles this *turn-around* configuration. In many cases, this equipment is called a digital *access* cross-connect system (DACS), but DCS is the more general term.

■ **M13** This is a DS-1 to DS-3 multiplexer. The main task of the M13 device is the multiplexing of up to 28 DS-1 trunks into a single DS-3 trunk. Of course, the same device can demultiplex as well. Not all 28 DS-1s are shown in the figure.

■ **DCS 3/1** DCS terminating DS-3 trunks and cross-connecting DS-1 trunks. The "3/1" designation is used to indicate that the DCS handles both DS-1s at 1.5 Mbps and the DS-3s operating at 45 Mbps. Note that there may be some DS-3s that do not emerge from an M13. These are *unchannelized* DS-3s that operate as a full 45 Mbps, not as 28 DS-1s or 672 DS-0s. As before, some DS-3s from some customer sites may terminate at other sites served by the same switching office. The DCS-3/1 equipment also handles this turn-around configuration.

■ **High-Order Mux** Another multiplexer device, but without cross-connect capabilities. The high-order mux combines multiple DS-3 signals into a single, high-speed signal. This is a purely proprietary device, without even a standard output format. Most of these devices combine 9 to 12 DS-3s on a single fiber optic link operating in the 1–2 Gbps range.

■ **Protection Switch** This device sends the high-speed transmission signal from the high-order mux out on one of two fiber optic transmission media. Naturally, this provides some protection from fiber cable damage and ensuing failure. This protection, in turn, depends on just how far apart the two fiber cables extend, what is known as *diverse routing*. Unfortunately, in the past, diversity was defined as a minimum separation of only 25 feet. This often translated to one fiber down one side of the road, and the other fiber on the other side of the road. Trenching operations across the road frequently severed both cables to the dismay of users and carrier alike. The protection switch is controlled by the operations system or automatically by the condition of the received optical signal. Often the switch is done automatically, but service must be restored manually.

- **FOTS** The fiber optic transmission system (FOTS) is the equipment that converts the electrical signal to an optical one and vice versa on the receiver side. It should be noted that the FOTS and the protection switch could be transposed. That is, the output of the high-order mux could transmit directly to a FOTS, which would now feed an optical protection switch. This has the added benefit of requiring only a single FOTS.

- ▲ **Operations** This is the nerve center for the operation of the entire transmission system, not just the simple link shown in Figure 10-1. All equipment alarms and performance statistics are relayed to this center and all remote control and provisioning is performed from here. Each piece of equipment in the system has a direct link to the operations center. Because each piece of equipment had its own unique brand of network management hardware and software, operations in this environment were complex and awkward.

SONET CARRIER SYSTEMS

SONET carrier systems can be contrasted with pre-SONET carrier systems in a number of ways; however, the easiest way is to contrast the equipment requirements and layout of the components. This section emphasizes SONET for ease of comparison to pre-SONET carrier systems, but the same ideas apply to SDH as well. A typical SONET carrier system is shown in Figure 10-2.

As can easily be seen in Figure 10-2, SONET carrier systems reduce the equipment complement required for a transmission system to a bare minimum. SONET architecture allows for a reduction in the complexity of the equipment that brings non-SONET digital information streams into the SONET system.

Now there are only the DS-1 and DS-3 DCSs left, and these still are needed to handle getting the T-carrier traffic onto and off of the SONET carrier itself. Of course, the DCS turn-around cross-connection functions for DS-0s, DS-1s, and DS-3s that originate and terminate at the same switching office are still needed as well. However, the bulk of the work done by the many pieces of separate equipment is done by the SONET multiplexer. High-order multiplexing, protection switching, and FOTS functions are all done in this single piece of SONET equipment. This is another example of the increasing power of electronic components over time.

Another area where SONET reduces complexity is in operations support. SONET standards also specify the OAM&P system for the entire transmission network. All SONET-compliant equipment must follow these standards and implement these functions and capabilities. This allows for a standard operations platform for control of the whole transmission system. Of course, the former links to the existing DCS equipment must be maintained, making the situation less than perfect.

SONET equipment also includes a standard set of craft interfaces for ease of maintenance and repair. Craft interfaces are those connection points where diagnostic equipment

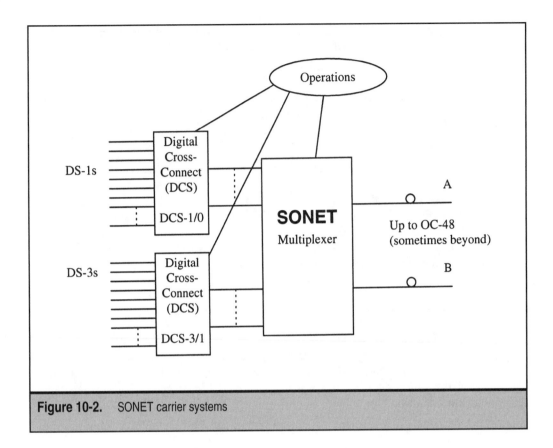

Figure 10-2. SONET carrier systems

can be attached by technicians performing on-site troubleshooting tasks. The terminals or consoles in the operations center where the technicians manage the equipment with software interfaces are also sometimes called a craft interface, but usually the term is reserved for attachments to the equipment itself.

A final advantage of the SONET carrier system is the capability of upgrading the carrier system to higher rates without the wholesale replacement of the transmission equipment. Certain functions of SONET remain the same regardless of the carrier rate. Therefore, the SONET multiplexer box in Figure 10-2 does not need to be replaced to allow the fiber links to operate at OC-48 (about 2.4 Gbps), instead of OC-12 (about 622 Mbps). After all, input ports from the DCSs, protection switching, and network management capabilities are exactly the same regardless of output speed. Typically, a simple processor board upgrade is all that is needed to change the output speed. Most port additions merely involve this simple card upgrade as well, yet another benefit of the spread of microprocessors to the transmission network. This simple upgrade process allows economical upgrades of carrier systems.

SONET/SDH IN ACTION

So far, the discussion of SONET/SDH networks in this chapter consists of replacing plesiochronous T-carrier or E-carrier components with SONET/SDH components: the overlay process. Of course, there is much more to SONET/SDH than T-carrier and E-carrier replacement, although this aspect of SONET/SDH should never be forgotten. SONET/SDH can do more, much more. Figure 10-3 shows the operation of a SONET/SDH link that takes more advantage of the unique capabilities of SONET/SDH.

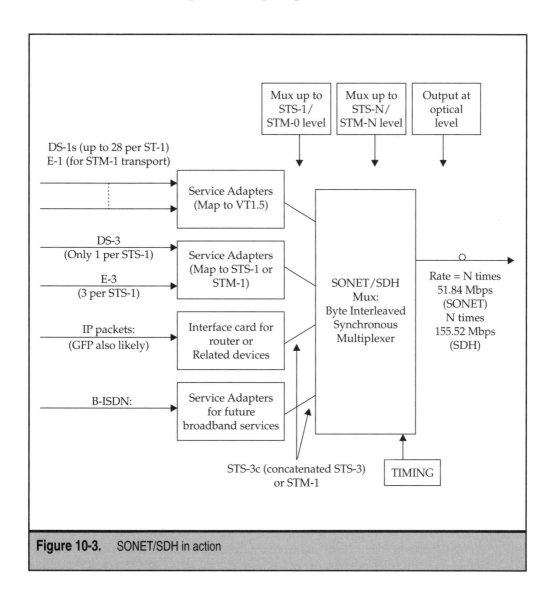

Figure 10-3. SONET/SDH in action

There is a lot happening in Figure 10-3, and requires a more detailed examination. On the left side of the figure, the main input possibilities for the SONET/SDH link are shown. Note that the familiar DS-1s, E-1s, DS-3s, and E-3s are still there, but more details are presented regarding their exact operations. For example, the output of an existing DCS-1/0 which is, after all, just another DS-1 (or multiple DS-1s), are aggregated and a group of 28 and sent into a *service adapter* which maps the 28 DS-1s into a single VT1.5. The concept of these service adapters also applies to SDH.

Although the service adapter in Figure 10-3 is shown as a separate physical device, it need not be and often is not. Most frequently, the service adapter is just a type of processor board or card put into a slot in the SONET/SDH equipment to handle that particular form of input and output. Vendors have their own names for service adapters, such as port adapters and interface cards.

Also shown on the left side of the figure is a single DS-3 or E-3 that may be connected to its own service adapter. Note the difference between adaptation to SONET/SDH of channelized transports such as 28 DS-1s and a single channel such as a DS-3. The 28 DS-1s require the use of VT1.5 (mainly for DS-0 payload visibility); however, a DS-3 can be transported directly inside a SONET frame through the process of asynchronous mapping. This may be desirable when the DS-3 in question is unchannelized itself, or when DS-0 visibility is neither wanted nor needed. Naturally, only a single DS-3 operating at 45 Mbps can be carried by an STS-1 link operating at 51.84 Mbps. Of course, similar equipment arrangements can be made for SDH structures.

The figure also shows that SONET/SDH is now often used to link Internet routers (or even large organizations' private routers). In many cases, there is little multiplexing done by the SONET/SDH equipment and router interfaces might run at OC-3c, OC-12c, and their STM equivalents all the way up to OC-768c. All of the traffic multiplexing (whether for voice, video, or data) is done by the routers combining the packet streams onto the high speed SONET/SDH fiber link. Usually the packets used are IP packets, and the framing is some form of POS. But today, GFP is possible as well. Using SONET/SDH links to connect routers has given new life to the SONET/SDH technology, which is sometimes written off as a needless complication to direct-to-fiber router interfaces. More on these aspects of SONET/SDH and routers will be discussed in the chapter on SONET/SDH issues later on in this book.

Finally, the figure shows a service adapter to be used for future implementation of broadband services, as defined as part of the entire SONET/SDH/ATM movement in the late 1980s. The breadth and scope of these standard broadband services are astonishing, but most people focus on services like switched digital video (SDV), color faxing, high-speed faxing, transport of ATM cells inside an STS-3c or higher speed frame, MPEG-2 compressed digital video, and a number of other services nearly impossible to provide without SONET/SDH. Standard broadband services should have their own service adapter(s), depending on the services provided.

It is important to remember that standard broadband services transported by SONET/SDH links must be handled by links operating at 155.52 Mbps or greater. The speed of 155.52 Mbps is the rate of STS-3/STM-1. Moreover, for broadband, the STS-3/STM-1 must be *unchannelized*, or rather *concatentated* in SONET/SDH talk, to operate at the full 155.52 Mbps. This is an STS-3c in SONET or, more commonly but less technically, an OC-3c. This is more of a problem in SONET than SDH, because a "plain" STS-3 consists of three STS-1s all operating at 51.84 Mbps. This may be what both the customer and carrier want, but not for the transport of standard broadband services. This is the exception that the service adapters will multiplex up to the STS-1 level.

Today, it is unlikely that any form of standard broadband services will ever be deployed. Broadband services will exist, of course, but are more likely to be based on IP routers than the ATM switches envisioned in the late 1980s and early 1990s. Outside of some tricky interoperability issues (which have a habit of occurring anyway), even with firm international standards in place, the basing of broadband services on IP routers should have little impact on customers or SONET/SDH.

Continuing through the figure, the major component is the SONET/SDH multiplexer itself. This device does the byte-interleaved multiplexing that distinguishes SONET/SDH. The SONET/SDH multiplexer also performs the electro/optical (E/O) conversion needed to send the SONET/SDH signal over the fiber itself. The BITS (a United States term) timing information is shown in the figure as well. One of the reasons that the service adapters may be separate devices, or even customer premises equipment (CPE), is because in most cases many payloads are combined, groomed, and multiplexed onto a single, high-speed SONET/SDH link at the switching office. For broadband, the service adapter should be CPE as well.

In any case, the SONET/SDH mux combines and grooms the input signals, channelized or not, in any combination, into a single SONET/SDH frame. The output of the SONET/SDH mux at this point will of course be some OC-n. This OC-N will operate at "N" times the basic SONET rate of 51.84 Mbps. It is quite common today for this SONET/SDH equipment in a switching office to operate at the OC-48/STM-16 rate of about 2.4 Gbps. However, higher speed equipment is possible and will become more common in the near future.

Finally, although not shown in the figure, it is good to remember that the output fiber is typically not a single strand of fiber. More often the output from the SONET/SDH mux consists of an "A" and "B" fiber pair with SONET/SDH protection switching performed on the part of the device or through some operations interface.

SONET/SDH AS TRUNKING

Figure 10-3 shows a typical SONET/SDH link on a SONET/SDH-based network in action. Any type of service, from voice to high-speed data to video, can be accepted by various types of service adapters on the SONET/SDH network. However, it is always good to keep in mind that SONET/SDH is basically a trunking technology. That is, SONET/SDH itself is neither capable of doing any switching based on traffic stream content, nor able to provide any of the neat features of fast packet networks, such as flexible and dynamic bandwidth allocation (also known as "bandwidth on demand"). When a SONET OC-1 pipe is filled with 28 DS-1s, which is the maximum, then the OC-1 is full.

This can be a problem when users are accustomed to seeing SONET (or SDH) services listed in the same categories as frame relay and/or ATM. If frame relay and ATM can give a user "bandwidth on demand," why can't SONET/SDH? Because frame relay and ATM run in switches as well as across links. SONET/SDH can be the link, access or trunk, technology for frame relay or ATM, but SONET/SDH can never replace frame relay or ATM switches by itself. The same argument can be made about SONET/SDH links hooked up to routers. The routers take care of packet multiplexing and bandwidth assignments, not the link.

SONET/SDH is a trunking technology and a "bit pipe" for traffic, no more and no less. Suppose a SONET multiplexer is used to provision three STS-1s operating at 51.84 Mbps on an STS-3 operating at 155.52 Mbps. The STS-1s are then sold to three customers. Each customer uses only a fraction of the 51.84 Mbps available at any one point in time. A fourth customer's traffic could easily be handled by the STS-3's speed.

However, that's it; there is no fourth STS-1 to sell. No more bandwidth can be provisioned on the STS-3 link, no matter how underutilized it may appear. The only solution is to add a frame relay, ATM, or some other kind of switch to the ends of the SONET link. Only then will features like flexible bandwidth allocation for "bursty" data applications help the service provider to support more than three users. Of course, this is not a trivial task. Figure 10-4 shows the situation with pure SONET.

Figure 10-4 shows that up to three STS-1s can be provisioned on an STS- 3. The math involved is simple but absolute. A fourth STS-1 cannot be added or sold.

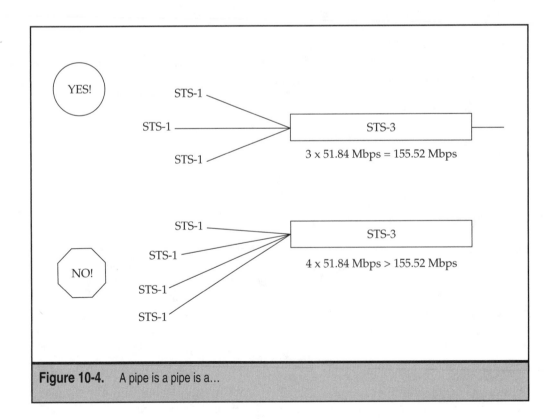

Figure 10-4.   A pipe is a pipe is a...

SONET/SDH LINKS IN SUMMARY

These last few sections have been full of alternative arrangements and complexities. This may be a good place to briefly summarize the operation of SONET/SDH equipment at the end of a SONET/SDH link.

All types of service adapters can map the incoming signal from one service type or another into the payload on a SONET/SDH link. New services and signal types (e.g., GFP) can be transported by adding new service adapters at the edge of the SONET/SDH network.

All inputs are eventually converted into the basic format of a synchronous STS-N or STM-N signal. Lower speed inputs, such as individual DS-1s, are first multiplexed onto VT 1.5s. In SONET, several synchronous STS-1s can be multiplexed together in either a single-stage or two-stage process to form an electrical STS-N signal. SDH would form an STM-N out of STM-1s, of course. Higher speed signals, such as those from newer broadband services and cells from ATM networks, can be carried on an STS-3c or STM-1 directly.

STS/STM multiplexing is performed by the byte-interleaved synchronous multiplexer. The bytes are interleaved in such a way as to provide "lower speed payload visibility" on the SONET/SDH-based network. No additional signal processing occurs, except for a direct conversion from the electrical STS/STM format to the optical OC format for fiber transmission.

SONET/SDH AND THE POWER BUDGET

As an experienced professional working in networking for many years, it came as somewhat of a surprise when discussing SONET/SDH networks and equipment that others did not have the same appreciation for the importance of the relationship between the fiber linking the pieces of SONET/SDH equipment and the gear itself. There was an awareness of what is called the *power budget* on a SONET/SDH link, but it seemed that this awareness was only at a very high level, and did not extend to the level of detail necessary for a rigorous analysis a SONET/SDH link's performance over time.

This section is an attempt to clarify the important (indeed, often crucial) relationships between things such as the transmitter power, receiver sensitivity, and just how much power is used pushing light through many kilometers of fiber. If there is not enough optical power at the end of the link, the error rate will be high, the link will intermittently fail, or the link will refuse to come up at all. There is more to the power budget consideration process than just figuring the light lost on the fiber span. Everything counts, from connectors to splices (both present and future, as we will see) to the safety margin added on, just because the world is often a cruel place that refuses to respect theoretical calculations, no matter how sincere.

What follows assumes that the SONET/SDH link is attenuation-limited, not dispersion-limited as in many DWDM links. This is almost always the case with SONET/SDH.

There are standard values for SONET/SDH transmitters and receivers, but unfortunately there is no "standard" fiber optic span in terms of physical structure and installation details. And the standards for transmitter power and receiver sensitivity for SONET/SDH are basically *minimums* that are only bare guidelines for vendors seeking to differentiate their equipment for everyone else's. In fact, some of the confusion over optical power often comes from the vendors themselves, who often treat the SONET/SDH span as a single long run of fiber. These factors make hooking up a SONET/SDH inter-face to many kilometers of fiber for the first time interesting, to say the least. This section is intended to take much of the mystery away from what happens between transmitter and receiver on a SONET/SDH link.

Decibels and the Power Budget

The SONET/SDH power budget is usually computed on a *span*. Here, the term "span" just means "the distance traveled without needing amplifiers or repeaters." The low at-tenuation (signal loss) and noise on fiber optic spans translate directly to longer spans, as pointed out earlier in this book.

But just how long can a fiber span be? The answer is always tied up in the concept of the optical *power budget*. This concept largely determines the distance of a fiber optic span. Power budgets are usually computed in decibels (dB). Decibels are the standard units used in all types of networks, and are familiar to every type of engineer, but can seem somewhat mysterious to non-engineers, even if they are network specialists. So, a few words of explanation about decibels are in order before proceeding on to the actual power budget calculations.

Decibels are used to represent ratios in a way that is more useful to an engineer than percentages. Decibels are used to represent signal to noise ratios (S/N), wireless system antenna gains, and (in the power budget) the ratio between input power and output (re-ceived) power. Because decibels are simple ratios, there are really no other units to the decibel, such as watts, as long as the ratio expresses the same units (which it always does). Decibels are a *logarithmic* scale, where 30 dBs represents a ratio of 1000 to 1. So a 60 dB power loss is one million (input) to one (output), and a loss of 3 dBs through a connector means that only half of the power that went in comes out.

Engineers routinely use decibels instead of percentages to measure power ratios, and all types of other ratios as well. This is because decibels are easily additive, while percent-ages are not. Consider a simple network where the output power on one link is piped di-rectly to the input on a second link. The signal power loss in milliwatts on each link is 90%, meaning one $1/10^{th}$ of the input power arrives at the output. What is the end-to-end signal power loss on the whole system? Adding percentages gives 90% + 90% = 180% which is nonsensical. Using decibels, the loss on each link is 10 dB, since an input to out-put power ratio of 10 to 1 translates to 10 dB. Ten to one is 10 raised to the first power (10^1), which is then multiplied by 10 to give decibels (*deci-* means 10). In decibels, the total end-to-end loss of the system is 10 dB + 10 dB = 20 dB which is correct, since $1/10^{th}$ of $1/10^{th}$ is $1/100^{th}$ or 10 raised to the second power (10^2). Naturally, not all ratios work out to neat whole numbers. For example, an 80% power loss in watts translates to a decibel loss of 6.99 dB ($1/5^{th}$ power received is expressed as 10 to the 0.699 power), a 33.3% power loss is 1.7 dB, and so on. The Bel, named to honor Alexander Graham Bell, was often so small that engineers had to add up too many fractions. So the whole unit was multiplied by ten to eliminate some of the fractions and gave science the "ten Bels" or decibel (dB). Decibels can just as easily be used to measure signal power loss (attenuation) or a power boost (gain) or almost any other ratio.

A small table giving the relationship between ratios, powers of ten, and decibels is shown in Table 10-1. Note that a 60 dB loss is not *twice* as bad as a 30 dB loss, but 1,000 times worse!

Ratio (P_{out}/P_{in})	Power of Ten	Decibels
	(Negative when P_{out} is smaller (loss), positive when P_{in} is smaller (gain))	
1/10	10^{-1}	-10 dB (1 ×10)
1/100	10^{-2}	-20 dB (2 ×10)
1/1,000	10^{-3}	-30 dB (3 ×10)
1/10,000	10^{-4}	-40 dB (4 ×10)
1/100,000	10^{-5}	-50 dB (5 ×10)
1/1,000,000	10^{-6}	-60 dB (6 ×10)

Table 10-1. Relationship between ratios, powers of ten, and decibels

That's about all there to understanding decibels. Really. And this is enough to understand everything about power budget calculations. Sometimes, SONET/SDH vendors even express the length of fiber span their equipment can support in terms of decibels, and not kilometers. This makes sense, because there is some variation in the attenuation per kilometer among all makers of fiber optic cable. So, for example, a SONET/SDH vendor might say "maximum system reach is…25 dB or approximately 86 km (53.4 miles)…" The dB figure is exact while distance is only approximate because not every fiber optic cable will have attenuation of 0.29 dB/km (25 dB / 86 km = 0.29 dB/km), although that figure is fairly common.

Power Budget Calculations

The power budget determines if enough light is present at the receiver to cover all the losses on the span. The receiver could be an optical amplifier or electrical regenerator, but in these examples, it is assumed that there is always another SONET/SDH network element at the end of the span. If the receiver is an optical amplifier, then the power budget computation is somewhat complicated, but valid nonetheless. The simplest case connects an optical transmitter with an optical receiver at the endpoint of a span. This is the configuration discussed in this section.

It should never be thought that when standards are followed and fiber span distances are spelled out in detail, that power budgets never enter the picture. All this means is that the standardizing body has considered the performance and operational characteristics of all of the span's components, and then translated these into maximum speeds and distances. For example, if a given specification recommends a semiconductor laser of a given power as the transmitter, an APD of a given sensitivity as the receiver, and the SONET/SDH equipment meets these specifications, this does not mean there are no power budget concerns. What standards for spans mean is that the organization issuing the standard have, by themselves or with others (such as equipment vendors), carefully computed worse-case power budgets, and knows that this configuration of transmitter and receiver will work over a typical span of this length or shorter, unless something unforeseen comes up.

However, there are always cases where the power budget needs to be understood in detail or even computed from scratch. There are times when standard configurations are pushed to the limit, or configurations where no standard yet exists. For these situations, which arise quite often when fiber optic links are used in newer configurations, firm understanding of power budgets is required.

But the overwhelming reason that the power budget today is so important is because of *dark fiber*. Dark fiber is fiber that has been run, but is not yet hooked up to the transmitter and receiver electronics. Now, if the whole reason that the dark fiber has been installed is to implement one of the standard fiber technologies, then the power budget consideration is less critical.

It is when the dark fiber can be used for a number of different purposes that the power budget is so important. Almost every company with right-of-way, from telecommunications companies to cable TV companies to local municipalities, have been busily running single mode fiber as fast as they can. Never mind that there is no immediate use or customer for the fiber, the bandwidth will not go to waste. In many cases, less than 10% of the total fiber capacity is "lit" (in use). For every $1 spent to install the dark fiber itself, it can cost up to $20 to fully provision a fiber link.

It is usually in the metropolitan area network (MAN) fiber arena that an understanding of the power budget is absolutely essential. MANs are typically defined as less than 100 km (about 60 miles) total distance end to end, but many MANs are much smaller. Typical MANs will cover no more than 20 or 30 miles and normally consist of a single span. When linking buildings or other facilities that are not so far apart, dark fiber is an attractive choice. A full service offering including all E-O conversion, transmitters, and receivers is, in many cases, more cost effective, the longer the distance end-to-end. This makes sense, because much of the expense of a fiber link is not the fiber optic cable, but the end electronics. But this also makes short fiber links quite expensive compared to their longer cousins. So, full-service offerings are often expensive for shorter distances and organizations with several facilities in a concentrated area. Lower bandwidth will save cost, but also decrease the effectiveness of the network. Dark fiber is attractive for all of these reasons.

Since dark fiber has no transmitter and receiver attached, the range of options open to the eventual user of the fiber is considerable, not only in terms of supported services and protocols, but in terms of the equipment. If the user can get the right to use dark fiber, then the user can supply the E-O conversion, transmitter and receiver. However, the more powerful the transmitter and the more sensitive the receiver, the more expensive the devices become. And even if a competitive service provider is used to provide the link, in many cases this service provider will rely on dark fiber from a third party.

The dilemma of the MAN solution is shown in Figure 10-5. A full service offering is too expensive, but if dark fiber is used, what transmitter and receiver should be used on the span? With a proper understanding of the power budget, dark fiber can be used with the proper transmitters and receivers for the span.

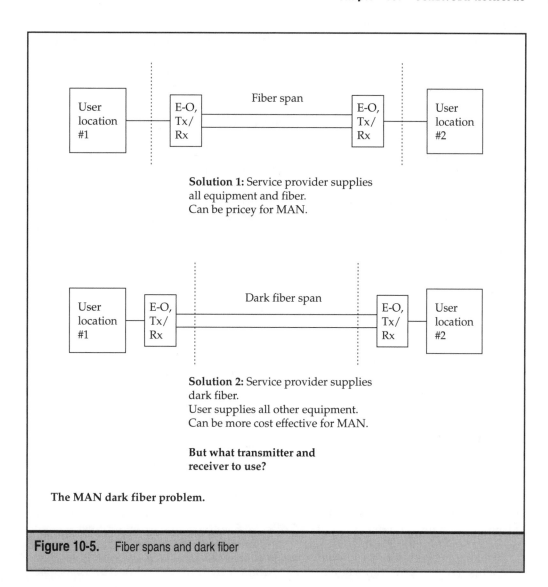

Figure 10-5. Fiber spans and dark fiber

Computing the Optical Power Budget

The basic idea behind computing an optical power budget is easy. If there is enough light left over, the span will work. If there is not enough light left over, the span will not work. As important as they are, optical power budgets are not difficult to compute. All that is needed is some simple skills like adding, subtracting, and multiplying. Power budget computations do not even require an understanding of decibels, but few practitioners would feel comfortable leaving power budget calculations up to such naïve implementers.

So the first step in computing the overall optical power budget is to figure out just how much light there is for the electronic devices on each end of the span to use. Two numbers are of paramount importance here. The first is the minimum transmit power and the second is the minimum receiver sensitivity. Note the use of minimums here, just to be on the safe side. These numbers are usually expressed in decibels with reference to one milliwatt of power, or dBm.

Both of these numbers are available only from the equipment vendor. Some equipment vendors try to be coy about these numbers, while others are quite open. Saying the equipment complies with "all relevant international and national standards" just forces interested parties to do more work. In this case, mystery is never an attraction.

The minimum transmit power is just what it says. This is the worst-case performance of the device. The transmitter is absolutely guaranteed to deliver that amount of power, and might give more. Some of the more coy vendors are fond of listing an "average transmit power" for the device, but this number is useless for power budget considerations and meaningless by definition. A power average means that sometimes the power is more, and sometimes the power is less. There is no "floor" to an average.

Using averages in a power budget is like packing clothing for a trip to New York. The average temperature is 50 degrees Fahrenheit (about 11 degrees Celsius). But if the trip is made in July (90 degrees F) or January (10 degrees F), the traveler will be very uncomfortable. If a power budget calculation is performed using the average power of the transmitter, what happens on the span when the power generated is considerably below the average listed? It most likely will not work.

The minimum receiver sensitivity is just as important. This is the minimum amount of light required to make the device operate properly. No average allowed here either.

If there were nothing more to the optical power budget, simple subtraction should give a valid result for the amount of light available. That is:

Light available = minimum transmit power – minimum receiver sensitivity

The only thing to remember is that the minimum transmit power and minimum receiver sensitivity are often supplied as negative numbers already. In some cases, the transmitter might include a semiconductor optical amplifier (SOA) and boost the signal right away, before the fiber itself, giving transmitter powers of 0 dB or even +2 dB. A typical minimum transmit power might be –5 dBm and a typical minimum receiver sensitivity might be – 26 dBm. That's okay, since minus a negative gives a positive, or

Light available = -5 dBm – (-26 dBm) = -5 dBm + 26 dBm = 21 dB (the difference is always just dB)

There is one small complication with this simple, overall power budget example. Fiber spans almost universally operate in *both* directions. The fibers are different, but probably of the same type, and the transmitter-receiver pair on both ends can also be different. This can happen when one organization links up to another. Each organization makes independent equipment choices, but of course the fibers still have to work in both directions.

So the rule for bi-directional operation is simple. Figure out the overall power budget in both directions, and then use whichever number is *smaller* for further calculations. It makes sense that if the smaller power budget works in one direction, then the larger budget will work in the opposite direction. However, this will only be absolutely true if the inbound and outbound fibers follow the same route between sites and are of the same construction. Fortunately, this is almost always the case.

For example, consider Company A and Company B. They wish to connect sites with SONET/SDH over a dark fiber link. Company A has chosen SONET/SDH equipment from Vendor A with minimum transmit power of –1 dBm and minimum receiver sensitivity of –26 dBm. Company B has chosen SONET/SDH equipment from Vendor B with minimum transmit power of –3 dBm and minimum receiver sensitivity of –28 dBm. Which available power numbers should be used?

Table 10-2 shows the answer. The power budget is computed twice, once from Transmitter A to Receiver B and then from Transmitter B to Receiver A (the "crossover" is critical for proper results). The lower number should be used, in this case 23 dB, from Vendor B to Vendor A.

If this is all there were to computing power budgets, the topic would not require a section on its own. However, the power budget calculation process has really just begun. All that has been done so far is to compute the *overall* available power on the span end to end. Light is not delivered direct from transmitter to receiver without additional losses. The next step is to consider these additional causes of signal loss (attenuation) into the power budget numbers derived so far.

Light is lost on a span due to a number of things. Any fiber, of course, has a characteristic attenuation per kilometer or mile. This is usually the largest component of loss of the span, and many organizations proceed only considering this loss. But there are also losses due to splices or connectors on the span. No fiber span is a straight shot of continuous fiber from transmitter to receiver. There are always connectors at the equipment ends and at any patch panels along the way, and simple splices on the fiber span in between. This situation is shown in Figure 10-6.

Vendor A		Vendor B	
Mimimum Tx power	-1 dBm	Minimum Tx power	-3 dBm
Minimum Rx sensitivity	-26 dBm	Minimum Rx sensitivity	-28 dBm
Power Budget #1		**Power Budget #2**	
Vendor A Tx	-1 dBm	Vendor B Tx	-3 dBm
Vendor B Rx	-28 dBm	Vendor A Rx	-26 dBm
Power available	-1 dBm –(-28 dBm) = 27 dB	Power available	-3 dBm –(-26 dBm) = 23 dB

Table 10-2.   Using the lower overall power budget

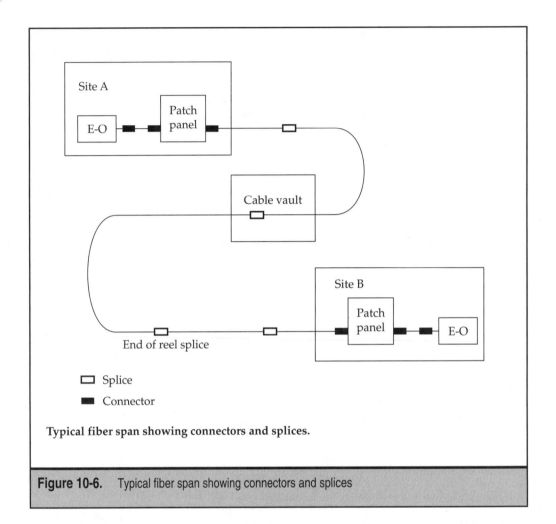

Typical fiber span showing connectors and splices.

Figure 10-6.   Typical fiber span showing connectors and splices

In the figure, sites A and B are to be connected by a fiber optic link. The E-O converter shown also includes the transmitters and receivers needed, although only one of the two fibers needed for simple two-way transmission is shown. This just makes the figure less cluttered. At each site, the E-O unit has a connector on the fiber tail end leaving the device. There is a connector on the tail end. This mates with another connector that takes the signal down the hall or to another floor on user fiber to a fiber optic patch panel. The patch panel, which is really just a "manual cross-connect" is not strictly necessary, but often used. The patch panel just makes re-arrangements (known as "adds, moves, and changes") of the building fiber cabling that much easier to do. Naturally, there are connectors of both sides of the patch panel as well. There is a short length of fiber jumper cable connecting the patch panel ports as well, but this is not shown in the figure. The patch panel is just a passive housing and adds no loss to the system. The network side of the patch panel at the entrance facility interfaces with the dark fiber itself. This arrangement is mirrored at site B. There could be additional connector pairs, but this scheme with three pairs on each end of the fiber is quite common.

The most important point about connector loss is that connector losses are always figured in connector *pairs*. There are six connector *pairs* in the figure, but 12 connectors total. But it makes no sense to compute the loss for *one* connector, since light will not flow end to end unless all of the connectors are mated. So it is connector pairs that count.

Once out of the building, the dark fiber is not usually connectorized. Rather, the fiber on the span itself is spliced. Splices are much more permanent than connectors, less bulky, and have lower losses than connectors as well. The figure shows four splices on the span between the two sites. There are usually three good reasons to splice fiber optic cable. All three are illustrated in the figure. First, fiber optic cable does not come on a 40 km (25 mile) reel. Most fiber reels are only a few kilometers or miles long. At the end of the reel, a splice is needed when installing the fiber. Second, there are places along the span where it is easier to splice fiber than other places. If there is an environmentally hardened cable vault made out of steel or concrete, this is a good place to put a splice even if the reel is not quite finished. Splices in a vault can last much longer than splices in the middle of a plastic conduit or between utility poles. Finally, fiber optic cable can be dug up ("backhoe fade") after installation or damaged during installation by related construction activities. These problems usually occur nearer to the connected sites, since construction activity is more intense in the immediate vicinity of a building. To fix the damage, splice the fiber.

There are many variations on this simple end-to-end theme. But all of the basics for power budget considerations are shown in the figure.

So the additional losses from the fiber itself, the splices, and the connectors have to be subtracted from the end-to-end power budget. Fiber optic cable attenuation across the span is the most serious cause of signal loss. Then come connectors. Splices contribute the least to power budget losses.

The typical numbers used in optical power budget computations are shown in Table 10-3. Keep in mind that these are example values, and should never be used instead of the actual numbers from vendors.

Unless information is available to the contrary, SONET/SDH equipment operates in the 1310 nm range. These losses are typically 1.25 dB/km for multimode fiber (rarely used outside of campus networks anymore), 0.4 dB/km for single mode fiber, and 1.0 dB/km for the newer non-zero, dispersion-shifted fibers (NZ-NSF) that are optimized not for SONET/SDH, but for DWDM systems operating around 1550 nm. The fiber cable vendor is the only reliable source for actual figures, and sometimes these numbers are stamped right on the cable jacket itself. For example, actual figures for single mode fibers range from 0.22 to 0.5 dB/km.

The standard value for connector pair loss is set by the Telecommunication Industry Association (TIA) at 0.75 dB per connector pair. Typical splice loss is only 0.1 dB per splice and can be compared to connector pair loss. Most fiber reels hold 6 km (3.75 miles) of fiber. The table also shows that when all is said and done, there should be at least 3 dB left over as a safety margin. Anything above and beyond this is just *excess power*. If there is a lot of excess power, however, perhaps some money could be saved by selecting a less powerful transmitter and/or a less sensitive receiver. Extreme excess power can result in receiver *saturation* (like

Typical multimode fiber attenuation loss at 850 nm	3.75 dB/km
Typical multimode fiber attenuation loss at 1310 nm	1.25 dB/km
Typical single mode fiber attenuation loss at 1310 nm	0.4 dB/km
Typical single mode fiber attenuation loss at 1550 nm	0.3 dB/km
Typical NZ-DSF attenuation loss at 1310 nm	1.0 dB/km
Typical NZ-DSF attenuation loss at 1550 (&1625) nm	0.25 dB/km
Standard value for connector pair loss	0.75 dB
Typical splice attenuation	0.1 dB/splice
Typical distance between splices (typical reel length)	6 km (3.75 miles)
Typical safety margin for power budget	1.7 to 3 dB

Table 10-3. Typical optical power budget values

shouting when a whisper is expected), and special external *attenuators* have to be added to very short spans to add some signal loss to the link. The safety margin is different than what is called the *operational margin* of the span. This requires some further explanation.

Consider the fiber span shown in the previous figure. To compute the operational margin and safety margin, two further steps are necessary.

First, it is necessary to allow for the additional span losses due to fiber attenuation, connector pairs, and splices. The fiber loss is multiplied by the distance of the span, and the others are added on a per component basis. For example, assume that the two sites are separated by 20 km (12.5 miles) of fiber and have six connector pairs and four splices, as shown in the figure. The fiber link will operate at the standard SONET/SDH wavelength of 1310 nm.

Optical power budget losses are always computed from largest loss contributor to smallest. That way, if the figure ever goes negative, which indicates that the transmitter and receiver selected will not work over the link, the entire power budget procedure can stop right there. So fiber losses come first, then connectors, then splices. This gives the *operational* margin of the span.

It is easy enough to multiple 20 times 0.4 dB to get a contribution of 8 dB loss from the fiber. There are six connector pairs. Exact connector pair loss figures can be provided by the connector vendor or installer, assuming any deviation from the normal value. Number of connectors is usually easy to determine from visual inspection at the building sites. Multiplying the six connectors in the example by 0.75 dB per connector pair gives an additional span loss of 4.5 dB.

Last come the losses due to splices. The hardest part here is for the user to determine just how many splices there are. Fortunately, the losses due to splices are so low that they are seldom the make or break factor in a power budget. Cable engineering and "as built" drawings should show splice number and location, but these might not be available for a user to consult (sometimes it seems as if they are not even available internally!). When in doubt, use this guideline: one splice per six km (3.75 miles) between sites, assuming a point-to-point fiber run (some vendors use very short run figures between splices, in some cases as low as 1 km or about 0.6 miles, but this seems excessive.) In this case, however, there are four splices. So 4 times 0.1 dB is 0.4 dB loss contribution due to splices.

The available power less the losses due to attenuation, connectors, and splices give the operational margin on the span. The process so far is shown on the optical power budget worksheet shown in Figure 10-7.

If the operational margin is negative, then there is not enough optical power to drive the link with the selected transmitter and receiver. But even if the operational margin number is positive, as it is here, there is more to be done. There are two more factors to consider in a complete power budget calculation.

Fiber cuts happen more frequently than is often assumed. Buried fiber is vulnerable to floods, simple cable dig-ups ("backhoe fade"), and more. Aerial fiber is at risk from tornadoes, hurricanes, earthquakes, fires, and even backhoe fade (the backhoe was being transported on a tall truck and snagged the cable). It is easy enough to splice the cable back together again, or use two splices to drop in a new fiber cable section, but this affects the attenuation on the span. So an estimate on the number of repairs made to the fiber over its lifetime is needed.

Longer spans are more at risk than shorter spans. There are just more places for bad things to happen the longer the span is. Most optical power budgets use five as the number of repair splices expected during the lifetime of the fiber span, which is usually considered to be 25 years and sometimes more. This is just another way of saying that a fiber cut should be expected every 5 years or so on a given span. The number can be adjusted up or down, of course.

Optical Power Budget Worksheet (Incomplete)

			Minimum transmit power:		-3 dBm
			Minimum receiver sensitivity:		-31 dBm
			Available power:		**28 dB**
Fiber loss:	20	km of cable x	0.4	dB/km	8.0 dB
Connector pair loss:	6	pairs x	0.75	dB/pair	4.5 dB
Splice loss:	4	splices x	0.1	dB/splice	0.4 dB
			Total losses:		**12.9 dB**
Operational margin (available power - total losses)					**15.1 dB**

Figure 10-7. Incomplete optical power budget worksheet

There are even a few other factors that affect fiber span performance. Temperature extremes are one important factor. Thermal expansion and contraction of connectors and some types of splices will cause additional losses over the lifetime of the fiber. There are other more arcane factors, but temperature and other considerations are usually lumped together in the *safety margin* entry in the power budget. The most common safety margin value used in optical power budget computations is 3 dB. In no cases should a value of less than 1.7 dB be used for a safety margin. Conservative organizations often use higher values, and organizations willing to assume greater risk can use smaller values for the safety margin.

When the final figure for the repair splices and safety margins are subtracted from the operational margin, the result should still be positive. The greater the value, the more confidence there is in continued link operation over time. If the value of the power budget is negative at this point, this means that the link might still work, but perhaps not after a few repair splices or during a heat wave. The failure might not be catastrophic, but even a modest rise in the bit error rate on the span at the wrong time can be disastrous.

Figure 10-8 shows the completed optical power budget worksheet for the example span configuration.

Optical Power Budget Worksheet (Complete)

		Minimum transmit power:		-3 dBm
		Minimum receiver sensitivity:		-31 dBm
		Available power:		**28 dB**
Fiber loss:	20 km of cable x	0.4 dB/km		8.0 dB
Connector pair loss:	6 pairs x	0.75 dB/pair		4.5 dB
Splice loss:	4 splices x	0.1 dB/splice		0.4 dB
		Total losses:		**12.9 dB**
Operational margin (available power - total losses)				**15.1 dB**
Repair splice loss:	5 splices x	0.1 dB/splice		0.4 dB
		Safety margin:		3.0 dB
Excess Power (operational margin - (repair + safety)):				**11.7 dB**

Figure 10-8. Completed optical power budget worksheet

In this example, the excess power is 6.6 dB. This is a little high (a figure of about 5 dB is more typical), but an indication of a link that should be expected to perform solidly. If the final result is more than 10 dB or so, a designer performing this power budget calculation might try to save money by selecting lower powered transmitters and less sensitive receivers. But without a detailed power budget calculation, this possibility would not even have been considered.

It is important to note the significant differences between the power available from a simple distance calculation (23 dB − 8.0 dB = 15 dB), a more accurate calculation including connectors and splices (15 dB − 4.9 dB = 10.1 dB), and the complete budget allowing for repairs and safety margins (10.1 dB − 3.5 dB = 6.6 dB).

SONET/SDH AND ATM

Much has been written in books and articles about the precise relationship between SONET/SDH and ATM. Some of this relationship has been mentioned earlier in this book. Because one of the last major points made in this book is that SONET/SDH is a trunking technology, and not a switching technology like ATM, or even frame relay, this is the place to discuss the relationship between SONET/SDH and ATM equipment more fully. There may be a false impression implied by this switching/transport split. The impression may be that there is no relationship between ATM and SONET/SDH. In fact, the ATM relationship with SONET/SDH is deep and precise. This section describes the relationship in more detail, and from an equipment point of view.

Today, this discussion might seem merely to be of historical interest. However, given the popularity of linking routers carrying voice, video, and data with SONET/SDH, this topic is still of importance. No matter what SONET/SDH can do, the high-level equipment (be it ATM switch or IP router) must still enhance the basic transport services than SONET/SDH provides. This section explains why.

As initially conceived in the late 1980s, SONET/SDH was intended to be the transport network for an ATM network. This network, with both ATM switches and SONET/SDH links, provides broadband services to users. If the broadband services comply with a series of services defined by the ITU, the switches are ATM, and the trunks are SONET/SDH, then the whole network complies with the international B-ISDN standard. Service providers once aimed toward providing full B-ISDN services because it was once thought that B-ISDN would ultimately be very popular with potential customers. Today, what broadband services there are (and probably will be) are provided by IP routers. The reason for combining ATM switching and SONET/SDH links into B-ISDN service networks in the first place was simple: Broadband networks are characterized by low and stable network delays to support applications that require interactive operations (e.g., voice and videoconferencing) and high bandwidths on access lines and trunks to support applications that are bandwidth-intensive (e.g., client-server networked graphics applications or videoconferencing). Note that some applications (e.g., videoconferencing) share the twin requirements of low and stable delays as well as high bandwidths.

Network delays have two causes. First, the bandwidth may not be sufficient for bits to travel quickly enough through the network from source to destination. Such applications are *bandwidth bound* and will not function properly without plentiful bandwidth. Obviously, this is SONET/SDH's function in a B-ISDN network: to supply the bandwidth needed for otherwise bandwidth-bound broadband applications. SONET/SDH's scalability certainly helps. When OC-3 operating at about 155 Mbps is not enough bandwidth, then perhaps OC-12 operating at about 622 Mbps will work.

Some applications are more affected by the delay through the network from a source to a destination. If the network delay were too high, or too variable and unstable, the application would not work. Voice is always a good example. The needs of voice in terms of bandwidth are quite modest by today's standards (it was not always so). The 64 Kbps needed for voice is easily provided by a number of means. Applications, which are more sensitive to low and stable network delays than bandwidth, are known as *delay-bound* applications.

Some delay can be eliminated with increased bandwidth, but most delay-bound applications, such as voice, will not benefit from more bandwidth. Voice at 128 Kbps will not operate better than voice at 64 Kbps when the delay is not low and stable enough to begin with. Thus, low and stable network delays are affected by more than bandwidth, of course. One of the two components that contributes to network delay is the propagation delay. No matter what the bandwidth is, the signal will take longer to reach a destination that is farther away than one that is close by. This is simple physics. When international voice traffic began to be routed over geosynchronous communications satellites, many argued that the delays were much too high for voice applications. In spite of concerns, nothing could be done about long uplink and downlink satellite delays.

The other component of network delay is nodal processing delay. Whether the network nodes are routers, central office switches, or something else, it takes a finite amount of time for bits to enter an input port on a network device and emerge from an output port. Here is where ATM fits in. A full discussion of ATM is far beyond the scope of this chapter and even this book. However, because ATM and SONET/SDH were once intimately related, a brief overview of what ATM switches were supposed to add to SONET/SDH links is not out of order.

ATM is the international standard for cell relay technology. In cell relay, data units are chopped into small pieces (the cells) to facilitate network node processing. The end result is more effective switching and multiplexing. ATM provides the low and stable nodal processing delay and, thus, the overall low and stable network delay needed for the whole spectrum of broadband services.

This relationship between ATM switches and SONET/SDH links is illustrated in Figure 10-9. Today, of course, because many routers are as fast as ATM switches, broadband services can be delivered on router-based networks. As a result, not only has ATM been "de-coupled" from SONET/SDH , but SONET/SDH has become more closely aligned with router-based networks.

SONET/SDH COMPONENTS

So far, most of the discussion of SONET/SDH equipment has revolved around single links, and (if the truth be told) only the sending side of these links. Not that receiving is

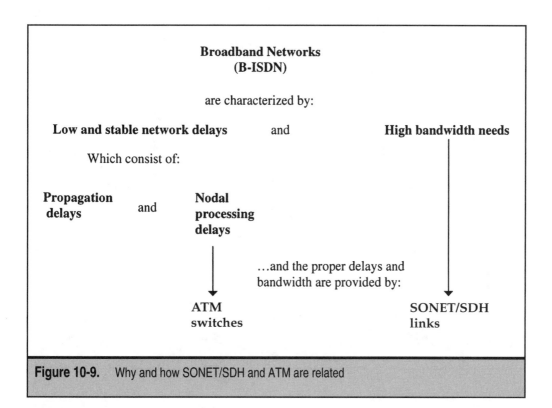

**Broadband Networks
(B-ISDN)**

are characterized by:

Low and stable network delays   and   **High bandwidth needs**

Which consist of:

**Propagation
delays**   and   **Nodal
processing
delays**

...and the proper delays and
bandwidth are provided by:

**ATM
switches**   **SONET/SDH
links**

Figure 10-9.   Why and how SONET/SDH and ATM are related

any more complex. Almost any figure in the text could have a mirror installed in the right-hand margin to reflect the fiber link and all to the left of it. For instance, all SONET multiplexers also demultiplex, and service adapters not only adapt from T-carrier to VT1.5s and the like, but also adapt VTs back to T-carriers.

However, SONET/SDH networks consist of more than single links. In fact, SONET/SDH networks can be quite complex. This chapter introduces the overall components of a typical SONET/SDH network, and the next chapter details their individual operation. Figure 10-10 shows the major components of a multiple-link SONET/SDH network, such as may be deployed by a major carrier in a metropolitan area today. SONET/SDH documentation refers to these components as *network elements (NEs)*.

Since there are more than single links involved in the network, clock distribution and network timing become all the more critical when DS-*n* cross-connecting is needed. Note that some of the SONET/SDH components form rings in the diagram. SONET/SDH rings have become a distinguishing feature of SONET/SDH, much as touch-tones have become a distinguishing feature of telephones. Both are simply a much better way of doing things. A full discussion of SONET/SDH rings deserves a chapter of its own. But this is the place to introduce them.

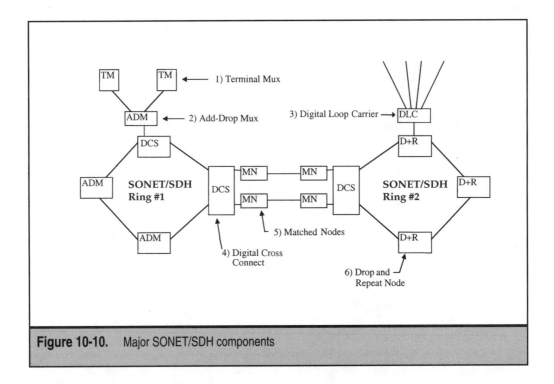

Figure 10-10.   Major SONET/SDH components

As shown in Figure 10-10, there are six main network elements that can be used in SONET/SDH networks. These are labeled in the figure as follows:

1. **Terminal multiplexer (TM)**   An "end point" device on the SONET/SDH network. This device gathers bytes to be sent on the SONET/SDH network link and delivers bytes on the other end of the network. TM is intended to be the most common type of SONET/SDH CPE, but ironically is still rare at the customer site. In many current SONET/SDH configurations, TM is still found in the serving office. The major exception is the router, which in most cases performs the role of the TM.

2. **Add/drop multiplexer (ADM)**   Really just a "full-featured" TM. However, it is more accurate to refer to the TM as an ADM operating in what is known as "terminal mode." This device usually connects to several TMs and aggregates or splits (grooms) SONET/SDH traffic at various speeds.

3. **Digital loop carrier (DLC)**   This SONET/SDH device is used to link serving offices with ordinary analog copper-twisted-pair local loops in order to support large numbers of residential users in what is known as a carrier serving area (CSA). The devices are also commonly referred to as remote fiber terminals (RFTs) by many vendors.

4. **Digital cross-connect (DCS)**   This SONET device can add or drop individual SONET/SDH channels (or their components in a VT/TU environment) at a given location. It is basically an even more sophisticated version of the SONET/SDH ADM.

5. **Matched nodes (MN)**   These SONET/SDH devices interconnect SONET/SDH rings. They provide an alternate path for the SONET/SDH signals in case of equipment failure. This feature is commonly known as signal protection, but it has a variety of other names.

6. **Drop and repeat nodes (D+R)**   These devices are capable of "splitting" the SONET/SDH signals and sending copies of bytes onto two or more output links. The devices will be used to connect DLC devices for residential video (or even voice) services.

Figure 10-10 shows a more complete network deployment that exploits the features of SONET/SDH. It should be noted that there are no "official" designations for the various pieces of SONET/SDH equipment shown in the figure. Although the figure generally follows terminology of such SONET/SDH equipment vendors as Nortel Networks, it is not uncommon for vendors to have their own names for the devices shown in the figure. However, the emphasis here is on function, not so much on the names themselves. The equipment found in this network is separated into central office terminal equipment and outside plant equipment.

The central office equipment consists of a SONET/SDH-compatible switch (i.e., digital cross-connect switch) and digital loop carrier equipment. This equipment forms the network terminating point for SONET/SDH transport services.

The outside plant equipment consists of a fiber ring topology for survivability, and access multiplexers to this ring. The multiplexers are functionally separated into two categories, TMs and ADMs. TMs are functionally similar to channel banks in the T-1 networks. They provide conversion from the non-SONET transmission media to the SONET format, and similar functions apply to the E-carrier environment. Add/drop multiplexers are an integral component of the SONET/SDH architecture; they allow access to the SONET/SDH transmission network without fully demultiplexing the SONET/SDH signal. Non-SONET/SDH signals may be added to or taken from the SONET/SDH transmission signal via this equipment. The differences between SONET/SDH TMs and ADMs is often only a difference in network location and in function, rather than a fundamental or intrinsic difference. Figure 10-11 shows the main difference between a SONET/SDH ADM and a SONET/SDH TM, which is really just an ADM operating in "terminal mode."

In Figure 10-11, the major difference between a SONET/SDH ADM and TM is that the SONET/SDH ADM has two network connections that must be equal. Channels, or even other OCs (as long as they are less than the network OC), can be dropped or added as desired. These channels or OCs can lead to SONET/SDH TMs, or course. On the other hand, a SONET/SDH TM has only one network connection. The "terminal mode" just multiplexes various bit rates onto and off of this single network link. The whole point is to allow (relatively) "low-speed" user access to the SONET/SDH network. The user access may be at DS-3 or even OC-3 rates, as long as the network link can handle the speed. Routers usually perform this TM role in many SONET/SDH networks.

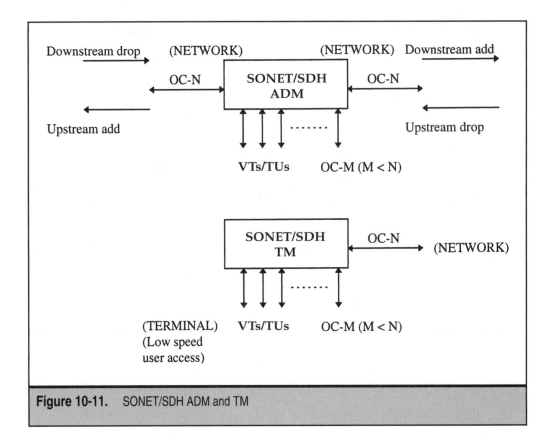

Figure 10-11.   SONET/SDH ADM and TM

Physically, the SONET/SDH equipment described is of quite modest size. The solid-state electronics that characterize SONET/SDH result in components that are typically rack-mounted in standard communications cabinets. The power requirements are correspondingly modest as well.

Figure 10-10 can be misleading in one sense. The figure implies that the six major SONET/SDH NE functions might all be found in all large SONET/SDH networks and with equal likelihood. In fact, the truth is just the opposite. The SONET/SDH NE packages offered by most equipment vendors are ADMs, first and foremost. Many of the other NE functions are added (or subtracted) from the basic ADM package to yield the required NE for the application. So an ADM with non-SONET/SDH interfaces and a single SONET/SDH interface becomes a SONET/SDH TM package. DCS functions can be added to the ADM in the same way, but this is seldom done. Most DCSs remain separate-gear-attached to the ADM. If the ADMs have multiple links used to connect their rings, this makes that pair of ADMs into Matched Nodes. In other words, the MN function is just another feature of the ADM.

The ADM/TM/MN package is normally what a SONET/SDH equipment vendor emphasizes, and that might be it. If a SONET/SDH DCS NE is offered, it is usually not an ADM feature, but separate equipment altogether. The D+R and DLC functions can be addressed in the same way. The D+R NE is a special ADM. The DLC is a special TM optimized for residential voice applications.

In summary, SONET/SDH equipment vendors typically offer a basic ADM package in larger or smaller sizes. Small ADMs make great TMs and DLCs with the correct line cards and associated functions. Larger ADMs make terrific D+R nodes and the MN function is usually just another interface card in the ADM unit to interconnect rings. So the same basic product line can be ADM, TM, D+R, MN, and DLC device all in one. It just depends on the precise hardware and software configuration. The odd NE out is the DCS. Cross-connecting has always been a complex function to configure and control. In SONET/SDH, it is often best to leave this task for more specialized equipment.

Oddly, this means that many a SONET ADM is busily extracting DS-1 and DS-0s from SONET frames just so the DS-1s and DS-0s can be sent over to the legacy DCS across the room for cross-connecting. The rearranged traffic then comes back to the ADM for re-packaging into outgoing SONET frames.

SONET/SDH Rings

A distinguishing characteristic of SONET/SDH links is their capability to be deployed in a ring topology and configuration. A full discussion of SONET/SDH rings and their various forms deserves a chapter all its own, but this section introduces the concept.

Fiber optic links, like T-1 twisted pair and most other digital transmission links, are inherently unidirectional and not full duplex. All SONET/SDH fiber links consist of a transmit fiber and a receive fiber, as T-1 links consist of a transmit-and-receive twisted pair. SONET/SDH ADMs have a minimum of four fiber interfaces: one for *upstream* and *downstream* transmission in each direction. What happens to these fiber links at the "ends" of a series of SONET/SDH ADMs? An interesting configuration results when the loose "ends" are looped around to form a closed loop, or ring, of SONET/SDH ADMs connected by the fiber links. This is shown in Figure 10-12.

In Figure 10-12, the simple SONET/SDH ring consists of several ADMs linked by their upstream and downstream fibers. The ring may span only a few miles, or stretch to literally thousands of miles, depending on its purpose. Note that there are still SONET/SDH TMs, with users attached to feed the ring with traffic. Naturally, the aggregate traffic from the TMs cannot exceed the capacity of the SONET/SDH ring.

What is the advantage of deploying SONET/SDH in a ring configuration? Quite simply, rings provide greater reliability by furnishing two separate paths for digital signals between the ADMs. Even this simple SONET/SDH ring provides what is known as path protection switching (PPS), or just *path switching*, between the SONET/SDH ADM nodes.

Typically, the two fibers are deployed in the same cable sheath and are laid in the same conduit. Consider what would happen if the fibers between SONET/SDH ADM #2 and ADM #3 in the figure were severed by some environmental disaster (e.g., flood, tornado) or even a construction "incident" (known in the industry as "backhoe fade"). Signals can still travel between ADM #2 and ADM #3 through ADM #1. There is still a fiber path between ADM #2 and ADM #3.

SONET/SDH rings allow for repairs to be made without disrupting customer service. This is a valuable capability that justifies the added expense of the extra fiber needed to

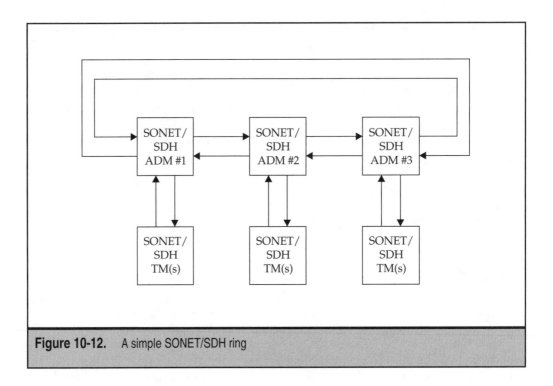

Figure 10-12. A simple SONET/SDH ring

"close the ring" of SONET/SDH ADMs. Today, it is rare to see a SONET/SDH deployment of any size without rings.

The essence of path switching is to provide an alternate path for signals; however, in this simple ring, spare capacity must be available on the links between ADM #3 and ADM #1 to allow for this capability. This only makes sense. There are more elaborate ring configurations that will be explored later. For now, it is enough to point out that the ring illustrated in Figure 10-12 is a 2-fiber, unidirectional, path-switched ring. SONET/SDH rings are often referred to as "self-healing" rings, although this may imply that the rings somehow fix themselves and restore failed links to service on their own. They do not.

Typically, a SONET/SDH ring consists of several ADM devices connected together in a ring topology. Users, however, get access into and out of SONET/SDH using TM devices. This is not a problem because ADMs can easily interface with one or more TM components at many points around the ring. ADMs take the form of a simple "card cage" with a power supply (often with backup) and several slots for interface cards. In their most basic form, ADMs have at least two cards for the input and output fiber links of the ring itself. Adding TMs to a SONET/SDH ring of ADMs is as easy as adding cards to card cage slots. Of course, when a device's slots are full, no more TMs (or other devices) can be added to the configuration.

It is also possible to extend the reach of a SONET/SDH ring by strategically placing another ADM not in the path of the ring, but off of an ADM on the ring itself. This allows for the connection of more TMs, of course. At the same time, however, this process cuts down on the amount of protection a customer receives from the SONET/SDH ring. It makes no difference to a customer if the carrier has a SONET/SDH ring or not when the problem is the single fiber link from the TM or ADM to the ring itself.

What would the typical components of a service provider's SONET/SDH ring look like when used to provide more reliable and efficient T-carrier or E-carrier circuits? Where would the typical components be located? Figure 10-13 shows such a layout, with the dual fibers rendered as single lines to reduce the complexity of the figure.

In Figure 10-13, the carrier has installed a SONET/SDH ring by linking the ADMs in several office locations (i.e., central offices and, in this case, a wire center). A wire center is a more centralized switching office for trunk connectivity to interexchange carriers and other long-haul network elements. The most common speeds used for this ring are in the 622 Mbps and 2.4 Gbps ranges, although higher speeds will become more common.

TMs that handle the customer interfaces to the SONET/SDH ring may be located in the carrier's space in a large office building, or even on the premises of a particularly large customer. Note that the customer's TMs may feed an ADM which, in turn, is connected to the ring. These ADMs are most often located in a co-location space servicing a large area, but may also be located in the carrier's space in an office building. Frequently, one building in an office campus complex holds the ADM for the whole complex, while individual buildings hold TMs. Of course, the campus ADMs may form a ring themselves.

TMs typically feed the ring at OC-1 (SONET) or OC-3 (STM) rates. There is not "bandwidth on demand" in action here, and the sum of the OC-1s and OC-3s cannot exceed the capacity of the SONET/SDH ring. Thus, when 24 OC-1s and 8 OC-3s feed an OC-48 SONET ring, the ring then is completely full because simple math shows that $(24 \times 51.84$ Mbps$) + (8 \times 155.52$ Mbps$) = (1.24416$ Gbps$) + (1.24416$ Gbps$) = 2.48832$ Gbps (or 2488.32 Mbps), which is the capacity of an OC-48 fiber link. It makes no difference whether the links are in a ring or not. As an aside, even at full capacity, the SONET ring may still provide some failure protection. This subject is discussed in more detail later in Chapter 14.

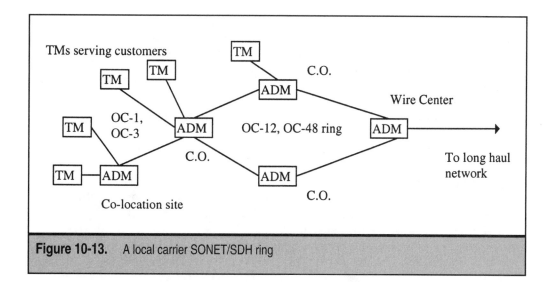

Figure 10-13.   A local carrier SONET/SDH ring

Rings in Action

What would a working SONET/SDH ring look like and be used for? SONET/SDH is not only for metropolitan areas. SONET/SDH rings exist in rural areas as well. To balance the perception of SONET/SDH rings in heavily populated city areas, this section describes a SONET ring in a more sparsely populated area in the United States, yet a ring that is still easily cost-justified.

A typical rural SONET ring may service about 300-square miles in the Midwestern United States. It would link five telephone exchanges and service perhaps 2,000 business lines and 6,000 residential subscriber lines. Operational costs are a big factor in rural areas. SONET rings can help to control these costs and at the same time maintain the high-service levels customers need and have come to expect.

The SONET ring is cost-justified as a way of simultaneously upgrading the copper feeder infrastructure while, at the same time, providing a basis for new high-speed services and controlling maintenance costs. Businesses are rapidly entering more rural areas around the country, and if the incumbent carrier cannot provide the services these new businesses need, they will look elsewhere.

The SONET ring itself may only span about ten miles. In addition to the protection features, the maintenance costs for SONET are about 75% less than the cost of copper cable upkeep. The TMs feeding the ring are placed around the 300-square mile service area. They provide both voice and data services at a variety of speeds. This keeps the cost of new circuits down and provisions service very quickly.

For example, a business ordering 100 new access lines does not need new cable laid from the serving office. The carrier links the new lines to the nearest TM, where software provisions the circuits. The carrier can provision the links in a day or so, compared to two weeks previously. The TM may combine some (or even all) of the features of a SONET DLC (sometimes called an RFT). In that case, any dial tone service, voice, or data, can be provided as well as leased lines. Sophisticated SONET equipment is just basically an *overlay network* superimposed over the existing copper network.

Network management takes advantage of the standard SONET set of features for this purpose. Outages are pinpointed more accurately, and service due to ring-link outages can be restored while most customers are unaware that there has been a problem. Naturally, the degree of protection depends on the details of the overall SONET ring topology; however, any SONET ring gives much better protection than does the most sophisticated T-carrier point-to-point links.

The same business may order 100 lines initially, expect 300 lines in five years, and 500 lines in seven years. The SONET infrastructure can support whatever growth is needed over time. The success of one business often encourages others in the same area. More customers mean more revenue for the service provider.

Eventually, the ring can grow to include more offices and an enlarged customer service area. A ten-mile ring could grow to 100 miles. TMs linked to rings by unprotected, single-fiber links could be "upgraded" to full ADM operation with new fibers to make the TM a node on the SONET ring itself. All of this can be managed with SONET-specific software and network management capabilities.

This chapter has opened up some new areas for further exploration. The example SONET ring just described should be examined even further. The operation of each piece of SONET/SDH equipment should be explored in more detail. Finally, the makers of various pieces of SONET/SDH equipment should be investigated to try to determine how each of them seeks to distinguish themselves in an area where, after all, everybody is selling the same thing.

CHAPTER 11

SONET/SDH Equipment

This section of the book began with an overview of SONET/SDH networks. Some of the various types of SONET/SDH equipment were identified and their overall function in a SONET/SDH network was outlined for each. There were six overall classes of equipment defined, but those classes are anything but hard and fast. Indeed, any new technology goes through an early phase of specialized equipment that can perform only one task. Already, SONET/SDH "equipment" has appeared in the form of simple interface cards for routers.

The one-task phase is followed by a phase that is characterized by rapid equipment evolution and convergence of form and function as an electronics evolution overtakes the standard. Consider early LAN bridges and routers that could handle only Ethernet or Token Ring, but not both.

Thus, even as the SONET/SDH standard evolves, there has been something of an evolution of equipment. This chapter explores the current state of the SONET/SDH equipment marketplace. It details the function of the various pieces of SONET/SDH equipment needed to build a SONET/SDH network. Finally, the chapter attempts to project the evolutionary trend of SONET/SDH devices.

First, a slightly different perspective of SONET/SDH devices is in order.

SONET/SDH NETWORKS

SONET/SDH networks can be roughly divided into four main parts. These four parts are not unique to SONET/SDH networks; they have been around for more than 100 years in one form or another. However, the application of SONET/SDH to the structure is the topic of discussion.

Although the topics and networks in this chapter could easily be applied to both SONET and SDH, the remainder of the chapter examines SONET only. That approach is used mainly for simplicity, but also because many of the network configurations shown are derived primarily from the distinctive architecture of the North American telephone network. That is not to say that similar networks could not be shown in an SDH context. But this chapter does not do so.

The four portions of SONET (or SDH) networks are distribution ("local access"), the feeder portion, the local backbone, and the long-haul backbone network. These sections vary in equipment capabilities as well as in scope. Figure 11-1 shows the overall structure of the four portions. From user to user, the parts may be reflected in the sense that user access must exist on each end of the network, and that the communication between local users close enough together may not involve all four portions of the SONET network; however, the general picture is clear.

The local distribution system takes signals to and from homes and businesses. The link to the home and business locations—frequently known as "the last mile" (although this can stretch to nearly five miles in many areas)—is not necessarily fiber. Most often, even in many business situations, the link is still coaxial cable or twisted-pair copper wire. With few exceptions, the link to the home is twisted-pair copper. SONET equipment interfaces with these traditional distribution systems at this point.

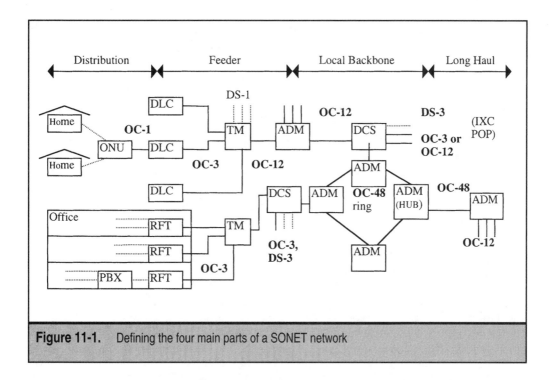

Figure 11-1.   Defining the four main parts of a SONET network

The next portion of the SONET network is the feeder system. Feeder links may stretch for miles as well, and generally cover whatever portion of the distribution system is not serviced by older coaxial or twisted-pair access links. As its name implies, the feeder system aggregates and distributes the SONET traffic from and to the distribution ("access") system. The feeder system equipment is characterized by modest SONET speeds and environmentally hardened casings and enclosures. This last aspect is due to the fact that, in most cases, SONET feeder equipment is deployed in the field in "outside plant" locations. As such, it is subjected to extremes of weather conditions and the rigors of all other environmental hazards. Nevertheless, few SONET rings are yet deployed at this level of the network, except in major metropolitan areas.

The third portion of the SONET network is the local backbone. Here is where SONET rings are first typically deployed by local exchange carriers (LECs). The ring may be a mixture of outside plant and office space ("inside plant") equipment locations. The ring protects the SONET network from single-point-of-failure concerns. This feature, however, is at least somewhat offset by the common single-link feeder systems that are the rule today. Local rings may span only 10 miles, or they may stretch to about 100 miles across. Generally, because they are found mostly in metropolitan areas, local rings of larger sizes are few and far between.

Finally, the local backbone (or backbones, because several may be linked at this stage) connects to a long-haul backbone. This last portion also may be furnished by the local service provider, but it is more likely to belong to one of the inter-exchange carriers (IXCs) in the United States such as MCI or Sprint or one of the dozens of others. The long-haul backbone may consist of more rings or single links, depending on the SONET or other capabilities of the IXC. These rings and links may span the country, or may even be international in scope.

THE DISTRIBUTION NETWORK

Taking a more detailed look at each segment in Figure 11-1 is instructive. The distribution ("access") network consists of two main types of users. These are business and residential (home) users. This split has endured since the early days of telephony, when the telephone was marketed as a business tool and sold only grudgingly to home users, who were used to writing letters instead. (One of the earliest residential telephone users was Mark Twain in Hartford in 1878. Twain was also said to have invented the complaint letter, since his protests about the service were so strident and frequent.)

In any case, the homes are serviced by special Optical Network Units (ONUs), which convert SONET fiber signals to ordinary twisted-pair copper-wire signals. The ONUs most often connect to the rest of the SONET network at low fiber speeds, and may not even support SONET. When ONUs do support SONET speeds, this is most often just an OC-1 at 51.84 Mbps. Of course, other speeds are possible, depending on the number and density of the homes served. ONUs are interesting enough devices in their own right, but their relationship to SONET is only peripheral; therefore, they will not be discussed further.

Naturally, if fiber were to extend right to the home in the so-called fiber-to-the-home (FTTH) or even fiber-to-the-curb (FTTC) configuration, the ONUs might be mounted on the side of the house (FTTH) or on the nearest utility pole (FTTC). Neither of these configurations is common yet and will be ignored for the remainder of the discussion.

In Figure 11-1, the ONUs interface with the first piece of SONET gear. This is the SONET digital loop carrier (DLC) device. DLCs are also called remote fiber terminals (RFTs) by many vendors and service providers, but the figure distinguishes between them. In the figure, DLCs are positioned as residential units supporting ONUs, while RFTs are depicted as business SONET customer premises equipment (CPE) devices. Neither designation is right or wrong; they merely reflect a possible system for telling them apart easily in this context.

DLCs interface with many ONUs, with the exact number varying among vendors and models. The link to the ONU may be non-SONET fiber, SONET fiber, or (most commonly) some form of T-carrier, such as T-1 operating at 1.5 Mbps. The evolution of this DLC (and RFT) is interesting as well.

The use of DLCs and RFTs in the local loop is the culmination of years of evolution in wire technology. Previously, miles of twisted-pair copper wires bundled together snaked their way across the landscape on poles, forming both the distribution and feeder portions of the networks. One pair of copper wires was needed for voice service and was dedicated for this purpose all the way back to the local serving office ("central office": CO).

In the 1960s, analog loop-carrier systems evolved to combine many voice signals on a few pairs of wires. These systems provided a way to gain back some of the pairs used for each and every user. Because of this, they came to be known as "pair gain" systems. By the 1970s, digital technology was used for the same purpose for a variety of reasons, not the least of which were improved quality and reduced costs. In 1979, AT&T, then the "national telephone company" in the United States, introduced the most successful DLC equipment, the Subscriber Loop Carrier 96 (SLC 96) device.

Network growth throughout the 1960s and 1970s led to a need to limit the amount of new copper placed in and above the ground. This has historically been the highest cost in the telephone system. The SLC 96 from AT&T (now from an AT&T component called Lucent Technologies) uses four pairs for transmission to the CO and four pairs from the CO, with a spare pair in either direction for control and backup.

The SLC 96 system supports up to 96 voice lines on the ten pairs of cable. In the case of a new housing development, an SLC 96 or equivalent is placed at a central point. From that point, individual copper pairs will be connected to the customers' homes and businesses. The system allows the network to support more customers with minimal deployment of new cable plant. This approach is ideal when providing telephony services. As service providers look to add new, higher bandwidth services, they will have to upgrade the local loop. After many years during which the local loop was viewed as having limited significance, it is now an area of great concern in upgrading to support a service network.

Thus, an SLC 96 uses four T-1 lines for 96 voice channels; the fifth T-1 provides protection switching. Without SLC 96, 96 homes would need 96 pairs of wires. Unfortunately, several design features made SLC 96 unsuitable as a long-term solution. Compromises in design were made to allow all types of copper pairs to interface on the user side, and more compromises were made to interface with all types of service provider switches. Throughout the 1980s, the quest was on to provide a better DLC solution than SLC 96.

In the 1980s, Bellcore defined a new, all-digital, T-1–based, loop-carrier system called *integrated digital loop carrier* (IDLC, Bellcore IDLC-TR303) just from digital and computerized serving offices. In the late 1980s, a migration strategy to SONET to deploy fiber-in-the-loop (FITL, an umbrella term for FTTH and FTTC and others as well) was defined by Bellcore as a supplement to IDLC-TR303.

Today, almost all DLCs or RFTs are SONET-compatible. As time goes on, more and more SONET fiber will extend nearer and nearer to the end user. This makes a lot of sense. SONET offers a consistent and standard interface for vendors between the access network to homes and the feeder network to the serving office. SONET will eventually allow for the replacement of copper cable with fiber, which provides enormous advantages. This will not be inexpensive, but it will definitely be worth the effort when the SONET fiber provides the basis for new broadband services.

Figure 11-1 also shows a business arrangement using SONET equipment. In this case, the device is shown as an RFT, but there are few differences between DLCs and RFTs. In the figure, the RFT at the business location can interface with a variety of other devices, from PCs with analog modems for data services, to the organization's PBX for voice services.

Naturally, digital interfaces can be provided for user data equipment as well. The RFT can often directly interface with the organization's building wiring when it meets current specifications. The RFT provides a digital path to the serving office at full SONET capacity. The RFT cabinet can provide a business or organization with interfaces operating at speeds from 64 Kbps (DS-0) to 1.5 Mbps (DS-1) to 45 Mbps (DS-3). Naturally, for DS-3 speeds, a SONET OC-12 instead of an OC-3 may be desirable.

The use of both DLC and RFT equipment will increase rapidly as SONET moves into the local loop and access portions of the network.

The Feeder Network

The feeder portion of a SONET network is the area where subscriber-line multiplexing technologies, such as SLC 96 are employed. In SONET networks, the key component of this portion of the network is the terminal multiplexer (TM). The TM is essentially a SONET add/drop multiplexer (ADM) running in "terminal mode." Previously in this book, the TM has been positioned more as a CPE device or piece of equipment that is SONET on one side and something else on the other.

However, the emergence of DLC and RFT equipment with SONET interfaces has changed that picture somewhat. In fact, it is probably just as accurate to say that newer DLCs and RFTs with SONET interfaces are now a kind of SONET TM, but people seldom talk this way. It is possible that a SONET TM may be located in a large office building, especially a building of 20 or 30 stories in a major metropolitan area. In many cases, however, vendors and needs for SONET TM and RFT equipment vary enough to make the retention of the distinction a wise idea. As time goes on, and prices decline, the DLC and RFT devices will merge rather quickly into one SONET TM device.

In Figure 11-1, the SONET TM can feed either a DLC for residential distribution or a RFT for business access. The biggest difference is usually in the speed. Even heavily populated residential areas can be fed from a few OC-3s, each providing more than 2,000 voice channels. Businesses, on the other hand, can easily require 2,000 phone lines in a modest-sized office building. Moreover, the needs of businesses for speeds beyond 64 Kbps and 1.5 Mbps will consume more bandwidth more rapidly. Thus, TM equipment deployed here would support up to OC-12 rates; each has the equivalent of some 8,000 regular telephone lines.

In a SONET network, the main job of the TM is to aggregate and groom the traffic passing through it. In large business applications, it is even possible to have a SONET TM as a true user CPE, although this application of the TM has been rare to date.

The TMs at the end of the feeder network for residential services are connected to ADMs, which are pretty much the same type of equipment as the TMs, with minor differences in operation and capabilities. The ADMs further adjust the traffic, because residential areas are characterized by many customers spread over a wide area. Although the links to and from these residential ADMs are typically OC-12s, this is not a problem. The density of residential users makes this feasible, and the speed of the link can always be increased if necessary.

Note that the business customers' TM usually links directly to an ADM on a ring. The amount of traffic to and from a business location makes it desirable to feed a SONET ring directly at the local backbone level. Typically, these links can operate at OC-12 speeds, although speeds of up to OC-48 are becoming more common.

Most links to and from the TM and ADM are linear links and not rings. There may be multiple links with diverse routing, but neither offers the same type of protection from failure that a SONET ring offers. Ironically, most service providers will play up the presence of SONET rings on the local backbone and never mention that the distribution network, and most likely the feeder network as well, is not deployed in a ring configuration. Naturally, this is where an outage affecting customer service is the most likely to occur and where environmental risks are the greatest.

The Local Backbone

The local backbone is where SONET rings are most often deployed. In Figure 11-1, the ring has only four SONET devices, but this is a representative ring only and should not be taken to imply any degree of precision. There may be 5, 10, or up to 16 ADM nodes on the SONET ring. The configuration depends on the size of the ring and the density of users in a given area. Typically, each node on the ring will be located in a serving office. This may be a telephone company local switch (CO, quite commonly), a cable TV company's headend, or almost any commercial office space. SONET equipment requires little special air handling or backup commercial power supplies, although both are a good idea.

Two main types of SONET equipment are employed on the local backbone ring. These are the ADM and the SONET digital cross-connect system (DCS). The differences between ADMs and DCSs, in terms of functions and features, are significant. The ADM can add and drop signals at a variety of speeds, including both SONET and T-carrier speeds. That is, the traffic that the ADM adds and drops is SONET traffic in SONET standard formats, such as STS-1 and STS-3c, and T-carrier traffic in DS-1 and DS-3 (or even other) formats.

A SONET DCS is sometimes called a "full-featured" ADM. The DCS can interface with other speeds and types of equipment, and also with the same forms of T-carrier traffic—such as DS-1, DS-3, and so on—as an ADM can. A DCS can also have SONET interfaces. Thus, a DCS can connect to an ADM or TM as well, as shown in Figure 11-1. Generally, a SONET DCS has more management features and interconnection options between the input and output ports than an ADM has. That is, instead of a passive dropping of DS-1 as in an ADM, a DCS can interconnect the DS-1 in the same device, a feature sometimes called "hairpinning" (as in "hairpin turn" on a road).

The main differences between SONET TM, ADM, and DCS are in the relationship between the interfaces supported. ADMs usually do not "hairpin" a channel in an outside-plant location. Their main task is to get the traffic to the serving office, where the DCS can decide where the channel should go. Of course, a DCS on a local backbone can still easily add and drop a DS-3 to a customer site or somewhere else. This direct DS-3–to–SONET-local-backbone arrangement is often used when a customer needs (and can afford) SONET service in a location where deploying TMs and RFTs does not make sense.

Consider a customer with many sites located in a downtown area, but one site (a factory perhaps) located some distance away. Obviously, the customer will want to have SONET-based services everywhere. The question is how to reach the factory site. The other sites may be served by RFT-TM-DCS arrangements to reach the local backbone. The factory may need to use an existing DS-3 to reach the ring. Without a direct DS-3 interface to the DCS on the ring, tying the factory into the rest of the network would be awkward and difficult.

A DCS is also typically used to carry traffic over SONET links to other service providers, such as IXCs or even local competitors. Naturally, other service providers may have their own SONET rings, but the link to their rings are still not commonly rings themselves. The speeds are usually at DS-3 (672 voice channels), OC-3 (about 2,000 voice channels), and even OC-12 (about 8,000 voice channels). The channelization does not need to be 64 Kbps units.

Most SONET rings operate at OC-48 speeds, with higher speeds in the near future. Some service providers are starting to fill their SONET rings with traffic, and some have looked at adding a second ring in the same geographic area. This is fine and even offers another measure of protection beyond that provided by a single ring; however, OC-96 and even OC-192 will be operational soon.

One other type of ADM is employed on the local backbone ring. This is the *hub* ADM, most often simply called a SONET hub, although it is not really a distinct piece of equipment. A hub also interfaces between other ADMs and DCSs. Typically, all of the links into and out of a hub operate at a very high SONET speed (usually OC-48), which helps to distinguish the hub device from other ADMs and DCSs.

The hub is used to interface to another SONET ADM device off the local backbone. The main purpose is to extend the SONET services into remote areas, usually further afield than those reached with feeder network arrangements. Alternatively, and as shown in Figure 11-1, the hub may feed an IXC's long-haul SONET links.

Admittedly, there is a fine line between ADM, DCS, and SONET hub. Some vendors have different products and names for all three. Others have only one product and rely on interface and configuration differences to tell them apart in the field. Still others deny that a difference exists. In the future, it seems certain that differences will become even more blurred. In a LAN environment, few distinguish between a bridge and a router today, but they began as completely different devices for completely different purposes.

The Long-Haul Backbone

The last piece of the SONET network is the long-haul backbone. Usually, this portion is controlled by the IXC, and may be accessed not only at SONET speeds (typically OC-3 or OC-12), but even at relatively modest DS-3 speeds. When T-carrier is employed, as in the case of DS-3, the nice feature is that the IXC is not required to have any SONET equipment available, although all of the T-carrier limitations are in force. Long-haul backbones have become a very active area of SONET deployment, where speeds of OC-48 (2.4 Gbps) and even OC-192 (10 Gbps) are common.

The long-haul network is not always ringed, but it may be. More IXCs are building their own SONET rings. As local competition becomes economically feasible on a facilities basis (local lines owned by the IXCs), the IXCs may begin to build their own local backbone rings, although some have criticized this as a needless duplication of effort.

All four portions of the SONET network—distribution, feeder, local backbone, and long-haul backbone—function together to bring the benefits of SONET to greater and greater numbers of customers. The only thing missing from Figure 11-1 to complete the picture would be a mirror image of the local, feeder, and distribution network somewhere across the country.

SONET COMPONENTS

The more one explores the world of SONET TMs, ADMs, and DCSs, the more those components come to be viewed as the same basic SONET device. What distinguishes them are the number and types of interfaces they support. Any input ports can be linked to any output ports with a little configuration effort.

What do the devices actually look like? Surprisingly, they are quite small and compact—another example of the triumph of electronics miniaturization. Those used to seeing a room full of cables, cabinets, and cross-connect panels for T-3 services are often astonished at the tiny size of SONET gear. A SONET ADM can easily be overlooked in a large equipment room, and SONET shipments have actually been misplaced in the shipping room. No one believed it could be so small.

Figure 11-2 shows the basic look of a basic SONET device. (SDH equipment looks exactly the same.) Although a lot of SONET equipment comes in its own cabinets, the basic component is the *card shelf*. These are just standard rack-mounted units that are installed in cabinet enclosures or a standard communications rack. Most are about one foot high, but smaller units are also sold. The key component of these units is the control module or main card—or whatever name the vendor chooses to give this component. This board contains the heart of the device, and usually controls whatever SONET or other type of processing is possible with the unit. It has its own power supply, and it interfaces with the backplane of the unit to allow communication between all other cards. This main board may be an integral part of the card cage, or a separate board. Separate boards can be changed without disturbing the rest of the device, which can be a plus.

None of the foregoing should be taken to mean that SONET equipment is not often found as seven-foot-high racks. It certainly is common to find standard telecommunications racks filled with shelves of SONET gear. But each shelf of the rack is capable of performing the tasks previously reserved for a whole room full of equipment. That is the point here.

The rest of the card shelf can be filled with interface cards to accomplish the task intended for the device. Most of the card slots are typically filled with various input and output ports with standard SONET connectors on them. Amazingly, standard SONET connectors are simply small plugs that are not much different in appearance from stereo

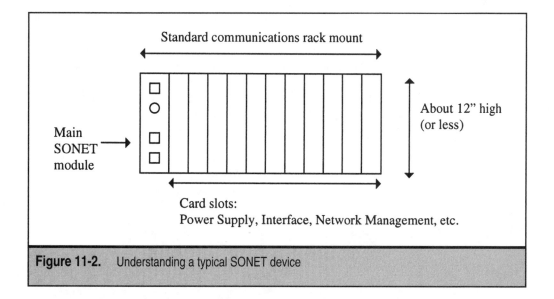

Figure 11-2. Understanding a typical SONET device

jacks. The types designated SC or ST are most common, although there are others as well. The more cards slots that are occupied, the greater the capacity of the device. When all of the slots are full, upgrades may be made by adding a second shelf if the vendor allows this. Otherwise, a complete box swap may need to be made.

Input and output ports are not the only types of cards that can fill the card slots. Other types can be used to add to the capabilities provided by the main board (often at a considerable price). One of the most common additions is a backup power supply for the unit. Such redundant power supplies can keep the unit functioning during extended AC power outages, sometimes for hours. The modest energy demands of electronic components means that the elaborate batteries and diesel generators of the recent past can be effectively replaced by battery packs not dissimilar to those used in laptops.

In many cases, even the main electronics board can be made redundant with a card. A special network management module may be available, but most units will include at least the standard SONET network management functions and interfaces on the main controller board. Some SONET network elements come with a "hot swappable" feature. This means that cards in the device can be removed for repair, or others installed, without shutting down or powering off the device. Needless to say, this is a very nice feature for repair-conscious service providers with a need to maximize availability.

Although specific vendors will add their own favorite features and functions to this basic package, the general appearance of SONET equipment will remain fairly consistent. The card cage and shelf approach is nearly universal.

SONET NETWORK ELEMENT FUNCTIONING

Most SONET equipment looks fairly similar, whether the device is an ADM, DCS, or some form of hub. However, differences do occur in what the SONET network elements do internally, or even externally. This section takes a closer look at what that statement implies.

All of the components of a SONET network are shown in Figure 11-3, with added details highlighting which arrangement is more suitable for business and which for residential customers; however, nothing precludes the use of either arrangement in a given area. As already noted, this chapter emphasizes SONET, but it should be remembered that SDH could follow the same arrangements if desired by the service provider.

Figure 11-3 shows a fairly complete network deployment that exploits the features of SONET. Such an arrangement would typically be employed by a local service provider, such as the local telephone carrier or even a cable TV company. Of course, this SONET network would not be limited to these entities; they are merely representative. Power companies have a lot of SONET in use internally, and there is little to stop them in the current regulatory environment from extending the SONET network to their customers.

The equipment found in this network is separated into CO terminal equipment and outside-plant equipment. Admittedly, these are telephone company terms, but they are widely understood and used in the industry and are not necessarily limited to a telephone networking environment.

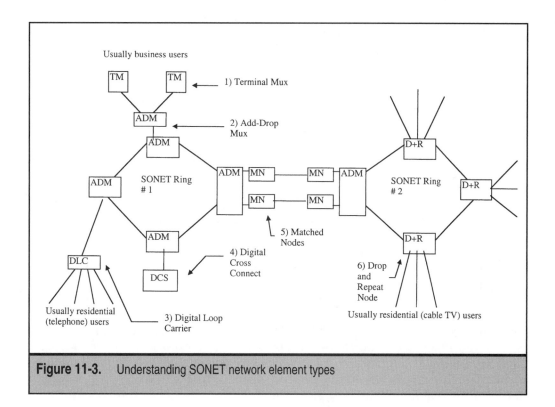

Figure 11-3.   Understanding SONET network element types

The CO equipment consists of a SONET-compatible switch (that is, DCS) and DLC equipment. This equipment forms the network terminating point for all SONET transport services. Note that the DCS is not capable of "switching" based on the bit configuration of the user's traffic. A traditional voice switch (for telephony services), router (for Intranet traffic or Internet access), or some other form of more exotic switch (such as ATM) must be used for this purpose. SONET is still very much a trunking—not a switching—technology.

The outside-plant equipment consists of a fiber ring topology for survivability, and access multiplexers to this ring. The multiplexers are functionally separated into two categories, TM and ADM. However, SONET standards specify that the TM is just an ADM operating in terminal mode. Nevertheless, in many cases, vendors manufacture and market two distinct products for these two network elements.

The TMs are functionally similar to channel banks in the T-1 networks. They provide conversion from some form of non-SONET transmission media and format, such as T-carrier, to the SONET format. The ADMs are the nuts-and-bolts component of the SONET architecture. The ADMs allow access to the SONET transmission signals, even down to the DS-0 level, without fully demultiplexing the SONET signal. Non-SONET signals may be added to or taken from the SONET transmission signal with ADM equipment.

As shown in Figure 11-3, and as mentioned previously, there are six main network elements that can be found in SONET. These are

1. Terminal multiplexer (TM)

2. Add/drop multiplexer (ADM)

3. Digital loop carrier (DLC)

4. Digital cross-connect system (DCS)

5. Matched nodes (MN)

6. Drop-and-repeat nodes (D+R)

It is important to remember that the names of some of the SONET network elements are covered by standards, but some are not. On the list, the names TM, ADM, and DCS are well defined and universally used. However, names like DLC, MN, and D+R have various names, depending on the manufacturer. The names used here were chosen for their descriptive powers and should not be taken to be an endorsement of any particular vendor or vendor's terminology. Whatever the name, the most important aspect of the device is where it is employed in the network and what function it must perform.

The access node into a SONET network is most often a multiplexer, of which there are two basic varieties. The simplest, the TM, acts as a concentrator of multiple DS-1 or DS-3 signals onto a SONET electrical STS-1/3 or optical OC-N backbone. It is analogous to the asynchronous M13 multiplexer that forms the entry point to the DS-3 digital signal hierarchy. The TM is usually a less redundant configuration of the SONET network element known as the ADM.

The next few sections take a closer look at the internal functioning of the SONET network elements.

Terminal Multiplexers

One major SONET network component is the TM. TMs are used to terminate several DS-1 signals and package them together as STS payloads. For example, one type may terminate 28 DS-1 signals, thus eliminating the need for today's M13 multiplexers. Yet another variety may terminate 84 DS-1 signals to generate an OC-3 signal.

Figure 11-4 shows the basic internal structure of a TM that takes 28 DS-1s, each with 24 DS-0s, and that produces an STS-1 frame for transport on an OC-1 fiber link. Note that multiple groups of DS-1s and multiple OC-1s may be produced in the same device, depending on the number of card slots available for use.

Figure 11-4 shows that a SONET TM consists of a number of T-carrier ports that use normal time division for multiplexing into SONET payloads. The payload is given the complete set of SONET overhead bytes and out it goes. The same happens in reverse on the receiving side. Note that each OC interface must have two fibers, one for input and one for output, as is common in digital systems.

The individual fibers are not shown in the figure. The TM is a point-to-point device; thus, even though there are actually two output paths, the two paths must terminate on the same SONET device. Two physically separate fiber cables would be used in between devices, with each cable containing a receive and transmit fiber pair.

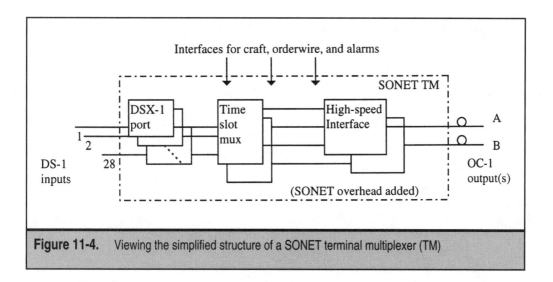

Figure 11-4. Viewing the simplified structure of a SONET terminal multiplexer (TM)

Although not strictly limited to point-to-point SONET links, it is unusual to find a SONET TM on a ring. This would essentially make it a DCS. Although no harm is done to the network, confusing TMs with DCSs (or even ADMs) is a misuse of terminology and is generally not acceptable.

In Figure 11-4, the "DSX-1" should not be confused with a "DCS." The digital system cross-connect for DS-1 (DSX-1) is not really a piece of electronic gear by itself, as is a DCS. A DSX interface is more likely to be a physical patch panel arrangement attached to a rack or even to plywood nailed to a wall.

▼ A DSX-1 is usually a series of terminated RJ-45 jacks that can be patched with jumpers to provide access to the SONET TM (the "DSX-1 Port" in the figure).

▲ A DSX-3 is usually a series of terminated coaxial cable jacks (less frequently, mini-SCSI connectors) that perform the same function for DS-3s.

The key is that DSX is an *interface*, not a piece of equipment. It is possible, but not necessary, to have a real DCS feeding a SONET TM DSX port. This arrangement can be confusing, and many people use the terms DSX/DCS interchangeably for this very reason.

The large amount of information that is transported by SONET devices requires a high degree of redundancy to eliminate any single point of failure with regard to the fiber path, internal circuits, and power supplies. There is too much information flowing too fast on the SONET links to permit a single point of failure. The concern for reliability is reflected in the architecture of the TM.

The controller in the main board plays a key role. The complexity of the SONET protocol favors the manufacture of a software-driven product whose configuration may be determined by software downloaded from a central network site. This type of design simplifies the addition of new feature sets at the modest expense of advanced microprocessors and large amounts of memory. Operating instructions stored in the device's memory provide the necessary intelligence for provisioning, monitoring, status reporting, and sounding alarms.

The DSX-1 ports (transceivers) provide access to the lower-speed channel signals. For example, up to 672 DS-0 signals are received in the form of 28 DS-1s by an STS-1/OC-1 multiplexer transceiver, which processes them in a form acceptable to the pair of redundant time slot multiplexers, or interchangers. There are usually four additional DS-1 channels: two are used for maintenance, and the other two for the European E-3 frame.

The time slot interchanger (TSI) in the TM handles the low-speed signals to a redundant pair of high-speed interfaces, A and B, which normally operate at the SONET STS-1/3 electrical or OC-N optical rates.

The high-speed interface controls the SONET output links. One of the two signals, either A or B, is transmitted; the other is held for back-up purposes. Sometimes the A and B signals could be transmitted simultaneously, and it could be up to the receiver which to use. The optical signal is simply "bridged" at the electrical level to transmit the identical signal from the TM. The unit monitors the in-service and the protection optical signals independently.

The TM has other specialized interfaces as well. The craft, orderwire, and alarm units are called the "COA" interface. The COA selects whether the A or B SONET link will be active. It also provides for craft (technician) interfaces and analog input and output (if used). Furthermore, it forwards the alarms to remote network operations centers.

As an aside, the clocking for TM is derived from either an internal source or a received optical signal. Figure 11-4 illustrates that a SONET TM can take existing DS-1s and DS-3s, whether employing DSX interfaces or not, and convert them to and from SONET.

SONET TMs, as well as ADMs, transport existing DS-*n* signals without alteration. The low-speed DS-*n* ports may feature bit- or byte-synchronous operation (locked VT) or asynchronous (floating VT) traffic. Bit-mode provides transparent transport of DS-1 traffic and is useful for transport of unframed DS-1 or framed DS-1 with ZBTSI coding. Byte-mode transports D4-framed DS-1s with superframe or ESF formats. When configured as a CPE, the TM becomes more specialized. There will be SONET multiplexer "gateways" (to use the term loosely) for Ethernet and Token Ring LANs, ISDN primary rate access, National Television System Committee/Phase Alternate Line TV, HDTV, switched multimegabit data service (SMDS) for metropolitan area network access, and ATM cells.

Add/Drop Multiplexers

The most crucial SONET (or SDH) network element is the SONET ADM. This is because the ADM is not only an essential piece of the SONET network, but also a kind of generic SONET device that can be easily modified to function as a TM, or as ring interconnection device, or as something else, as the deployment scheme requires.

The ADM is a device unique to SONET with no real counterpart in the T-carrier world. The SONET ADM replaces the back-to-back M13s and associated manual DSX patch panels often used for DS-1 cross-connections. The ADM is a fully synchronous, byte-oriented multiplexer capable of adding and dropping DS-*n* signals onto one of two SONET links. DS-*n* payloads within the OC-N signal also may be added or dropped within the ADM. A simplified schematic of the ADM is shown in Figure 11-5.

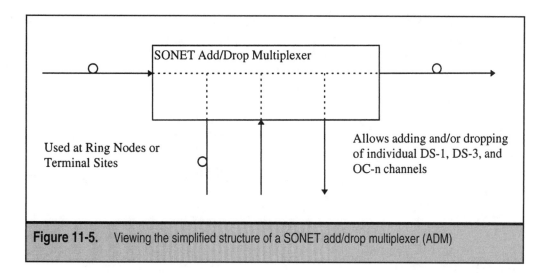

Figure 11-5.   Viewing the simplified structure of a SONET add/drop multiplexer (ADM)

Figure 11-5 shows less internal detail than does the previous TM figure. The emphasis here is on overall function, but all of the internal components still exist. SONET ADMs have two sets of OC interfaces instead of one, as in the TM. ADM features include DS-*n* electrical low-speed interface, OC-N high-speed optical interface, DS-*n*/OC-N multiplexing, DS-*n* add/drop, and the usual COA management interfaces.

The input and output STS-1/3 interfaces may be used to interconnect within an equipment frame or building site, but are unable to drive signals any significant distance over copper wire. For this, an electrical-to-optical conversion must be added to the ADM. Then, the customary SONET physical distance rules apply. For example, a 1,310 nm fiber optical signal may span 26 miles before requiring regeneration.

The ADM interfaces with existing T-carrier equipment and facilities by way of multiple DS-*n* electrical interfaces. The incoming payloads that are not terminated by the ADM pass through to the OC-N interface on the other side of the ADM. Each DS-*n* interface may drop information from an incoming OC-N and add information to the outgoing OC-N. Therefore, an ADM may add or drop DS-1 signals within an STS-1 or STS-3. In addition, some ADMs are capable of performing DS-0 "groom and fill," which aggregates and distributes normal voice channels. This goes beyond the specifications of the SONET ADM, as described in the Bellcore technical references, which require ADM capabilities only at the DS-1 level.

SONET ADMs can be placed in series along a SONET path, so that non-SONET signals can be added to or dropped from the STS payload. ADMs operate bidirectionally and component terminations can be accomplished from either direction. Typical implementations of ADM employ protection-switching capabilities, providing an active fiber and a standby fiber (A and B), as do TM devices.

ADMs can also be equipped with time-slot interchangers to allow digital cross-connect functions between channels, carried as part of the STS-1 payload. When optioned in the add/drop mode, ADMs eliminate the need for back-to-back multiplexers—the common solution in T-carrier networks—which allow component signals to be broken. Of course, the ADM then effectively becomes a SONET DCS.

The SONET DCS can be an add-in card to the overall SONET equipment rack. Alternatively, the DCS can be a separate physical device altogether.

Because they interface with both the optical network and the conventional electrical network, ADMs need a low-speed electrical interface and a high-speed optical interface. Additionally, they require remote- and local-operation interfaces, a data communications channel for section overhead functions, local nonvolatile memory for backup purposes, and distribution frame and facility maintenance capabilities.

SONET Digital Loop Carrier

The SONET DLC, like any loop carrier, is designed to minimize the assignment of dedicated plant to the loop environment. The IDLC combines the features of intelligent remote digital terminals (RDTs) and a feature of the digital switch called integrated digital terminals (IDTs). The two are connected by a digital transmission facility. The primary goal of this architecture is to allow RDTs from one manufacturer to efficiently and transparently interface with switches (that is, IDTs) from another manufacturer, thus providing a generic interface of sorts.

For most of the digital revolution, the local loop has remained primarily analog. This will have to change with the introduction of fiber into the local loop and the deployment of SONET DLC equipment. The current local telephone environment is ideal for SONET DLC deployment.

Within a local access and transport area (LATA) are numerous central offices ("wire centers"), which serve specific communities of interest. Wire centers handle trunk-to-trunk traffic, such as for IXC access. A wire center may be a 50,000-line CO or even what is known as a community dial office (CDO) serving only a few hundred subscribers. Service may also be provided by remote terminals (RTs) which give commercial tenants of a distant office park access to the telephone network.

The SONET DLC may be considered a concentrator of low-speed services before they are brought into the local CO for distribution. In remote areas, SONET DLC is actually a system of multiplexers and switches designed to perform concentration from the remote terminals to the CDO and, from there, to the CO.

Typically a multiple-port OC-3 device, the SONET DLC provides direct access to its constituent DS-0 signals. Available DLCs support up to 2,016 DS-0s and are most economical where the general demand falls between 200 and 2,000 lines. The DLC supports both voice and data traffic and can provision, test, and maintain POTS, coin phones, multiparty lines, digital data service (DDS), and ISDN basic and primary rates.

The SONET DLC is really a short-term transition device meant to increase the capacity of the existing network until fiber is deployed almost everywhere. Initially, the DLC acts as a helpful addition to the CO switch by terminating an OC-N data stream and applying its embedded DS-0s to the switch. As the switch is upgraded or replaced by a SONET-capable switch or SONET DCS, the DLC may be moved elsewhere in the carrier serving area (CSA).

A SONET TM may be deployed at the customer premises, but the SONET DLC is intended for service in the CO or a controlled environment vault that belongs to the service provider. Bellcore describes the generic IDLC, which consists of intelligent remote digital terminals (RDTs) and digital switch elements called IDTs, which are connected by a digital line. The IDLCs are designed to more efficiently integrate DLC systems with existing digital switches.

To facilitate the introduction of SONET, additional feature groups have been included that provide needed functions. Feature set B provides such features as remote test unit (RTU) integration, remote switch unit (RSU) integration, and integrated network access (INA).

Feature set C, called the IDLC/SONET interface, generates savings in the areas of facility terminations, information transport, and operations expense. These savings come from a variety of sources, most notably reductions in DSX terminations, intra-office cabling, and signaling terminations (for example, framing circuits, timing signals).

SONET Digital Cross-Connect Systems

The SONET DCS is a key component in any SONET network that performs cross-connecting. Sometimes called a "full-featured" ADM, the DCS does it all.

One major difference between a DCS and an ADM is that the DCS can be used to interconnect a larger number of STS-N links. The DCS can be used for grooming (that is, consolidation and segregation of) STS-1s or for broadband traffic management. For example, a DCS may be used to segregate the high-bandwidth traffic from the low-bandwidth traffic arriving from a SONET ring and to send that traffic to a high-bandwidth switch (for example, ATM switch for videoconferencing) or to a low-bandwidth switch (for example, voice switch) as appropriate. This is the synchronous equivalent of DS-3 DCS and supports hubbed network architectures.

In the serving office (and eventually perhaps even on the customer premises), SONET digital cross-connects will replace much if not all of the current back-to-back multiplexer configurations. This will reduce the number of intermediate distribution frames and manual jumper panels between these frames. Cross-connection is a fundamental capability of SONET standard network element descriptions. SONET DCSs will be used at the serving office to terminate an OC-N and to separate switched and non-switched DS-n traffic. Then, any switched DS-0s can be brought into a standard voice switch. Even better, by taking advantage of the SONET remote configuration features and capabilities, selected signal streams can be cross-connected even at intermediate sites as needed, although most often DCSs will be deployed on centralized office-based rings.

The SONET DCS can become a large interoffice hub. This type of DCS is capable of managing large amounts of traffic transmitted at the DS-1, DS-3, or OC-N rate, and primarily at OC-3 through OC-12. DS-1 cross-connections will be performed by cross-connecting floating virtual tributaries (VTs). Higher-level cross-connects will use electronically alterable memory maps to groom the various VT signals into OC-N signals.

Figure 11-6 shows, in some detail, the structure of a SONET DCS. (SDH equivalents could also exist, of course.) The detail is warranted in this case, owing to the position of the SONET DCS as a "superset" of the ADM and TM devices. The DCS still physically looks like a shelf or card-cage device, but the figure shows a more functional organization. The DCS must still contain a SONET timing module and will have other interfaces used for craft, operations (technically, operations support systems: OSS), and alarms. There will usually be some backup functions included for memory, power supply, or other redundancy.

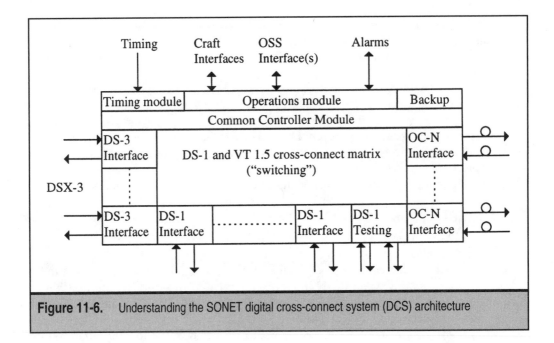

Figure 11-6. Understanding the SONET digital cross-connect system (DCS) architecture

Besides the internal operation of the DCS, Figure 11-6 also shows the common controller module that oversees the general functioning of all input and output operations. A variety of interfaces are supported, from DS-1 through DS-3, and up to some large OC-N level. To be accurate, the DCS actually "terminates" the DS-3 or OC-N and is not merely an interface; however, the term "interface" is more common among equipment vendors. In practice, the availability of SONET "level n" service depends on the availability of SONET DCS equipment from a vendor to support it.

The heart of the SONET DCS is the cross-connect matrix. This is how signals from an input port are "switched" (or more accurately, "shuffled") to an output port. Some models can cross-connect all the way down to the DS-0 level, but this is uncommon. All ports can be connected through the matrix.

Whether the input and output signals are at the same level or whether they aren't (SONET or T-carrier), the cross-connection is simple owing to the synchronous nature of SONET. Grooming and hubbing are automatic and transparent.

There are two distinct types of SONET DCS. They are called the broadband DCS and the wideband DCS. They differ in the types and speeds of signals that they cross-connect.

BROADBAND DCS

The SONET broadband DCS (B-DCS) is similar in function to a DS-3 cross-connect in the T-carrier hierarchy. The B-DCS has interfaces for full-duplex SONET signals, as well as for DS-3 clear-channel interfaces. The basic function of the SONET B-DCS is to make two-way cross connections at the DS-3, STS-1, and even the concatenated STS-N (STS-Nc) levels.

The B-DCS, thus, provides a type of "switching" at STS level and the DS-3 level. However, this "switching" is not the same as would be encountered in a voice or frame relay switch. Rather, the "switching" is similar to that achieved by DS-3 cross-connect. In other words, no input-to-output port correspondence is established by the content of the information stream. The correspondence is established strictly by configuration software and remains in effect until changed.

Figure 11-7 shows the general architecture of a B-DCS. The B-DCS provides transparent cross-connections between DS-3 interfaces (actually, the DSX-3) and between DS-3s and OC-Ns terminating at the DCS. In this context, the term "transparent" (or clear channel) means that any DS-3 at the normal DS-3 rate can be cross-connected, whether it is carried synchronously or asynchronously on the SONET network. Thus, the B-DCS does not need to frame align on the incoming DS-3 unless enhanced performance monitoring is required.

The B-DCS is also fully capable of performing STS-1 and STS-Nc cross-connections. As a consequence, the B-DCS must be able to frame-align with the incoming OC-N. It must also identify and gain access to the desired constituent SPE. Again, this terminates the OC-N because the SONET transport overhead is terminated and processed. The B-DCS is also capable of generating new outgoing OC-N with valid transport overhead. In general, this makes the DCS a natural focal point for gathering network information and performance statistics.

WIDEBAND DCS

The main distinction between a B-DCS and a wideband DCS (W-DCS) in SONET is that the W-DCS goes "deeper" into the digital hierarchy, all the way down to the VT level. In this respect, the W-DCS resembles the DCS-3/1 cross-connect in T-carrier. Thus, a W-DCS terminates SONET and DS-3, while cross-connecting DS-1 signals. The W-DCS DS-1 and DS-3 interfaces meet the DSX-1 and DSX-3 specifications.

Figure 11-8 shows the general capabilities of a SONET W-DCS. The picture is simple but accurate. Note the presence of the DS-1 speeds into and out of the device. The ability to

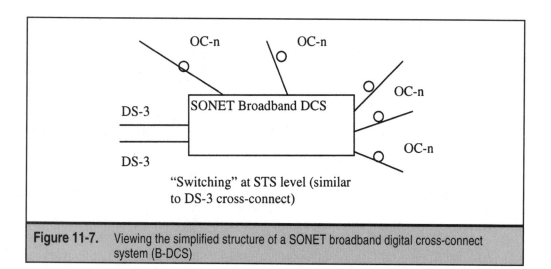

Figure 11-7.   Viewing the simplified structure of a SONET broadband digital cross-connect system (B-DCS)

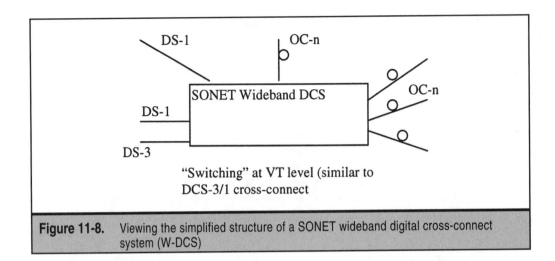

Figure 11-8.   Viewing the simplified structure of a SONET wideband digital cross-connect system (W-DCS)

cross-connect at the DS-1 level (actually, the VT1.5 level) makes the W-DCS a very useful SONET device.

In Figure 11-8, it is easy to see that the major difference between a broadband and a wideband SONET DCS is that the W-DCS operates at the VT1.5 level. This device, therefore, is similar to a DS-3/1 DCS because it accepts DS-1s and DS-3s, and is equipped with optical interfaces to accept optical carrier signals. The W-DCS is suitable for DS-1–level grooming applications at hub locations, such as the serving office. One major advantage of deploying W-DCSs is that less demultiplexing and multiplexing are required because only the affected VTs are accessed and switched.

The W-DCS network element cross-connects floating VT1.5 signals between OC-N terminations and transparent (clear-channel) DS-1 signals. It also provides transparent DS-1 cross-connections between DS-1 interfaces and DS-3/OC-N terminations. The term "transparent," when applied to cross-connection, means that the W-DCS is capable of handling any type of signal at the normal DS-1 rate. This is important because DS-1s come in a variety of framing formats and digital coding methods. The W-DCS should be able to handle them all, regardless of the nature and format of the signal.

Transparency is accomplished by framing the incoming OC-N to identify and access the target VTs. This, in turn, requires that the section, line, and path overhead of the OC-N be regenerated from scratch using new STS-1 SPEs and new section, line, and path overhead. This is why both the broadband and wideband SONET DCS network elements are said to "terminate" the signals.

SONET Matched Nodes

SONET networks today are usually deployed in a ring architecture, but how do the rings talk to each other? That is, how can a set of users on one ring communicate with a set of users on another ring? One of the most common ways of accomplishing this is with a set of SONET matched nodes. It should be noted that this device is usually some form of

SONET ADM deployed in a pair for this task. Some vendors and service providers consistently refer to the arrangement by the "matched nodes" name and, therefore, it requires a few words of explanation.

In a SONET matched-node configuration, whether the devices used are sold as "matched nodes" or not, SONET traffic is duplicated and sent over two links between two rings. One of the signals is selected and continues around the rings. If the connection were lost between the active nodes, the backup link could be used until repairs were made. Figure 11-9 shows this use of matched nodes, which are also used in SDH rings.

Rings are not even necessary for matched-node operation in SONET. Even on point-to-point links, as is common from a customer site TM to a ring ADM, matched nodes can be used. Matched nodes can be configured with diverse routing to the serving office, separate entrance facilities at the customer site itself, backup and redundant power supplies, and even totally redundant common control electronics to make SONET links almost impervious to failure, except in the most catastrophic circumstances.

SONET Drop-and-Repeat

Like matched nodes, SONET drop-and-repeat nodes are not really separately defined SONET network elements. Matched and drop-and-repeat nodes both add a feature to duplicate traffic received on a input port and send the same bits out of two (or even more) output ports. They are merely SONET devices deployed in a certain configuration to accomplish a specific task.

When SONET is deployed in a ring architecture, drop-and-repeat SONET equipment provides alternate routing for traffic passing through a higher-level ring from lower-level feeder rings. For instance, several OC-12 rings with traffic from individual customer sites may be linked with an OC-48 ring configured as a regional ring. Figure 11-10 shows this use of SONET drop-and-repeat devices.

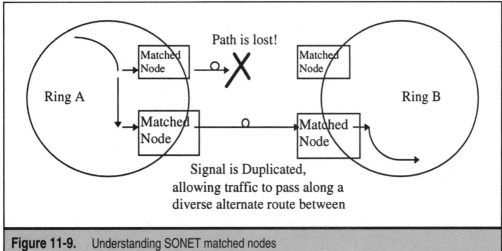

Figure 11-9.   Understanding SONET matched nodes

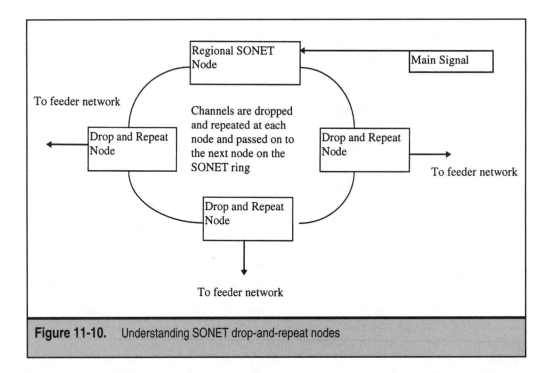

Figure 11-10. Understanding SONET drop-and-repeat nodes

At each node in Figure 11-10, a drop-and-repeat SONET device "drops" the signal off of the main ring onto the feeder ring, yet keeps a "copy" of the signal going around the main ring. This is obviously one giant step beyond the simple "matched-node" configuration. Drop-and-repeat also applies to SDH nodes.

In multiple-node distribution applications (such as cable TV), drop-and-repeat is essential. One transport channel can efficiently carry traffic between multiple distribution nodes. For example, when transporting cable TV video signals, each programming channel can be dropped (that is, delivered) onto the feeder ring at the node and yet repeated for continued delivery around the ring. Not all of the bandwidth (that is, program channels) need be delivered at all of the nodes. Channels not dropped at a node are passed through, without physical intervention, to the other nodes.

The SONET drop-and-repeat feature is sometimes called "drop and continue." It depends on the service provider and equipment vendor. In each case, operation is identical.

The equipment terminology used in this chapter is exclusively for SONET, but SDH networks frequently use the same terms—especially ADM and TM. The correct SDH network element (NE) terms were introduced in Chapter 9 and are repeated here for convenience.

In SDH, the following NEs are typical TM/DCS/DLC devices, using their proper SDH names:

▼ Low-order (or low-speed) multiplexer

■ Wideband cross-connect system

▲ Subscriber loop access system

SDH specifications offer examples of SDH ADMs (multiplex section terminated equipment devices: MSTEs) called

▼ Optical line terminals

■ Radio terminals

■ High-order multiplexers

▲ Broadband cross-connect systems

CHAPTER 12

SONET/SDH Equipment Providers

his last chapter in this section of the book takes the major categories of SONET/SDH network elements established in the previous chapters and examines just how equipment vendors have implemented them. It is one thing to investigate standards and bit configurations and frame structures, but it is quite another to take these specifications and build them into products that people will buy to implement SONET/SDH networks. SONET/SDH equipment must embody the SONET/SDH standards. This chapter looks at how and to what extent vendors have achieved this. At the end of this chapter, a section new to this third edition examines how two major router vendors—Cisco Systems and Juniper Networks—have implemented SONET/SDH as router interfaces for use in corporate networks and the Internet.

All of the SONET/SDH product offerings from the major SONET/SDH vendors are explored in this chapter. Every attempt is made to compare products "apples to apples." However, some vendors are more open than others regarding information about their products. Many make a lot of detailed information, including pictures and technical specifications, available at a product web site accessible over the Internet. Many are more guarded about exact product functioning in many key areas; therefore, a complete feature-by-feature comparison is not possible.

Another area of concern that is not addressed in this chapter is interoperability. It may sound strange that products based on exactly the same set of international standards would be anything but interoperable, but this is not always the case. This is an especially sensitive issue in SONET, one of the major goals of which was to provide the interoperability that the T-carrier hierarchy lacked. However, not all standards are rigid.

The term "interoperability" encompasses more than simple bit communication. The term is more inclusive. How are issues such as network management, troubleshooting, and even performance monitoring handled when two or more vendors' products are included in a SONET/SDH ring? These are the basic concerns of interoperability, beyond the respect for standards. Unfortunately, these compatibility issues cannot be explored in this chapter for the simple reason that not many unbiased and wide-ranging studies have been made on SONET/SDH product interoperability issues.

This chapter makes an effort to be as up-to-date as possible. In all cases, web sites are given for further research to determine the latest information from a vendor. The major difference between the products listed in the first edition of this book and in the second edition was the shift in product emphasis from SONET to SDH as vendors moved into an international market arena. Between the second edition and the third edition, the shift has been toward making SONET/SDH equipment part of an equipment vendor's overall "optical networking strategy." This latter shift includes not only continued and enhanced support for SONET/SDH, especially with regard to OAM&P and provisioning capabilities, but also added features supporting WDM and DWDM, MPLS (multiprotocol label switching), better IP/Ethernet integration, and use of various quality-of-service mechanisms for voice and video over IP (such as RSVP, the resource reservation protocol).

As a result, I now tend to classify equipment into "traditional" SONET/SDH network element gear and "newer" devices that happen to include SONET/SDH interfaces on their DWDM boxes or IP routers and wireless broadband equipment, or some even more exotic combination of networking equipment. There is nothing wrong with buying an "optical bandwidth manager" and getting SONET/SDH, of course. If nothing else, this shows the stability and popularity of SONET/SDH. But the use of SONET/SDH by only a few IP router vendors (Cisco Systems and Juniper Networks) will be examined in this chapter. When SONET/SDH is combined with other optical networking capabilities, such as DWDM, these other packages for SONET/SDH will remain beyond the scope of this chapter. The question of just how much of a SONET/SDH network element this new breed of hybrid gear might or might not be is a matter open to debate.

THE BIG SIX

When it comes to direct vendors of "traditional" SONET/SDH network element equipment, there are really six major vendors. Combined, they account for most of the SONET/SDH network elements sold to date. The six vendors are (in alphabetical order) Alcatel Network Systems, Fujitsu Network Transmission Systems, Lucent Technologies (formerly AT&T Network Systems), NEC Transmission, Nortel Networks, and Tellabs. This is not meant to slight other vendors with SONET/SDH products on the market. For example, ADC, Marconi, and Siemens make fine SONET/SDH products as well. The prominence given to the selected vendors is just an acknowledgment of the current market mix and an attempt to narrow the focus of the chapter.

A little overall information on each of the vendors follows. Readers concerned with corporate histories and present financial status are referred to each vendor's web site. As mentioned, all of the vendors once highlighted their SONET/SDH products at their web sites. It is surely a sign of the times that now all of them prominently feature DWDM and other optical networking products. SONET/SDH equipment is now sometimes buried in the back pages of the site, or is just not available except on the DWDM platform. This is a sign of changing times and of the maturity of SONET/SDH equipment.

Alcatel Network Systems *(www.alcatel.com)* became a major player with the acquisition of the transmission-product arm of Rockwell in 1991. Rockwell was a respected supplier of asynchronous transmission gears and products, but was very conservative when it came to new technology. Alcatel inherited the conservative outlook of Rockwell and brought out SONET/SDH products on a leisurely schedule, supporting OC-3 initially and OC-12 later. Alcatel now has a SONET/SDH product line that includes OC-3, OC-12, OC-48, and OC-192 support, with OC-768 on the way. Alcatel is known for its fiber ring products as well as digital cross-connects for SONET/SDH networks.

Fujitsu Network Transmission Systems *(www.fnc.fujitsu.com)* was the first company to market Phase 2 SONET-compliant products. Phase 2 included such welcome features as the data communications channel protocol (but not yet a full message set) and network management systems, among quite a few others. Fujitsu has been a pioneer in SONET/SDH ring products, as well as user-friendly adjunct software packages that make it easier to add features and functions to SONET/SDH networks and products.

Lucent Technologies *(www.lucent.com)*, formerly AT&T Network Systems, was formed after a reorganization of AT&T into more realistic lines of business and was billed as the beneficiary of the technologies developed by Bell Labs. The idea that the best SONET products should come from the organization that invented the laser in the first place is a powerful marketing tool. Lucent/AT&T was late to enter the SONET ring market, but its latest ring products are second to none. Although Lucent has fallen on some hard times lately, it remains at the forefront of DWDM and optical networking. Lucent products now include SDH support, of course.

NEC *(www.nec.com)* is another company that has excelled in SONET/SDH ring products. Its products specialize in providing high capacity in a small package. NEC has generally introduced products at higher and higher speeds as quickly as the speed becomes economically and technically feasible. Its products are quite versatile, with special emphasis on reliability. Of all the vendors discussed in this chapter, NEC has changed its offerings the most between the second and third editions of this book.

Nortel Networks *(www.nortelnetworks.com)*, formerly Northern Telecom, arose from the former Bell Canada system after 1984. Nortel Networks is notable for its wide range of SONET/SDH products. It has also emphasized the network management services that are needed to configure and run SONET/SDH networks. Nortel Networks has also been quick to support higher and higher speeds. The products are tightly integrated and easily upgraded or reconfigured as network requirements change.

Tellabs *(www.tellabs.com)* was founded in 1975 and might seem out of place rubbing shoulders with giants like Alcatel, Fujitsu, and Nortel Networks. Nevertheless, Tellabs has been successful in the SONET field by finding its niche and exploiting it well. Tellabs has evolved from being a maker of analog-based telecommunications products; it has become a truly global equipment provider, mostly in the field of digital cross-connects. Lately, Tellabs has begun to branch out into the main lines of SONET/SDH gear, and has even begun to offer optical networking gear and integrated switching. But the main strength of Tellabs remains its line of digital cross-connects.

Several of the companies mentioned here should be singled out in this book for reasons other than their SONET/SDH products. Alcatel promptly and gladly provided otherwise hard-to-get product information. And Nortel Networks' numerous white papers and product information provided a welcome basis and framework for this book as a whole.

It should also be noted that of the companies on the SONET/SDH list, Alcatel and Lucent are the two that make complete product lines that can be integrated seamlessly with their SONET products. In other words, from cross-connect to long-haul ADM, both Alcatel and Lucent can provide equipment that does it all.

PRODUCT OVERVIEW

Each of the major vendors introduced in the preceding section makes a number of SONET/SDH products, in both the "traditional" network element arena and the "newer" field of optical network bandwidth managers or similarly named products. And each one of these SONET/SDH products comes in several sizes and has various features. So, before detailing the key aspects of each individual product, it might be a good idea to briefly look at some of the products from each of the vendors. A representative sampler of these products will then be detailed.

Table 12-1 is certainly not meant to imply that the vendor in question makes only these SONET/SDH products and no others. Such a claim to be exhaustive would be quickly out of date anyway. Again, the intention is to profile a sample of the vendor's SONET/SDH products as representative of the line as a whole. All of the information in the table comes directly

Vendor	Product	SONET/SDH Device*	OC-N/STM-N Levels Supported
Alcatel	1631 LMC	W-DCS, ADM, TM	OC-1, OC-3, OC-12
	1630 GSX	W-DCS	OC-1, OC-3
	1603 SM	ADM, TM	OC-3, OC-12
	1677 SONET Link	B-DSC, ADM, TM	Up to OC-48 and OC-192
	1603 SMX	ADM	OC-3, OC-12, OC-48
	1603 SE	ADM, TM, DCS	OC-3
Fujitsu	FLASH-192	ADM, B-DCS	OC-12, OC-48, OC-192
	FLASHWAVE 4300 (FLASH-600 ADX)	ADM & ATM	OC-3, OC-12, OC-48
	FLM-2400 ADX	ADM, TM	OC-48
	FLM-150 ADX	ADM, TM, DCS	OC-3, OC-12
	FLM-600 ADM	ADM, TM	OC-12
Lucent	DDM-2000	ADM, TM	OC -1, OC-3, OC-12
	FT-2000	ADM, TM	OC-48
	SLC-2000	DLC	OC-3
NEC	Vista	ADM, TM	OC-3. OC-12
	SMC-150	ADM, TM	OC-3 (STM-1)
	SMC-600	ADM, TM	OC-12 (STM-4)
	SMC-2500	ADM, TM	OC-48 (STM-16)
Nortel	TN SDH family (TN-16X, etc.)	ADM/ TM, D+R, MN	STM-1, STM-4, STM-16, STM-64
	S/DMS transport node (SONET)	ADM, TM, D+R, MN	OC-3, OC-12, OC-48, OC-192
Tellabs	FOCUS LX	TM, DCS	STM-16
	5500	DCS	OC-3, OC-12
	6500	DCS & ATM	OC-3, OC-12, OC-48

* Main product use. Does not imply that product cannot be used in other SONET/SDH network situations.

Table 12-1. A Sampling of SONET/SDH Products from the "Big Six" Vendors

from the individual vendors' web sites. The intent is mainly to give an overall flavor of the range and depth of each vendor's SONET/SDH offerings. Some of the packages are small enough to sit on a desktop. Others are full seven-foot-high equipment racks. Table 12-1 is only a starting point for exploring the world of SONET/SDH equipment.

Some of the products are listed as either SONET or SDH devices. In many cases, the same equipment can be configured for both SONET and SDH interfaces. Some products listed are mainly SONET devices geared for the North American market, while others are mainly SDH devices with more international appeal. The list should not and cannot be used to imply that a particular vendor is more focused on SONET or SDH.

Table 12-1 must also include equipment with some DWDM and other optical networking features and capabilities. Many vendors have begun to package SONET/SDH ADMs and DWDM equipment together to provide multiple SONET/SDH links on one fiber. The list can no longer be restricted to pure SONET/SDH equipment without optical multiplexing considerations.

Even a quick look at Table 12-1 shows that all varieties of SONET/SDH network elements are well represented, especially ADMs. It is worth mentioning that many of the vendors' ADM products can be configured to perform other SONET/SDH equipment functions, such as drop-and-repeat and matched nodes. In many cases, the matched-node ring interconnectivity can be handled by a digital cross-connect or by even more imaginative uses of SONET/SDH devices. Note that two of the products listed, the Fujitsu FLASH-600 ADX (now the FLASHWAVE 4300) and the Tellabs TITAN 6500 Multiservice Transport Switch, combine SONET/SDH functions with ATM cell switching.

THE PRODUCT SAMPLER

Before exploring some of the products in more detail, a few words of caution are needed. First, when researching or comparing SONET/SDH products (or any other networking products, for that matter), there is no substitute for up-to-date information. As current as books or articles may wish to be, vendors are constantly introducing new products, updating old ones, and prototyping the next generation of SONET/SDH equipment on an ongoing basis in labs and even in the field. It would not be surprising, therefore, that specifics might differ on some products. The vendor's web site should always be consulted for the latest information. This section is intended as a general guide, not as a substitute for vendor product information.

Second, because the information presented is of a general nature and is, after all, dependent on the openness of the vendors themselves, there is a variation in the amount of detailed information available. Although every effort has been made to present information in a balanced manner, in some cases this has not been possible. Just because a product profile does not specifically mention a feature like remote memory backup and restore, this does not necessarily mean that this feature is unavailable.

Usually, the differences in product coverage are in matters of detail, and so the overall value of this section should not be diminished. Not every vendor, for instance, will list all ring configurations or component requirements for their products. But this would not add substantial value to a general description of the products, at least from a general perspective.

No effort has been made to evaluate the products in terms of absolute or even relative value or utility. Only two products from each vendor are profiled. No pricing information is included. This is not a marketing pitch. Marketing is not the intention of this chapter.

ALCATEL NETWORK SYSTEMS

Alcatel Network Systems is based in Richardson, Texas. This is the headquarters for Alcatel Telecom's North American organization. A leading supplier of telecommunications products and services, Alcatel provides basic network infrastructure to all major long distance carriers, to virtually every local telephone company in North America, and to private networks, government agencies, and telecommunications systems throughout Central and South America and the Pacific Rim.

Alcatel has been around for more than 100 years. Alcatel Network Systems is part of Alcatel Alsthom, of Paris, France. In North America, the roots go back to Rockwell International's Commercial Telecommunications Group, the Collins Radio Company, and ITT. Alcatel's product line has evolved a lot between the second and third editions of this book.

Alcatel 1631 LMC Wideband Digital Cross-Connect

The Alcatel 1631 LMC Wideband Digital Cross-Connect System (W-DCS) provides good performance. It is the highest capacity and density SONET-based W-DCS available. The 1631 LMC is modular and has a growth path while in service that allows the system to start small and expand as needed up to the current capacity of 3,360 DS-3 ports. The device is very compact, with 240 DS-3s or STS-1s in five bays, which include one matrix rack connecting all the others, and one administrative rack. This supports up to 96 DS-3s or 896 DS-1s per rack. There is support for UPSR (unidirectional, path-switched rings), and plenty of them. Up to 1,120 OC-3 rings or 280 OC-12 rings can be configured, an impressive number, all with full DCC processing.

The participation in rings allows networks with the 1631 LMC to survive catastrophic failures such as fiber cable cuts. Several advanced features make it easy to reroute critical traffic with minimal impact on end users. Because the 1631 LMC is such a valuable survivability and restoration tool, it has also been designed for robust operation. It can be coupled with the more conservative protection schemes for the electrical DS-3 and STS-1 interfaces.

Key features of the 1631 LMC include 3,360 DS-3 capacity; STS-1, DS-3, OC-3, and OC-12 ports; and T-1 conferencing capability. The 1631 LMC also supports concatenated connections (STS-3c, STS-12c) and firewall protection during network failures or storm conditions, and is intended as an M13 replacement.

Alcatel 1677 SONET Link

Despite the name, the 1677 SONET Link is more than just a SONET product. This high-density, optical bandwidth management platform includes SONET features, of course, but it also contains high bandwidth optical networking capabilities. The 1677 SONET Link consolidates multiple networking functions such as SONET add/drop multiplexer (ADM), digital cross-connects (DCS), and DWDM into a single, compact platform, which helps to lower the cost of ownership.

The 1677 SONET Link provides carriers with a full range of service offerings, such as full redundancy and interfaces from DS-3 to OC-192 and Gigabit Ethernet, making the device a carrier-class system.

The key features of the 1677 SONET Link are the wide range of interfaces to support existing SONET services and future optical services, nice port density for grooming in a small space (320 Gbps per shelf), scalability from 1 to 4 Tb of traffic per rack, and support for SONET virtual concatenation. The grooming can be done at the STS-1 level, and the Gigabit Ethernet provisioning offers SONET protection to the Gigabit Ethernet traffic.

FUJITSU NETWORK TRANSMISSION SYSTEMS

Fujitsu has been around since 1935 in Japan as the Communications Division of Fuji Electric Company. Starting in 1949 Fujitsu was listed on the Tokyo stock exchange, and in 1953 they began to make radio communications equipment. In 1954 the company manufactured the first commercial computer made in Japan (the FACOM 100). The combination of communications and computers proved to be a fruitful one, and Fujitsu grew rapidly throughout the 1960s. Overseas offices opened in New York (1967) and California (1968), eventually growing into Fujitsu America.

The company has a laboratory that researches supercomputing and artificial intelligence applications and semiconductor technology, as well as communications products. Fujitsu is now a huge multinational organization specializing in communications systems, computers, and other electronic devices. In the field of communications, Fujitsu makes digital switches, satellite communications equipment, and, of course, all manner of SONET/SDH fiber-based systems, although the emphasis at Fujitsu is always on SDH. The latest products deal with DWDM and optical bandwidth management, and so Fujitsu recently renamed a number of products to bring them under the FLASHWAVE optical networking umbrella. For instance, the FLASH 600 is now the FLASHWAVE 4300.

Fujitsu FLASH-192

The FLASH-192 is a member of the Fujitsu family of SONET/SDH products. All are designed to service the high-capacity transport applications of interoffice networks and inter-exchange carrier networks. The 10 Gbps speed of the FLASH-192 provides very high speed SONET/SDH transmission bandwidth.

The emphasis here is on high capacity and economies of scale. When compared with multiple lower-rate transmission systems operating in parallel, a single FLASH-192 system offers significant equipment and fiber optic cable savings for transporting 192 STS-1 circuits.

The FLASH-192 unit consists of basic shelves that can be mixed and matched, along with appropriate plug-in units, to support all of the following SONET network configurations (and their SDH equivalents):

▼   **OC-192 Transport**   The shelf contains high-speed optical line units and system control equipment and integrated OC-48 optical tributary interfaces.

■   **OC-48/12 Transport**   The shelf provides a combination of OC-48 and OC-12 optical tributary interfaces.

- **FLM 2400 ADM (as tributary shelf)**   The shelf provides OC-12, OC-3, EC-1, and DS-3 tributary interfaces.
- ▲ **OC-192 Routing Complex**   This provides fully flexible cross-connect matrix, unrestricted time-slot assignment, and STS-1 cross-connect granularity with hairpinning. (Hairpinning is the cross-connecting of tributaries so that they go right back out of the SONET device.)

The FLASH-192 has a number of more technical advanced features. For example, the unit supports in-service upgrades of the embedded OC-48 multiplexers while the device is operational. The FLASH-192 supports the Fujitsu FLM 2400 ADM (point-to-point or ring) and other point-to-point OC-48 equipment.

The device has the same basic shelf support in all configurations—terminal mode or drop-and-repeat, or as a ring ADM. Several survivable SONET ring architectures are supported, and the FLASH-192 allows full access and routing of individual signals within the entire STS-192 bandwidth, including hairpinning between tributaries running at STS-1, STS-3c, STS-12c, and STS-48c speeds. A full range of network management and integrated operations capabilities are supported, including software downloads of new features and enhancements, and remote memory backup and restore.

The FLASH-192 has an onboard Stratum 3 timing source and features synchronization status messaging as well as DS-1 BITS primary and secondary clock output and input. The unit is extremely compact, with OC-48 tributaries in one-half of a seven-foot bay and OC-12 tributaries in a single seven-foot bay. The total transmission capacity is 129,024 DS-0–equivalent channels. The tributary interfaces run at OC-48 or OC-48c (from the OC-192 transport shelf); OC-48 or OC-48c and OC-12 or OC-12c (from the OC-48/12 transport shelf); and OC-12 or OC-12c, OC-3 or OC-3c, STS-1 and DS-3 (from the FLM 2400 ADM).

Fujitsu FLM 2400 ADM

The FLM 2400 ADM is Fujitsu's OC-48 line rate multiplexer. The device is flexible enough to construct a multitude of network configurations, especially when used with the FLASH-192. The FLM 2400 can be used to build point-to-point networks, linear chains, several types of rings, and even interconnected rings. The FLM 2400 ADM can also be used to construct a high-capacity SONET or SDH backbone that interworks seamlessly with the FLM 150 ADM, the FLM 600 ADM, or other Fujitsu products, to create transport networks that are very versatile.

The networks are easily managed because Fujitsu supports complete OAM&P access from a single location across the entire SONET/SDH product line. To accommodate all the new broadband services, the FLM 2400 ADM transports concatenated STS-3c (STM-1) and STS-12c (STM-4) payloads.

The FLM 2400 supports several ring configurations. Each configuration has advantages, depending on traffic distribution. With Fujitsu, all are supported from the same shelf, so that you can upgrade from one type of ring to another in-service—a very nice capability. In some configurations, Fujitsu's OC-48 rings also give you the flexibility to expand to OC-192 in-service.

Other features of the FLM 2400 include software downloads to allow easy feature upgrades using industry-standard protocols. Also, remote memory backup and restore can be used for quick provisioning of similar nodes or for disaster recovery. In some ring configurations, an automatic squelch table update can be used to prevent misconnected traffic during ring-failure scenarios.

The FLM 2400 has several possible configurations. The unit can run in terminal mode, as a hub, as a linear add/drop, as several ring configurations, and as a matched node. In-service upgrades can change the FLM 2400 from TM to ADM, from linear ADM to ring, from OC-48 terminal to OC-192 terminal, and from OC-48 ring to OC-192 ring.

Interfaces supported include OC-48, DS-3, EC-1, OC-3 or OC-3c, and OC-12 or 12c, again with their SDH equivalents. In terms of capacity, the unit supports 48 STS-1s in various flexible combinations: time-slot assignment, 48 STS-1s, 16 STS-3s or STS-3cs, 4 STS-12s or STS-12cs, or combinations of these.

LUCENT TECHNOLOGIES

Lucent Technologies builds on the developments of Bell Labs. It has locations and offices or distributors in more than 90 countries and territories around the world, and Bell Labs facilities in 17 countries. Headquarters is in Murray Hill, New Jersey.

Lucent Technologies was created in 1996 as part of the AT&T decision to split into three separate companies. Lucent Technologies combines the systems and technology units that were formerly a part of AT&T with the research and development capabilities of Bell Labs.

Lucent started out as a Fortune 40 company. It provides both hardware and software for worldwide telecommunications networks. Lucent also builds local networks, business telephone systems, and consumer telephones that access global networks. They also make the microchips and related components needed to run a host of products and systems, from digital cellular phones and answering machines to advanced communications networks.

Lucent now markets a number of SDH products under their WaveStar line, such as the WaveStar ADM 16/1 for STM-16 to STM-1 applications. However, so that this section remains about the same size as the others, only Lucent's major SONET products are profiled here.

Lucent DDM-2000 OC-3/OC-12/OC-48 ADM

The DDM-2000 OC-3 and OC-12 ADM product is Lucent Technologies' main entry in the SONET race. The DDM-2000 also supports OC-48, but this section will focus on the OC-3 and OC-12 models. The OC-3, OC-12, and OC-48 models share many characteristics, and the DDM-2000 for OC-12 is basically an enhanced version of the OC-3 product in terms of interfaces. For example, the maximum number of DS-3s supported in the OC-3 model is 3, while the number supported in the OC-12 model is 12 (as one would expect).

In terms of tributary support, the DDM-2000 with OC-3 interfaces supports DS-1, DS-3, STS-1, or OC-3c tributaries, while the OC-12 supports DS-3, STS-1, and OC-3c. Note that the DDM-2000 for OC-12 does not support a direct DS-1 interface. There is also the DDM-2000 Fiber-Reach product that is an OC-1 SONET multiplexer.

Both product models offer continuous DS-3 performance monitoring and full-line protection switching. The protection extends to full one-to-one protection in point-to-point configurations. There are downloadable software updates available for each model, and the equipment itself is fully protected by a variety of redundancies.

The DDM-2000 product can be deployed in point-to-point or ring configurations. In a point-to-point deployment, one DDM-2000 would typically reside in a central office, with the other end housed in a remote terminal location, perhaps attached to a SONET digital loop carrier (DLC), which Lucent also makes, of course. The DDM-2000 can also be used in a hubbing and add/drop configuration, which takes full advantage of the ADM capabilities of the product. In a hubbing arrangement, a DDM-2000 at a remote location can extend an OC-3 or OC-12 to other remote locations, all feeding DLCs for hubbing and grooming purposes. But the real power of the DDM-2000 product naturally comes into play when SONET ring configurations are deployed.

The products are surprisingly lightweight and compact, especially the DDM-2000 with OC-3 interfaces. In some cases, the full shelf is not even needed and may be covered with a faceplate to keep out dust and other contaminants.

Lucent SLC-2000 Access System DLC

A recent check of Lucent's web site turned up little more than a mention of the SLC-2000. (SLC stands for "subscriber line carrier," and the 2000 is Lucent's common designation for a lot of their SONET products.) However, just for the sake of variety, this section will profile this product as a representative of a DLC SONET offering. Many of them are still around. A SONET DLC device interfaces residential subscriber lines (local loops) onto a SONET carrier system. Obviously, the DLC is a key piece of equipment for SONET deployment, yet one that gets little attention and generates little excitement.

The SLC-2000 can handle 768 subscriber ports, more than enough for a good-sized neighborhood, even when second lines are considered. Most DLC products come in twos, one for the central office (known as the COT, "central office terminal"), and one for the remote site where the local loops are serviced (known as the RT, "remote terminal"). In the SLC-2000, identical plug-in boards are used for both the COT and RT—a nice feature. The RT can be handled with remote inventory and diagnostic software—another real plus.

The SLC-2000 can support up to 28 DS-1 lines, all of which can be powered ("wet T-1") and protected by backup links. This capability is attractive for business districts and applications where DS-1s are common. The SLC-2000 has an integrated OC-3 interface and a compatibility mode that complies with the Bellcore specifications for DLC SONET products (TR-008 and TR-303).

NEC OPTICAL

NEC, a Japanese company, originally known as Nippon Electric Corporation, first established a U.S. presence when it opened a sales office in New York in 1963. Since that time, NEC has broadened its operations in the United States by expanding into manufacturing, research, and software development operations, by employing 7,000 people, and by establishing extensive marketing, sales, and service networks nationwide with revenues exceeding $5.9 billion.

NEC is a huge, sprawling operation, with a web of U.S. subsidiaries and diverse operations. NEC's SONET products are part of its first U.S. subsidiary, NEC America. This organization oversees a vast range of telecommunications operations from its Melville, New York, headquarters, including manufacturing in Hillsboro, Oregon.

NEC develops, manufactures, and markets a complete line of advanced communications products and software for public and private networks, as well as SONET/SDH products. NEC America also oversees software and hardware development operations for various communications technologies in Dallas, Texas; Herndon, Virginia; San Jose, California; and Hillsboro, Oregon.

NEC's equipment has evolved rapidly between the second and third editions of this book. Now, SONET and SDH products are presented at NEC's web sites in two entirely separate ways. SDH equipment can be found at *www.nec-optical.com*. SONET gear is found at NEC's American subsidiary, Eluminant *(www.eluminant.com)*. NEC sometimes writes this as eLUMINANT. NEC Eluminant Technologies' SONET equipment is marketed under the Vista name, and so this section will also explore one of NEC's more advanced SDH products.

NEC Vista

Billed as an "intelligent access multiplexer," NEC's Vista SONET equipment is really a shelf-wide package that can be used in three different places in a network. Vista is a SONET broadband access system specifically designed for both access network and local loop transmission of any TDM, IP, frame relay, or ATM traffic. Vista can be used as a traditional SONET multiplexer, ready to support any terminal or unidirectional path switched ring (UPSR) application, without additional software or hardware add-ons. A GUI-based management system makes configuration easy. In addition to standard interfaces needed for legacy OC-3/12 SONET speeds, Vista also supports HDSL, span-powered T1s, DS-3 transmux capability, as well as traditional DS-1 and DS-3. Optical extensions eliminate the need for additional service shelves in collocation, customer-premises, and remote-cabinet applications. Vista also has an optional 10/100BaseT (LAN) interface capabilities, a nice feature.

Vista will support any traditional M13 application with full grooming capabilities as a feeder to backbone fiber trunks. Vista has an in-service upgrade capability to OC-12 when access network growth demands additional bandwidth, ensuring full access to all the added bandwidth at every node without any interruption in service. Vista's bandwidth management features also include hairpinning and drop-and-continue, which permit efficient use of facilities and provide support for most dual-homing applications.

As a SONET ADM, Vista supports UPSR (per GR-1400) and can act as a DS-3 access multiplexer or optical hub with interconnected rings. Operations systems support (OSS) includes TL-1/X.25, TCP/IP CO (connection-oriented) LAN over 10BaseT (very nice), and OSI section DCC for multi-vendor interoperability. There are methods for local or remote provisioning, software download through a LAN or modem, and nine addressable "housekeeping" alarms.

Optical interfaces include OC-3 at 1310 nm, OC-3 at 1550 nm (for WDM), OC-12 at 1310 nm and 1550 nm, and OC-3L 1550 nm. The unit has dual DS-1 BITS inputs, supports synchronization status messages (SSMs), and can act as an M13 multiplexer as well. Vista was designed to be flexible, and whether the system is placed in a central office environment, interfacing with a voice or ATM switch, or in an outside remote environment, Vista can provide scalable and easy-to-provision loop network services.

NEC SMS-2500A

Because NEC uses the Vista umbrella for all its SONET products (and there is really only one SONET product with various configurations), this section profiles an SDH piece of equipment, the SMS-2500A.

The SMS-2500A (there is also an SMS-2500C) provides high-quality, reliable networking through the use of two-fiber or four-fiber MS-SPRing (the SDH term for rings), long-haul interfaces, and the massive tributary capacities needed for backbone networks. Internetworking for a variety of services is available through the use of VC-4-4c concatenations. The device can directly interwork with DWDM equipment, giving a fully managed solution, including services, network elements, and customers.

The SMS-2500A is an STM-16 (2.5 Gbps) multiplexer. It can adapt to high-transmission networks such as backbone networks, all with MS-SPRing protection. The SMS-2500A offers tributary interfaces for constructing the backbone network, including STM-1, STM-4, and add/drop capacity for STM-16. A cross-connect facility is included to allow support of fully flexible non-blocking digital link functions. The SMS-2500A supports a "colored" optical STM-16 interface for effective integrated photonic networks using DWDM.

There is multiplex section protection (MSP) for point-to-point and linear bus systems, as well as interworking rings consisting of multiple two-fiber and four-fiber rings linked by STM-1 and STM-4 tributary interfaces. Broadcast applications can use the drop-and-continue function at the STM-16 level. There is a VC-4 cross-connect function, and the device supports VC-4-4c concatenation. Management is integrated with other NEC products.

NORTEL NETWORKS

Nortel Networks (Northern Telecom) marked its 100th anniversary in 1995. The company has evolved from small Canadian telephone equipment supplier to global player in networks. Nortel Networks really began in 1880 when the Bell Telephone Company of Canada was formed, based on Alexander Graham Bell's Canadian patents. The Manufacturing Branch especially prospered, and in 1895 it was incorporated as a separate company, called Northern Electric and Manufacturing Company Limited.

For the next 50 years, Northern Electric essentially manufactured equipment based on designs and processes licensed from AT&T's Western Electric. It made products primarily for the use of the Bell Telephone Company of Canada, although it did make some consumer electronics products. The Western Electric association endured until 1956, when Western Electric had to terminate its patent and licensing relationship with Northern Electric. Northern Electric created its own research and development facilities known as Bell Northern Research, based in Ottawa, Ontario; and, in 1959, it established Northern Electric Research and Development Laboratories in Ottawa, Ontario.

Since 1984, Nortel Networks has grown into a respected global supplier of products, with large markets in Canada, the United States, Japan, and elsewhere around the world. It is certainly a sign of the times that Nortel Networks' "first generation" OC-192 and STM-64 equipment is being replaced by more DWDM-friendly devices. However, this does not apply to all of Nortel's SONET/SDH products, of course.

Nortel Networks TransportNode (TN) SDH Family

The second edition of this book fondly profiled the Nortel Networks S/DMS AccessNode FST (for "full services terminal") SONET DLC. The section concluded that it is "worth remembering that OC-192 and gigabit links are fine, but the success of SONET is also dependent on backward-compatibility and just how easy it is for everyone to take advantage of SONET benefits." Sadly, the Nortel DLC is on its way out. But in its place, this edition profiles the Nortel SDH family of products, and this family is certainly a worthy substitute.

The Nortel SDH product family—all from the "TN family," such as the TN-64X—supports speeds from an E-carrier's 2 Mbps all the way up to 160 Gbps. The family includes a range of SDH multiplexers (compact SDH Access, STM-1, STM-4, STM-16, and STM-64) and cross-connects (STM 4/1, STM 4/4), some with WDM capabilities, and all using Nortel Networks management. The integrated network management provides a nice feature in most markets, so that there is a common platform for operations around the globe.

Nortel Networks SDH solutions provide compliance with all SDH standards, enabling interconnection with multi-vendor networks, with an installed base of more than 30,000 network elements. The products are designed and manufactured with a high degree of reliability for high availability, and they enable service providers to offer broadcast-quality video services, a very demanding application.

As an example of the SDH TN family, consider the TN–16X. This product is intended for network backbone traffic, delivery of services to high-volume urban core areas, and other bandwidth-intensive applications. In addition to an STM-16 interface, the TN-16X is equipped with DWDM capabilities to maximize available bandwidth. The TN-16X has a common sub-rack, and provides hub, ADM, ring, and TM functions. The platform is equipped to handle a range of electrical (34 Mbps, 140 Mbps, STM-1) and optical (STM-1, STM-4) tributary interfaces. In addition, the TN-16X STM-4 optical interface supports the transmission of concatenated VC-4-4c payloads typically used by IP and ATM equipment.

The platform is fully scalable and is seamlessly compatible with other Nortel Networks SDH products. In addition, the DWDM STM-16 interfaces are compatible with other Nortel Networks DWDM products and can be used without the need for wavelength translators. Compliance with SDH and OSI standards permits the TN-16X to interwork with optical equipment supplied by other vendors.

The TN-16X offers a full complement of network survivability features. At the tributary level these include 1:n (where $n = 1$–8), 1+1, and matched nodes for automatic protection switching. At the aggregate level STM-16 1:n, 1+1, and MS-SPRing protection are supported. For added reliability, the STM-1 and STM-4 protection switching systems operate independently of each other.

As demand increases, the DWDM interfaces on the TN-16X support 32 DWDM wavelengths, which translate to 80 Gbps of bandwidth on a single fiber pair, in 2.5 Gbps increments. The operational costs associated with the TN-16X can be minimized by using the remote maintenance and administration available through Nortel Networks software.

Nortel Networks S/DMS TransportNode OC-192

Nortel Networks poses a special challenge when it comes to SONET products. The company has been so influential in the whole SONET industry in terms of white papers, configuration information, and just product information in general, that many of the ideas in this book, and much of the terminology itself, owes a great deal to the Nortel Networks influence on SONET thought.

In any case, the Nortel Networks S/DMS TransportNode OC-192 product offers up to 20 Gbps of optical tributary access (two OC-192s) in a single frame, along with very sophisticated management. The S/DMS system is a single SONET network element that can groom, segregate, and transport literally millions of telephone calls at once. The S/DMS can be configured in protected linear or ring configurations and offers centralized control through a graphical network management system.

The S/DMS TransportNode OC-192 system is a member of a family of products that runs at other SONET speeds as well: OC-3, OC-12, and OC-48. The S/DMS OC-192 interworks fully with its OC-3, OC-12, and OC-48 counterparts, as well as with other S/DMS products. It also provides centralized, end-to-end network management.

The S/DMS terminates up to 20 Gbps of tributary bandwidth. These tributaries can be fed to the transport side or looped back into other tributaries. The S/DMS can be used to provide service with full-survivability form self-healing, interconnected rings using the matched-nodes feature of the S/DMS. There are linear deployments as well, all with OC-48, OC-12, or OC-3 tributaries with bandwidth management down to the STS-1 level.

The S/DMS TransportNode OC-192 system is a fully modular system in a single seven-foot rack. There are shelves and plug-in circuit packs that make it easy to deploy varying amounts of capacity, allowing for easy expansion. A typical OC-192 version of the S/DMS would include these elements:

▼ **Control shelf** The control shelf supports the breakers and power units for all battery connections and a number of controller components. The controller units support such functions as timing interfaces, SONET overhead processing, the orderwire and DCC channels, the hard drives (and even tape drives) for software downloads and backups, various remote access ports, and Ethernet ports.

■ **Local craft access panel** A panel provides the user interface for local craft personnel. There is an RS-232 port for a VT-100 asynchronous terminal, a handset and headset jack for orderwire communications, and alarm cutoff switches.

■ **Main terminal shelf** The main terminal shelf is the heart of the SONET transport system. It is used in all configurations to support OC-192, OC-48, OC-12, and OC-3 interfaces. The top half of the terminal shelf contains the interfaces for the OC-3, OC-12, and OC-48 tributaries. The bottom half contains the OC-192 line interface.

- ■ **Extension shelf (optional)**   The optional extension shelf can house additional OC-12 and OC-3 tributaries. It can also support optional OC-192 line interfaces if desired.
- ▲ **Environmental control panel**   The environmental control panel houses the fan units that control the temperature of the main terminal shelf. The control shelf above is cooled by convection.

The S/DMS can be deployed in a wide range of configurations. In linear configurations, the S/DMS can support 10 Gbps on the active OC-192 and another 10 Gbps of traffic on the protection link. In rings, the same 20 Gbps of capacity can be deployed and protected by the rings.

TELLABS

Tellabs, which was started in 1975, is headquartered in Lisle, Illinois. The company ended its first year with a total of 20 employees and sales well under a half-million dollars. Since then, Tellabs has carved out a position in the digital cross-connection field that has resisted all assaults from larger and older companies with telecommunications roots going back more than 100 years. Originally specializing in analog products, Tellabs is now a $2-billion global company. Tellabs recently began to explore more of the SONET/SDH market than just digital cross-connects, but that specialty remains the core business.

Tellabs acquired two new companies toward the end of 1999. Alcatel's Co Communications businesses in Europe became Tellabs Denmark, and, in the United States, Tellabs acquired NetCore Systems.

Tellabs 5500 DCS

The Tellabs 5500 family of digital cross-connects (formerly marketed under the TITAN 5500 name) is intended for all types of telecommunications networks, including local telephone service, long distance networks, wireless networks, private networks, the emerging networks in corporate America, and networks around the world. The 5500 family has many applications in a wide variety of network types and has very good migration capabilities. Telecommunications competition makes it imperative for service providers to upgrade networks to cut costs, improve quality, and guard against equipment obsolescence. Telllabs systems satisfy the demand for new wideband and broadband services. They also allow remote and centralized performance monitoring and control.

The Tellabs 5500 system replaces multiple pieces of co-located equipment such as digital patch panels (DSX), M13 add/drop multiplexers (ADM), fiber-optic terminals (FOTs), and the manual wiring techniques previously used to cross-connect and manage DS-3, DS-1, and OC-N transmission facilities. A Tellabs 5500 system can grow from a small Tellabs 5500S system with 32 ports to 1,024 DS-3/STS-1–equivalent ports and beyond, in increments of 32 DS-3/STS-1 equivalents. The growth path is accomplished without manual reentry of database information and without rendering current equipment investments obsolete. With a single architecture that can grow far beyond 1,024 ports, large hub locations no longer have to co-locate or tandem smaller DCSs that have reached maximum capacities.

The maximum configuration of the Tellabs 5500 is quite impressive. It can cross-connect a maximum of 2,048 DS-3s, with 57,344 DS-1s representing 1,376,256 DS-0s.

Tellabs FOCUS LX

It is fitting to end this product sampler with the Tellabs FOCUS LX SDH ADM and DCS because this product demonstrates a growing trend in SONET/SDH product offerings. SONET, after all, is just the North American version of SDH. So SDH is the true global market, not SONET. Many vendors of SONET equipment therefore also offer a more or less complete line of SDH. Tellabs Denmark has emphasized the growing SDH market with this product.

The FOCUS LX system—which is part of the FOCUS family of access, transport, and optical systems—is a multi-purpose synchronous digital hierarchy (SDH) multiplexer with integrated cross-connects. To simplify the design and use of a transport network, the FOCUS LX system is compact and flexible, and this reduces the network's deployment costs. The FOCUS LX system can be deployed in all parts of a transport or access network. The system's main strength is in regional and metropolitan networks, where operators can use its cross-connect functions for efficiency.

The FOCUS LX is an all-in-one network product that works with other Tellabs broadband access solutions to deliver integrated voice, high-speed data, and video-on-demand services to a subscriber. The modular architecture of the FOCUS LX system supports in-service upgrades. The system offers add/drop multiplexing directly from 2.5 Gbps (STM-16) to 2 Mbps (VC-12) and handles standard electrical and optical SDH, plesiochronous digital hierarchy (PDH), and ATM interfaces. In addition, its advanced pointer handling and resynchronization capabilities provide mobile telephony operators with a low-jitter, 2 Mbps data line for direct base transceiver station synchronization, which enhances quality of service.

The FOCUS LX system complements Tellabs' FOCUS SDH and DWDM products that are used by major service providers worldwide. The FOCUS products provide a full range of integrated solutions for access and transport/optical network applications, supporting managed access speeds ranging from 64 Kbps to 622 Mbps and transmission speeds up to 10 Gbps per channel. Like all FOCUS SDH and DWDM network elements, the FOCUS LX system is managed by a carrier-class network and element management system, which offers graphical network configuration and fault management.

SAMPLE PRODUCT CONFIGURATIONS

It is one thing to outline product features and interfaces. But it is quite another to see just how all of the vendors' offerings might fit together to solve a network problem with SONET or how typical SONET deployments might look on a smaller or even a larger scale. ("Small scale" means a fairly restricted area, as opposed to multi-city or even nationwide SONET network deployments. "Large scale" includes rings or links that might span the country.)

In this section, each of the vendors' products is shown in a typical SONET deployment. This does not imply that the product is necessarily limited to the configurations shown, especially when it comes to SONET rings. An example showing a small SONET network does not rule out the use of the product for large SONET networks. The aim is to be informative with respect to the various SONET network elements, rather than to be exceedingly precise about all of a product's features and possible scenarios.

Alcatel Network Systems

Alcatel makes a full range of SONET/SDH-based products, including wideband and broadband DCS products and SONET/SDH ring devices that operate at a range of speeds. Figure 12-1 shows how several of these products can be used to implement SONET ring configurations.

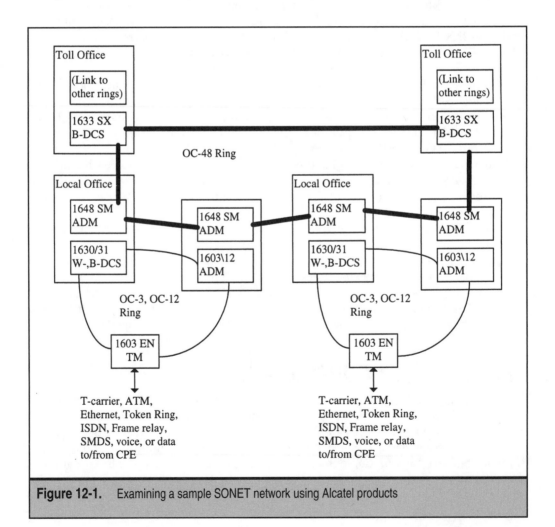

Figure 12-1.   Examining a sample SONET network using Alcatel products

In Figure 12-1, the Alcatel 1603 TM (a form of SONET terminal multiplexer) concentrates and interfaces with customers' T-carrier, OC-3, Ethernet, Token Ring, ISDN, frame relay, SMDS, or even ATM voice and data equipment. This traffic is carried back to the serving office by an OC-3 or OC-12 ring with 1603 SM (for OC-3 or OC-12) SONET ADMs. Once at the serving office, the signals can be cross-connected with the 1631 LMC wideband cross-connect network element. Other Alcatel cross-connect products can be used here as well.

Traffic—especially voice traffic—can be switched or transferred to a higher-speed SONET ring for longer hauls. Typically, a 1648 SM would be used for this purpose, running at the OC-48 rate. Further signal processing can be done at this level with the 1633 SX broadband DCS.

Although not shown in Figure 12-1, feeding an OC-192 ring for covering even larger areas is possible. Also, the Alcatel product line includes a network management system for all of the components.

Fujitsu Network Transmission Systems

Fujitsu makes a very respected line of SONET/SDH TM and ADM products for implementing all types of SONET/SDH rings (the emphasis is on SDH). The line culminates with the FLASH-192 product, which is both powerful and versatile. Figure 12-2 shows how the FLASH-192 product can be used to implement a very high speed SONET ring. Naturally, there are more Fujitsu SONET products in the whole family, but the FLASH-192 can incorporate the functionality of many of them as optional modules.

In Figure 12-2, the FLASH-192 network elements, variously equipped, handle the SONET ring links running at OC-192 and, at the same time, cross-connect down to the OC-1 (STS-1) level. Network management is a key part of the product as well.

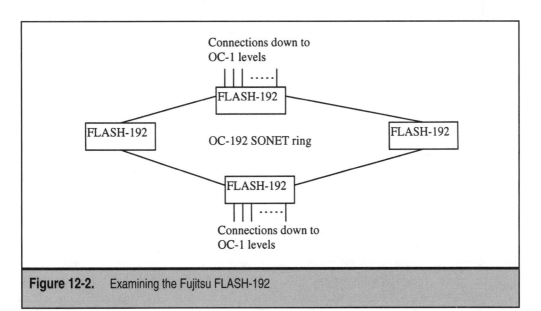

Figure 12-2. Examining the Fujitsu FLASH-192

Lucent Technologies

Lucent (AT&T) makes a full line of SONET/SDH-based products. In the interests of variety, Figure 12-3 shows a rather simple arrangement of Lucent equipment. Of course, the DDM-2000 can also be used in a wide range of ring configurations, and the WaveStar line can be used for SDH situations.

In Figure 12-3, the Lucent DDM-2000 is shown in a typical metropolitan area service configuration. In this point-to-point configuration, the OC-3 ADM is located at a central office location. The DDM-2000 has DS-1 links to a digital cross-connect—in this case, a DACS IV-2000—and a digital central office switch. Another OC-3 ADM is located at a large customer location, perhaps an office building. Two sets of fibers are employed, a working pair and a protection pair. In the building, the ADM connects to a remote terminal linked to telephones, again with a DS-1 interface. To service other floors, fiber cables can be installed to other remote terminal units through another DS-1 interface. The OC-3 links can deliver other than voice services, of course. Video is only one such possible application.

NEC Transmission

NEC makes a range of SONET/SDH products well-suited for ring configurations. Figure 12-4 shows a modest but complete SONET ring application using NEC's Vista products.

In the figure, the Vista gear operating in terminal mode (Vista can also be configured as an ADM) gathers traffic from user devices at customer locations. The traffic can be HDSL, DS-1, or DS-3. These can be linked by OC-3 links to the ADMs at the serving offices. The Vista ADMs can be linked to an ATM switch, or to an passive optical network (PON) serving other customers at DS-1 or Ethernet speeds.

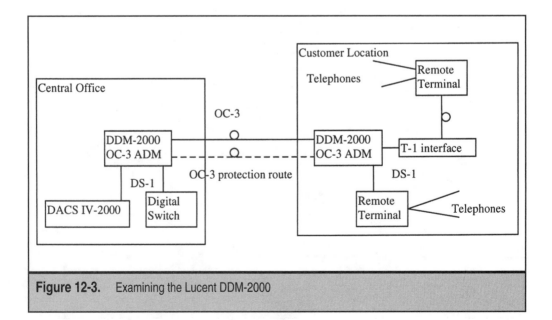

Figure 12-3. Examining the Lucent DDM-2000

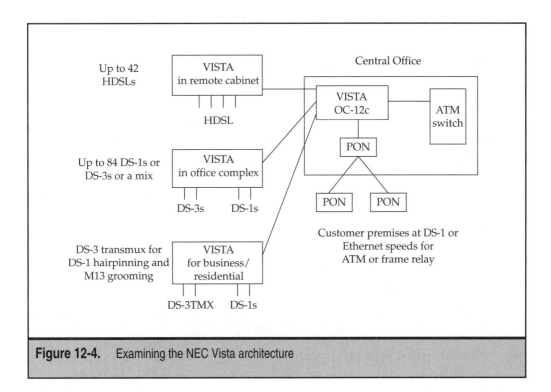

Figure 12-4.   Examining the NEC Vista architecture

Nortel Networks

Nortel Networks' prominent position when it comes to SONET need not be commented on further. Figure 12-5 shows a SONET ring implemented on the Nortel Networks S/DMS products.

In the figure, the S/DMS products, which run at a range of SONET speeds, are configured in an OC-12 or OC-48 SONET ring. Each can deliver up to 48 DS-3s or the equivalent (on an OC-48 ring) to cross-connect or digital switches. Other configurations are also possible, naturally, and the traffic capacity can be handed off to other rings or other SONET terminal multiplexer arrangements.

Tellabs

Tellabs has staked out its claim in a major SONET/SDH category: easily upgradeable digital cross-connects for SONET or SDH. Tellabs has also begun to explore the marketplace for new offerings in the optical networking arena. This section will emphasize Tellabs' traditional role, however.

Figure 12-6 shows how the Tellabs 5500 DCS and the Tellabs 5320L (not previously mentioned) can be used to create a mixed SONET/SDH environment with both T-1 and E-1 traffic. While there are other Tellabs products that could be used in this configuration as well, the figure is intended to showcase the role of the Tellabs 5500 and 5320L.

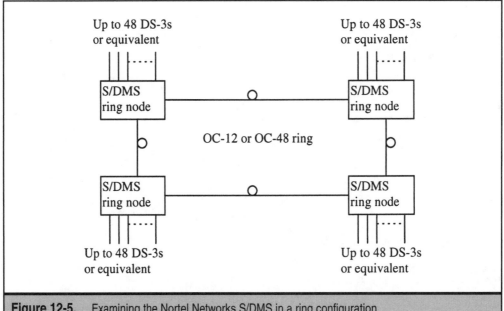

Figure 12-5. Examining the Nortel Networks S/DMS in a ring configuration

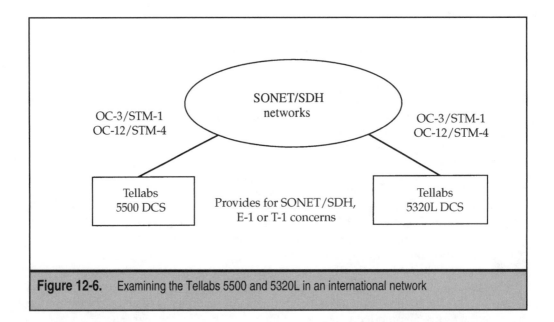

Figure 12-6. Examining the Tellabs 5500 and 5320L in an international network

DWDM EQUIPMENT PROVIDERS

This book is about SONET/SDH, and this chapter has been about SONET/SDH equipment providers. However, the recent trend in the optical products industry has been to emphasize DWDM products, often at the expense of the older, more mature SONET/SDH product line. Already, several of the vendors in this chapter have begun to phase out SONET/SDH-only products in favor of products that integrate SONET/SDH with DWDM. This is only natural; vendors are always keeping an eye on the future. The global Internet and World Wide Web have made it clear that the future belongs to international networks and higher-bandwidth solutions. This translates to DWDM. It is often difficult to navigate the SONET/SDH vendors' web sites and isolate the SONET/SDH equipment details, because many of the newer offerings are intimately tied to optical networking and DWDM.

A full survey of DWDM equipment to the depth provided on SONET/SDH equipment would not only take many pages, but be far beyond the scope of this book. Interested readers are referred to books emphasizing this aspect of networking.

ROUTERS AND SONET/SDH

SONET/SDH is primarily a product of the world of telephony. That is, SONET/SDH was originally invented by telephony service providers as a trunk network to carry more and more voice conversations, and with an eye toward a future of broadband services based on adding some video and data services to the telephone. Early examples of "videophones" and "data terminals" were designed to transition customers to a world of advanced services. The service providers were to remain the same, however: the same faithful telephone companies that provided your dial tone.

All of this changed with the Internet and World Wide Web. A huge SONET/SDH market developed for Internet service providers (ISPs) seeking the most bandwidth they could get, not for telephone calls, but for shuttling IP packets among a collection of IP routers. This is what packets over SONET/SDH was all about: making SONET/SDH more router-friendly. In many cases, the corporate entity using the SONET/SDH links was still the telephone company, but in its new role as ISP. And even voice made its way into the packet world as voice over IP (VoIP).

SONET/SDH cannot ignore the role of the router in the modern network. This fact has already been pointed out in the earlier chapters of this book. What has not been pointed out is that, in many cases, the role of the SONET/SDH TM network element is not played by a device packaging low-speed tributaries and containers onto higher-speed SONET/SDH links, but by a simple router trying to blast as many packets per second as possible over concatenated (unchannelized) SONET/SDH links. The routers still link to SONET/SDH ADMs arranged in rings, and might even support voice in the form of VoIP (and packetized video as well).

Figure 12-7 shows today's use of routers in conjunction with SONET/SDH rings.

The main point is that SONET/SDH products have evolved to the point where a router line card (also called an interface card) can implement one or more SONET/SDH ports into the router itself. Not all ADM functions are supported, of course, but most of

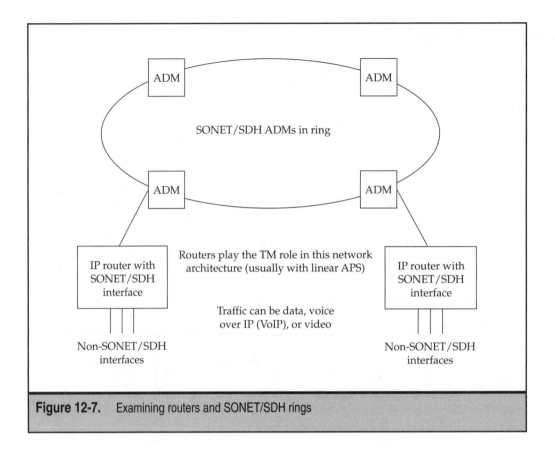

Figure 12-7.   Examining routers and SONET/SDH rings

the basics of a SONET/SDH TM are there, ready to carry packets from one router acting as a TM to another across a SONET/SDH ring maintained by a SONET/SDH service provider. SONET/SDH equipment has thus been reduced from a rack full of equipment to a form of "SONET/SDH on a chip" suitable for ISP ease of use.

This chapter will therefore close with a look at how two major router vendors—Cisco Systems and Juniper Networks—implement and support SONET/SDH interfaces. It the future, the router might even take on the role of the ADM ring, if current plans to implement IP-based resilient packet rings (RPRs) are completed as IEEE 802.17.

Cisco Systems

Cisco Systems (*www.cisco.com*) more or less invented the router, and Cisco routers still often form the foundation of the Internet and most corporate, education, and government networks around the world. Cisco provides more than just routers. It offers a line of products for transporting data, voice, and video within buildings, across campuses, or around the world.

Cisco was founded in 1984 by a group of computer scientists from Stanford University (in the same building where SUN Microsystems and SGI began). Since the start, Cisco engineers have been prominent in advancing the use of IP as the basic language to communicate over the Internet and in private networks. No matter how many products Cisco makes, be they for switching, voice and video over IP, optical networking, wireless, storage networking, security, broadband, and content networking, the focus remains IP.

SONET/SDH LINE CARDS

Cisco Systems makes line cards for SONET and SDH frame formats alike, in both concatenated and channelized versions, mostly for the higher-end Cisco 7300- and 12000-series routers. The older line cards have fewer ports at a given speed, and the newer line cards have four or even eight ports, although the architecture limits the number of ports at higher SONET/SDH speeds. All SONET/SDH configurations use POS, except when packets are carried by tributary or container rates such as E-1 or DS-0.

In keeping with the earlier SONET/SDH product samplers, this sampler section is not intended to be an exhaustive list of Cisco's SONET/SDH line cards. But it is representative. Only SONET/SDH is included, but Cisco makes ATM line cards that also run at the OC-3/STM-1 and OC-12/STM-4 rates. Because routers essentially perform a TM role in a SONET/SDH network, only a simple list is needed for this product sampler.

Channelized SONET/SDH line cards include these:

▼ **PA-MC-STM-1**   The PA-MC-STM-1 is a single-port, but multi-channel, port adapter for the Catalyst 6000 and Cisco 7200 routers. It can be configured to carry 63 E-1s or 256 DS-0s.

■ **Two-port STM-1/OC-3 Channelized E1/T1 Line Card**   This card is for the Cisco 12000 series. When configured for SDH, each STM-1 port is capable of supporting channelization into either 63 independent E1s through VC-12 ETSI mapping, or 84 J1 (Japanese T1 equivalent) channels through VC-11 ANSI mapping. When configured for SONET, each OC-3 port is capable of supporting channelization into 84 independent DS1 (VT1.5 mapping or CT3 mapping into STS-1) channels. Fractional E1 and individual 64 Kbps channels can also be configured, up to the limit of 105 channel definitions per STM-1 (35 per TUG-3). Fractional T1 and individual DS-0 channels can also be configured, up to the limit of 35 channel definitions per STS-1 (210 per line card). Each card supports up to 126 E1s when configured for SDH operation, 168 DS1 connections when configured for SONET operation, or any combination of 64 Kbps/DS-0, fractional E1/T1, or E1/T1 interface definitions, up to the limit of 210 per line card.

■ **Four-Port Channelized OC-12/STM-4 POS/SDH ISE Line Card**   This card supports SONET/SDH framing and provides DS-3/E3 aggregation for the Cisco 12000 series router. For SDH, AU-3 and AU-4 mappings are both supported. The card offers flexible channelization. For example, the line card can be channelized to have DS-3 interfaces only or a combination of OC-3 and DS-3 interfaces, as long as the SONET/SDH framing boundaries are respected.

■ **16-port Channelized OC-3/STM-1 to DS-3/E3 Line Card**   This card supports SONET/SDH framing and provides DS-3/E3 aggregation for the Cisco 12000 series router. For SDH, AU-3 and AU-4 mappings are both supported. The card offers flexible channelization. For example, the card can be channelized to have DS-3 interfaces only or a combination of OC-3 and DS-3 interfaces, as long as the SONET/SDH framing boundaries are respected.

- ■ **OC-12–DS-3 Line Card**   The OC-12–DS-3 line card provides 12 channels of DS-3 multiplexed over a single 622-Mbps OC-12 port. It interfaces through an ADM with other DS-3 line cards at DS-3 line rates in a configuration that usually consists of a Cisco 7200 series router and a Cisco 7500 series router configured with the POET port adapter, or a third-party T3 DSU such as Digital Link, Larscom, or Kentrox.

- ▲ **CHOC-12/STS-3 IR-SC Line Card**   This card provides the Cisco 12000 series product line with four streams of STS-3/STM-1 service through a single 622-Mbps OC-12/STM-4 interface.

Concatenated SONET/SDH line cards for Cisco routers are simpler:

OC-3c/STM-1   These line cards come in 4-, 8-, or 16-port versions, but the higher the speed, the fewer the ports. The router makes a difference too, and there are four-port OC-12c/STM-4c and OC-48c/STM-16c cards for the Cisco 12000, but only single-port versions for the Cisco 7300.

The Cisco concatenated POS cards run at these speeds:

- ▼ OC-3c/STM-1
- ■ OC-12c/STM-4c
- ■ OC-48c/STM-16c
- ▲ OC-192c/STM-64c.

Only the highest-end Cisco routers can keep up with the OC-192c/STM-64c packet stream. Cisco's major router architectures will struggle with OC-768c/STM-256c.

CONFIGURING SONET/SDH ON A CISCO ROUTER

This section cannot be a complete guide to configuring SONET/SDH links on a Cisco router, especially the channelized line cards. Instead, this section concentrates on the SONET/SDH overhead bytes that can be configured when using POS. This section is generic for POS, but details may vary across platforms and line cards.

Generally, the default framing is for SONET, and the default frame size for packets is 4470 bytes, but this can be adjusted from 64 bytes to 15,360 bytes. "Keep alives" are generated every 10 seconds, but can be set to between 0 seconds and 32,767 seconds. The clock source for the link is for loop timing, but the internal Stratum 3 clock can also be used.

The path overhead C2 signal label value defaults to 207 decimal, or the experimental 0xCF value for IP inside PPP (or Cisco HDLC). Any value from 0 to 255 is valid, but only a few make sense. Various BIP and related alarm thresholds can be configured, and alarm reporting can be set as well, a very nice feature. The J0 trace byte value defaults to 1 in some Cisco documentation, and to 0xCC (204 decimal) in others. The allowed range is 0 to 255. The J1 trace byte defaults to host name, interface name, and IP address for SONET. The SDH J1 trace default is empty. Neither default can be changed. The line cards support linear APS and can scramble the payload (the default is no scrambling). The setting for the s1s0 bits (also called the "ss bits") in the pointer byte is allowed, but newer line cards can derive these from the framing type (they have meaning only in SDH and are set to decimal 2 or bit values 10).

Juniper Networks

Juniper Networks *(www.juniper.net)* was founded in 1996 and seemed to come out of no-where to capture a large portion of the core router market, mostly at Cisco's expense. Juniper Networks essentially reinvented the high-end router along the lines of a more hard-ware-driven approach, with custom chipsets and very small form factors. The M-series router remains the core product, and variations for wireless and cable TV service providers also exist.

Compared to other vendors, Juniper Networks remains tightly focused on the router marketplace. But no routers are faster or bigger in terms of packets per second or aggregate throughput than Juniper Networks routers.

SONET/SDH PHYSICAL INTERFACE CARDS

Just as with Cisco cards, physical interface cards (PICs) from Juniper Networks can be used for SONET or SDH. But in contrast to cards from Cisco Systems, which are router-specific, many Juniper Networks PICs for SONET/SDH can be used in *any* Juni-per Networks router. Even the smallest Juniper Networks router, the M5 (about the size of a large server), can have up to 16 OC-3c/STM-1 ports or 4 OC-12c/STM-4c ports, giv-ing an impressive 2.5 Gbps aggregate speed. There are the usual port-number variations depending on speed, but only the fastest OC-192c/STM-64c PICs are used in the high end of the product line. In addition, high-port-density PICs can run only concatenated and do not allow for channelization. All SONET/SDH configurations use POS, except when packets are carried by tributary or container rates such as E-1 or DS-0.

The Juniper Networks SONET/SDH PICs are these:

▼ **Two-port and four-port OC-3c/STM-1 PIC**   The two-port version is for the smaller M5 and M10 routers. These PICs operate in concatenated mode only.

■ **One-port and four-port OC-12c/STM-4 PIC**   The one-port version is supported on all M-series routers and operates in both concatenated and channelized modes. The four-port version is for the M40, M160, and T640 routers only, and operates only in concatenated mode.

■ **One-port and four-port OC-48c/STM-16 PIC**   The single-port version is for the M10, M20, and M40 routers, and supports both concatenated and channelized modes. The four-port version for the high-end T640 operates in concatenated mode only.

▲ **One port OC-192c/STM-64 PIC**   This PIC is for the M160 and T640 routers. It operates in both concatenated and channelized modes.

Support for OC-768/STM-256 should be easier to accomplish on the unique, hard-ware-based architecture of the Juniper Networks router than on routers from Cisco Systems.

CONFIGURING SONET/SDH ON A JUNIPER NETWORKS ROUTER

This section cannot be a complete guide to configuring SONET/SDH links on a Juniper Networks router, especially for channelized PICs. Instead, it concentrates on the SONET/SDH overhead bytes which can be configured when using POS.

Configuring a SONET/SDH link on a Juniper Networks router can be somewhat complicated. There are chassis parameters that can be set (mainly for SONET or SDH framing), overall options (such as APS and overhead byte values), physical parameters that can be set for the overall properties of the link (such as encapsulation type), and logical parameters mainly for IP addresses.

The default framing is for SONET, but this can be changed to SDH, of course. The setting of the s1s0 bits (also called the "ss bits") in the pointer byte is automatic and is decimal 2 or bit value 10 for SDH. The default PIC mode is concatenation for POS, but this can be changed for the PICs that are capable of supporting channels.

The default timing on the link is to use the internal Stratum 3 clock on the PIC (line timing), but external (loop timing) is allowed. The default frame size for packets is 4470 bytes, but this can be adjusted upward to 9,192 bytes. "Keep alives" are generated every 10 seconds, but can be set to between 1 second and 32,767 seconds.

The router can set values for many SONET/SDH overhead bytes. These include configurable values between 0 and 255 for orderwire E1 (default is 0x7E), the idle value, the SSM S1 byte (in the range 0 to 255), the user F1 and F2 bytes (in the range 0 to 255; default 0x00), and the Z3 and Z4 (user F3 and path APS K3) bytes (in the range 0 to 255; default is 0x00).

The path overhead C2 signal label value is determined automatically and is generally 207 decimal (0xCF) or 22 decimal (0x16) depending on the length of the frame-check sequence configured. The J1 trace byte can be set to any string up to 15 characters, usually the router name. In SDH framing, the router can be configured to increment to Z0 byte (unused J0 positions) for older SDH ADMs that expected this (similar to the old C1 overhead byte behavior). The default is not to increment the Z0 byte. The PICs support linear APS and can scramble the payload (the default is to use scrambling).

PART IV

SONET/SDH Advantages

Previous parts of this book have mentioned the distinct advantages that SONET/SDH has over other forms of transmission networking. However, the discussions to this point have been scattered throughout earlier sections on the SONET/SDH technology and have compared SONET almost detail by detail with alternatives such as T-carrier. (The same arguments apply to E-carrier for the most part.) This section attempts to be more systematic about the advantages that SONET/SDH offers to both customers and service providers. It thoroughly examines SONET/SDH as more than just a way to overcome the limitations of the existing digital hierarchy, proceeding on a higher level than the bits and bytes of the technology itself.

In fact, this section goes beyond a simple recitation of SONET/SDH benefits for customers and service providers. Many of the benefits of SONET/SDH are in the area of more efficient and comprehensive network management. Service providers are concerned with much more than simple network management, naturally. As well as being managed, networks must be provisioned, configured, and run on an ongoing basis. In the United States, service providers refer to the daily tasks involved in running a network of transmission links as "operations." The record-keeping work, which involves tasks such as billing the customer, is known as "administration." The repairs and preventive maintenance that keep the system up and running are referred to as "maintenance." Finally, the tasks initially required to configure and set up the network are known as "provisioning." Usually, these tasks are all gathered under the umbrella acronym of OAM&P (operations, administration, maintenance, and provisioning).

The OAM&P tasks are often called OAM (for "operations, administration, and maintenance," or just "operations and maintenance") in various standards documents. This does not imply that OAM involves no provisioning or administration; the term OAM&P is just more explicit.

The investigation into SONET/SDH OAM&P leads into a fuller discussion of SONET/SDH rings than was given earlier in the book. More details on the actual equipment and the operations used in forming and running SONET/SDH rings are presented.

This part of the book contains four chapters.

Chapter 13 looks at the overall advantages that SONET/SDH brings to customers and carriers alike. It points out that, although SONET/SDH was essentially invented to overcome the limitations of existing digital hierarchies, especially in the United States, SONET/SDH has gone far beyond this simple goal. SONET/SDH possesses many advantages beyond the obvious technological ones. Owing to the emphasis in Chapter 13 on the digital hierarchy in the United States, most of the chapter deals with SONET; SDH is mentioned only in passing. But once again, this approach avoids needless repetition. What can be said of SONET in this regard easily applies to SDH.

Chapter 14 explores in detail one of the major SONET/SDH advantages: the distinctive OAM&P procedures and protocols built into the SONET/SDH overhead. The amount of overhead defined in SONET/SDH has been a perceived liability in the past, but newer concerns for reliability have made SONET OAM&P a crucial part of modern networks. Chapter 14 also explores the handling of alarm conditions in the SONET/SDH environment. Alarms signal the presence of error conditions and thus are an important part of SONET/SDH OAM&P.

Chapter 15 explores the application of OAM&P to SONET/SDH network management. It also has some words about the older SONET-based network management protocols, such as TL1, the newer SDH methods firmly based on ITU and OSI specifications, and the latest discussions about basing SONET/SDH OAM&P protocols on the TCP/IP protocol stack. This only makes sense in a world dominated by the use of SONET/SDH for Internet and Web links, the migration of voice to a voice-over-IP (VoIP) environment, and the use of IP for video distribution by global cable TV companies.

Chapter 16 details methods for addressing the increased concern for reliability in networks. It considers traditional linear protection arrangements and then examines a key feature of SONET/SDH: the deployment of formerly point-to-point transmission systems in ring architectures. Not all SONET/SDH rings share a common architecture or method of protection switching. Chapter 16 explores the architectures of two-fiber and four-fiber ring configurations, and compares their characteristics.

CHAPTER 13

Customer and Carrier Advantages

To view SONET/SDH as simply the natural evolution of T-carrier and E-carrier onto fiber-optic cabling is tempting—that is, seeing SONET/SDH not so much as an improvement on older carrier systems, but just "different" owing the fiber-optic media employed. However, that view would be an oversimplification at best, and a complete miss of the whole point of SONET/SDH at worst.

Technologies like SONET and SDH are *developed.* They do not, like species, evolve, adapting themselves to environmental conditions. It is hard to say that the modern horse is an "improvement" on older versions the size of a dog, or that humanity "improves" the ape (especially given the misery that humankind seems fated to unleash upon itself and others). On the other hand, it is easy to list the precise ways that SONET/SDH improves upon older carrier systems. This is because development is intentional, and the technology in question is always guided by its creators along a certain path. Of all the ways in which SONET and SDH could have been appeared, it is no coincidence that SONET/SDH happens to overcome a number a limitations under which T-carrier and E-carrier labored. SONET (especially) and SDH took their final forms with the goal not only of using fiber-optic cables, but also of improving the way that former carrier systems used equipment and links. SONET/SDH has distinct advantages over older methods.

Some technologies benefit mainly the users and customers that employ it. The technology may require a large expense on the part of the service provider to deploy, with marginal financial benefit. In some cases, only the intangible benefits of increased user satisfaction may result; but, hopefully, those intangible benefits lead to increased revenues in the long run. One is tempted to put ISDN in this category. Upgrading analog copper local loops and central office switches to support ISDN proved to be enormously expensive, and yet customers were unwilling to pay much more for ISDN services which, in most cases, still amounted to making telephone calls. The interest in ISDN for Internet access has not changed the basic economic perception of ISDN on the part of carriers. It can still be seen as an expensive upgrade that does not substantially increase revenues in proportion to the expense. As a result, ISDN was unavailable in many areas in the United States, and DSL was developed by the service providers to overcome the limitations of ISDN with regard to economics and equipment for Internet access.

Other technologies benefit mainly the service providers and carriers that deploy it. The benefit of the new technology may be invisible to users and customers. If users and customers benefit at all, the benefit is indirect and typically in the form of vague assurances of "improved service," without more specifics. One is tempted to put things such as T-carrier in this category. The first deployments of T-carrier increased the trunk capacity of service providers and carriers, but had little-to-no direct impact on, or benefit to, users. Carriers would often imply that the new digital trunks made telephone calls sound better, but with at least two noisy analog local loops on both ends, the claims lacked the credibility they might otherwise have had if the local loops had been digitized as well. (That change would have to wait for ISDN—which means that, in many cases, customers are still waiting.)

Happily, SONET/SDH falls into neither of those categories. SONET/SDH benefits customers and carriers alike. The benefits have virtually assured that SONET/SDH would be both widely and quickly deployed and would be simultaneously popular with customers. This has proved to be the case in all areas where SONET/SDH is available.

Naturally, the benefits of SONET/SDH fall into a number of categories. Some involve cost reductions in deploying network equipment to support a given mix of services; others involve the possibilities of enhanced service; still others involve a direct reduction of customer network costs. Various features of SONET/SDH contribute to each of these benefits.

While what is true of SONET in this regard is also true of SDH, differences appear in the details. Some of the differences are due to the differences between T-carrier and E-carrier. For example, E-carrier, with its signaling channel in the frame itself, never suffered from the need for T-carrier to limit the digital channels used for data to 56 Kbps until newer methods were developed to overcome this limitation. The rest of this chapter therefore concentrates on SONET; SDH is mentioned only peripherally. This is not to slight SDH in any way. It just acknowledges that SONET came first and deserves a "pioneer preference" in this regard.

KEY BENEFITS OF SONET/SDH

Table 13-1 lists the key benefits of SONET/SDH for customers and service providers. It reflects much information compiled by Nortel Networks, but differs in many details from similar tables.

Feature	Reduced Capital Expense for Network Builders	Customer Cost Benefit	Service Provider Revenue Benefit
1. In multipoint configurations			
A) Grooming	Yes		
B) Reduced back-to-back muxing	Yes		
C) Reduced DSX panels and cabling	Yes		
2. Enhanced OAM&P			
A) OAM&P integration		Yes	
B) Enhanced monitoring capabilities		Yes	Yes
3. Integrated operations and monitoring		Yes	
4. New service offerings			Yes
5. Optical interface ("mid-span meet")	Yes	Yes	Yes

Table 13-1. Key Benefits of SONET/SDH

The table is organized into the five main features of SONET/SDH. Each of these major features produces a benefit, either in terms of reduced capital expenditure for the service provider building SONET/SDH networks, reduced cost (service pricing) to customers using SONET/SDH networks, or enhanced revenue streams for service providers. The new revenue sources usually (but not always) come from new data services that SONET/SDH can support. When a feature provides a particular benefit in one of the three categories, that category is given a "yes" entry. Sometimes a feature provides benefits in all three categories.

What follows, as mentioned earlier, deals mainly with SONET. As applied to SONET, the five main features listed in Table 13-1 include these:

1. **Multipoint configuration** SONET, unlike many T-carrier networks with strictly point-to-point links, is frequently deployed in multipoint configurations. This means that several sources of SONET bits can be combined or distributed without terminating the digital stream to recover and process the constituent signals (such as the basic 64 Kbps channel structure). This combination of multiplexing and cross-connecting in the same device is known as "hubbing." Many SONET ADM devices are deployed at hub locations to perform this essential SONET task. The SONET device that combines these functions is called the hub, and the whole process is generally known as "grooming." At the hub, traffic is combined or distributed to "spurs." Grooming is not unique to SONET equipment. Much the same was, and is, done with T-carrier devices; however, SONET grooming requires much less equipment, reducing the need for linking multiplexers and digital cross-connects back to back (as in T-carrier), and reducing the need for cabling between T-carrier terminations and patch panels (that is, DSX panels). All three aspects of SONET multipoint configurations help to reduce the capital expenditures needed to build and deploy SONET network links. Grooming can concentrate traffic and service more customers with fewer links than would otherwise be needed without grooming. The reduced need for back-to-back multiplexers indicates that one SONET device can perform the functions of several pieces of T-carrier equipment, saving even more capital. The reduced need for cabling and DSX cross-connection panels not only saves money, it also reduces the need for personnel to run the network and makes more efficient use of those staff members who have responsibility for day-to-day operations.

2. **Enhanced OAM&P** SONET not only greatly enhances the OAM&P capabilities available in digital systems such as T-carrier, it also tightly integrates them into all SONET network elements. The integrated and enhanced OAM&P overhead and procedures built into all SONET equipment help in two ways. First, the integration means that service providers can offer SONET at lower cost to customers. Second, the enhanced performance monitoring (in terms of errors) can not only lower customer costs, it can also offer a means for service providers to earn more revenue by making more efficient use of SONET network resources. This is possible because the integrated SONET OAM&P can help to service more customers with the same physical resources.

3. **Concurrent operations and monitoring**   "Operations" is the process of configuring and running a network on a daily basis. Monitoring is needed to make sure that everything is running correctly. Ideally, monitoring should result in a network that detects and corrects problems even before users are aware of them. Without close monitoring, a network operations center can resemble a hospital emergency room, with stressed-out network "doctors" scurrying around to treat the most serious cases, while many painful but less life-threatening situations go untreated. Because the OAM&P capabilities in SONET are enhanced and integrated, service providers can consolidate their operations centers and minimize the equipment needed to support a community of users. Although not necessarily a true cost reduction in terms of capital, this consolidation helps to lower the user costs for SONET.

4. **New service offerings**   The huge amounts of bandwidth in a SONET network (in the gigabit range) can support new services that were unheard of in the T-carrier world. Business services will likely lead the way. Businesses needing bandwidth to support video applications or even modest-speed LAN interconnections at 4, 10, or 16 Mbps are often at a loss for affordable solutions to their networking problems. With SONET, however, routine support of video, 100 Mbps LAN interconnections, color faxing, and the like will become commonplace. Customers will gladly pay for reliable support for these needed services. Many of the new services will be tied to asynchronous transfer mode (ATM) networks and B-ISDN deployments. Therefore, the customer is not directly buying SONET, but an ATM or B-ISDN service delivered on a SONET network. However, the continued reliance of customers on private network solutions will enable a service provider to immediately increase revenues as a direct result of SONET.

5. **Optical interface ("mid-span meet")**   The final benefit is one that is frequently considered to be the biggest and most important—but is really less of an issue as time goes on. In the near future, it will seem as unusual to think that fast-fiber systems were once mostly proprietary as to contemplate the fact that routers from different vendors would once not work together at all. But the "mid-span meet" is still important. The fact that SONET is a standard that defines a standard optical interface means two things. First, it is possible to deploy SONET in a multi-vendor environment with a transmitter from one vendor and a receiver from another. Second, it is even possible to have two separate service providers operating the two ends of a SONET link. The responsibility is divided in "half" at the middle of the link—hence the term "mid-span meet." In the current environment of deregulation in both long-distance and local areas, this standard optical interface is an important one. The existence of a SONET standard optical interface provides benefits in all three areas defined in Table 13-1. There is less outlay of capital because equipment procurement can easily and confidently be put out to competitive bid. Customers can do the same with customer premises equipment. Finally, service providers can potentially earn more revenue because competition keeps equipment prices down for everyone.

This list of SONET benefits could easily be extended in both dimensions, and of course it includes SDH (with slightly different details in some categories). The ones chosen for the table are those deemed to have the greatest impact. (That choice also makes it possible to avoid repeating long sections of text with slightly different terms.) The remainder of this chapter takes a more detailed look at SONET in each of the categories.

SONET MULTIPOINT GROOMING

The practice of "grooming" refers to the consolidating (that is, combining) or segregating (that is, distributing) of traffic to make the most efficient use of existing facilities. With grooming, traffic to and from customer sites is carried on lower-speed SONET links, such as OC-3, and can easily be serviced with a higher-speed SONET link. In some cases, the link may even operate at the same speed as the links to the customer site.

Although grooming is not unique to SONET, an argument could be made that *efficient* grooming is. Grooming was certainly done in T-carrier networks. In a simple form, T-1s carried on two T-3s from several customer sites could be "groomed" onto a single T-3, assuming that the total number of active T-1s was not greater than 28. Without grooming in this situation, two T-3s would be required all the way to a cross-connect location or other processing site. However, grooming was often inefficient and difficult to do in T-carrier, given the limited range of speeds available and the amount of T-carrier equipment needed to groom.

SONET, with its simple and step-like series of defined speeds, always offers numerous opportunities to groom as signals make their way through the service provider's network. Figure 13-1 shows the general process of grooming in a SONET multipoint link configuration.

In the figure, network traffic that originates at node A must travel to node D, but so must the traffic from node B. The easiest way to accomplish this is to establish two direct point-to-point links from node A to node D and from node B to node D. This was typically how the situation was handled on T-carrier networks. With SONET, however, instead of connecting two links from each site (node A and node B) to node D directly, it is easy to establish an intermediate hub site (node C in the figure) to handle both traffic streams.

Before SONET, T-carrier could also accomplish this same thing with extensive multiplexing equipment at node C. This usually meant at least one pair of T-carrier multiplexers at the appropriate speeds, manual DSX patch panels, and a digital cross-connect system (DCS). This substantial equipment requirement made it much more difficult to establish a site with adequate floor space, power supplies, and so on, for T-carrier hub sites. With T-1s and T-3s to deal with, it was not unusual to backhaul a large amount of traffic to already established sites for hubbing and grooming purposes. The term "backhaul" indicates that a service provider must take links from a local user to a remote site to provide a service or to accomplish what is desired. For example, with T-carrier links, much of the traffic from node D to node A may actually have been routed to node B, where the required grooming equipment had been located previously for other purposes. Node C—if it existed and played any role in the T-carrier network—was just

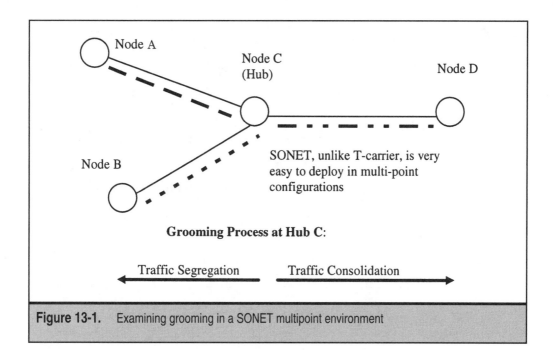

Figure 13-1.   Examining grooming in a SONET multipoint environment

another link in the chain in spite of its central geographic location. No one wanted two hubbing sites and the associated expense.

When it comes to grooming or hubbing, SONET can combine or distribute traffic at the STS-1 or VT1.5 level, right down to the individual DS-0, with ease. SONET grooming is not limited to mere manipulation of anonymous bit streams. Grooming in SONET can involve *service* combination or distribution as well. For example, with a SONET hub, video-conferencing links, data, or even voice channels can simply and effectively be handled to or from a number of sites.

Thus, with SONET, grooming is much more efficient. In fact, SONET gear is so compact that in the future no need may exist for a "central office" or other location for switching or grooming! A central office was needed to provide a controlled environment for sensitive and bulky switching equipment. Why switch every 1,000 lines with one switch when a central location with one switch could handle 10,000 lines?

SONET microprocessor-based equipment is robust and inexpensive, especially in comparison to a central office. Why not move as much of the handling of signals as close to the customer as possible? Although this may make a lot of sense, there has been little incentive—financial or otherwise—for service providers, especially local exchange carriers, to decentralize the central office. However, this may change as many former exclusively voice-oriented service providers are forced to compete with cable TV service providers for the residential high-speed Internet access market. The closer DSL central office equipment gets to the customer, the better.

REDUCED BACK-TO-BACK MULTIPLEXING

The integration of SONET equipment into compact, shelf-mounted, microprocessor-driven units has led to a revolution in the way that signals are combined onto fiber links. An array of multiplexers are no longer needed to build up high-speed bit streams to funnel onto fiber links. SONET marks the culmination of a long evolution in multiplexing equipment. With SONET, integrated devices reduce the need for the back-to-back multiplexing equipment that T-carrier required to build up fiber-optic speeds to access individual DS-1 or DS-0 streams. Equipment cost is dramatically reduced.

The SONET evolution has been a long time coming. Before SONET, multiplexing environments that combined signals onto fiber links were usually known as "fiber terminals." Figure 13-2 shows the evolution of pre-SONET fiber terminals to SONET.

In Figure 13-2, fiber-optic terminal evolution is shown in three stages, culminating with SONET. Initially, multistage multiplexing devices, each with its own footprint and

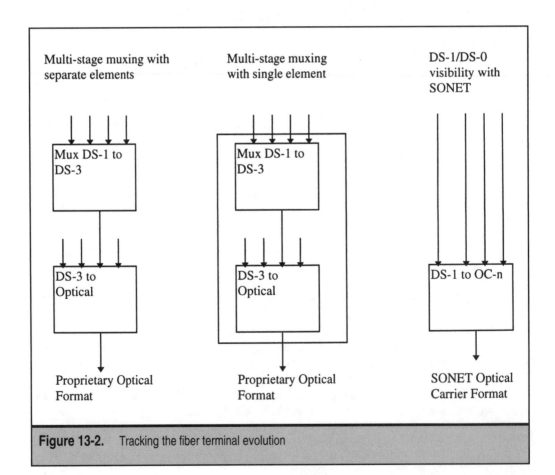

Figure 13-2.   Tracking the fiber terminal evolution

power supply, were needed to combine the standard number of 28 DS-1s to the DS-3 level. Then, several DS-3 electrical signals (the exact number varying from vendor to vendor) were combined and converted into a proprietary optical format.

The next stage in the evolutionary process was to combine the multiplexing into a single device with a smaller footprint and reduced power requirement. However, the output was still a proprietary fiber-optic signal format. Moreover, great care had to be exercised when routing the links. The provider had to avoid multiplexing and demultiplexing the fiber signal too many times in too many places for cross-connection purposes, because the total electronics package needed (with its associated capital cost), had to be installed wherever a DS-1 signal was processed. The net result was a fair amount of backhauling to the adequately provisioned sites.

Finally, on the far right portion of the figure, the SONET arrangement is shown. With SONET, not only is the equipment footprint and power requirement quite small, but the whole process is done at once with dedicated microprocessors. The optical signal produced is standard SONET, not proprietary. This prevents vendor "lock-in" and means that SONET signals can be groomed practically anywhere, not just at large switching nodes (typically, central offices). The synchronization features in SONET also mean that only the required STS or VT signals (with the needed DS-1s or DS-0s) need to be demultiplexed, not the entire bit stream, as in T-carrier.

It is this development of the SONET fiber terminal that leads directly to the reduced need for back-to-back multiplexing in SONET. A roomful of equipment has become a small unit that can be deployed in many more types of locations.

T-carrier systems are characterized and even dominated by the presence of back-to-back terminal and multiplexing equipment because the "asynchronous" or plesiochronous T-carrier hierarchy is inefficient and less cost-effective for other than simple point-to-point configurations. The contrast is stark. Figure 13-3 compares how DS-1s are accessed in a fiber T-carrier environment and in a SONET environment.

The upper part of the figure represents a T-carrier fiber environment; the lower part represents how SONET accomplishes the same task. In T-carrier, a fiber-optic termination system (FOTS) is needed to convert the optical signal to an electrical one. The FOTS is naturally a separate piece of equipment. The recovered DS-3 signal (known as a "DSX-3" signal at this point, because it flows with simple electrical coding on short cables) must next be demultiplexed to access the 28 DS-1s inside. These signals follow a format known as DSX-1, similar to the DSX-3 format in the sense that the coding is simple, and the distance, quite limited.

Once recovered, the required DS-1s are dropped or inserted (or both). These DSX-1s are typically taken to a DCS or some other switching device (as shown in Figure 13-3). The DCS is usually a separate device. The recovered and processed signals must be recombined into a single DSX-3 and reconverted to an optical signal at the outgoing FOTS.

Contrast this with SONET. Using SONET, all of the above is accomplished in one simple, cost-effective package. A SONET add/drop multiplexer (ADM) can convert the optical signal, access the bit stream, drop and insert the required DS-1s, DS-3s, or even STS-1s, and recreate the optical signal. Nothing could be simpler.

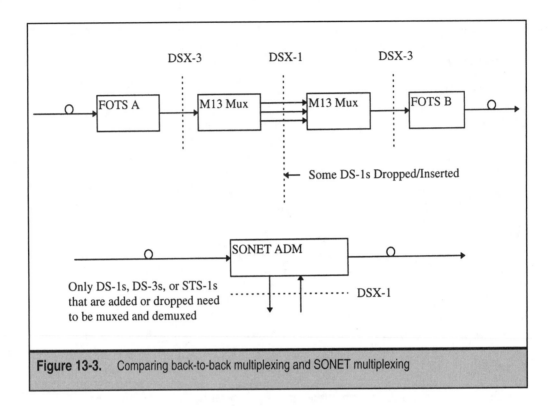

Figure 13-3.   Comparing back-to-back multiplexing and SONET multiplexing

In fairness to T-carrier, it should be noted that the SONET ADM does not actually cross-connect the required signals. In both cases, a DCS or some other switching device must be used when cross-connecting is needed. SONET would require the use of a wideband or broadband DCS (W-DCS or B-DCS). The DCS could still be a card on a SONET shelf, of course. In neither the T-carrier nor the SONET portion of the figure is the DCS device shown.

Reduction of Cabling and DSX Panels

In the FOTS-dominated pre-SONET world of T-carrier multiplexing and demultiplexing, a huge amount of equipment was needed to transport a signal from one end of a network to another. Even in extremely simple cases where an electronic DCS or switch was not needed, the expense could be considerable, because when a number of DS-3s simply need to be combined or distributed to or from several sites (the process known as "grooming"), an array of spaghetti-like coaxial cables "nailed up" on DSX-3 cross-connect panels is needed.

The setup is expensive, in terms of panels, bays, cables, and the labor needed to install them. Congested cable racks not only take up a lot of floor space, they all pose a mild hazard to employees through job-related injuries (for example, cable trips). Obviously, a financial incentive exists to dispense with the need for this labor-intensive environment when possible.

Fortunately, SONET provides the answer. Figure 13-4 contrasts the use of DSX panels for grooming with the use of a SONET hub for the same purpose. The hub is

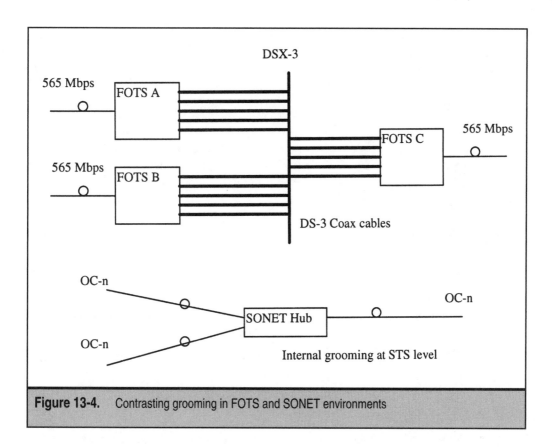

Figure 13-4. Contrasting grooming in FOTS and SONET environments

just another form of SONET ADM; no special SONET network element needs to be deployed for this purpose.

Physical cross-connect panels, such as those used in T-carrier grooming, are undesirable for a number of reasons unrelated to expense. First, the cables are exposed to environmental damage due to dust, moisture, and other hazards. Second, considerable harm can come to the cables through accidents (for example, careless cleaning personnel) or even sabotage (for example, disgruntled employees). No technical expertise is needed to cause damage by ripping cables from a piece of plywood, whether with a misguided broom or well-placed screwdriver. All in all, DSX panels are bulky, expensive, and difficult to maintain. They are an inelegant solution to a common problem.

Moreover, the most common high level of the T-carrier hierarchy was DS-3. Grooming usually consisted of trying to groom underutilized DS-3s—in terms of the number of "live" customer DS-1s they carried—with traffic from more highly utilized DS-3s. In other words, even though the fiber links might operate at 405 Mbps or 565 Mbps, the difference was in terms of the number of DS-3s (groups of 28 DS-1s) they carried, not in the speed of the DS-3s themselves. This did not make much sense, but it *had* to work that way. There was really no place to go above DS-3!

SONET hubs can combine the ADM capability with the DCS capability in the same box. This arrangement offers a cost-effective method of doing grooming *internally*, within the hub itself. The need for back-to-back equipment is reduced, except when needed to interface with existing T-carrier equipment. Of course, a SONET network element used for this purpose can still be a pure ADM, with a separate (usually preinstalled for T-carrier) DCS used to cross-connect signals at the hub site.

The availability of various OC levels also makes the whole grooming process more efficient. Instead of trying to squeeze a number of DS-1s from various DS-3s into an outgoing DS-3 or group of DS-3s, the signals from various OC-3s (for instance) can be groomed onto a single OC-12 or even an OC-48, depending on the amount of the traffic at the hub, or the number of sites served.

ENHANCED OAM&P

OAM&P (sometimes just OAM, especially to the ITU) is the art of controlling and managing an extended physical network consisting of many links in diverse geographic locations. OAM&P procedures were added haphazardly in the T-carrier hierarchy, due to a lack of any way to integrate such procedures with the link itself. The result was a system of OAM&P that had modest capabilities and many limitations.

SONET integrates and enhances OAM&P, mostly through the inclusion of dedicated overhead bytes reserved for the purpose. This section compares and contrasts OAM&P in T-carrier with OAM&P procedures in SONET.

OAM&P in T-Carrier Networks

In a T-carrier network, OAM&P was, and still is, an adventure. Separate operations systems are required to provide a centralized, single-ended maintenance group. This means that craft technicians need to be trained and well-versed in using a variety of management packages and techniques for providing OAM&P functions for FOTS equipment, DCS devices, and so on. The craft technicians must have access to a number of terminals, because no single package can monitor the network as a unified whole. If the terminal is in use, the problem has to wait.

Each vendor of a specific device (for example, FOTS or DCS) makes use of different management software. For example, any encryption devices used for secure communications links (heavily used in military and large corporate environments) must use the vendor's proprietary management software.

Even worse, a separate data-link network is used to connect the network operations center terminal to every manageable network element. Some service providers, making a virtue of necessity, extolled the "diversity" that the separate OAM&P network supposedly provided. But this was an illusion. Typically, the *links* are used for internal OAM&P, but the physical path is shared with the traffic-bearing links. When one is lost, so is the other.

In T-carrier, both the bandwidth and the types of information available to the network operations center are extremely limited. The extended superframe format (ESF) for DS-1 and the C-bit parity for DS-3 help the situation, but not much. The problem of not having enough overhead in the DS-1 and DS-3 frame structures for OAM&P purposes continues.

The limitation is not merely an annoyance. It has real impact on users and service providers alike. Because of this limitation, even a simple OAM&P function, such as performing a bit error rate test (BERT), is awkward. Suppose a customer has complained about end-system retransmissions. A possible cause is an elevated BER on the link. The customer may even suspect that a contract or tariff is being violated with regard to these errors. Performing the BERT requires taking the link out of service for the duration of the test. Frequently, the BERT shows that the link is fine. Usually, the problem turns out to be at one of the endpoints, making it the customer's responsibility. The link, however, still had to be taken out of service and tested intrusively.

OAM&P in SONET

With SONET, the art of controlling and managing an extended physical network consisting of many links in diverse geographic locations for OAM&P is much simpler than in T-carrier networks. SONET makes all of the OAM&P procedures an integral part of the SONET standard. In other words, a SONET network cannot be built without these OAM&P pieces as well. With SONET, network management operations systems are not added on; they are built in. The SONET standard uses much more bandwidth and has much more bandwidth available for OAM&P than does T-carrier. Moreover, the information available is very sophisticated.

SONET OAM&P makes use of the same physical network as the SONET links facilities themselves. At the network operations center, a local-area network is included in the SONET OAM&P standard to allow multiple terminals (actually, just PCs with OAM&P software) to access the same information instead of accessing a specific terminal, as in T-carrier. None of the OAM&P methods are proprietary in SONET.

This means that OAM&P in SONET networks is at once easier to implement and much more robust than the OAM&P procedures in T-carrier networks. One connection can reach all manageable network elements. SONET OAM&P also includes a means of remotely provisioning and configuring SONET network elements. This means that SONET networks can easily be centrally maintained, reducing travel expenses for maintenance personnel. Additionally, the substantial amounts of information available in the SONET overhead allows for much quicker troubleshooting and detection of failures before the network degrades to unacceptable levels.

For instance, SONET uses a non-intrusive BERT, rather than the intrusive type needed in T-carrier. SONET uses three distinct bit-interleaved parity (BIP) check fields. BIP checks exist for the SONET payload, as well as for the line and section levels. The three BIP fields allow for a simple and quick table lookup and conversion from BIP error counts to BER levels. Because all SONET equipment must record the BIP errors, the BER for the last hour, day, or even week can be easily determined from a central location without disturbing the link or the service to users.

Consider the customer situation outlined earlier. A customer complains about end-system retransmissions. As before, one possible cause is an elevated BER on the link. However, instead of needing to take the link out of service to perform a BER test, the technician can employ the craft interface for OAM&P to gather up-to-date BIP counts from the devices in question. The OAM&P software can then perform a simple table lookup and, in seconds, present the BERT result in both directions to the technician. This process can be made as easy as possible for the technician. Network maps, point-and-click interfaces, and graphs can make the task much more efficient.

INTEGRATED OPERATIONS AND MONITORING

Not only are the OAM&P procedures in SONET a great improvement over the methods of performing the same activities in T-carrier networks, but SONET integrates operations and monitoring capabilities. This can be a tremendous advantage when the twin concerns of operational tasks and performance monitoring are considered.

Figure 13-5 illustrates the general needs of OAM&P procedures in T-carrier. The figure shows a relatively simple arrangement of FOTS and DCS devices, with a multiplexer thrown in for variety.

The figure shows the complexity of OAM&P in a T-carrier network. In T-carrier, separate craft technician interfaces (that is, terminals or PCs) are needed to access the various types and pieces of equipment in the network. This can be a problem when, for example, two technicians are trying to troubleshoot two separate problems involving FOTS devices on the same parts of the network. Usually only one PC is capable of gathering FOTS OAM&P information from FOTS devices with one particular vendor. When one technician is using it, the other must wait—and so must the customer.

Even worse, with T-carrier OAM&P procedures, the overhead OAM&P communications can't pass through the various systems. In other words, no FOTS information can pass

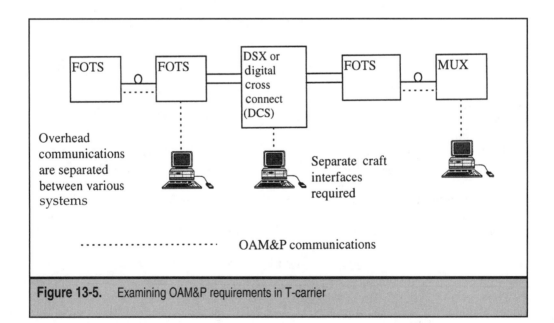

Figure 13-5.   Examining OAM&P requirements in T-carrier

through a DCS to reach the operations system in the network management center. A separate data link to each of the devices or portions of the network is required.

The overall effect is that OAM&P in T-carrier networks is disjointed and complicated. Craft technicians typically scurry around to locate the correct terminal or section of the room, wasting time and effort.

Contrast this with the SONET-integrated OAM&P shown in Figure 13-6. The SONET philosophy of OAM&P is to integrate network management functions within the SONET network.

In SONET, control of various network elements or physical portions of the network does not require separate terminals. Because the concept of the LAN is included in the operations system at the network operations center, there is no need to run around trying to find the proper terminal (which may be in use to diagnose another problem) or section of the room before settling down and trying to troubleshoot the problem. Moreover, because the standard SONET overhead is passed along from device to device on the SONET network, regardless of type, the craft technician can just as easily manage a FOTS device as a DCS from one network link to almost any SONET network device.

This philosophy makes OAM&P in SONET not only more efficient, but more cost-effective. Craft technicians need to be trained only in SONET, not in a diverse array of devices with varying OAM&P techniques and procedures.

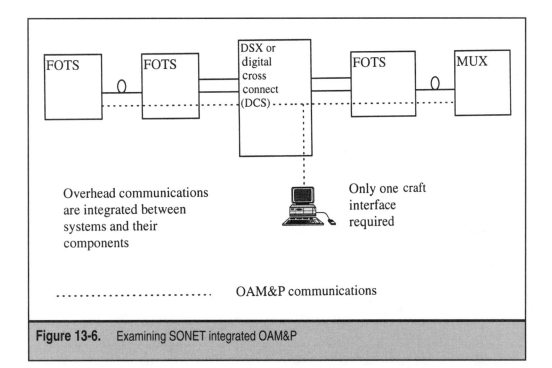

Figure 13-6. Examining SONET integrated OAM&P

NEW SERVICE OFFERINGS

The types of things that users seek to do with their networks today have been dealt with earlier in this book. They need not be repeated in detail here. Briefly, the kinds of services that users are beginning to demand in quantity all share at least one major characteristic: They require more bandwidth than ever before.

One of the most important benefits of SONET is its ability to position the network for carrying revenue-generating services unheard of before. Nothing but SONET will be able to handle the new bandwidth-hungry services and applications that users are beginning to deploy in a networked environment. SONET, with its modular, service-independent architecture and enormous bandwidth, provides vast capabilities in terms of flexibility of services. SONET supports high-speed packet-switched services, such as Internet access, LAN interconnection, and video applications. Figure 13-7 shows how SONET will enable service providers to support these new services, while supporting higher numbers of older applications and services as well. Note that the growing trend to do voice, video, or data over IP is of no concern to SONET. SONET carries IP packets, ATM cells, and other forms of data with ease.

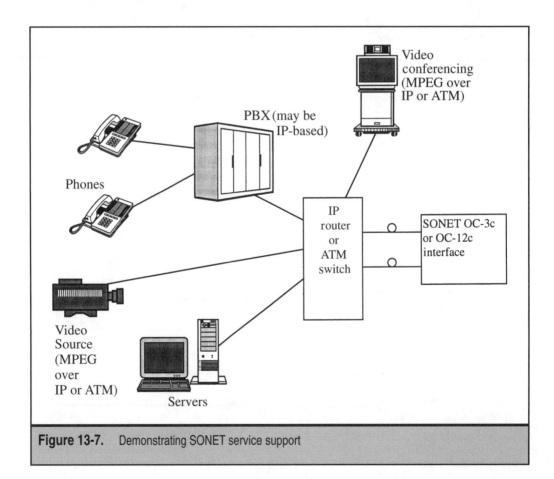

Figure 13-7.   Demonstrating SONET service support

Figure 13-7 shows a variety of devices and services that may be used by applications and users in many organizations today. Users need public or private carrier services to support these applications. In the figure, a SONET OC-3c or OC-12c transport can be used to handle the traffic generated by an ATM switch or IP router. Organizations still need voice service support from the PBX, and SONET can support these traditional services with ease. But SONET can also help to support newer services such as video (represented by the camera), videoconferencing, and high-speed LAN interconnection.

Users in large business and more specialized areas have begun to explore the potential of applications and services that involve the use of interactive multimedia. These combine the traditional use of networks to distribute data with the ability to add video or audio to the application. Advanced graphics and other niche applications, such as computer-aided design and manufacturing, also play a role. None of this rules out higher-speed pure data uses, such as bulk file transfers or remote server backup.

Many of these broadband services could use ATM, a fast packet-switching technique using short, fixed-length packets called "cells." When used to transport ATM cells, SONET (and SDH) fills the payload with a cell stream that can be generated and switched as necessary. Because of the bandwidth capacity it offers, SONET/SDH is a logical transport network for ATM. In fact, the ITU specifies SONET/SDH as the transport for ATM cells in a B-ISDN network. Even if IP renders ATM totally obsolete (as it just about has today), SONET/SDH can still be used as a transport for IP packets using POS.

Yet ATM still has characteristics that IP struggles to provide. In principle, ATM is quite similar to other packet-switching techniques—but ATM switches cells, not packets. However, the details of ATM operation are somewhat different. Each ATM cell consists of 53 "octets" (bytes). Of these, 48 octets make up the user information field ("payload"—but different from the SONET payload), and 5 octets make up the header. The cell header identifies the "virtual path" to be used in switching ("routing") a cell through the network. The virtual path defines the connections through which the cell is switched to reach its destination.

An ATM-based network is bandwidth-transparent, which allows handling of a dynamically variable mixture of services at different bandwidths, as shown in Figure 13-7. ATM also easily accommodates traffic of variable speeds.

An example of a service that identifies the benefits of a variable-rate interface is that of a video code-based service. The video signals can be coded into digital signals and packetized within ATM cells. The number of cells generated per second may vary, reflecting the variable speed. Because ATM is essentially packet switching with fixed-length packets, support for this variable bandwidth requirement is easy.

During periods of low activity within the field of view of the video camera, the rate of packet transfer decreases. Hence, an average rate of transfer may occur below the maximum capacity of the virtual path. During peaks in activity, the rate of packetization exceeds the normal maximum. This "flexible bandwidth allocation" (sometimes called "bandwidth on demand") is a feature that distinguishes ATM networks from the fixed, channelized bandwidth that SONET/SDH links provide, as large as that is.

SONET/SDH and ATM, therefore, can work hand-in-hand to make more effective use of SONET/SDH bandwidth to support more customers than when SONET/SDH is used to provide point-to-point private lines. This is because a customer having an OC-3 link (for instance) operating at 155 Mbps may use it to support three DS-3s operating at 45 Mbps each. However, each of the DS-3s must terminate at some device or multiplexer on the customer premises. The multiplexer can aggregate lower-speed signals onto the DS-3, or a single device can use the full bandwidth of a DS-3 by itself. Once all of the high- and low-speed channels are assigned, the SONET link is full.

ATM is different in the sense that all of the devices combined can dynamically share the entire 155 Mbps of the OC-3 link. The combined traffic from the low-speed devices generates a lesser number of cells, and the single device formerly linked to a full 45 Mbps DS-3 generates a greater number of cells, but all of the cells are multiplexed onto the OC-3c *based on the instantaneous bandwidth demand from those devices.* SONET/SDH helps to meet peak bandwidth demands while supporting many average demands for service at the same time. It should be noted that many of these ATM characteristics are shared by IP networks as well. But ATM has more explicit behavior in this regard. IP router-based networks often struggle to deal gracefully with traffic flows that vary widely. Both IP and ATM can form an upper switching and multiplexing layer for a SONET/SDH link, and the future favors IP over ATM.

High-speed LAN interconnection is another area in which service providers have used SONET/SDH to support new services. LANs represent a special challenge for private-line buyers. LANs are characterized by high speeds: usually at least 10 Mbps, in many cases 100 Mbps, and upwards of 1000 Mbps (1 Gbps), with 10 Gbps promised soon. However, LANs are also characterized by extremely restricted network distances, usually much less than 1,000 feet or so (although Gigabit Ethernet breaks many of these rules and sometimes seems like SONET/SDH itself in terms of distances). The challenge is to link LANs with net- work links that are long enough and yet have enough bandwidth to enable LAN users to per- form common tasks efficiently across these interconnected LANs. Leased private lines have no distance limitations, but readily available and easily affordable private lines typically top out at 1.5 Mbps, the DS-1 speed. This is a mere fraction of the modern LAN speeds and can act as a significant bottleneck when interconnecting LANs.

In some cases, a customer with particularly deep pockets may purchase a SONET OC-1 operating at 51.84 Mbps, or even an OC-3c operating at 155 Mbps. SONET speeds help to alleviate the LAN–to–private-line bottleneck. Even when a customer cannot afford the full 155 Mbps bandwidth represented by an OC-3c, the customer may be able to buy a DS-3 on SONET at a price lower than otherwise possible if the DS-3 were not delivered on SONET.

SONET will help to make DS-3s available in areas where no private lines at speeds greater than DS-1 have ever been available.

OPTICAL INTERCONNECT: "MID-SPAN MEET"

Because of varying and proprietary optical formats among vendors' T-carrier products, it is not possible to optically connect one vendor's fiber terminal device to another. For example, one manufacturer may use a line rate of 570 Mbps, while another uses a line rate of 565 Mbps. Both may package the same number of DS-3s in the aggregate signal, but the number of over- head bytes and their exact format do not match. This precludes any interconnection effort.

Figure 13-8 shows the two benefits provided by SONET optical mid-span meets.

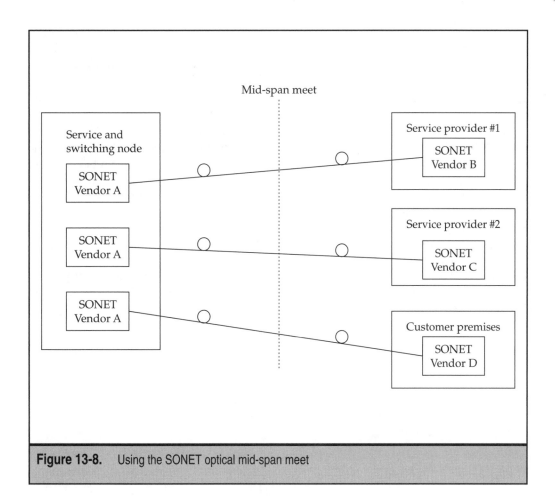

Figure 13-8.   Using the SONET optical mid-span meet

As can be seen in the figure, a major advantage and benefit of SONET is that it allows for a mid-span meet for multi-vendor compatibility. Today's SONET standards contain definitions for fiber-to-fiber interfaces at the physical (photonic) level. These low-level aspects of the SONET standards determine the optical line rate, wavelength, power levels, pulse shapes, and coding for bits on the fiber link. Above the physical transmission level, the current SONET standards also fully define the frame structure, overhead, and payload mappings. Other aspects of the standards continue to be developed to define the messages in the overhead channels to provide increased OAM&P functionality.

SONET allows optical interconnection between network providers regardless of who makes the equipment. The network provider can purchase one vendor's equipment and conveniently interface with SONET equipment from other vendors at either the various carrier locations or the customer sites. Users may now obtain the OC-N equipment of their choice and "meet" with their network provider of choice at that OC-N level.

Figure 13-8 shows how one service provider can easily decide to deploy one vendor's equipment and still interface optically with others. For instance, the two service providers use SONET equipment from other vendors. However, a mid-span meet is still possible. In the figure, a customer using a direct SONET interface (for ATM, perhaps) also can have another vendor for the customer premises equipment. The need for a mid-span meet was one of the original goals for developing the SONET standard. As formerly tightly regulated telecommunications environments around the world are turned into more competitive marketplaces, this benefit is extended to SDH as well.

It is fair to say that the benefits of SONET/SDH far outweigh the expense of deploying SONET/SDH links. This chapter presented a more or less systematic look at all of the benefits. However, it is still worth remembering that these are only the major benefits and the ones upon which most people would readily agree. As time goes on, what were once perceived as incidental aspects of SONET/SDH may emerge as major benefits in their own right. For instance, T-carrier, once perceived strictly as a way to deliver digitized voice for internal carrier use, turned out to be virtually indispensable for sales to customers interested in building private networks. It took the upheaval of divestiture in the United States in 1984 to give carriers the incentive to tariff and offer T-carrier to the general public. The success of T-carrier in this added role came as somewhat of a surprise to some who still saw DS-1 as only the vehicle to deliver DS-0 links at 64 Kbps to scattered users. Perhaps SONET/SDH will emerge as ideal for situations yet undreamed.

CHAPTER 14

SONET/SDH OAM&P

One of the most powerful features of SONET/SDH is its built-in standards for operations, administration, maintenance, and provisioning (OAM&P). Also know as OAM in the international context, OAM&P involves all aspects of day-to-day operations and fault detection in all parts of a SONET/SDH network. When things go wrong, it is up to the SONET/SDH OAM&P procedures to detect the problems and enable network operators to put them right. Even when things do not go wrong, operations involves configuration issues, performance statistics, and so forth.

It is tempting simply to equate OAM&P with "network management," but OAM&P involves more than the traditional activities associated with a network management center. SONET/SDH OAM&P is not only network management, but also customer support, trouble tracking, performance evaluation, configuration management, technical support, and billing.

Considering only SONET for the moment, the difficulties of performing OAM&P in T-carrier networks have been dealt with earlier and need not be repeated here. Briefly, T-carrier networks were characterized by inadequate bandwidth for OAM&P needs, lack of OAM&P standards, and even many key operational areas where OAM&P procedures were lacking (at worst) or haphazard (at best). This apparent oversight in T-carrier was not due to a lack of interest or attention. At its inception, T-carrier pushed the envelope of equipment performance to such an extreme that simply making the network work was enough of an accomplishment at the time. It was only after more exacting OAM&P needs became apparent that T-carrier added what functions it could. Limiting factors included the absence of adequate overhead bits to carry OAM&P information, alarms, and commands; and the inability of equipment to handle the processing needs required by extensive performance statistics and monitoring information.

SONET (and SDH) labors under no such restrictions. SONET/SDH overhead is more than adequate to handle the most sophisticated OAM&P indicators and messages. SONET/SDH equipment has internal processors that are powerful enough to deal with all internal, functional monitoring. This chapter is a systematic approach to SONET/SDH OAM&P, including all major aspects of this key SONET/SDH feature.

SONET/SDH OVERHEAD REVISITED

Repeating a discussion of the full structure of the SONET/SDH overhead bytes is not necessary here. It is enough to point out that some of the overhead bytes at the path, section, and line levels in SONET (and the path, multiplex section, and regenerator section levels in SDH) are used for OAM&P procedures. Several bytes are constantly monitored by receiving SONET/SDH equipment for fault detection; others are used to try to correct and compensate for the errors.

Although the structures of the SONET and SDH frames differ, the names of the functional overhead bytes are the same for the most part. It is enough, therefore, to examine only the SONET overhead bytes in detail, without regard to their actual position in the frame structure. Figure 14-1 shows the overhead bytes at the SONET line, section, and path levels. The bytes directly relating the SONET OAM&P are these:

▼   **A1/A2 framing bytes**   The framing bytes are monitored by receiving equipment to ensure frame alignment. Loss of these expected framing bit patterns for 3 ms results in a loss-of-frame (LOF) error condition or alarm.

■   **D1, D2, and D3 bytes**   The D1, D2, and D3 bytes comprise the SONET section (and SDH regenerator section) data communication channel (DCC) at 192 Kbps. They are used to convey network management messages between a SONET/SDH operation system (OS) and the particular section/regenerator device being managed.

■   **H1/H2 pointer bytes**   The pointer bytes are monitored by receiving equipment to locate the SONET synchronous payload envelope (SPE) and SDH containers. The values must be valid. If invalid pointer values are received for 8, 9, or 10 consecutive frames (all three values are allowed), the result is in a loss-of-pointer (LOP) error condition or alarm.

■   **K1/K2 automatic protection switching (APS) bytes**   The APS bytes are monitored by receiving equipment to determine when the SONET or SDH signal should be switched to an alternate physical path. Specific bits within these bytes are also monitored to detect line (SONET) or multiplex section (SDH) error conditions.

■   **D4 through D12 bytes**   The D4 through D12 bytes comprise the SONET line (and SDH multiplex section) DCC at 576 Kbps. They are used to convey network management messages between a SONET/SDH OS and the particular SONET/SDH line/multiplex section device being managed.

■   **S1 synchronization byte**   The synchronization byte can be used to determine the source and stratum level of the SONET/SDH timing clock. As such, it can be used for troubleshooting and reliability purposes.

■   **M0/M1 byte**   The M0/M1 byte is used to convey error count information to the sending device. At the STS-1 level, only bits 5–8 of the M0 byte are used. At the other STS levels, the M1 byte exists. At the STS-768 level, all bits of the M0 and M1 bytes are used for the counts. In SDH, the STM-1, STM-4, and STM-16 frames have only an M0 byte. STM-64 and STM-256, with more possible errors to report, use both an M0 and an M1 byte.

■   **C2 signal path byte**   The signal path byte is monitored by receiving equipment at the end of a SONET/SDH path (typically the terminal multiplexer) and is used mainly to determine if the path is used, or equipped (has a sender and receiver). If equipped, the signal path byte gives the format of the content of the SONET/SDH payload (for example, ATM cells, virtual tributaries).

▲   **G1 path status byte**   The path status byte is monitored by receiving equipment at the end of a SONET path (typically the terminal multiplexer) and is used to convey bit error rate information to the sending device at the other end of the SONET/SDH path (usually the sending terminal multiplexer).

A1 Framing	A2 Framing	J0 Trace	J1 Trace
B1 BIP-8	E1 Orderwire	F1 User Channel	B3 BIP-8
D1 Data Com	D2 Data Com	D3 Data Com	C2 Signal Label
H1 Pointer	H2 Pointer	H3 Pointer Action	G1 Path Status
B2 BIP 8	K1 APS	K2 APS	F2 User Channel
D4 Data Com	D5 Data Com	D6 Data Com	H4 Multiframe
D7 Data Com	D8 Data Com	D9 Data Com	Z3 Growth
D10 Data Com	D11 Data Com	D12 Data Com	Z4 Growth
S1 Synch Status	M0 REI-L	E2 Orderwire	N1 Tandem Conn.
Column 1	Column 2	Column 3	(Position varies)
	Section and Line Overhead		Path Overhead

Figure 14-1. Reviewing the SONET overhead bytes

Additional monitoring features and alarm conditions are built into the virtual tributary (VT) levels of SONET and virtual container (VC) levels of SDH. However, because they add little to an understanding of SONET/SDH OAM&P functions, the VT/VC-level OAM&P functions are not detailed in this chapter.

Most of this chapter makes repeated reference to the SONET/SDH overhead bytes and the conditions that they represent. It is important to realize that it is one thing for a SONET/SDH network element (NE) to realize that the framing has been lost from the upstream device; it is quite another thing to make this information known to a network management center or to automatically take action until human intervention is possible.

The very speed of SONET/SDH links makes network management especially challenging. Even if some corrective action could be taken in a second (which would be much faster than humanly possible), some 10 Gb of information (presumably user information) would be lost at the OC-192 or STM-64 rate.

WHAT COULD GO WRONG?

What could happen on a SONET/SDH network that would affect users?

SONET/SDH standards define several major failure conditions and their associated alarm indicators. The alarm indicators are used to inform SONET/SDH network operations that the failure exists. The names of the events often differ between SONET and SDH, but the essentials are the same. SDH also differentiates between the higher path or higher-order path level (the SONET path level) and the lower path or lower-order path level (the SONET VC level). SONET and SDH names for events are both given here when they differ. Most of the details about these alarms are provided by the basic ANSI SONET specification. The basic ITU SDH documents on these signals are not as detailed. The major failure types and indicators include these:

▼ **Loss of signal (LOS)**   An LOS failure occurs when a SONET/SDH receiver detects an all-zeros pattern for 19 ± 3 μs or longer, or when the power of the received signal drops below a preset threshold. LOS indicates that the upstream transmitter has failed. However, owing to the nature of SONET/SDH fiber links, the reverse path to the upstream device may still be operational. The condition is cleared when two consecutive valid frames are received. (Although loss of light (LOL) is not a separate SONET/SDH alarm, a lot of equipment also generates the LOL alarm. LOS occurs below a signal threshold, and LOL occurs below an optical receiver threshold. The LOL should therefore occur after an LOS alarm. In other words, with LOS, there could be light on the link, but not enough. With LOL, light is essentially gone.)

■ **Loss of frame (LOF)**   The absence of a valid framing pattern for 3 ms (called severely errored framing) leads to an LOF failure condition. The condition is cleared when two consecutive valid A1/A2 framing patterns are received.

■ **Loss of pointer (LOP)**   An LOP failure condition occurs in the absence of valid H1/H2 pointer bytes in 8, 9, or 10 consecutive frames, or when there are "excessive" new data frame (NDF) values. The condition is cleared when three consecutive valid pointers are received. Of course, to SONET, this is LOP at the path level (LOP-P). SDH calls it an administrative unit LOP (AU-LOP).

■ **Alarm indication signal (AIS)**   The AIS condition can occur in response to one of the earlier-described failures (LOS, LOF, or LOP). It indicates to other SONET/SDH devices that a failure condition exists. The AIS, which consists of all 1 bits in the SONET/SDH frame, is sent downstream from the NE device that has detected the failure. The link in that direction has basically failed and contains invalid information. AIS can occur at the SONET line (AIS-L) or SDH multiplex section level (MS-AIS), the SONET path (AIS-P) or SDH administrative unit level (AU-AIS), or the SONET virtual tributary (AIS-VT) or SDH tributary unit level (TU-AIS). At the SONET/SDH path level, an AIS-L/MS-AIS can trigger an AIS-P as well. The AIS-P is generated by three frames of all 1 bits in the H1, H2, and H3 bytes, plus the payload. (The SDH equivalents behave in the same way, but have more complex rules.) One frame of a valid pointer with NDF is enough to remove the AIS-P/AU-AIS.

■ **Remote defect indicator (RDI)**   RDI was formerly called the far-end receive failure (FERF). It is another type of failure indicator condition. RDI is sent upstream from the NE device that has detected the failure, because the failure may not have affected this return link. It is related to APS at the line/multiplex section level (RDI-L in SONET and MS-RDI in SDH). RDIs also exist at the path/higher-path level (RDI-P and HP-RDI) and the virtual tributary/lower-path level (RDI-V and LP-RDI). The SONET RDI-L is generated within 100 ms of an LOS, LOF, or AIS-L failure. Clearing is within 100 ms of the defect removal, as long as the RDI-L has continued for 20 frames. RDI-L is indicated by a bit value of 110 in bits 6-7-8 of the K2 APS byte. The SDH equivalents behave in the same way. The RDI-P is more complex and is carried in bits 5-6-7 of the G1 byte. Five conditions can cause a SONET RDI-P (detailed in Appendixes B and C); only two of them have standard triggers and clearing events. For an AIS-P and LOP-P trigger, the RDI-P is generated within 100 ms, must continue for at least 20 frames, and is cleared within 100 ms of the clearing trigger. RDI-P is a little different in SDH, as shown in Appendixes B and C. There is also a SONET path payload defect indicator (PDI-P) to indicate problems with virtual tributaries. It is carried in the C2 signal label byte. These values are also listed in the appendixes.

▲ **Remote error indicator (REI)**   RDIs can be sent upstream in response to several events on an incoming fiber. In contrast, REIs are produced in direct response to received coding violations detected by the status of the B2, B3, and V5 BIP bytes, and are reported upstream by the M1, G1, and V5 bytes. So an REI is just a simple report of the number of BIP errors received. REIs are found at the line/multiplex section level (REI-L in SONET and MS-REI in SDH). There are also REIs at the path/higher-path level (REI-P and HP-REI) and the virtual tributary/lower-path level (REI-V and LP-REI). This latter REI ignores the effects of the forward error correction (FEC) in the STS-48 (STM-16), STS-192 (STM-64), and STS-768 (STM-256) frames. The FEC, owing to its nature, can "compensate" for BIP errors that would otherwise count in the REI.

Each of the first three items on the list is a major failure condition. The last three items on the list are indicators of the error condition sent and received by the SONET NEs. Naturally, because the failure affects more than a single fiber link or piece of equipment, these indicators may be going off into the great bit bucket, but they must be generated anyway.

Other failure conditions are defined in the SONET/SDH standards. These include simple equipment failure, which covers a variety of situations and is more or less implementation dependent, because one vendor's equipment may differ radically from another's. (For example: Backup power supply on? Cooling fan failure?) Equipment failures fall into five categories: service-affecting (SA), non-service-affecting (NSA), critical (CR), major (MA), and minor (MN). A failure state for loss of synchronization, defined as a loss of timing source for periods of 100 to 1,000 seconds, also exists, as does failure of the protection switching and DCC. These last two failures have still not been fully defined in SONET standards.

A number of virtual tributary signals and alarms occur only when problems affect the non-SONET/SDH portions of the network, and they are not detailed here. The emphasis here is on SONET/SDH-related conditions.

The relationship between the major failure states detected and the alarm indicators raised and sent in consequence is an important one. Figure 14-2 shows, for a simple SONET point-to-point link, the relationship between the failure states and alarm indicators. For the sake of completeness, VTs are shown, although they are not discussed in detail.

In the figure, several types of SONET equipment are shown to illustrate the differences in SONET section, line, and path. The figure shows two types of paths, the STS path from SONET TM to TM, and the VT path from one digital loop carrier to another. Failure detection is illustrated by a solid circle, and the error indication procedures are represented by open circles.

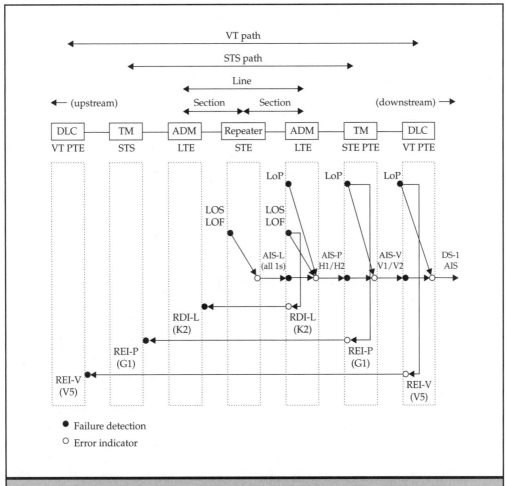

Figure 14-2.   Understanding failure states and alarm events

Note than a failure condition is detected by one piece of equipment or another and results in one or more error indications being sent both upstream and downstream on the network. In each case, the actual SONET mechanism used to convey the error indication is shown. For instance, the RDI-L is conveyed by the value of some K2 overhead byte bits, the REI-P is in the G1 path status bits, and so on. Sometimes multiple failure events result in the same indicator. For example, a SONET ADM detecting a LOS, LOP, or LOF failure may respond with an AIS-L in the H1/H2 bytes to a downstream device.

Generally, the detection of LOS, LOP, or LOF at the section-terminating equipment (STE, a SONET regenerator in Figure 14-2) or line-terminating equipment (LTE, a SONET ADM in the figure) will cause the generation of alarms on that NE's output port to the downstream NE (the right portion of the figure). An LOS detected at the repeater would generate an AIS-L consisting of all 1's to the LTE (ADM). The ADM would, in turn, generate an AIS-P to the path-terminating equipment (PTE). At the same time, the ADM would indicate an RDI-L to the upstream LTE (ADM) using the K2 byte. The other PTE devices would indicate REI-P using the G1 byte in the upstream direction as well.

A major goal in the SONET/SDH failure and indication system is to avoid "cascading" error conditions wherein a single failure can generate an ever-increasing circle of error indications. The rules for detecting failures and raising indicators in SONET/SDH try to prevent cascades. Despite of these efforts, SONET/SDH is famous for "alarm storms," where one trigger event can raise enough alarms to keep network operators very busy. As equipment is consolidated into single, large devices, this threat only gets worse. The key is to find the *lowest-level* alarm and deal with that. For example, clearing SONET line-level alarms should make all of the path-level error conditions clear up on their own. Newer network management software can be of help as well.

OAM&P SIGNALS AND LAYERS

SONET/SDH OAM&P functions include all aspects of trouble detection, repair, and service restoration. To support these functions, SONET/SDH contains a number of alarm surveillance operations designed to detect a problem or a potential problem. However, before the specific surveillance operations are explored, the terms "state," "indication," and "condition" should be defined further.

In SONET/SDH, the term "state" describes any occurrence in the network that must be detected. A SONET/SDH NE enters a specific state whenever a particular occurrence is detected and leaves the state when that occurrence is no longer detected. The detection of an occurrence may lead to an alarm condition being signaled by the NE device, which is called an "indication." An indication represents the presence of a certain condition in the NE. Indications are not always reported; rather, they made available for later access by an OS. Others indications can be reported immediately, either as an alarm or a non-alarm indication.

As mentioned previously, the purpose of the AIS is to alert *downstream* equipment of a problem that an *upstream* NE has detected. Various types of AISs are reported by the various SONET/SDH layers. Figure 14-3 shows the relationships between the SONET layers and various AIS indications. This is a SONET example, but much of what follows in this section is also true of SDH.

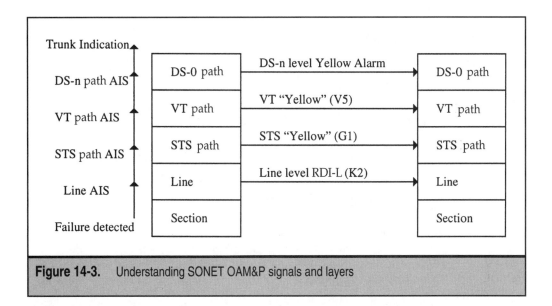

Figure 14-3.   Understanding SONET OAM&P signals and layers

Figure 14-3 shows the interplay between the two aspects of OAM&P. The higher layers and peer layers in downstream devices must both be informed of the failure condition. In the left portion of the figure, vertical arrows and their associated OAM&P indications show how higher layers are informed. The purpose of these AIS conditions is to inform the downstream SONET NE peer process of a failure. In the figure, the failure is detected at the section level (not atypical), and the AIS flow is upstream to downstream. The vertical arrows indicate that the following events are taking place:

1. The upstream SONET STE (for example, a regenerator) detects an error condition. The event triggers a line AIS, informing a downstream LTE (for example, an ADM) of the failure.

2. The upstream STE now informs the downstream PTE (for example, a TM) of the failure by generating an STS path AIS and an STS "yellow alarm," using the path G1 byte.

3. The upstream STS PTE detects the line AIS or path AIS. Once this happens, the upstream STS PTE informs the downstream STS PTE of the failure, using either a VT path AIS (in the V5 byte), a DS-3 AIS, or even a DS-0 AIS, which are also all 1's, depending on the specific payload in the STS SPE.

4. If the payload is transporting DS-n signals, then the SONET NE informs a downstream NE of the failure or termination of the DS-n path using the proper DS-n AIS.

The figure also shows some "yellow" signals between some layers. Readers familiar with T-carrier will recognize the term "yellow alarm" instantly. In the SONET world, a "yellow alarm" loosely applies to the error indications sent to the OS in response to a "red alarm," which is the failure initially detected. In SONET, the RDI-L is the SONET line-layer maintenance signal, and SONET "yellow signals" are STS and VT path-layer signals. As in T-carrier, yellow signals can be used for trunk conditioning (recovery procedures) or by a downstream terminal to inform an upstream terminal to initiate trunk conditioning on the failed circuit. These yellow signals are used for troubleshooting and trouble isolation.

The position of the horizontal arrows in Figure 14-3 illustrates the events described next. These events now flow not upstream-to-downstream, as the vertical flow shows, but rather downstream-to-upstream:

1. A downstream LTE informs an upstream LTE about a failure along the downstream SONET line level using the line-level RDI-L (in the K2 byte).

2. The downstream PTE informs an upstream PTE that a downstream failure indication has been declared along the STS path using the STS path "yellow alarm" (in the G1 byte).

3. Because virtual tributaries are in use in this example, the downstream VT PTE informs the upstream VT PTE that a failure indication has been detected along the downstream VT path using the VT path yellow (in the V5 byte).

4. Any required DS-n T-carrier yellow-alarm signals are generated either due to the failure or for any DS-n paths that are terminated using the proper T-carrier DS-n yellow alarm. A detailed description of T-carrier yellow alarms is not necessary here because the emphasis is on the SONET portion of the link.

The series of steps just described can be summarized in a simple illustration. The important events occur between SONET NEs, typically ADMs. Figure 14-4 shows a simple

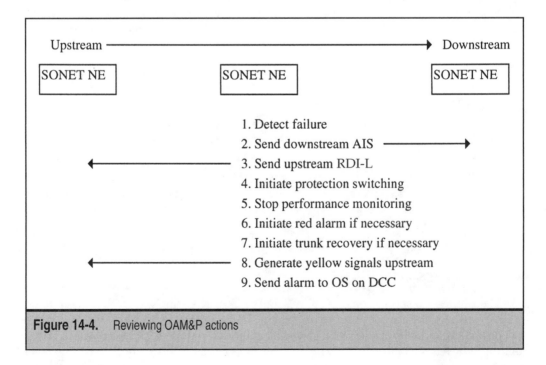

Figure 14-4.   Reviewing OAM&P actions

example of how SONET NEs react to a failure. This example is a brief summary and is not intended to show all possible situations; thus, the figure shows only a few actions from among the wide range of possibilities. The exact types of AISs, RDIs, and yellow signals sent between the NEs will vary, depending upon the nature of the failure.

Two other common OAM&P terms appear in Figure 14-4 and should be explained. They are "red alarm" and "performance monitoring parameters." In SONET, a red alarm is generated when an NE detects a failure state that persists for 2.5 seconds or longer, or when the NE is experiencing continuous but intermittent failures. Performance monitoring statistics include parameters such as the number of bit interleaved parity (BIP) check errors. The collection and recording of these performance monitoring parameters are suspended during the handling of a failure state. This only makes sense: The SONET NE has more pressing things to do, and there is no meaningful performance to report anyway.

USERS AND OAM&P OPERATIONS

In a perfect world, SONET/SDH OAM&P surveillance and monitoring would be so good that users would never have to worry about outages or detecting SONET/SDH path-level failures. In the real world, however, users still want to feel that they have some control over the network, especially because SONET/SDH is, at heart, basically just another kind of private line.

Figure 14-5 shows an example of several OAM&P information flows in relation to user equipment. The term "flow" just means that a sequence of OAM&P indicators is "flowing" between the SONET network elements. Again, this example uses SONET terminology.

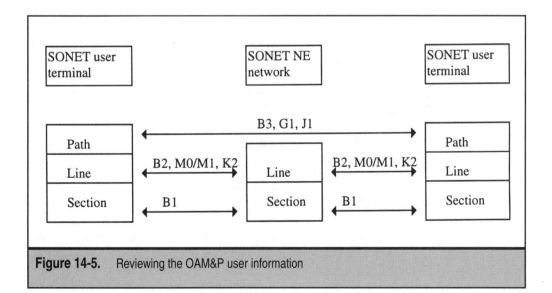

Figure 14-5.   Reviewing the OAM&P user information

In the figure, the user device terminates the SONET section, line, and path. The user thus has access to the following SONET overhead bytes and the OAM&P information that they contain:

▼ **J1** Verifies a continued connection between the two user devices at the ends of the SONET network (user must have access to the SONET TM).

■ **G1** Indicates an STS yellow signal when byte content includes a 1 value in bit 5.

■ **B3** Using even parity, gives the BIP-8 calculation results on the previous STS SPE.

■ **B2** Using even parity, gives the BIP-8 calculation results on the previous line overhead and STS-1 envelope.

■ **M0/M1** Informs the originator of the error counts from the B2 calculations using a count in the M0 octet of an STS-1 in an OC-1, or the M1 octet of third STS-1 in an OC-N.

■ **K2** Indicates an RDI-L and that the line is entering a LOS or LOF state. The RDI-L is indicated by the bit value 110 in bits 6-7-8 of the K2 byte.

▲ **B1** Using even parity, gives the BIP-8 calculation results on all bits of the previous STS-N frame.

So OAM&P in SONET involves a SONET NE monitoring and processing the arriving overhead bytes at the section, line, path, and even virtual tributary levels. Based on the overhead processing at the receiving NEs, certain *events* can occur that raise alarm conditions in the detecting NE. These events are covered generally by the blanket alarm indication signal (AIS) *state* in the detecting NE. The AIS state is used to convey events *between* SONET layers in the same NE device. (AIS *signals* convey this information to other NEs.) Sometimes, events are relayed downstream to other NEs and upstream to the sending NE (assuming that the upstream link is still capable of functioning) in other overhead bytes. In the case of section overhead, the upstream device could be a regenerator, but no events are reported to regenerators. The events typically report some error condition, but the severity of the error is not always catastrophic, as has already been pointed out.

The end result is a kind of cascading series of events generating AIS conditions all the way through the SONET layers. Now, some events can take place at the path overhead (payload) level, for instance. The SONET frames are formed and arriving just fine, but the payloads inside are hopelessly confused and error-ridden. This event will generate alarms only at the STS path and virtual tributary levels, if the payload is used for VTs.

Figure 14-6 shows the complete structure of all SONET events (except for a catastrophic and complete loss of signal) and the overhead bytes monitored to detect them. With a few exceptions, the structure also applies to SDH, but this example uses SONET terms exclusively. The whole figure is quite complex, and so a few words of explanation are in order.

At the bottom of the figure are the overhead bytes that are processed to detect the event. The event detected by each set of overhead bytes is shown at the end of an arrow. An upward arrow represents events that are detected either by processing the bits in the overhead byte(s) or by monitoring the bits within the byte for a certain pattern (which means that the event is actually occurring at the *upstream* device). The six downward arrows represent events that are reported to other SONET NEs by bit patterns within the indicated overhead bytes. The triangle symbol represents a logical process by which any

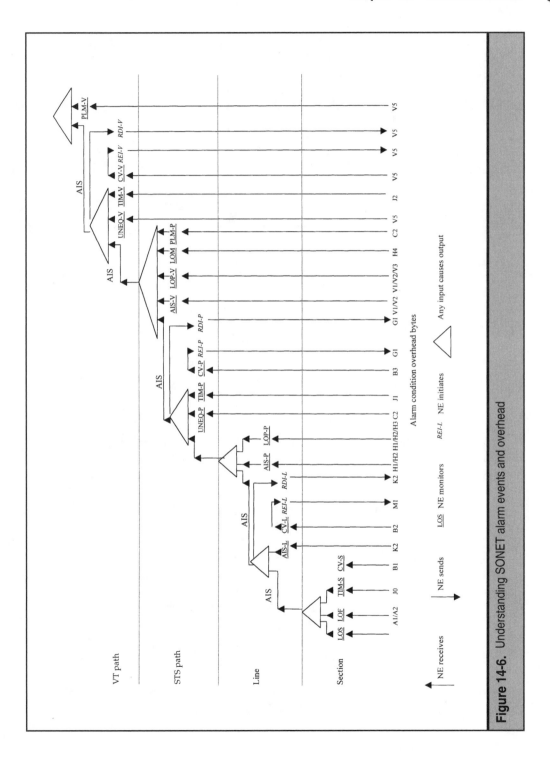

Figure 14-6. Understanding SONET alarm events and overhead

event entering any triangle symbol from below generates the actions indicated at the top of the triangle. The process is hard to talk about in the abstract, but easier to understand by examining some concrete examples.

For example, suppose that a SONET NE detects coding violations (BIP errors) at the line level based on the processing of the B2 bytes. This CV-L event will result in a remote error indication at the line level (REI-L) being reported to the sending NE in the M1 byte. Table 14-1 shows the details of how the overhead bytes are used to detect and report events.

One further, more complex, SONET example will suffice to show how SONET events and alarms are related.

Overhead Byte	Event	Meaning	Comment
N/A	LOS	Loss of signal	All 0's or below receive threshold
A1/A2	LOF	Loss of framing	Framing pattern lost
J0	TIM-S	Regenerator section trace identifier mismatch	The J0 network address of the regenerator is not what was expected
B1	CV-S	Coding violation, section	BIP errors detected in B1 byte
K2	AIS-L	Alarm indication signal, line	Bit positions 6-7-8 of K2 byte set to 111
B2	CV-L	Coding violation, line	BIP errors detected in B2 byte
M1	REI-L	Remote error indication, line	Count of B2 BIP errors
K2	RDI-L	Remote defect indication, line	Bit positions 6-7-8 of K2 byte set to 110
H1/H2	AIS-P	Alarm indication signal, path	Excessive new data flags (NDFs)
H1/H2/H3	LOP-P	Loss of pointer, path	Invalid payload pointer values
C2	UNEQ-P	Unequipped STS at path level	All 0's in C2 byte, no user information
J1	TIM-P	Trace identifier mismatch at path level	The J1 network address at the path level is not what was expected

Table 14-1. SONET events and their meanings

Overhead Byte	Event	Meaning	Comment
B3	CV-P	Coding violation, path	BIP errors detected in B3 byte
G1	REI-P	Remote error indication, path	Bit positions 1-4 of G1 report B3 error count
G1	RDI-P	Remote defect indication, path	Bit positions 5-6-7 of G1 code path events
V1/V2	AIS-V	Alarm indication signal, VT	Excessive VT pointer adjustments
V1/V2/V3	LOP-V	Loss of pointer, VT	Invalid VT pointer values
H4	LOM	Loss of multiframe, VT	Loss of 0 through 3 payload sequence in H4
C2	PLM-P	Path label mismatch, path	Payload content is not what was expected
V5	UNEQ-V	Unequipped VT at VT level	Bit positions 5-6-7 of V5 set to 000
J2	TIM-V	Trace identifier mismatch at VT level	The J2 network address at the VT level is not what was expected
V5	CV-V	Coding violation, VT	BIP errors detected in bit positions 1-2 of V5 byte
V5	REI-V	Remote error indication, VT	Bit position 3 of V5 reports any BIP error
V5	RDI-V	Remote defect indication, VT	Bit position 4 of V5 reports any VT event
V5	AIS	Alarm indication signal, VT	Alarm reported to affected tributary

Table 14-1.   SONET events and their meanings *(continued)*

Consider a SONET NE that suddenly finds that the J1 path trace byte (which contains the network address of the SONET NE that packaged the payload) does not match the address expected in the link configuration table. This change is consistent—not the result of random bit errors—perhaps because a new PTE has been turned up and is not configured properly. The receiving PTE will realize that this trace identifier mismatch at the path level (TIM-P) has occurred. This realization triggers two actions at the path level.

First, this remote defect indication at the path level (RDI-P) is conveyed in the G1 path status overhead byte to the PTE across the SONET link or links that packaged the SONET payload. The RDI-P is conveyed in bit positions 5-6-7 of the G1 byte. Now, the sending SONET NE knows that the receiving PTE does not like the network address in the J1 byte or that the path is "unequipped" (the receiving PTE is not configured to do anything with the payload contents).

Second, the TIM-P event triggers an AIS to the virtual tributary level above (as, obviously, do a lot of other things). This AIS triggers a remote defect indication at the virtual tributary level (RDI-V) since, obviously, all of the assumed VTs in the payload are affected by the J1 mismatch (or unequipped status of the receiver). Table 14-1 shows that this RDI-V is conveyed to the sender by setting bit 4 of the V5 byte in each VT. Naturally, an AIS condition is also raised to whatever device is attached to the VT at the receiver.

By using Figure 14-6 and Table 14-1, almost any SONET event and alarm can be traced from cause to reaction, and from layer to layer. This is true of even the complex messaging protocol used in the automatic protection switching (APS) K2 byte.

As mentioned earlier, most of the SONET events and alarms also correspond to SDH events and alarms. But differences do exist. For example, SDH sets the codes in bit positions 5-6-7 of the G1 path status byte in response to slightly different triggering events. Readers concerned with international paths are referred to the base ITU and ANSI specifications.

Note that the event triggered by the J1 path-level event resulted in no events at the line and section levels. This is as it should be, given that the line and section SONET NEs are happily shuttling perfectly valid (to them) SONET frames end to end along the path. The hierarchy of events is enforced from bottom to top. A section-level event therefore triggers AIS at all SONET levels above, because lines, paths, and virtual tributaries should all be informed of a lower-level failure or alarm condition. Similarly, a section-level LOF triggers RDIs at the section, path, and VT levels. As mentioned, SONET networks were plagued by "alarm storms" where one low-level event triggered an avalanche of alarms that swamped network operations centers and obscured the low-level root cause. Improved network management software and systems have better isolated the lower-level events that result in higher-level alarm conditions.

SDH events and alarms are quite similar to those in SONET. The differences are mostly of terminology. Table 14-2 shows the SDH equivalent terms for the SONET events and alarms shown in Table 14-1.

Overhead Byte	Event	Meaning in SONET	SDH Equivalent
N/A	LOS	Loss of signal	LOS (same)
A1/A2	LOF	Loss of framing	LOF (same)
J0	TIM-S	Regenerator section trace identifier mismatch	RS-TIM (regenerator section TIM)

Table 14-2. SDH equivalent events and alarms

Overhead Byte	Event	Meaning in SONET	SDH Equivalent
B1	CV-S	Coding violation, section	RS-BIP (regenerator section CV)
K2	AIS-L	Alarm indication signal, line	MS-AIS (multiplex section AIS)
B2	CV-L	Coding violation, line	MS-BIP (multiplex section CV)
M1	REI-L	Remote error indication, line	MS-REI (multiplex section REI)
K2	RDI-L	Remote defect indication, line	MS-RDI (multiplex section RDI)
H1/H2	AIS-P	Alarm indication signal, path	AU-AIS (administrative unit AIS)
H1/H2/H3	LOP-P	Loss of pointer, path	AU-LOP (administrative unit LOP)
C2	UNEQ-P	Unequipped STS at path level	HP-UNEQ (higher path unequipped)
J1	TIM-P	Trace identifier mismatch at path level	HP-TIM (higher path TIM)
B3	CV-P	Coding violation, path	HP-CV (higher path CV)
G1	REI-P	Remote error indication, path	HP-REI (higher path REI)
G1	RDI-P	Remote error indication, path	HP-RDI (higher path RDI)
V1/V2	AIS-P	Alarm indication signal, VT	TU-AIS (tributary unit AIS)
V1/V2/V3	LOP-V	Loss of pointer, VT	TU-LOP (tributary unit LOP)
H4	LOM	Loss of multiframe, VT	TU-LOM (tributary unit LOM)
C2	PLM-P	Path label mismatch, path	HP-PLM (higher path PLM)
V5	UNEQ-V	Unequipped VT at VT level	LP-UNEQ (lower path unequipped)
J2	TIM-V	Trace identifier mismatch at VT level	LP-TIM (lower path TIM)

Table 14-2. SDH equivalent events and alarms *(continued)*

Overhead Byte	Event	Meaning in SONET	SDH Equivalent
V5	CV-V	Coding violation, VT	LP-BIP (lower path BIP)
V5	REI-V	Remote error indication, VT	LP-REI (lower path REI)
V5	RDI-V	Remote error indication, VT	LP-RDI (lower path RDI)
V5	AIS	Alarm indication signal, VT	TU-AIS (tributary unit AIS)

Table 14-2. SDH equivalent events and alarms *(continued)*

TIMING PROBLEMS

Next to outright equipment failures, perhaps the most vexing thing about SONET/SDH networks is timing problems. Such problems not only affect the synchronous nature of SONET/SDH, but make it impossible to cross-connect lower-speed channels or to retrieve the information inside the SPE.

When timing problems occur, a SONET/SDH network administrator or manager can look for two possible causes. First, the NE may be at fault. Second, the timing network feed may be faulty. Fortunately, tests can be run with a variety of test sets with standard pointer-shifting patterns to determine where pointer adjustments are being made on the SONET/SDH network. With a simple strategy of working back toward the upstream NEs, the tests can reveal the location where pointer adjustments are not being made. That is, that particular NE will have proper pointer adjustments on the input side, but not on the output side.

Once the trouble area has been pinpointed, the next step would be to determine whether the clock recovery board in the NE is faulty, or whether the BITS timing has somehow failed, isolating the NE.

SONET AND T-CARRIER NETWORK MANAGEMENT

Once again, this section uses a SONET example for the sake of simplicity.

Surveillance is the practice of monitoring a network for various alarm ("alert") conditions that may arise. Alarms typically are used by NEs to notify the OS of some critical problem or condition in the device, or on the links between the devices. Just as in SONET, LOS and AIS are examples of alarm conditions on T-carrier networks.

When it came to network management, T-carrier faced all of the same problems that SONET did. But SONET greatly improved on the surveillance capabilities of T-carrier. This section explains how and why.

Obviously, it is not enough to have just one network element aware of a problem. Networks consist of many devices that are not always under the control of a single service provider. For example, a T-carrier network may have had a DS-3 link between an inter-exchange carrier and a local access provider, each with their own fiber-optic terminal system (FOTS). Presumably, the mid-span meet is not a problem, because the same vendor's equipment is used on both ends of the link between them. When a problem occurs on the inter-exchange carrier's network, it would be nice if the local access provider had a way of determining this as quickly as possible, because users will typically call the local access provider once the end-to-end bit stream is lost owing to the failure. Figure 14-7 shows how this process would work in a T-carrier situation.

In Figure 14-7, if a FOTS link were lost within the inter-exchange carrier's network, then the symptom would be the loss of the incoming bit stream to the local access provider. The link is still there physically, of course, but nothing on the fiber makes sense. Each vendor will have a proprietary "keepalive" signal between the devices to let the "downstream" device know that the link is still there, but the information bits are now gone owing to the upstream failure.

The loss of the data stream triggers the generation of an AIS signal (usually called "blue alarm" in T-carrier) to other downstream devices. This suppresses a flurry of other downstream alarms at lower levels. The blue alarm process allows the local access provider to quickly determine that the problem is not a local one.

No standard mechanism exists for informing the *upstream* devices in the inter-exchange carrier network that the problem even exists, and it is these devices that must ultimately deal with the problem. Of course, the local access provider at the other end of the link will be fielding calls from local end-users and will not have a clue as to what is going on.

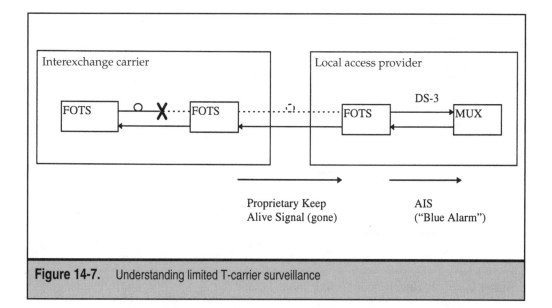

Figure 14-7.   Understanding limited T-carrier surveillance

Now consider how SONET improves the situation. T-carrier networks are limited in the types of alarms that they support (yellow, red, and blue) and in the way that alarm conditions are propagated throughout the network to all upstream and downstream devices affected. This limitation is necessary for two reasons: first, to prevent the original failure from propagating a storm of other alarms from downstream devices through the network; and second, to alert the other devices to the problem in case recovery steps need to be taken (for example, protection switching).

As illustrated in Figure 14-8, SONET is much more robust than T-carrier when it comes to network surveillance. In the figure, a SONET hub at an inter-exchange carrier site has a failed OC link from a SONET terminal multiplexer (TM A) to the SONET hub. The SONET hub combines add/drop multiplexing (ADM) and digital cross-connect system (DCS) capabilities. Naturally, the SONET hub at the local access provider's site will want to notify the downstream devices about the failure.

With the full SONET overhead to employ, an STS path AIS (AIS-P) condition is raised and passed along. The SONET TM B device will respond with a downstream DS-3 AIS. But SONET does more. On the upstream path, which is not affected by the failure, an "STS path yellow alarm" (technically the remote error indicator, REI) condition is raised and passed along to the SONET TM at A. In addition, the SONET hub raises its own "STS line RDI" (RDI-L in the K2 byte) to let the SONET TM at A know that the problem exists.

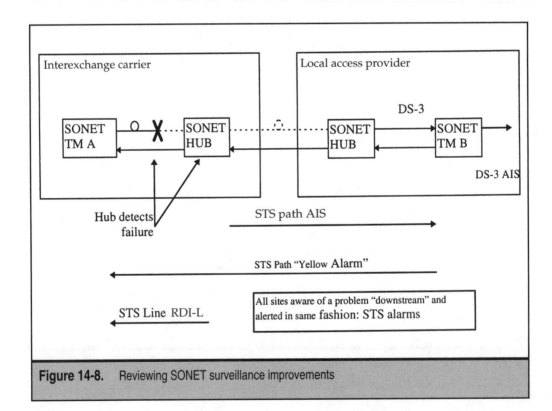

Figure 14-8. Reviewing SONET surveillance improvements

The key here is that all sites are aware of the problem upstream and downstream, and all devices are alerted in exactly the same way: STS alarms are in the SONET overhead. Note that the alarms flow between various service providers as well.

Now, SONET/SDH overhead alarms and indications are simply the tools that can be used to manage a SONET/SDH network. The actual SONET/SDH network management application and protocols are another thing altogether. Previous editions of this book included network management in this chapter on OAM&P. But recent developments in the area of SONET/SDH network management mean that the topic deserves a chapter of its own.

CHAPTER 15

SONET/SDH Network Management

S ONET/SDH networks are today in a period of transition when it comes to network management issues. The ISO and ITU-T have both been working for a number of years on the development of several open systems interconnection (OSI) network management standards. These standards revolve around the common management information service element (CMISE), which defines the network management applications, and the common management information protocol (CMIP), which is the protocol used for CMISE applications.

Unfortunately, when SONET deployment began, neither CMISE nor CMIP was in any condition to be used for production equipment and networks. Early SONET vendors were forced to adapt the standard transaction language 1 (TL1) interface from Bellcore (now Telcordia) for SONET use.

Not that vendors of early SONET equipment were reluctant to implement TL1 network management for their products. Quite the opposite. Many saw CMISE/CMIP as a needlessly complex solution to the situation of a technician sitting at a network management terminal sending messages to, and receiving alarms from, a SONET NE. And that was entirely the point of the whole TL1-versus-something-better debate: if a human being is sitting at one end, then TL1 is fine. But once SONET (and SDH) networks grow to the point where the whole network of NEs cannot be seen and managed from a single terminal with a human operator, the quest for some other form of network management becomes critical.

A little later, this chapter will consider what is needed to perform OAM&P in modern SONET/SDH networks. For some, TL1 is still the answer. For others, TL1 may remain, but only as part of an evolutionary transition to something better. (What that something better might be is still debated as well.) For now, the time has come to take a closer look at just what the original attractions and limitations of using TL1 are.

TL1 is a widely used telecommunications management protocol (although it has been assailed on all sides in recent years). Many of the advantages of TL1 in relation to its rivals mirror the advantages of the IETF's Internet protocol (IP) compared with the more full-blooded international standard protocols proposed in the 1980s. TL1 is not an international standard. However, it is vendor-neutral, simple, extensible for new features, and easy to implement. TL1 can be used to manage almost any vendor's product. As a result, service providers do not always have to buy proprietary implementations for each vendor's gear.

However, the biggest drawback of TL1 is that it was originally intended as a "man–machine language," with a human sitting at a terminal where the operations support system (OSS, or just OS in modern SONET/SDH management) is located and the managed NEs at the end of communications links. It was typical to have a separate link to each managed device. Today the need is for OSS components themselves to be computer programs that can gather information and perform routine diagnostics automatically, and then both store the results in a historical database and post the results to a web site for authorized personnel to see.

Network management languages and the protocols they use first appeared in the 1970s, once networks with intelligent (that is, computerized) NEs began to appear in force. These NEs could not only flash a red light on their outer covers when disturbed, but also actually send data messages to a central OSS. Technicians did not have to be in the same room as the equipment at all times. The OSS center could be almost anywhere. This was a much more efficient arrangement, and the human operator could even compose messages—such as configuration commands or diagnostic commands to the device—at the terminal.

By the early 1980s, the ITU had looked at this whole arena in telecommunications and established the Z.300 series for user interfaces to the human-and-computer OSSs. Like many ITU umbrella specifications, the Z.300 series was a mass of choices and options, and did not directly address interoperability issues. So, two vendors of a similarly acting and configured M13 device (for example) could be totally compliant with Z.300 and yet still require service providers with both devices to buy, use, and maintain two totally separate-looking and separately-acting network management hardware and software.

Bellcore (now Telcordia) came along in early 1984 to bring research and development skills to the newly emerged local pieces of the former Bell System—the RBOCs. Bellcore decided right away to specify a standard language to control transmission network devices (SONET/SDH was still a few years off)—mostly T-carrier. Most of TL1 was essentially done in January 1985, but pieces were added throughout 1985 and 1986. Firmly based on Z.300, TL1 embraced both language and messages to and from all telephone company devices except the circuit switches themselves. (Interoperability and standardization was much less of a problem in the highly conservative voice central office marketplace.)

The TL1 specifications had no interest in the lower-layer protocols used to shuttle the TL1 message to and from NEs. After all, a direct link usually ran from the OSS to the NE. Packets and frames were of little importance except as packages for the TL1 messages and higher-layer protocols. But in 1988, Bellcore specified CMIP as the protocol to be used for TL1 messages. No one really worried, because there were TL1 implementations that used CMIP, X.25, or even IP as the protocol package for the TL1 messages.

There was nothing wrong with using TL1 for SONET. After all, TL1 worked. However, TL1 suffered (and suffers) from several limitations that are a problem in modern networks.

First, TL1 defines a human–machine interface. This means that TL1 expects a human being to be sitting at a terminal in an operations center somewhere, typing commands and looking at alarms.

It is not easy today to realize how awkward the type-and-read interface is. SONET NEs used to ship with about three linear feet of documentation. Most of it went into a bookcase or cabinet and stayed there. But one document was never far out of reach: the TL1 command book for the particular device.

Figure 15-1 shows what a typical SONET alarm looks like on an OSS console. The message shown in the figure reports a loss-of-pointer (LOP) condition at an ADM on a SONET OC-12 ring. Of course, a command to the SONET NE in distress had to be composed, formatted, and properly typed using just as arcane a format. Some point-and-click front ends were built for TL1-based OSSs, but all they did was interpret and formulate the TL1 on behalf of the human operator. Besides limiting the range of possible TL1 messages to the programmed ones, these graphical front ends did not enhance TL1 in any way.

In the simpler world of T-carrier, this was not a particular problem. However, in the SONET/SDH world, things happen faster. Not only are line rates higher, but protection switching should happen in milliseconds. SONET/SDH would be better served by machine–machine interfaces. A management process (or program) running in a SONET/SDH

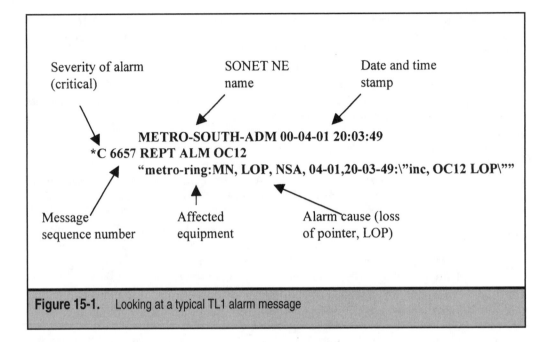

Figure 15-1. Looking at a typical TL1 alarm message

NE could then communicate with a management process (or program) running on a computer in the network management center. Decisions could then be automated to a large extent, and the whole task made more efficient. CMISE includes such interfaces.

The second problem is that TL1 is a Telcordia and T-carrier specification. As international links become more common, a transition from TL1 to some form of CMISE would be welcome. Indeed, this is happening now. CMISE/CMIP use is also the foundation for the ITU's standard for telecommunication management networks (TMNs). TMNs will be discussed more fully a little later. However, this chapter needs to further explore some aspects of TL1, owing to its position in SONET networks today.

As if TL1 and CMISE were not enough, a third contender for network management protocols emerged. The activities of the Internet Engineering Task Force (IETF) represent a major effort into network management standards. The current importance of the Internet need not be commented on here. The emphasis is on network management and the Internet.

Initially, the Internet was organized and built through the ARPANET research project that originated in the United States. In 1971, the Defense Advanced Research Project Agency (DARPA) continued the work of the earlier organization. DARPA's work in the early 1970s led to the development of the "transmission control protocol/Internet protocol," known as TCP/IP. Early network management techniques on the Internet were proprietary and quite spotty. In the last ten years or so, the Internet has assumed the lead in setting standards for network management for all devices, not just Internet routers and other components. This is the "simple network management protocol" (SNMP).

SONET vendors and service providers are faced with the difficult decision of sticking with TL1 or migrating to something else. But what else? SNMP or CMISE? Both are open standards, but only CMISE is currently an official ISO and ITU-T international standard. Yet, SNMP is much more widespread, especially as the Internet has grown around the world. Recently, proposals have come forward to implement IP as the transport network protocol of choice in the SONET/SDH DCCs. The latest proposals will be discussed later in this chapter, because IP can transport management information for TL1, SNMP, and even the latest SONET/SDH NEs with ease.

Of course, the key goal of all network management standards is to develop an integrated set of procedures and standards that will work equally well across various vendors and products. Both CMIP and SNMP offer enhanced network management services beyond the simple alarms and indications of the basic SONET/SDH OAM&P. Both CMIP and SNMP can run inside the SONET/SDH DCC bytes to actually try to manage failed equipment. For example, DCC can be used to run a remote diagnostic on the failed device and to report the results to the network manager.

Both SNMP and CMISE use one other component that is essential for all standard network management systems. This is called the management information base (MIB). Despite its name, the MIB is a *database of network management information* that resides in the managed device. Technically, the MIB is just a description of the database fields and contents. When implemented in a piece of equipment, the MIB becomes the managed object, but the term MIB tends to remain the preferred name for the whole object, not just the piece of paper describing the object. "MIB" therefore defines the content and structure of a database, common among similar network devices, that provides information about the managed network elements.

With either SNMP or CMISE, the managing process (or the management workstation) is responsible for directing the actions of the managed system. The managed system contains a remote manager as well as the managed objects defined by the MIB. This remote manager is called the "agent." The agent is software that is responsible for receiving the network management messages from the managing process and for making sure that proper access control measures are taken regarding the managed objects (so that no one can hack into the device). The agent controls the logging ("recording") of network management information as well. Through the use of a process known as "trapping" in SNMP, or as an "event-forwarding discriminator" in CMIP, the agent makes decisions about whether messages are to be returned to the managing process. The agent software typically resides in a SONET/SDH NE.

Although a one-to-one relationship may exist between a managing process and a managed system, in the sense that the managing process always manages all managed systems, the managing process itself can also be an agent (managed) process. No restriction applies to the roles that these two entities play. In fact, the roles can even be exchanged. Most commonly, a device with modest processing capabilities and a small agent will be managed by another process which, in turn, is an agent managed by the network management system. This arrangement is called "proxy agent" because the device is not directly managed by the network management system.

OSI AND SNMP NETWORK MANAGEMENT LAYERS

Although OSI and SNMP network management standards share many features, such as agents and MIBs, they are quite different in implementation. The keyword in SNMP is *simplicity.* Ironically, the simplicity in SNMP that led to its widespread acceptance eventually proved to be a liability. SNMP had little in the way of security, authenticity, or even reliability. These features have been added to form an SNMP version 2 (SNMPv2), but the effort has been marred by political infighting among Internet factions and so has languished. The net effect was to turn SONET/SDH vendors towards OSI and CMIP.

OSI Network Management

If SNMP seems overly simple for complex network management tasks, then OSI seems overly complex for simple network management tasks. The OSI network management model is consistent with the overall OSI application layer architecture (layer 7 of the OSI reference model).

In addition to defining a series of MIBs (managed objects), OSI also provides several standards for the use of these objects in widely used management areas. The OSI management standards help in defining standard ways to solve common network management problems. Several system management functions (SMFs) are designed to support the more generalized system management functional areas (SMFAs). These include fault, configuration, accounting, performance, and security management. Each of the SMFs, together with the standard objects, can contribute to one or more of the functional areas.

Several standards address the set of SMFs. They range from low-level object management to more involved application areas, including security, accounting, and performance. They provide a useful framework for the use and application of the base MIB object classes, which are defined in the X.721 standard. The standardized SMFs include these:

▼ **Object management (X.730)**   These basic object manipulation services consist primarily of the CMIS functions, and do not contain extensive object models.

■ **State management (X.731)**   These standard mechanisms for managing object states include state monitoring, state-change notifications, and state-change commands.

■ **Relationship management (X.732)**   The relationship management standard defines ways in which managed objects relate to one another.

■ **Alarm reporting (X.733)**   The mechanisms for alarm reporting include mechanisms for alerting objects that an alarm has been detected, and defining the characteristics of the alarm. They can be used to better understand the nature of the condition causing the alarm.

■ **Event management (X.734)**   The event management set of mechanisms controls the distribution of event reports within the management system.

■ **Log control (X.735)**   Log control provides mechanisms to manage the recording of management information.

- ■ **Security alarm reporting (X.736)**   Security alarm reporting defines services for managing event-forwarding discriminators that are responsible for managing the flow of security-critical alarm notifications.

- ■ **Confidence and diagnostic testing (X.737)**   The testing standard defines services and objects used in managing the execution and reporting of system tests.

- ■ **Summarization function (X.738)**   The summarization function provides a framework for defining, generating, and scheduling summary reports of system information.

- ▲ **Workload monitoring function (X.739)**   This function manages and controls workload monitoring services.

Additional ISO standards have been developed to address management functions, including testing and metrics. Additional SMFs may be included in the already significant list of standards as the technology matures.

The object management SMF defined in X.730 identifies the services that a management entity should provide for handling system objects. These services have been mapped into two sets. The first are "pass-through" services that are directly mapped from the CMISE services. The second are "direct" services that provide additional services beyond the basic CMISE services. The "pass-through" services map directly to the MIB object creation, deletion, action, set, get, and event report primitives in CMISE. The "direct" services consist of additional reporting services, and are specified to assist managers in tracking the changes that can occur to objects within a managed system.

Obviously, OSI network management is quite comprehensive, although somewhat complex, especially when compared with SNMP. The attraction of CMISE for SONET is the ready implementation of an international standard that applies to SDH as well as to SONET.

Of course, these standards all rely on the underlying OSI application-layer standards. Figure 15-2 shows those standards.

In the figure, the systems management application service element (SMASE) creates and uses the protocol data units (PDUs) transferred between the management processes of the two devices (manager and managed). These PDUs are called management-application PDUs (MAPDUs). The SMASE messages have to do with the five OSI SMFAs of fault, performance, configuration, accounting, and security operations. The CMIP protocol supports these messages through the use of a common set of procedures (for example, Get, Set, Create).

The SMASE component may use the communications services of various application services elements (ASEs) or the CMISE. The use of CMISE implies the use of CMIP with either ROSE or ACSE, which are just other elements of the full OSI application-layer standards.

The purpose of the ACSE is to set up an association ("session") between the managing process and the agent. Once the session has been established, ROSE is used by the SMASE and CMIP to invoke operations and exchange information between the managing process and the agent.

Figure 15-3 shows an example of the OSI SONET/SDH message flow for OAM&P operations (using only SONET for simplicity). Here, a managed system (usually the agent

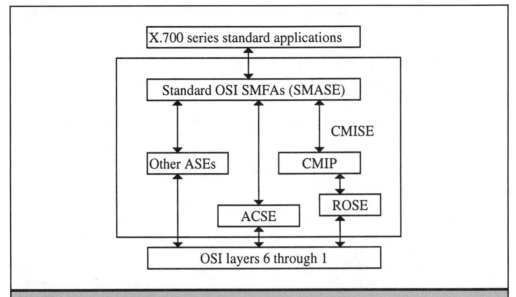

Figure 15-2.   Understanding the network management layers in OSI

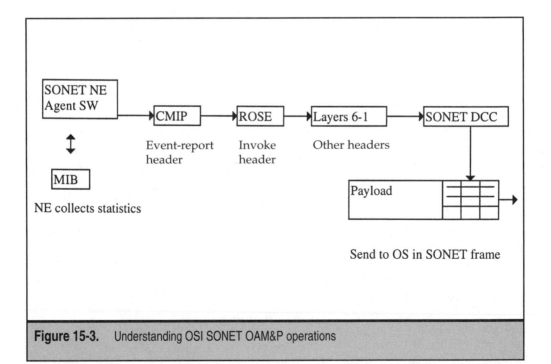

Figure 15-3.   Understanding OSI SONET OAM&P operations

software in a SONET NE) is prompted or is configured to periodically send some performance statistics to the OS at the management center. The statistics are application-independent; what the SONET link is transporting is therefore immaterial. There are reports on traffic conditions at the NE, error reports, and so on. Next, the statistics are passed to CMIP, which creates a header to indicate just what type of network management message is being conveyed to the peer CMIP process at the OS site. This header also informs the companion CMIP how to process the traffic—in other words, what type of message CMIP is to generate as a result of the network management message. In the example in the figure, an "event report"–type message is generated.

Next, the performance information and the CMIP header is passed to the ROSE process, which creates an Invoke header and appends this header to the data unit originated from the upper layers (in this case, the performance data itself and the CMIP header). This header is used by the peer ROSE process at the OS site to determine its actions. The action could be to report back to the SONET NE about the success or failure of the operation at the OS, or other possible actions.

Next, the information is passed down through the lower OSI layers of the SONET equipment. When the SONET layer receives the information, it places these bytes into the DCC bytes of the SONET header. It could also use the payload of the SONET envelope, but the DCC method is preferred and more common, so that the payload area is kept available exclusively for user traffic. Finally, the information is sent to the OS management process in a series of SONET frames, because each DCC (line or section) accommodates only a few bytes per frame.

At the OS computer, the entire encapsulation process is reversed. The various layers of headers and data units are passed up through the OSI layers to reach the performance and management process itself, which processes the data in accordance with some standard SMFA application-specific requirement.

SNMP Network Management

Compared to OSI, the layering for SNMP atop the Internet suite is much simpler than the OSI suite. The SNMP protocol forms the foundation for the Internet network management architecture. However, the network management applications (the Internet equivalent of the OSI SMASE) are not defined in the Internet documentation and specifications. In SNMP, these applications consist of vendor-specific network management modules such as fault management, log control, security, and audit trails. It is somewhat ironic that SNMP is everywhere but has no standard network management applications, while OSI network management applications are standard and full but OSI implementations on which to run them are difficult to find. Figure 15-4 shows the utter simplicity of the SNMP architecture.

As shown in the figure, SNMP is used by the custom-written (usually by the SONET/SDH equipment vendor) network management application and uses the (connectionless) user datagram protocol (UDP). The wisdom of using a connectionless protocol for network management has repeatedly been questioned. Much can be learned simply from whether a connection to a network device can be made or not. SNMPv3 (SNMPv2 suffered from many problems) allows the use of TCP, which is connection-oriented. In

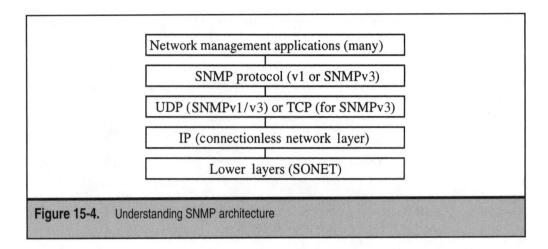

Figure 15-4. Understanding SNMP architecture

most cases with SNMP, UDP is used, and IP packets form its messages. IP uses the lower layers of the Internet protocol suite, which correspond to the data-link layer and the physical layer of OSI.

When SNMP is used for network management, its function (along with some other software) is to keep track of the ongoing SNMP operations (get/set) between the agent and manager processes. UDP directs the messages to and from the network management ports, and IP packet addresses are used to identify and route the traffic being exchanged between the two devices.

The figure merely shows some generic "network management applications" running on top of SNMP. None of these is defined by Internet standards, but nothing would stop an enterprising vendor from rewriting all of the SMFA standards for SNMP. To date, however, most vendors have chosen to gear their network management products closely to their specific products, making them difficult to use in a multi-vendor environment. This situation will slowly change as OSI network management becomes more common in SONET/SDH devices.

The interactions of the SNMP layers with respect to SONET are almost identical, if simpler, than the OSI sequence. In this case, SONET maps the SNMP management traffic instead of OSI into the DCC fields or the payload area for transmittal to the management process or OS site.

STANDARD OBJECTS AND SONET/SDH

It should come as no surprise that standard technologies such as SONET/SDH demand standard network management techniques such as SNMP and OSI. The major benefit of any standardized network management package is the ability to identify, in an unambiguous manner, any SONET/SDH network element. Actually, this is true of *any* network element. The network management standard must therefore include an associated naming and identification registration program.

The concept of a registration program simplifies internetworking between various network components that need to know about the locations and identities of other network components (such as network management software applications). The concept is similar to the telephone system: a standard, worldwide agreement exists on a hierarchical numbering scheme. (After all, a telephone number is just a network address of a device on the voice network.)

Both SNMP and OSI use a standard object "tree" for designating devices and the MIB objects that they contain. ISO maintains the international registration and naming hierarchy. Vendors can apply for a "branch" in this standard naming and numbering scheme. Once registered, a customer with any standard network management package, SNMP or OSI, can access a SONET/SDH network management database MIB. The objects (fields) in the database are easily obtained by using the names (the object identifiers) that are registered and known by the network management software vendor and, thus, by the customer as well.

OAM&P, the ITU, and SONET/SDH

SONET/SDH OAM&P is more than failure conditions and alarm indications. Moreover, SONET/SDH network management is more than SNMP or OSI protocols. What should be done with the information gathered? What exactly is the OS doing in a SONET/SDH network? This section looks at the overall architecture for SONET/SDH OAM&P at a higher level.

The overall architecture for OAM&P communications is defined by the ITU, an agency of the United Nations based in Geneva, Switzerland. Figure 15-5 shows this architecture, known as the telecommunication management network (TMN) architecture. TMN provides

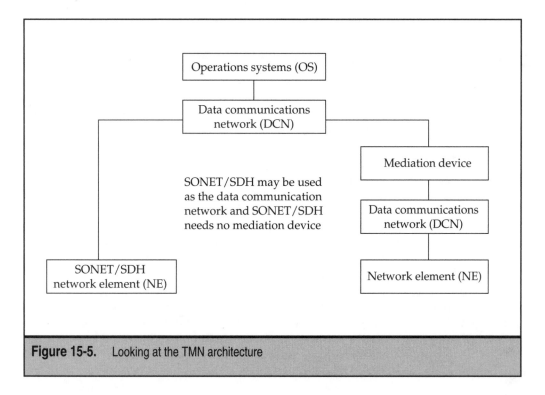

Figure 15-5. Looking at the TMN architecture

a framework for an OS at a network operations center to communicate with the physical network, and a way for alarms to be sent to the OS. The OS, a sophisticated collection of software applications, manages the entire network.

An OS can gather information from the network, along with alarm, status, and performance statistics. The OS can perform remote testing and network configuration tasks. An OS may also maintain information about facilities available for provisioning and may be used to provide resources by sending remote commands to enable or disable links.

As shown in Figure 15-5, the OS communicates with the managed NEs either directly over a data communications network or through a mediation device. A mediation device can support multiple functions (such as consolidation of data links for devices in a specific geographic portion of the network—east coast/west coast, for example), convert between proprietary OAM&P techniques for different devices, and handle a collection of performance statistics into a database.

The advantage of SONET/SDH is that *no separate data communications network is needed*, because all SONET/SDH OAM&P information flows on the network itself in the SONET/SDH overhead bytes. Because SONET/SDH is entirely standard, *no mediation devices need be used*, unless the service provider wishes to continue to provide for separation of OAM&P functions (east coast/west coast). This paragraph describes an ideal situation, of course, and assumes that the entire network management protocol stack complies with TMN. Unfortunately, network management for SONET/SDH devices has proven to be in a constant state of flux, as vendors equip the NEs with the latest modules based on the latest SNMP applications and even more Internet-related software layers than ever before. All this does is make the use of IP over the DCCs look more attractive than ever before.

TELECOMMUNICATIONS MANAGEMENT NETWORK

TMN is a huge, sprawling topic all on its own. Only those pieces of TMN of direct concern for SONET/SDH are discussed here. The discussion is not intended as a TMN tutorial, of which many are available both in print and on the Internet and Web. It is enough to note that the most important component of TMN for managing SONET/SDH NEs is known as the Q3 (often, "q3") interface. The Q3 interface is used for communications between a TMN-compliant operations system (OS) and various other components of TMN. Only devices equipped with a Q3 interface can be directly managed by the TMN OS. The Q3 interface specifies that TMN messages using CMISE are sent between NEs using CMIP. In other words, TL1 has no place in Q3.

So, how can legacy SONET/SDH NEs that still understand only TL1 be managed? Two ways are possible. First, the SONET/SDH NE can employ a Q-adapter (QA) which understands Q3 on the OS side but still understands TL1 internally. A QA could therefore be placed in each and every TL1-managed SONET NE to make it compliant with TMN. This is usually a combination hardware and software upgrade (sometimes, a firmware upgrade) that many SONET/SDH vendors have made available for their NEs.

Second, the Q3 interface can terminate at a TMN mediation device (MD). Not only can the MD employ TL1 to directly manage non-TMN compliant NEs with a QA, but the MD can also be used to gather regional information about a cluster of SONET/SDH NEs (perhaps a SONET/SDH ring). The MD can therefore act as a kind of proxy for the TMN

OS. The MDs support Q3 on the OS side and can use various Q*x* interfaces to communicate with QAs or NEs directly. The important thing is that any non-Q3 interface be shielded from the view of the OS itself. The use of these interfaces is shown in Figure 15-6.

Some SONET/SDH vendors began experimenting with MDs as well as QAs for their NE devices. However, the MD became a real issue, because the presence of the MD introduced a massive database-and-associated-processing burden without offering any real incentives. MDs complicated and congested the network. Many vendors therefore emphasized the direct QA approach for TMN compliance. QAs exist today for all four of the common variations employed to convey TL1 messages to and from SONET NEs, as shown in Figure 15-7.

In spite of the complexity of QAs in and of themselves, the emergence of TMN at least has offered the potential for service providers to buy SONET/SDH NEs from different vendors using different transport protocols for TL1 messages and to manage them all through one OS.

It is important to realize what the SONET/SDH DCCs really do for network management. Because of the presence of a full OSI stack in each NE, the SONET/SDH network becomes more or less a router network for network management purposes. The routing tables in each SONET/SDH NE are all updated by a routing protocol, so that the SONET/SDH network (for example, a SONET ring with attached SONET TMs) maps its

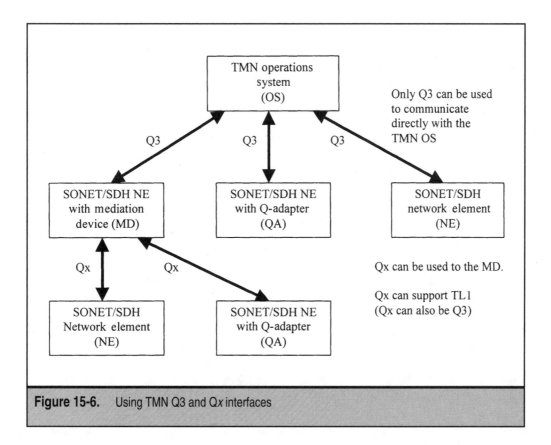

Figure 15-6.   Using TMN Q3 and Q*x* interfaces

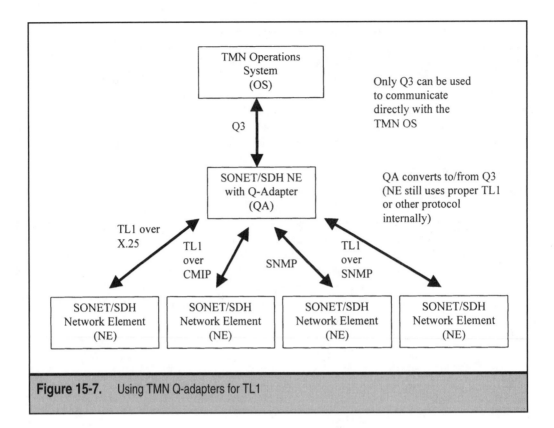

Figure 15-7. Using TMN Q-adapters for TL1

own topology automatically. All the OSS (or OS, to be TMN-compliant) operator needs to do is inject a message into one SONET/SDH NE directly. The DCCs then take care of routing the message to the proper NE. The same goes for the reply. Figure 15-8 shows how a SONET/SDH ring could be managed today.

The figure shows a simple ring with two ADMs located in a central office (CO), and three other ADMs on the ring itself. The remote ADMs have TMs attached to gather and groom customer traffic. Each CO-based ADM has a network management module with a common 10Base-T Ethernet interface. Two network management workstations are attached to the LAN. This arrangement provides redundancy for workstations and ADMs alike.

Now, suppose that an operator needs to reconfigure the VTs on the TM in the upper right of Figure 15-8. No other links (or direct links) are needed with SONET. The operator simply points to and clicks the icon of the TM on the screen. The proper TMN or TL1 message is packaged up ("put yourself into configuration mode," perhaps) and is sent across the LAN to the ADM that is active for network management. (No need exists to duplicate the message.) This ADM has a complete routing table to shuttle such messages around the ring. The message is inside a packet, just like everything else today. The message inside the packet is routed onto the ring in the DCC (because the TM is a

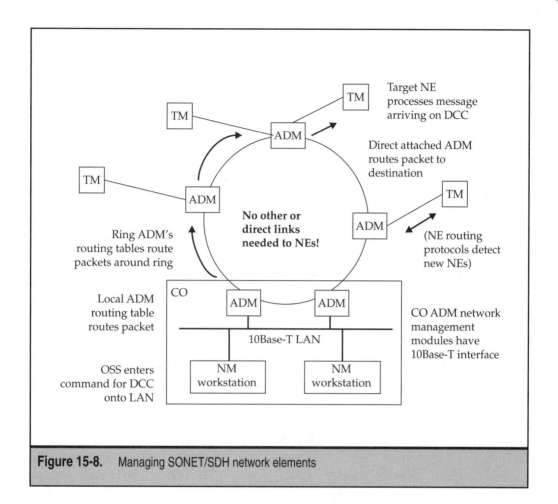

Figure 15-8.   Managing SONET/SDH network elements

line-level device and, even though the complete operation will affect the VT paths, the line DCC is used). Intermediate ADMs have their own DCC routing tables, and so the packet is passed through to the ADM directly attached to the target TM. Once the TM recognizes its own packet address, the packet is parsed, and the message is retrieved. Any reply is routed through the collection of NEs in the same way. Routing protocols running between the NEs take care of automatically detecting and mapping the topology of newly installed TMs and ADMs.

Nothing would stop the LAN shown in Figure 15-8 from being extended to other rings and locations in a variety of ways. A remote OSS could therefore easily gain access to the SONET/SDH ring as well. But local management is often provided by a standard and common 10Base-T Ethernet LAN interface built into the network management module of at least one of the SONET/SDH NEs. Both network management and access to the ring should be redundant.

Security is not much of a concern, at least not to the extent seen on other packet networks such as the Internet, because users have no access whatsoever to the SONET/SDH overhead bytes. Remote access ports on the management LAN *are* vulnerable, though, and steps must be taken to reduce the threat of unauthorized access by that means.

SONET Management Information: 1/7/24

All SONET equipment, whether repeater or terminal multiplexer (TM) or add/drop multiplexer (ADM), must comply with certain Bellcore (Telcordia) network management standards for SONET. SDH has its own similar specifications, of course. This section uses only SONET as an example.

All SONET equipment must gather network management information and performance statistics for the previous hour, previous 7 hours, and previous 24 hours (known as "1/7/24"). This information is stored in the equipment itself and is gathered by some centralized network management center on a periodic or as-needed basis. What follows is basic management information. The amount of information that SONET NEs must gather has been extended. Interested readers are referred to relevant Telcordia and ANSI specifications for current details, which have grown as equipment sophistication and power have increased over the years.

What are SONET network managers able to monitor in this fashion? The information goes above and beyond alarm conditions, of course, which are usually the triggers for some network management activity that gathers performance statistics.

At the photonic level, network managers can check the current laser bias relative to some initial, tuned installation value. Lasers tend to "age." They need periodic preventive maintenance, usually every six months or so, for continued acceptable performance. The received optical power also must be monitored to detect a sudden drop, which could indicate either tampering or damage to the fiber.

At the section level, there are three things to monitor. First, the section BIP is checked for coding violations (CVs), which are the "bipolar violations" of SONET. Up to eight CVs may be recorded per BIP, of course. Next, out-of-framing (OOF) seconds are recorded. These are seconds during which framing is lost. Naturally, the framing will normally be lost only for a fraction of a second. Finally, the number of errored seconds (ESs) is tracked, which is defined as seconds with more than one CV.

Most of the monitoring action takes place at the line level. In addition to the CVs and ESs (defined this time on the line BIP), the line level records severely errored seconds (SESs), which are defined as x number of CVs in a second (where x is greater than 2, but not universally defined). In addition, the line level watches the H1/H2 pointer justifications, which cannot exceed one in every four frames. All of the preceding are also stored in SONET equipment for the previous six days, showing their importance.

The line level also records unavailable seconds (UASs), during which the SONET equipment is simply out of action. Degraded minutes, which are minutes containing one or more SESs, used to be monitored, but are no longer. Finally, the time during which the SONET link or ring is "protect switched" is recorded.

At the path level, SONET equipment keeps many of the same statistics as are kept at the line level, with a few differences. CVs, ESs, SESs, and UASs are based on path overhead BIPs, of course. Instead of watching H1/H2 pointer justification, the path level keeps track of VT pointer justifications when the VTs are allowed to float as the SPEs are allowed to. (Usually, VTs are locked.) Also, the path equipment does not record "protection switched" duration.

Table 15-1 summarizes all of the foregoing information. Many SONET NEs today are capable of capturing and storing much more information, as already noted. The table contains only the basics.

All of the SONET equipment network management information, from coding violations to protection switch duration, is useful only when made available to network managers in the network management center. The problem is to allow network managers to query the SONET equipment in some standard fashion, regardless of vendor or service provider, to gather the information either periodically or as needed. The results of the queries may be used to diagnose problems, verify tariff compliance, or just to maintain a historical database.

Fortunately, the section and line DCCs have been set up by ANSI and the ITU-T for exactly this purpose. The SONET section DCC (D1–D3 bytes) operates at 192 Kbps, and the SONET line DCC (D4–D12 bytes) operates at 576 Kbps. Both DCCs are supposed to follow the full seven-layer OSI-RM at all layers. Many of these standards exist in both OSI and ITU-T "flavors," often referencing exactly the same text with different titles. Of course, unique pieces do exist.

For instance, at the data-link layer, the DCCs employ the ITU Q.920/Q.921 link-access protocol for the D channel (LAP-D). The default option is to be "acknowledged information transfer service" (AITS), which will perform error recovery of each SONET link; however, implementers are required to offer unacknowledged ITS (UITS) as an option. This helps to speed things up on the DCC.

Information Type	Photonic Level	Section Level	Line Level	Path Level
Laser bias	Yes			
Optical power received	Yes			
Coding violations		Yes	Yes	Yes
Out-of-frame (OOF) seconds		Yes		
Errored seconds (ESs)		Yes	Yes	Yes
Severely errored seconds			Yes	Yes
Pointer justifications			H1/H2	VT
Unavailable seconds (UASs)			Yes	Yes
Protection switch duration			Yes	

Table 15-1.   SONET Management Information

At the network layer, the DCCs employ ISO 8473 connectionless network layer protocol (CLNP), which is the ISO "version" of IP. DCC information is independently routed network by network. The transport layer uses ISO 8073/8473, which describes a connection-oriented protocol similar to TCP. In fact, it describes a method of supporting the ISO transport protocol 4 (TP4) over CLNP, which forms the ISO version of TCP/IP (TP4/CLNP).

Considerable DCC documentation overlap occurs at the higher layers. At the session layer, the DCCs use ISO 8073/8473 or X.215/X.225 (or both), which provides sessions to recover from transport failures, and access validation for security. At the presentation layer, the DCCs use ISO 8327 or X.216/X.226 and X.209 (or both), which describes a common abstract syntax notation language (ASN.1) for internal data representation differences on the part of SONET equipment vendors, as well as compressed ASCII coding and encryption.

The application layer uses the usual ACSE/ROSE/CMISE/SMASE network management structure. ISO 8650 or X.217/X.227 (or both) defines the ACSE, ISO 9072 or X.219/X.229 (or both) defines the ROSE, and ISO 9595/9596-1 defines the CMISE for SONET. These pieces essentially define only an API for network management, not the network management application itself.

TMN brings order to this apparent chaos. Vendors can run TL1 in the F1 "user channel," which is totally in line with TMN.

TMN Issues

In a perfect world, TL1 or other management software for SONET/SDH would evolve to TMN, and that would be that. Vendors could migrate from using TL1 messages inside X.25 (or SNMP, or whatever) by using QAs and, eventually, native TMN messages and protocols. However, just as TMN is reaching a certain level of maturity, and TL1 is entrenched enough to have considerable momentum on its side, the whole arena of telecommunications network element management is being challenged by changes in the way that OSSs are deployed and used. The new challenges arise from usage considerations surrounding Java, the common object request broker architecture (CORBA), the Microsoft distributed component object model (DCOM), and the Web itself. Certainly nothing in TMN addresses any of these newer concerns. Some commentators have gone so far as to pronounce TMN dead on arrival, with the CORBA Internet bus model being the most logical successor.

Much of the interest in using Internet-related software concepts in SONET/SDH has come from the use of SONET/SDH links to connect routers, of course. ISPs, even when subsidiaries of telephone companies, do not often have the same elaborate network management centers required to run telephone networks—especially regulated telephone networks. Why not be able to reach out over the Internet (in a secure fashion, of course) and get a webpage-like report on a SONET/SDH NE linking router? In fact, why deprive telephone company users of SONET/SDH the same capability? It makes a lot of sense, and makes the push for IP over DCCs that much stronger.

Some of these issues are quite complex. This section is intended to at least introduce some of the latest ideas in the OSS field.

It seems premature to write off TMN entirely right now. But the whole point is that it looks more and more as if TMN alone cannot accomplish all of the goals that service providers are seeking to achieve with their networks today—especially when the use of SONET/SDH on the Internet is concerned. The root of the issue is not TMN itself (or even TL1). The messaging protocol used in all OSSs simply has to be understood by each end of the conversation. The issue revolves around the infrastructure used at the lower layers of the OSI model to shuttle these messages through the distributed communications network linking all the NEs.

Distributed computing and networking today means the Internet protocol suite—TCP/IP. CORBA adds distributed object computing support to TCP/IP. A full discussion of exactly what an "object" is or is not would fill many pages, and such a discussion is not important here. What is important is that objects can be defined once for some purpose (for example, network management) and then be used over and over again in other forms to provide information in billing, provisioning, order tracking, and related systems. Many vendors of SONET/SDH equipment today include CORBA, the webpage markup language (HTML), and even the newer XML in their NEs. It only makes sense to adapt the SONET/SDH DCCs to make these tools easier to use over the network itself, in the DCCs.

As an example of the attraction of objects today, consider the case of simple error tracking for a VT provisioned on a SONET ring. The bit error rate (BER) promised to the customer is a key part of the tariff or contract. But how can a customer possibly know what the BER on a SONET VT is?

One way is to take the alarms or errors recorded by the NEs involved (which go to the OSS, of course) and post them in mutated form to a web site. Customers could access the web site in a secure fashion and yet see only the VTs that belong to them. All of the appearances of the BER-VT association—from network management workstation, to web site, to customer web browser—are just varying manifestations of the same object. CORBA uses TCP/IP to connect all these instances of the BER-VT object. No conversions are necessary to relate all of these views of what is, at heart, exactly the same information. CORBA allows the website software application to communicate with the OSS software regardless of vendor or support for TMN. (Purists might cringe at this extremely simple example of objects, but the underlying reasons for object use are the important thing here.)

It is somewhat ironic that CORBA is used mostly with the UNIX (and now LINUX) operating system and the C++ programming language, when many clients and servers on the Internet today are Windows-based devices. Microsoft's DCOM supports DOS and Windows, as well as Visual Basic (now adapted nicely for network use). The common link between the two environments is Java, which both DCOM and CORBA buses can use—although Microsoft also employs ActiveX in place of Java in many places, and many Windows applications are written in C++. (These are just generalizations based on most common usage.) Figure 15-9 shows the lure of CORBA and, to some extent, DCOM to enhance or at least complement TMN.

In Figure 15-9, all traditional NE management functions are still performed with either TMN or TL1. But in addition, CORBA is used to enable direct access by Java or C++ to this information. For example, an application could be written, using CORBA, to allow a field technician with a personal digital assistant (PDA) or browser-enabled cell phone to access the information stored at the network management workstations connected to a SONET ring. For back-office operations, a simple DCOM-to-CORBA bridge could be written to enable direct access from any Windows browser to this information. For re-

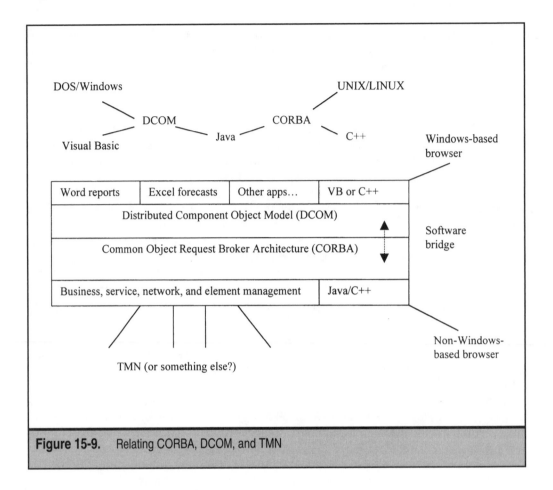

Figure 15-9.   Relating CORBA, DCOM, and TMN

ports, modeling, and so on, all of the TMN or TL1 information could be directly piped as objects into spreadsheets or reports. Commentators with a bleak view of the TMN future see little point in duplicating CORBA/DCOM capabilities into a TMN-compliant distributed network. Better to extend the CORBA/DCOM structure into the TMN/TL1 world of SONET/ SDH NE management with TCP/IP, they argue.

Now, "object servers" can be added to TMN and TL1 as well, to take information from NEs and massage it for web sites and servers. But CORBA is complex enough and powerful enough that extending it to the OSS rather than interfacing the OSS to it can be considered.

After all, most service providers today use CORBA (and more) on their own IP networks for e-commerce and e-business. They have all the hardware, software, security, procedures, and so on, because this environment is already in place and running. Why attempt to recreate the whole thing just for TMN purposes? TMN can even be redefined to ride atop CORBA rather than CMISE/CMIP. It all seems to boil down to a question of what is more useful and valuable to the organization as a whole: CORBA or TMN?

What would happen to TMN in a CORBA/Web world? TMN is really three things. First, TMN is a layered model with systems for viewing the network as a series of business, service, network, and element management pieces. This is all part of the TMN distributed approach. Second, TMN establishes a system of common names and relationships for all of these items, from NEs to trouble tickets. That piece is mainly for vendor independence. Finally, TMN establishes a distributed network for connecting the consistently named and addressed components. Some of the TMN standards rely on OSI (CMISE/CMIP) for this last piece.

Perhaps it would be a good idea to replace OSI with CORBA in the third component of TMN. Why reproduce (for instance) the security already in place for CORBA with un-proven, untested, and redundant OSI security?

The next step would be to replace TMN and OSI naming and addressing with IP/Internet naming and addressing. Why not give everything an IP address and be done with it? And once those two pieces of TMN fall to the IP empire, why not just do everything with some secure version of SNMP?

And so the debate rages. Some argue that, as long as TMN clings to its first mission on business and element and other types of management, TMN is not dead, just changed. Others see any crack in the TMN architecture as a whole as a sign that the whole edifice will come crashing down. Still others see TMN as providing a certain level of security, be-cause if it is based on something other than IP, no one will bother to hack into it! (But ob-scurity never stopped hackers from getting into telephone company management systems such as LMOS, the loop maintenance operating system.)

IP AND THE SONET/SDH DCCS

Officially, both SONET and SDH use the same seven-layer protocol stack for conveying network management information of all kinds across the SONET/SDH DCCs. Recently, a simplified protocol stack based entirely on TCP/IP tools and applications has been pro-posed as a replacement for the whole structure. Figure 15-10 shows the protocol is to be used at each layer in both TMN and the IP proposal.

Some of the ISO protocols have been mentioned previously in this chapter, but are re-peated here for the sake of completeness. Readers are presumed to be familiar enough with the IP protocols to need no detailed explanation. (The information is readily available from numerous other sources anyway, although the same is not true of ISO protocols.)

Although IP application layer tools such as SNMP are included on the IP side of the fig-ure, it is important to note that they are not used directly for SONET/SDH network man-agement. Despite the inclusion of an "application layer" in both ISO and IP, SONET/SDH network management actually starts with custom-built *user* applications that employ ap-plication-layer protocols such as CMISE or SNMP for gathering the information used by the network management software at the user end. These are not shown in the figure. Oddly, few user applications, such as FTP or Telnet, exist at the application layer.

In any case, the ISO application layer consists of CMISE, described in ISO 9595 and 9596. This just means that the application is for network management and follows those rules. To help CMISE out, there is also ROSE for remote access rules ("run this and tell me

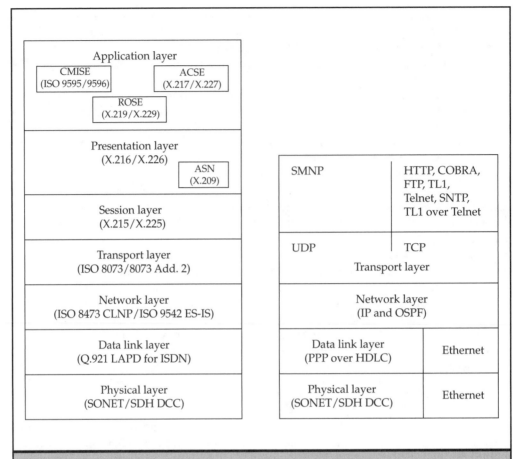

Figure 15-10. Comparing ISO protocols and IP over the SONET/SDH DCCs

what happens," and so on), specified by ITU recommendation X.219 and X.229. (Many ISO documents are duplicated by ITU specifications, and vice versa.) The ACSE, X.217/X.227, handles connections (associations to ISO) across the DCCs.

The presentation layer is described by X.216/X.226, and also includes X.209, which establishes rules for representing data on the network, a method known as ASN.1. This method is needed because many NEs represent a number (for example) as 16, 32, or 64 bits internally. So when a network management application asks for a count of BIP errors (for example), how many bits should be expected from the NE? ASN.1 establishes rules such as, if the number reported has a range of 0 to 255 defined, then only eight bits should be sent. Different chips also store bit strings in different (reverse) order. ASN.1 clears this up as well (and so on).

The session layer, which contains a sort of history of the connection ("three things sent, one more to go"), can be used for recovery if the connection is severed. The specification is X.215/X.225. (The double specifications come from the fact that many of these standards come in connectionless and connection-oriented flavors.)

The transport and network layers are the heart of the network protocol. ISO uses the connection-oriented transport-layer protocol (ISO 8073 and its Addendum A), similar to TCP, and the connectionless network layer protocol (ISO 8473), similar to IP. The routing protocol used at this layer to find NEs in the network is called end system–to–intermediate system (ES-IS), defined in ISO 9542.

The data-link layer wraps a frame around the network layer packet. In ISO, these frames are called "link access protocol—data channel" (LAPD) frames, defined in ITU Q.921 and originally used for ISDN.

The physical layer is just the SONET/SDH line and section DCCs, of course.

Now compare the IP protocol stack proposed for use as part of the all-important Q3 interface (defined in ITU Q.912 and Q.913). Recall that only Q3-compliant devices can be managed directly by the OS, without the need for complex (and expensive) QAs. The application layer (again without the actual user applications written on top) now consists of SNMP or a variety of standard and common tools such as HTTP (the transfer protocol for web pages), CORBA for objects (built into many SONET/SDH NEs already, but hard to fit into the ISO application layer), TL1, Telnet (for direct NE access), SNTP (the simple network time protocol, so that the operations workstation can set its time from the network), and even TL1 over Telnet for continued human–machine dialogs.

Transport is by connectionless UDP (SNMP) or the more robust connection-oriented TCP. The network layer is IP, with the OSPF routing protocol. Framing is provided by PPP over HDLC (the Internet's point-to-point protocol within a high-level data link control frame) or even Ethernet LAN frames (useful for the network management LAN in the operations center). Finally, physical connectivity is by the SONET/SDH DCCs or by the Ethernet LAN.

Given the nature of the newer generation of SONET/SDH NEs, the IP over DCC proposal makes a lot of sense. More and more SONET/SDH links will lead to routers, not telephone switches. Moreover, the proposal explicitly states that this arrangement should not limit the applications used. Anything that runs over IP could be used to manage SONET/SDH equipment, as long as it works. IP has already found its way into the ITU definitions of what a data communication network (DCN) for network management information can be, a major step for a protocol that until recently was never considered an international standard. This proposal simply acknowledges IP (and its related protocols) as the most widely used protocol in the world.

Figure 15-8 in this chapter showed how TMN-compliant protocols could be used to manage SONET/SDH equipment. Figure 15-11 shows how the figure remains the same, and yet changes dramatically, when IP over the DCCs is used.

The problem with the original TMN proposals is that the protocols used for the Q3 interface have to be developed especially for this use, are complicated and expensive, and are not readily available. All of this changes when IP is used. The proposal does not force changes to older equipment, but it allows IP to be used in newer devices. Many people in the industry, especially equipment vendors and ISPs, hope that IP will find a place in the world of SONET/SDH network management as soon as possible.

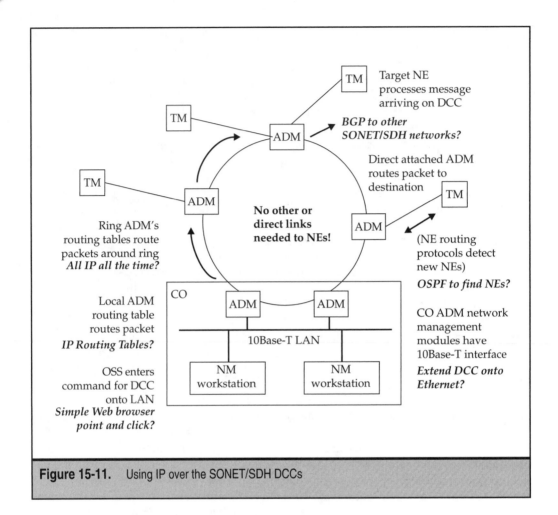

Figure 15-11. Using IP over the SONET/SDH DCCs

TRANSACTION LANGUAGE 1

This chapter closes with a more detailed look at the TL1 language. Why bother? First of all, until something better comes along, TL1 will occupy an important place in SONET network management. There is even a place for TL1 over Telnet in the IP-over-DCC world, as already pointed out. Second, even a brief description of TL1 outside of the source volumes of documentation is extremely rare. Even if not all readers gain an insight into TL1 operation from this section, it is worthwhile to present. This section provides a fairly detailed overview of TL1, based on the older, but simple, Bellcore documentation (SR-NWT-002723). Because of its source, this section on TL1 makes no mention of SDH.

TL1 is a generic Telcordia (at the time, Bellcore) human–machine interface for managing network devices. Because the decision was made quite early in SONET deployment to support a transition in SONET to OSI management based on CMISE, only a limited set of TL1 messages was defined for SONET NEs. This message set allows service providers to establish services, to monitor their SONET networks for failures, and to perform common and needed administrative tasks associated with maintaining and monitoring NEs in all simple SONET architectures. These include point-to-point, daisy-chained, tree, and path-switched ring architectures.

TL1 for SONET has these two objectives:

1. Provide Bellcore's own view of how equipment vendors and suppliers can map SONET requirements to specific TL1 messages, and support implementation of these specific requirements.

2. Identify the distinctly SONET requirements for which TL1 messages do not exist. In some cases, the existing TL1 message set was extended to support the requirement. In other cases, new messages were needed.

To perform their tasks, network OSs generate, accept, and process subsets of the existing generic TL1 message set. Equipment vendors and suppliers use a process known as the "operations systems modification for intelligent network elements" (OSMINE) process to get support for non-generic TL1 messages in Bellcore-compliant OSs.

Why bother with TL1 at all? Bellcore hoped that if suppliers supported TL1, the SONET NEs used by a typical service provider would be more consistent in security, provisioning, and maintenance with regard to OS/NE interface implementations. This level of standardization would result in reduced operational expenses for deployment of these products in a SONET network. If everyone were to follow these OS/NE message mappings, it might also reduce the expense of determining conformance to the proposed Bellcore requirements in product testing and analysis.

The focus of TL1 messages for SONET is on SONET NEs. This is hardly surprising. However, although it identified the core set of TL1 messages applicable to SONET NEs, Bellcore did not mean that all NEs must support the entire TL1 message set. The TL1 message set for a particular NE depends on the interfaces and the supported features. A good example is the conditional requirement for exercising protection switching. If that feature were not provided, then the associated messages should not be needed. All additional features in an NE, or features for which messages have not been provided, may need other TL1 messages beyond the basic set.

SONET ADMs are often configured in several types of self-healing rings to provide network survivability. These NEs require special equipment functionality and the exchange of additional information across the OS–NE interface.

There are also two specialized applications for SONET technology for which TL1 messages have been defined. These are the fiber in the loop (FITL) and integrated digital loop carrier (IDLC). For the sake of completeness, the TL1 messages required by FITL and applicable for IDLC systems are mentioned in this section.

TL1 Messages

All of the TL1 messages were developed to provide a generic human–machine interface language between NEs (where the machine is) and OSs (where the human is) and to meet a series of generic network operations requirements. All of these TL1 messages fall into two general categories: command–response and autonomous.

In the command–response type of messages, the OS initiates commands (usually by simply typing them or clicking on an icon) for the NE. The SONET NE must provide a response message. Conversely, an autonomous message is initiated by the NE to inform an OS about its current status. In this case, the OS does not itself respond to an autonomous message.

The TL1 language has its own semantics and syntax, but it is not necessary here to detail the full language. Instead, the emphasis here is on the use of TL1 for SONET networks.

The specific TL1 message set for OS–NE communications falls into two general areas: memory administration and network maintenance. Both are defined in four Bellcore (Telcordia) TRs. These are security administration messages (TR-TSY-000835), memory administration messages (TRNWT-000199), surveillance messages (TR-NWT-000833), and access and testing messages (TR-NWT-000834). Moreover, some messages applicable to SONET can be found in two TAs: memory administration messages (TA-NWT-000199), and surveillance messages (TA-NWT-000833).

In TL1, security is technically part of memory administration because it deals with databases resident in the memory of an NE. The security messages are found in separate Bellcore documents because the security database is independent of the data necessary for functional operation of the NE. That is, security—as important as it is—is an issue separate from TL1.

In spite of efforts to cleanly divide TL1 areas, there are instances of overlapping functions between memory administration and network maintenance, and between network maintenance and testing. For instance, setting performance monitoring parameter thresholds (for example, for errors) has aspects of both memory administration and surveillance functions. TL1 messages, therefore, are provided in both areas.

However, significant differences exist in the messages defined for these two areas. For example, the "EDIT" TL1 command in memory administration can change all parameters for a specific entity. That is, all parameters (for example, monitored threshold values, the state of service) are changed with this single EDIT command. In contrast, the network maintenance messages have specific commands for more individualized functions. The SET-THRESHOLD command, therefore, sets performance monitoring threshold values, the RESTORE command places an entity into service, and so on.

SONET and TL1 Messages

A good place to begin a discussion of TL1 messages is with security. All TL1 messages involve security considerations. Security in TL1 involves, among other things, logins, passwords, and security options to control access to NEs and their management databases. The TL1 security requirements focus on both the users of the NE and the security administrator for the NE. A login ID is the "name" by which an NE recognizes a valid user. The claimed identity of the user can be verified by several methods in TL1, including passwords. Security options may restrict a user from executing specific TL1 commands, or may limit the user to only certain fields and parameters within a specific command. Security may also restrict the operations channels and resources available to users. Security features also include tools to provide audit trails, to administer the security database, and to perform consistency and reliability checks. TL1 security requirements apply to all SONET NEs in all configurations. TL1 security messages can be originated by one of three sources: the user, the administrator, or the NE itself.

In TL1, "users" are defined as those people (or systems) that have a need to access the NE for performing OAM&P functions on the NE itself or on the overall network. A SONET NE should recognize these TL1 user commands:

▼ **ACTIVATE-USER**   The ACTIVATE-USER command permits a user to initiate a session with an NE. It supports the important requirement of user identification.

■ **EDIT-PID**   The EDIT-PID command allows users to modify their personal identifiers.

▲ **CANCEL-USER**   The CANCEL-USER command permits the user to terminate the session. It is part of the session, or system access, control.

Other related user functions, not explicitly required for SONET NEs, allow users to examine their permissions and the security permission levels associated with the current channel used for TL1 messages.

For a SONET NE to recognize the user's IDs, the NE must have a database. Thus, the requirement that SONET NEs administer user IDs also requires those NEs to support the security messages designed for the security administrator. This only makes sense.

TL1 Messages and Memory Administration

In TL1, the messages that fall into the class known as "memory administration" are those that control and access the data in the databases of a SONET NE. The databases contain configuration information critical to the services provided by the NE, such as cross-connection information. The messages also contain information involved in maintaining a certain level of quality for the services. Thus, memory administration deals with the addition, deletion, editing, and retrieval of information resident in the NE database. In addition to the manipulation of this information, memory administration also deals with memory backup, restoration, and system administration.

The most common command verbs used in memory administration are ENTER, EDIT, DELETE, and RETRIEVE. These are abbreviated in TL1 as ENT, ED, DLT, and RTRV, respectively. In the following discussion, the abbreviated verb forms are used. There also are references to "data dictionaries." The TL1 data dictionaries contain information about the SONET NE that memory administration commands can access. In this way they resemble MIBs (but not completely).

A TL1 message consists of a verb and one or two modifiers: the identity of the SONET NE for whom the message is intended, the entity to which the message is addressed, and the data to be transmitted. The modifier often identifies the data dictionary, which contains the list of permitted keywords and values. SONET data dictionaries in TL1 include these:

▼ **OCn**   Optical carrier at line rate n (where n = 1, 3, 9, 12, 18, 24, 36, or 48). This specifies keywords for the SONET section and line levels at an optical interface.

■ **ECm**   Electrical carrier at line rate m (1 or 3). This specifies keywords for the SONET section and line levels at electrical interfaces.

■ **STSn**   Synchronous transport signal, where n is the rate of the STS synchronous payload envelope (n = 1, 3c, or 12c). This specifies keywords for the SONET STS path level.

- ■ **VT***x*   VT, where *x* is the rate of the VT synchronous payload envelope (*x* = 1, 2, 3, 6, or 6C). This specifies keywords for the SONET VT path.

- ■ **ULSDCC**   The upper-layer section data communication channel specifies upper-layer protocol stack parameters for the section DCC.

- ▲ **LLSDCC**   The lower-layer section data communication channel specifies lower-layer protocol stack parameters for the line DCC.

There also are TL1 messages for SONET that do not require data dictionaries. These messages include the modifiers TADRMAP and OSACMAP and are for the verbs ENT, ED, DLT, and RTRV. All of these may be used by what are known as "gateway NEs" to route TL1 messages while OSs and NEs are in transition from TL1 to OSI/CMISE network management.

PROVISIONING SONET NES WITH TL1

TL1 memory administration messages support "provisionable parameters." These are associated with the SONET DCC protocols, linear automatic protection switching, and information associated with performance monitoring and alarm reporting that involve provisioning.

Much OAM&P activity concerns the establishment of SONET "termination points." These determine just where a SONET link begins and ends. With TL1, SONET termination points may be created by use of the ENT-OC*n* and ENT-EC*m* commands for an optical facility or electrical facility, respectively. As an example, the ENT-OC*n* command may be used to specify that the DCC is or is not supported on this link, that the OC-N is or is not a protection line, and whether the OC-N is the working line in a protection group, and so on. Another key function of these commands is to specify whether the interface is drop (customer or tributary) side or line (trunk) side. These commands could also indicate the type of board in the NE (if the equipment can contain multiple cards, each with different SONET OC-N characteristics).

The TL1 messages ED-OC*n*, RTRV-OC*n*, and DLT-OC*n* allow a user to modify, retrieve, and delete information about the identified OC-N, respectively. A similar set of messages exist for the EC*m* view.

Payload Mappings and Cross-Connections

When it comes to terminations, SONET NEs (especially TMs, naturally) can support a variety of types of facility terminations, both SONET and non-SONET. Non-SONET signals, or tributaries, are mapped into a SONET VT payload for transport through a SONET network. SONET DCSs and ADMs also permit the service provider to define the SONET time slot in which the signal is transported.

The payload mapping and cross-connect functions are both accomplished with a single TL1 command. This is ENT-CRS-*rr* (that is, enter–cross-connect–rate). When the cross-connect requires payload mapping, the *rr* modifier identifies this non-SONET rate. Each of the data dictionaries for non-SONET tributary rates (DS-1, DS-1c, DS-2, DS-3) contains the keyword "MAP" and a list of permitted values. The value assigned to the MAP keyword defines just how the signal is to be mapped and gives the content of the appropriate path-level signal label in the SONET payload. Of course, messages to cross-connect non-terminated VT-SPEs or STS-SPEs through a SONET NE, such as an ADM, do not require mapping information. Mapping information is required only at path terminations.

Both SONET ADM and DCS NEs have several optional features for which supporting TL1 messages have not been provided. For example, an optional feature is one-way broadcast services. This feature is not supported in the limited TL1 message set, and will probably have to wait for OSI/CMISE.

PROVISIONING AND SECTION DCC

With TL1, an individual OC-N (or ECm) can be provisioned to indicate support of the DCC. The section DCC uses a full OSI seven-layer protocol stack. Each layer of the stack contains user-obtainable parameters. The protocol stack has been divided into two data dictionary views in TL1. These are the lower-layer section data communications channel (LLSDCC) and the upper-layer section data communications channel (ULSDCC). Only layers 1 and 2 are covered in the LLSDCC view; the remaining layers (3 through 7) are in the ULSDCC. The ENT-LLSDCC, ENT-ULSDCC, ED-LLSDCC, and ED-ULSDCC commands are used to create and modify parameters at each of the layers, including the address at each layer. The state of the DCC can be controlled by means of the EDIT command and the primary state (PST) keyword. The RTRV-ULSDCC and RTRV-LLSDCC can retrieve the current value of any keyword or the entire data dictionary.

Until such time as both the OS and SONET NE support CMISE interfaces, the NEs may communicate with the OS via TL1. To this end, the gateway NEs must maintain the additional data required to route automatic messages to the intended OS and to route OS messages to the correct NE beyond the gateway. Using the TL1 commands ENT-OSACMAP, DLT-OSACMAP, ED-OSACMAP, and RTRV-OSACMAP, the network service provider can maintain, at gateway NEs, a table that maps an OSI application "context identifier" (similar in function to a network address) to the X.25 network address of the OS.

During the transition period from TL1 to OSI/CMISE, a gateway NE also needs information to correctly route messages from the OS to the NE. The command modifier TADRMAP may be used with ENT, ED, DEL, and RTRV to maintain a table that maps the TL1 target identifier (TID) of an NE to its network address (NSAP, the OSI equivalent).

TL1 and System Backup

The use of TL1 allows the OS to provide for "memory administration" of a SONET NE. This is like a system backup function for the memory of the NE, which contains configuration information, of course. The NE must notify the OS of any automatic or local changes to this configuration. Some documents call these "hidden updates." The REPORT-DATA_BASE_CHANGE message, which is abbreviated as REPT-DBCH, provides this function.

What kinds of changes would be reported?

Assume that a maintenance person has made some changes to the local database. These changes may include changing the service state of a termination point (using the

REMOVE and RESTORE TL1 messages), changing the threshold of a performance parameter (SET-TH), or changing a cross-connect (ENT-CRS, DLT-CRS, ED-CRS).

TL1 Network Maintenance and Alarms

With TL1, the service provider defines which SONET maintenance signals (for example, AIS, RID-L) are to be reported to the OS. The service provider must decide whether they are reported as alarms or events. The choice is made in the TL1 network maintenance environment by way of the SET-ATTRIBUTE command. The SET-ATTRIBUTE command is also used to define how other trouble conditions (failures, for example) are to be reported. The choices are critical, major, minor, and whether or not they are service-affecting.

In some cases, the network provider may want an NE to report environmental conditions at the NE's location to the OS. The SET-ATTRIBUTE-ENVIRONMENT command is used for this purpose.

All SONET NEs are required to detect and automatically report incoming signal failures, which only makes sense. These are serious problems. Thus, the SONET NE must support the autonomous REPORT-ALARM, REPORT-EVENT, REPORT-ALARM-ENVIRONMENT, and REPORT-EVENT-ENVIRONMENT messages. The first two of these messages also report trouble in the incoming SONET APS protocol channel (bytes K1, K2 in the line overhead).

The TL1 message set fails to account for one area in the SONET requirements: the ability for the user to specify the amount of time a defect should be present (or absent) before the NE reports the failure (or the clearing). These are covered by SONET standards.

In addition to all of the autonomous messages, the NE must allow the network service provider to retrieve current alarms and other information about the state of the NE. The RETRIEVE-ALARM and RETRIEVE-CONDITION messages meet these needs. Other commands are used to retrieve the attributes of the various trouble conditions and to suspend and resume the generation of autonomous messages from the NE. If an OC-12 SONET link has just been installed and has no live service on it, but a bad connection is causing trouble reports every ten minutes, the network managers may want to suppress the messages until all work is complete and live traffic is present. The INHIBIT-MESSAGE command is used for this purpose. The condition is enabled again with the ALLOW-MESSAGE command. Also, the SET-ATTRIBUTE and RETRIEVE-ATTRIBUTE TL1 messages tell an NE to set or retrieve the notification codes associated with a specified event. The event is a TL1.

Monitoring Performance

The point of monitoring the overhead in a SONET signal is to provide early detection of trouble and to help isolate the cause of customer-reported complaints. For this to be effective, the service provider must be able to set threshold values for the various performance parameters, and the NE must be able to report threshold-crossing messages automatically when the thresholds are crossed. The SET-THRESHOLD and REPORT-EVENT messages provide these two functions. The monitored types (designated as in the TL1 messages) supported by these commands were expanded for SONET.

Just to show how important TL1 monitoring can be, the following list gives the TL1 for each defined SONET performance parameter and the name of the parameter.

TL1	SONET performance parameter	Monitoring Status
LBCN	Laser bias current normalized	(R)
LPTN	Optical power transmitted normalized	(CR)
LPRN	Optical power received normalized	(CR)
CVS	Coding violation count – section	(CR)
CVL	Coding violation count – line	(R)
CVP	Coding violation count – path	(R)
ESS	Errored second count – section	(CR)
ESL	Errored second count – line	(R)
ESP	Errored second count – path	(R)
SESS	Severely errored second count – section	(CR)
SESL	Severely errored second count – line	(R)
SESP	Severely errored second count – path	(R)
SEFS	Severely errored framing seconds	(R)
PJC	Pointer justification count	(O)
PSC	Protection switching count	(R)
PSD	Protection switching duration	(R) in revertive systems
UASP	Unavailable seconds – path	(R)

(R) = monitoring required.
(CR) = monitoring conditionally required.
(O) = monitoring constitutes a performance objective.

Naturally, pointer justification counts are unique to SONET, and the associated performance monitoring is identified as an objective. Note that there is no special differentiation between STS line and VT path pointer justifications. This has been a source of some concern.

All SONET equipment must recognize these TL1 performance monitoring messages:

▼ INITIALIZE-REGISTER

■ SET-PERFORMANCE MONITORING MODE

■ RETRIEVE-PERFORMANCE MONITORING SCHEDULE

■ RETRIEVE-PERFORMANCE MONITORING

■ RETRIEVE-PERFORMANCE MONITORING MODE

■ RETRIEVE-THRESHOLD

■ SCHEDULE-PERFORMANCE MONITORING REPORT

■ ALLOW-PERFORMANCE MONITORING REPORT

▲ INHIBIT-PERFORMANCE MONITORING REPORT

Protection switching is initiated by an excessive BER or a degraded signal, in addition to failures. The level of bit errors that determines an excessive BER and a degraded signal BER level can be set by the user, as may be expected. The SET-THRESHOLD command accomplishes this by specifying the monitored types as BERL-LT for excessive BER and BERL-HT for degraded signal.

All SONET NEs are required to support performance monitoring. Performance monitoring is not turned on or off, but autonomous messages can be suppressed with the INHIBIT-MESSAGE command.

TL1 and Testing Process

TL1 documentation defines several functions to help with trouble isolation on carrier networks, but these were not included in the TL1 subset for SONET. In particular, TL1 surveillance messages have not been defined for these circumstances:

- ▼ Activation of a corrupted BIP
- ■ Retrieval of optical power transmit and optical power received values
- ■ Examination of path trace byte at non-path-terminating NEs
- ▲ Retrieval of the content of the STS path or VT path signal label

However, things are not as bleak as they may seem. The existing TL1 DIAGNOSE and DIAGNOSE-DETAIL commands are generic and may be used, in some cases, to meet the needs of SONET networks in this area. The ability to initiate either a facility or terminal loopback is provided in the OPERATE-LOOPBACK command. The REMOVE and RESTORE messages are used to change the service state of the line being looped back.

This final section on TL1 may seem overly long. However, the intention was to give the reader an appreciation of how TL1 operates, and at the same time, provide a feel for what TMN systems in a SONET network must be capable of doing. Whether TL1 of OSI or SNMP, all TMN systems can be quite complex.

CHAPTER 16

SONET/SDH
Protection and Rings

Rings are perhaps the most distinctive feature of SONET/SDH networks in general, and the most obvious way that SONET/SDH differs from T-carrier or E-carrier. The ability of SONET/SDH to be deployed in a ring architecture, rather than strictly in point-to-point or multipoint architectures, has become the defining feature of SONET/SDH to date. This chapter explores all aspects of SONET/SDH rings, from the very basics to the more sophisticated timing methods that must be employed with ring architectures.

SONET/SDH rings can be a few miles long and span a few city blocks, or can stretch to literally thousands of miles and span continents. Rings of international proportions are not uncommon, and small rings in downtown metropolitan areas are no longer news-worthy. SONET rings are built in the United States by local exchange carriers (LECs), competitive access providers (CAPs) or competitive local exchange carriers (CLECs), and inter-exchange carriers (IECs). In the rest of the world, rings of ADMs are usually called MS-SPRings (multiplex section-shared protection rings), but their structure, function, and provisioning closely resemble those of their SONET cousins. For a while, MS-SPRing documents simply referenced the ANSI specifications on SONET rings, but now ETSI documents addressing solely SDH exist. In addition to public network rings, a few totally private SONET/SDH rings exist, intended for the exclusive use of a single organization. All share the distinctive protection characteristics of SONET/SDH rings.

Still, SONET/SDH rings come in many shapes and sizes, having varying characteristics that make them suitable for one application or another. This chapter deals with all such characteristics.

The discussion begins by outlining the incentive for building SONET/SDH rings in the first place. After all, protection switching was not unknown in T-carrier and E-carrier networks. However, although the linear protection afforded by T-carrier and E-carrier was adequate for less exacting applications and services in terms of reliability and connectivity, the newer applications and services delivered by SONET/SDH networks place new demands on protection and reliability schemes.

All of the variations involved in SONET/SDH ring architectures are then detailed, and a summary compares the major architectures. Finally, the chapter concludes with several sticky issues that come to the fore in circular telecommunications networks. The two most critical issues concern SONET/SDH ring delays and timing. Both are given full treatment at the end of the chapter.

SONET/SDH RINGS AND FIBER FAILURES

SONET/SDH links are sometimes deployed in a straight point-to-point fashion. Of course, SONET/SDH has distinct multipoint features that make grooming and hubbing much easier for service providers. But, despite those multipoint configuration capabilities, a SONET/SDH network may still consist of a number of point-to-point trunks and links. The problem with single point-to-point links is that the entire path of the link is vulnerable to failures.

When it comes to talk about SONET/SDH rings, there is no doubt that SONET blazed the trail for SDH ring deployments. This section therefore addresses SONET rings first. To best appreciate what SONET rings bring to networks, it is a good idea to examine the situation that fiber optic networks had to deal with *before* the systematic deployment of SONET (and, later, SDH) rings.

Table 16-1 and Figure 16-1 show the causes of fiber-optic facilities failures in the United States from July 1, 1992, through June 30, 1995—a three-year period. The statistics come from the Alliance for Telecommunications Industry Solutions (ATIS). This information may seem somewhat dated, but the point is exactly that, before 1995 or thereabouts, service outages from fiber failures were common. Statistics obtained since the deployment of large numbers of SONET/SDH rings (after 1995) would not reflect the same conditions, because SONET/SDH rings offer high degrees of service protection.

Obviously, if some way could be found to allow service to continue uninterrupted when a fiber cable fails, the benefit to users and providers alike would be big. The elimination of 75% of all fiber-system failures would be a huge plus for SONET (and SDH, of course). Reliability terms in tariffs and contracts would be respected, public utility commissions would field fewer complaints, and the risk of the service provider being fined for inadequate service would be minimized.

The only question is the form that the protection should take. T-carrier was not immune to cable cuts and the like. Protection switching began with T-carrier. Perhaps a quick look at protection in non-SONET/SDH architectures may help to convey a better appreciation of the contribution that SONET/SDH rings make to overall network reliability.

NON-RING AUTOMATIC PROTECTION SWITCHING

It has already been noted that automatic protection switching (APS) did not begin with SONET. Many T-carrier links were protected from cable failures by APS systems as well. Typically, these were not employed with the T-carrier links in a ring architecture; they just sought to back up one or more links with spares in a rather simple fashion.

The first form of protection switching used in T-carrier was known as "1:7 switching." This is usually spoken as "1-by-7" or "1-to-7" protection switching. There is nothing magical about the number "7" here. The second number can theoretically be any number, and the

Failure Cause	Percentage
Fiber cable dig-ups (backhoe fade)	51%
Fiber non-dig-ups (aerial/electronics)	24%
Other causes or equipment	15%
Digital cross-connects	7%
Synchronization timing	2%
Internal power components	1%

Table 16-1.   Causes of fiber-optic system outages

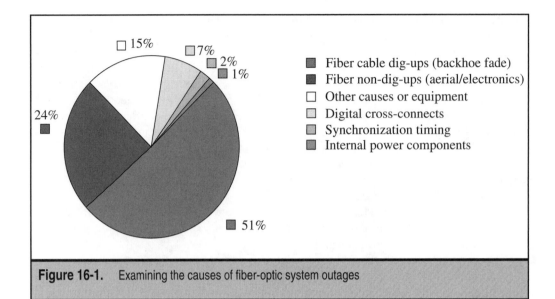

□ 15%   □7%   ■ 2%   ■ 1%   24% ■   ■ 51%

■ Fiber cable dig-ups (backhoe fade)
■ Fiber non-dig-ups (aerial/electronics)
□ Other causes or equipment
□ Digital cross-connects
■ Synchronization timing
■ Internal power components

Figure 16-1.   Examining the causes of fiber-optic system outages

general form is 1:n protection switching. The term just means that n active links are protected by one standby link in case of a failure. The reason that $n = 7$ is so common had to do with the way that T-1 was initially deployed in the United States.

It was common practice in T-carrier networks to install the transmitting and receiving equipment in standard seven-foot-high communications bays with 21-inch-wide shelves. Eight transmitting and receiving modules ("boards") could fit on an individual shelf. Usually, seven of these were used as active ("working") links. The eighth formed the protection ("standby") link in case any of the others failed, giving the now familiar 1:7 APS pattern. Figure 16-2 illustrates the scheme.

There is nothing wrong with 1:7 protection switching. It is still common and respected. In some cases, two full shelves could be protected this way, giving a maximum n of 14. In fact, many SONET products allow the 1:7 form of APS. Most commonly, the protection switching from a working link to the protection link is quick and automatic. No human intervention is required. In telephony applications, a brief disruption in the conversation is all that is encountered. Even in newer applications, such as live football and other sports broadcasts over DS-3 facilities, a quick "freeze" of the picture that quickly clears is the only symptom of this APS procedure. Typically, restoration of service from a protection link to a formerly working link after repairs have been made requires human intervention, but the initial APS engagement does not.

APS permits the network to react to many conditions that threaten service quality. Not only outright failures can trigger APS action, but also errors and overall poor signal quality. If an interface board were to fail, the net effect would still be a failed link. A network manager could also initiate APS on a link, switching it from a working facility to a standby facility for routine maintenance, bit error rate testing (perhaps due to user complaints), and so forth. Most commonly, APS operations occur as a result of lost connections or poor signal quality on connections in imminent risk of failure.

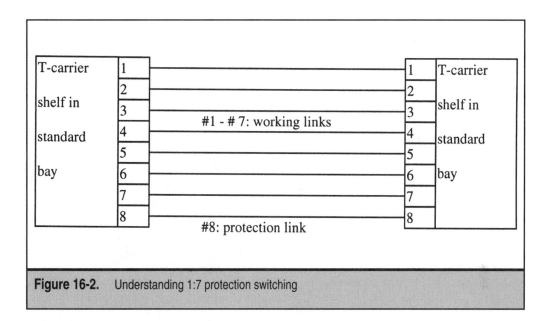

Figure 16-2.   Understanding 1:7 protection switching

Not that 1:7 protection switching is perfect. Far from it. Consider what would happen when more than one link of the working group is lost at the same time. Why should this happen? Frequently, several links follow the same physical path (and sometimes use the same feeder bundle or cable sheath) for a considerable portion of the entire link—perhaps even the entire distance. This means that when one link in a shelf is lost, generally several are lost at the same time.

Obviously, it is impossible to protect many T-carrier working links with a single protection link. Only one working signal can be diverted to the protection link at a time. A big step forward came as the cost of operating T-carrier facilities came down and the cost of the end interface electronics also dropped. These cost reductions eventually resulted in the deployment of 1:1 protection-switching methods in high-risk areas (usually major downtown metropolitan areas where constant street work was common). In 1:1 protection switching, every working link has its own individual backup link.

In general practice, the 1:1 protection operation involves sending the same signals over both links. That is, a single set of information is duplicated and sent over both the working and the protection path, a process known as "bridging." Both signals are constantly monitored for signal quality by the receiver, and the higher quality signal is chosen for live reception. The active link is reverted ("switched") to the standby link when the connection is lost or when signal quality degrades beyond a predetermined threshold. Naturally, a network management center is notified when the automatic protection switch takes place. The return to the active link is usually done manually, but it can also be automatically "revertive" if the equipment allows this mode of operation.

To make 1:1 protection switching function ideally, it would be nice to have the working and protection links taking different paths from source to destination; however, this is not always possible. Cable sheaths may be separate, but the cables may use the same conduit. Even if the conduit is different, the feeder route may be shared by several conduits, owing to right-of-way or congestion considerations. This "banking" of conduits from various service providers on the same feeder route is a common practice in many parts of the United States, and not merely in urban areas.

Figure 16-3 shows some of the options available to service providers, often used in T-carrier networks. In most cases, extra cost to the customer is involved in one or more of these route-diversity situations.

In the figure, several types of 1:1 protection-switching options and possibilities are illustrated. In its simplest form, 1:1 protection switching offers physical diversity, because a single business location is connected to a central office for T-carrier services with the same feeder, but the links are placed in two different cable sheaths. Obviously, if road work or a backhoe were to disrupt services on the working link, the protection link would almost invariably be lost as well.

The next step is to use the fact that feeders are typically laid in a grid pattern. The same feeder (in the sense that the feeder links the same source and destination) can be routed in two separate conduits for as much of the route as is feasible. Note two things about such 1:1 protection switching. First, "route diversity" is usually defined as 25 feet

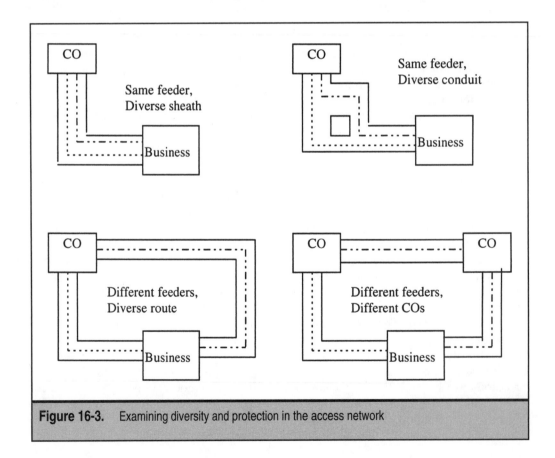

Figure 16-3. Examining diversity and protection in the access network

of separation between the conduits. Therefore, one conduit usually goes down one side of the road, and the other goes down the other side of the road. Second, the "entrance facilities" for the two buildings (the places where cable actually enters and leaves the premises) are still weak links in this arrangement. Many construction trenches stretch all the way across a street, and many building projects (for example, new steps, sidewalks, ramps) can affect entrance facilities. Thus, this type of protection is better, but is still lacking in many respects.

Next, it is possible to try to separate everything—that is, feeders, conduits, entrance facilities, and even central offices. Feeder routes in densely populated areas are good candidates for this scheme. The arrangement offers customer and carrier alike the best 1:1 protection switching available. This solution is not cheap; it is not even possible in many circumstances. Usually, separate entrance facilities are too expensive and require too much effort (including architectural and structural waivers and easements) to make total separation possible. Of course, the diversely routed cable must still be laid to the serving central office, as shown in Figure 16-3.

All of these 1:1 protection variations are quite common in T-carrier. All are lacking in one way or another. But when SONET was standardized, the SONET architects noticed something quite interesting about this last "loop-shaped" protection arrangement. What if the architecture were extended to include not just one customer site, but many? What if two or more serving offices were routinely included in the network design? What if, instead of access networks, backbones were created by linking serving offices of varying kinds, such as central offices, wire centers, and toll offices?

To make a long story short, this indeed was what was allowed and actually encouraged with SONET. Thus, SONET rings were born as a means of standardizing and extending 1:1 protection switching to its logical conclusion. (SDH inherited this SONET capability, naturally.)

SONET/SDH protection switching is not restricted to ring configurations, of course. The SONET/SDH K1/K2 overhead bytes used for APS are not restricted to rings. In fact, linear APS has become common when extending SONET/SDH from a ring of ADMs to routers acting as TMs. When an IP router is acting as a TM, the SONET/SDH links are mostly used for packet-over-SONET/SDH (POS) links to carry packets to another router acting as a TM somewhere else on the SONET/SDH network. Owing to the increased importance of routers in the world of SONET/SDH, a few words about extending SONET/SDH ring protection to router TMs with linear protection will be added at the end of the SONET/SDH ring discussion.

SONET/SDH RING BASICS

This section discusses basic configurations and automatic protection switching in SONET/SDH rings. The discussion begins with definitions of SONET and SDH rings, and continues by exploring various types of ring architectures and by examining the definitions of "unidirectional" rings and "bidirectional" SONET/SDH rings. It then examines the differences between two-fiber and four-fiber SONET/SDH rings, between "ring switching" and "span switching," and between these forms of SONET/SDH ring APS.

Application examples of bidirectional line-switched ring (BLSR) and unidirectional path-switched ring (UPSR), and the definitions and functions of the SONET K1 and K2 overhead bytes for ring APS applications are discussed in some detail. Although SDH rings will also be mentioned later in this chapter, this section begins with a full discussion of SONET ring architectures.

A "SONET ring" is defined as a collection of more than two SONET network elements ("nodes") forming a "closed loop." That is, the SONET NEs are linked together, with the last being connected to the first to form a closed loop. Each NE is thus connected to *two adjacent nodes* by means of at least one set of fibers operating in duplex fashion (that is, one transmit, one receive).

A SONET ring provides many benefits to the service providers that deploy them and to the customers served by them. They can provide redundant bandwidth for protection, redundant network equipment and electronics, and, in most cases, both. A SONET ring is commonly known as a "self-healing" ring, but the term is a little misleading because it implies that the rings somehow repair themselves. However, because services on the ring can be automatically restored following a failure or degradation in the network signals, the term has stuck.

SONET rings are intended to bring "loop diversity" to networks. Although ring architectures are distinctive in SONET, recall that loop architectures and the protection they provide have been around since the T-carrier days. The preferred way of creating loop diversity in SONET rings is to deploy the fiber links with three major guiding principles: use separate fiber sheaths, use separate conduits, and take different physical routes from source to destination. This provides what is known as "route diversity" in SONET rings. Loop diversity can also be achieved by using separate fiber sheaths and separate conduits, and by taking the same physical route from source to destination—but this is not the preferred method. Known as "structural diversity," this latter method is not quite as good as true route diversity, but may be the only form of SONET ring diversity feasible in many metropolitan areas.

It should be noted that both forms of loop diversity (route and structural) often must still use the same entrance facilities in either a central office (the cable entrance facility, CEF) or a business location (also known as the "entrance facility"). Route diversity is defined as separate routes from the first diverse point achievable at the source to the last achievable diverse point at the destination. Usually, but not universally, this is within a few hundred feet of the endpoints. In any case, route diversity is much better than no diversity at all.

SONET RINGS

There is more to SONET protection than just rings. SONET links can form linear, point-to-point architectures feeding SONET rings, and a form of protection switching exists for each of these basic architectures. Linear protection will be discussed more fully later in this chapter. The emphasis in this section is on SONET rings.

In the early days of SONET, ring types were varied. They also differed widely in their features and characteristics. This is understandable, since no one really had any idea of the best way to build and operate a SONET ring. Today, specifications have more or less settled down into four major types of SONET rings, although many vendors of SONET (and SDH) equipment will typically emphasize and document only three major ring types. However, for the purposes of illustration and to distinguish the characteristics and features of the operation and protection offered by the various ring types, all four types of SONET rings are explored here.

It should be pointed out that SONET/SDH equipment vendors make little to no distinction today between their products intended for SONET and their products intended for SDH. Generally, what works and is feasible for a SONET ring also applies to SDH. Even today, a lot of ETSI documentation on SDH rings references ANSI documents that technically apply only to SONET rings. The few distinctive features of SDH rings, especially with regard to terminology, are presented in a later section of this chapter.

The four types of SONET rings discussed in this section are:

1. Unidirectional line-switched rings (ULSR), which are seldom mentioned in newer documentation.

2. Unidirectional path-switched rings (UPSR), which are the primary type of unidirectional ring in use today.

3. Two-fiber bidirectional line-switched rings (2F BLSR), which are a very common ring type, especially in areas where fiber availability is at a premium or four fibers are not called for.

4. Four-fiber bidirectional line-switched rings (4F BLSR), which are also very common, especially in fiber-rich areas such as the United States, and for large rings with many miles of fiber spans that are at risk from any number of environmental hazards.

The ring terminology and definitions reveal something about the ring operation and protection characteristics. ANSI provides standard definitions for all of the terms. In any form of protection switching, linear or ring, key concepts are the idea of "working" traffic or fibers or spans or signals, and "protection" traffic or fibers or spans or signals. This just means that some of the fibers on a ring are used for the transport of live ("working") voice or video or data traffic, and that other fibers are often used for the transport of a "copy" of that traffic for "protection" purposes. Protection fibers do not have to carry a copy of the live traffic. They can even be used for "unprotected" traffic, increasing the useful capacity of the ring. But having a copy of the live traffic more or less instantly available on the ring minimizes customer service interruptions. The practice of duplicating traffic bits for protection switching purposes is known as "bridging." When a SONET NE has to select from the alternate of the two sources available for bridged traffic, this is known as "switching." The key SONET ring concepts are therefore bridging and switching. SONET NEs arranged in a ring are called "ring nodes." Current specifications limit the number of SONET NEs on any fiber ring to 16 ring nodes, owing to limitations in the field sizes of the K1/K2 bytes (four bits).

Unidirectional and Bidirectional

When it comes to rings in particular, the rings might be unidirectional or bidirectional. According to ANSI, working traffic for a bidirectional stream of traffic (almost universally the case) will always "travel around the ring in the same direction ..." on a unidirectional ring. By convention, working traffic flows clockwise (CW), and protection traffic flows counterclockwise (CCW) in figures. In other words, each connection between users on the SONET ring "uses capacity along the entire circumference of the ring." In contrast to unidirectional rings, bidirectional rings carry working traffic so that "both directions of a bidirectional connection travel along the ring through the same ring nodes, but in opposite directions."

These definitions of unidirectional and bidirectional apply to SONET rings only. Unidirectional and bidirectional protection switching also exist, but their definitions have nothing whatsoever to do with ring operation. Unidirectional protection switching is a *linear protection* SONET concept that ANSI essentially defines as using a backup fiber when a fiber in one direction on a span fails, leaving the working traffic on the original fiber unaffected. Bidirectional protection switching (the more common form) moves traffic from *both* the failed fiber and its unaffected return fiber to a protection pair of fibers. This difference between unidirectional and bidirectional *rings* and unidirectional and bidirectional *protection switching* was, based on feedback received, a source of confusion to readers of previous editions of this book. The present clarification should prevent further confusion over the use of the terms "unidirectional" and "bidirectional" in this chapter. It should be clear from the context when the unidirectional/bidirectional ring type or unidirectional/bidirectional protection switching type is meant.

Figure 16-4 shows the major difference between unidirectional and bidirectional rings. In each case, the normal operational traffic flow is shown. The unidirectional ring happens to be a path-switched ring, but this will become clearer in the following section. This applies to SONET and SDH rings alike.

Note the usual convention for working and protection traffic in unidirectional rings: clockwise for working and counterclockwise for protection. The six ring nodes are shown with only one bidirectional connection, perhaps a virtual tributary, from a TM on ADM node B to a TM on ring node E. Traffic from B to E flows through nodes B-C-D-E, always on the working fiber. Traffic from E back to B flows through ring nodes E-F-A-B, always clockwise, always on the working fiber. The figure also shows that each source sends traffic in *both* directions over the working and protection fibers (protection traffic and fibers are shown as dotted lines, a convention followed throughout this chapter). This is bridging. Likewise, each receiver has a choice of two incoming signals: the signal on the working fiber and the signal on the protection fiber. Unless the working signal is clearly inferior to the protection signal, the receiver always selects the working signal. Working traffic therefore always exhibits the asymmetrical delays characteristic of unidirectional rings as working traffic flows around the ring.

The figure also shows the normal operation of a bidirectional ring. Again, fibers carrying working traffic are shown as solid lines and arrows, and fibers carrying protection traffic are shown as dotted lines and arrows. Bidirectional rings also bridge and switch when ring protection is required. This time, however, the working traffic always follows the

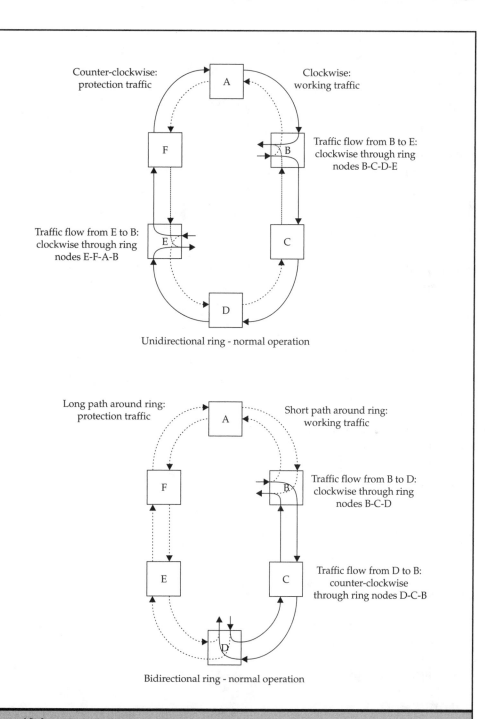

Figure 16-4. Examining normal operation of unidirectional and bidirectional rings

shortest path around the ring, in opposite directions (obviously, a special case occurs when both ring nodes are equidistant). In the figure, working traffic from ring node B to ring node D flows through B-C-D, and the return path is through ring nodes D-C-B. Note the use of both fibers for working traffic, and the availability of the "long path" around the ring for protection. Sources and receivers bridge and switch as before. The details of how these types of rings behave when fibers fail will be fully explored later in this chapter.

Line Switching and Path Switching

In addition to being unidirectional or bidirectional with regard to traffic, a SONET ring can provide line switching or path switching. These types simply refer to the SONET level at which the protection switching takes place, but fundamental differences in operation do exist.

As might be expected, path switching takes place at the SONET path level and the "trigger mechanism" is found in bits present at the SONET path overhead level. Essentially, a path-switched ring consists of two counter-rotating rings, clockwise for working traffic and counterclockwise for protection, and the channels on one ring protect the channels on the other. Path-switched rings operate using a simple 1+1 protection scheme borrowed from linear protection mechanisms. Incoming signals are permanently bridged at the ring node closest to the source and sent in both directions around the ring to the ring node closest to the destination. One of the two signals is selected as the working path, as long as the choice is consistent at all ring nodes.

Line switching uses the K1/K2 APS line-level overhead bytes directly to perform switching at the SONET line level. Only the line level of the user channel connection is affected, and no real change is visible to users at the path level. Line-level overhead includes signals for failure conditions and messages to coordinate line protection switching between adjacent ring nodes ("where is the protection channel?"). Line protection always takes place between adjacent ring nodes.

Two-Fiber and Four-Fiber

Finally, SONET rings are distinguished by the number of fibers that connect each node on the ring. There are two-fiber rings and four-fiber rings. All unidirectional rings are two-fiber rings, more or less by definition. On any fiber failure, two-fiber rings have no choice but to ring switch and seek to continue the connection the "other way" around the ring.

Four-fiber rings are capable of another type of protection switching in addition to the usual ring switching at the line or path level. Four-fiber rings can perform "span switching." Span switching is just a form of 1+1 bidirectional linear protection applied to rings with four fibers available between every ring node. One fiber pair is for working traffic; the other pair is for protection traffic. Span switching just moves over to the protection pair before needing to ring switch and alter the traffic flow around the ring itself. Span switching is therefore local between two adjacent ring nodes.

Ring Applications

So three or four types of SONET rings can be built and deployed. But when and where is each type of ring most appropriate? This issue goes above and beyond simple two-fiber or four-fiber considerations. Naturally four-fiber rings (with the addition of span switching) would be preferable to using two-fiber rings exclusively. But, at some point, the added cost and complexity of doubling the fiber requirement might reach the point of diminishing returns. In other words, four fibers everywhere is often complete overkill.

Fortunately, experience with fiber rings over the years has defined the roles best suited for the different types of SONET rings fairly well. Keep in mind that the discussion that follows offers only guidelines, and any type of SONET ring is better in all cases than a simple network of point-to-point links.

In general, unidirectional rings are used to cover smaller geographic areas than are bidirectional rings. The always-in-the-same-direction aspects of unidirectional rings—borrowed from older ring technologies such as fiber distributed data interface (FDDI)—makes the delay experienced by end users asymmetrical. In other words, the delay for outbound traffic is not equal to the delay for inbound traffic. Usually, this is not much of a problem, *as long as the ring is kept small enough*. In this case, "small" means total fiber miles and has nothing to do with the number of ring nodes. Delays on networks consisting of a series of point-to-point links are symmetrical more or less by definition, of course.

Asymmetrical network delays can be a problem in voice, because once the end-to-end delay is high enough, mostly due to the length of the circuit carrying the speech, echo cancellers are required to avoid annoying speech repetitions in the speaker's ear. A full discussion of the details of modern electronic echo cancellers is beyond the scope of this book. It is enough to point out that, in the vast majority of cases, echo cancellers are devices that are assigned in pairs, one in each direction, on a voice call. On a SONET unidirectional ring, it could be the case that an echo canceller is required in one direction (the one with the higher delay), but not in the other. There is no easy way to do this, because the need for echo cancellers is usually determined strictly by the source and destination telephone numbers or locations for the call. (Even mobile telephone users are still rooted to a particular cell tower at one or both ends.) Using echo cancellers in both directions for all telephone calls on a ring is wasteful and expensive. To eliminate the need for echo cancellers on the ring, the size of the ring in terms of distance must be limited. (This ignores special cases such as combined ring–point-to-point calls, but the main point is still valid: asymmetrical voice delays can complicate echo cancellation.

Data is usually more forgiving of asymmetrical delays, except (of course) when a data protocol such as TCP/IP is used to carry streaming voice or video. This is more common today that many people assume. And when delays reach a certain point, TCP windowing ("flow control") parameters might need to be adjusted, often at both ends, although the problem may exist in only one direction on the ring. Naturally, the same size limitation avoids TCP timeout performance issues as well.

What this all boils down to is that unidirectional rings are often said to be most suitable for access networks where distances are limited to a few miles in a major metropolitan or downtown area. This generalization is acceptable, as long as the reasons behind it are firmly kept in mind.

Sometimes (especially in older documents), it is claimed that unidirectional *path* switching is most suited for rings where most traffic flows from user sites to a central service-provider location, such as a telephony central office. On the other hand, the same sources often claim that unidirectional *line* switching is best suited for highly distributed, almost random connectivity patterns, such as those seen in a metropolitan fiber ring connecting user sites directly to one another for router-to-router connectivity. This position is normally justified by pointing out that path switching should be more effective when the paths all terminate at a central site. This point is discussed a little more when ring types are being detailed later in this chapter. Most unidirectional rings today are UPSRs, mostly owing to their deployment by traditional telephone companies that are firmly built on the central-office model.

Bidirectional rings are always line switched, so that the path-or-line argument loses steam. Bidirectional rings have symmetrical delays, making them suitable for rings and spans of any size at all (again with the ring-node limitation of 16 always kept in mind). The traffic distribution pattern is of secondary importance. Not only can bidirectional rings be used for small access networks, they are also suitable for the largest SONET rings that can be built. Naturally, few huge SONET rings spanning a country (or even several countries) gather traffic and connections to a central site. The distributed nature of the connections on a bidirectional ring is therefore always a consideration that tilts this type of ring toward line-switched operation. Access networks can also use bidirectional rings, of course.

Once a ring design decision in favor of bidirectional operation is made, only the consideration of two fibers or four fibers remains to complete the overall architectural design. There are two advantages to having four fibers between every ring node instead of two. First and foremost, the ability to perform span switching is a nice feature on rings with spans that might be hundreds of miles long, running over mountains and across valleys in isolated areas. Repairs to broken fibers in these areas might be slow. The longer the span, the more can go wrong. And two-fiber rings are always "broken" into two segments or more by multiple concurrent failures. Four-fiber rings can absorb multiple failures and still remain intact, assuming that the protection pair of fibers between two ring nodes that have span-switched from the working fiber pair are not the exact other fibers that have failed. Route working/protection diversity helps prevent this situation.

Also, having twice the fiber on the rings can effectively double the capacity on the ring. This might seem like a tautology, but the situation is complicated by the practice of allowing rings to carry unprotected traffic on the channels to be used when protected working traffic is switched. A special class of unprotected, extra traffic is non-preemptive unprotected traffic (NUT). Although this traffic is not protected from fiber failures, it cannot be preempted (replaced) by any other traffic on the ring when failures occur. Customers usually pay less for unprotected traffic on the ring, or more for protected traffic, depending on the rate structure of the service provider. Four-fiber rings are common on long-distance rings built by inter-exchange carriers in the United States (Sprint more or less invented the four-fiber BLSR), but scarce outside of such fiber-rich infrastructures.

Table 16-2 shows the typical role for each of the SONET ring types.

Keep in mind that, in most cases, more than one type of ring is a viable solution. The table reflects typical ring-type applications.

Ring Type	Application
ULSR	Access networks (small metro rings) with distributed traffic patterns
UPSR	Access networks (small metro rings) with site-to-central office traffic patterns
Two-fiber BLSR	Access or regional or national networks with distributed traffic patterns and minimal fiber availability
Four-fiber BLSR	Access or regional or national networks with distributed traffic patterns and high fiber availability

Table 16-2. SONET rings and their applications

REAL-WORLD SONET RINGS

Not much has been said yet about SONET ring speeds and protection features. Generally, SONET rings start at OC-3 speeds, but many service providers begin their ring structures at OC-12 and above. SONET rings are referenced and characterized by their line-side ring fiber speeds and architectures: an OC-12 UPSR, an OC-192 BLSR, and so on.

A lively debate revolves around the length a time a SONET (or SDH) ring must perform a protection switch. The intervals of 50 ms and 60 ms are most often mentioned in specifications and vendor documentation. Some sources also break this down into 10 ms to detect a failure on the ring, and then 50 ms to react to the failure and restore service. This is literally faster than the blink of an eye, given that humans take a leisurely 100 ms (1/10th of a second) to blink. The story goes that Sprint, essentially the inventor of the four-fiber BLSR, would demonstrate the effectiveness of ring protection by putting streaming video over a demonstration SONET ring and then having a technician cut a pair of fibers between two of the ring nodes with an axe. The point was supposed to be that, after a short freezing of the picture, the video continued uninterrupted. The problem was that few people ever saw the effect. Most witnesses from the telecommunications industry, raised in the world of vulnerable point-to-point links, actually closed their eyes or looked away from the dreaded axe, and most missed this very valuable demonstration of the value of SONET and SDH rings.

The limiting factor in protection switch speed is not the APS protocol itself. Fifty milliseconds is enough time to send 400 SONET/SDH frames. The limitation is the ability of SONET/SDH NEs to detect and react to APS messages rapidly, while at the same time performing their usual framing and switching operations 8,000 times per second on multiple ports. It is totally unreasonable to expect ring connections to restore themselves in 60 ms, especially on large unidirectional rings.

Ring nodes that simply pass traffic through can react faster than ring nodes that must loop or switch traffic around when failures occur. The best discussion of ring-protection switching delay is in Appendix D of the ETSI APS specification (ETS 300 746). This assumes 16 ring nodes on a ring with a total of 746 miles (1,200 km) of fiber connecting the (unidirectional) ADM ring nodes. The fiber transmission delay is assumed to be 5 μs (0.000005 seconds) per kilometer, and the time it takes each node to validate an APS message is three frame times (0.375 ms or 0.000375 seconds). The processing delay for frames passing through a pass-through node is given as the variable Dp, and the processing delay through a switching node that must do more is given a multiplication factor of J. So delay through these ring nodes, such as those adjacent to the failure, is $J \times Dp$.

Those interested in the details should consult the document itself. A formula for the worst-case scenario is included. The role of each node is added to the final time interval. The calculation begins with the time it takes for the detecting node to react to a fiber cut ($J \times Dp$). Then comes the time it takes for the 14 other ring nodes to pass the APS message the long way around the ring: $14(Dp + 0.000375)$. Next comes the time to process the message at the ring node at the other end of the break: $(J \times Dp) + 0.000375$. This node must generate a response in $J \times Dp$, and this response is passed back along the ring ($14 \times Dp$). Finally, the message is received and processed at the originating node: $(J \times Dp) + 0.000375$. The long-path transmission delay, applied twice (once in each direction) is $15/16(1200 \times 0.000005)$.

When these times are added, the ETSI formula for unidirectional ring protection switch delay (Sd) is given as

$$Sd = (4J \times Dp) + (28 \times Dp) + (16 \times 0.000375) + 2(15/16(1200 \times 0.000005)) \text{ seconds}$$

$$Sd = 4(J \times Dp) + 28Dp + 0.01725 \text{ seconds}$$

If $J = Dp$, so that it takes twice as long to process a protection message as to just pass it along through a ring node, then the equation takes the form

$$Sd = 4x^2 + 28x + 0.01725 \text{ seconds}$$

This curve is fairly steep, and the protection delay grows rapidly for even small values of x (Dp). For example, if $Dp = 500$ ms, then the switch takes place in about 16.4 seconds. But if $Dp = 100$ ms, then the protection switch happens in a little less than 3.4 seconds. With a protection nodal processing delay of 50 ms, the protection switch around this worst-case ring is a little less than 1.5 seconds.

BLSRs are supposed to be optimized with the K1/K2 APS bytes to protect basic STS-1 connections, but of course this protects whatever virtual tributaries are inside the STS-1s. The point here is that no SONET ADM ring node on a BLSR is going to shift VTs around among STS-1s to protect traffic. If the STS-1 keeps flowing, that's good enough for the ring.

However, UPSRs do not react at the STS-1 level. Path switching relies on path-level overhead to detect and react to failures on the ring. Virtual tributaries have to have their own mechanisms to respond to error conditions.

This opens up the whole issue of how all of the major ring types detect and respond to failure conditions. The next sections detail the protection operation of each major ring type. In all cases, only a few simple fiber failure conditions will be considered. Things are much more complex when ring architectures and errors are more complicated, such as when UPSRs are connected to or with a BLSR, or when an entire ring node is lost. The specifications detail many of these scenarios, which is why they are more than 100 pages long; this chapter addresses only the basics.

UNIDIRECTIONAL LINE-SWITCHED RING PROTECTION

As already mentioned, ULSRs are not often mentioned in vendor or standards documentation (ULSRs rate a mere six sentences in the ANSI specification). But they still exist, at least in theory, and are worth exploring if for no other reason than to contrast them with UPSRs, especially with regard to protection characteristics.

Figure 16-5 shows a ULSR in normal operation and after a fiber failure between ring nodes C and D. It assumes that (as is almost always the case) both fibers have been cut.

Note that sources place working traffic only on the working fiber. They do not bridge signals onto both fibers, nor do receivers have to choose between different signals. This is not necessary on a ULSR. In fact, the protection fiber is just "there," and the ring nodes do not even pass the signals through on the protect fiber. No live signals are found there anyway. Normally, the connection traffic flows from B to E through ring nodes B-C-D-E. Return traffic flows through ring nodes E-F-A-B.

The figure also shows what happens when the fibers between ring nodes C and D fail. Note that only traffic from B to E is affected, not traffic from E to B. The ring nodes adjacent to the failure, nodes C and D, know about the failure right away and loop traffic from the working fiber onto the protection fiber. The other ADMs on the ring, nodes A and F, must now pass the protection traffic through (usually called "switching to through mode").

Finally, note that the sources and receivers still function exactly as before. It is the ring that changes. Naturally, the K1/K2 APS bytes coordinate all this activity, but the exact configuration of the ring is essential. Each node on a ULSR must know whether it is adjacent to the failure, local to the source and receiver, or a through node on the ring.

UNIDIRECTIONAL PATH-SWITCHED RINGS PROTECTION

UPSRs, while still unidirectional in nature, operate and protect differently than do their ULSR relatives. As mentioned earlier in this chapter, all path-switched rings operate using a simple 1+1 protection scheme borrowed from linear protection mechanisms. Source signals are therefore permanently bridged at the ring node and sent in both directions around the ring to the ring node at the destination. One of the two signals is selected as the working path, usually the clockwise, working fiber, unless this fiber is broken.

Figure 16-6 shows a UPSR in normal operation and after a fiber failure between ring nodes C and D. It is assumed that (as is almost always the case) both fibers have been cut.

As in the case of the ULSR, this UPSR failure affects only the traffic from ring node B to E, not traffic from E to B, which continues as before. In contrast to ULSRs, which must coordinate activities among many ring nodes, UPSR protection and recovery is local and swift. Because the traffic is present both on the working and protection fibers at all times, the senders continue bridging just as before, as shown in the figure. Receivers, which are always comparing the signal quality on the working and protection fibers through the SONET (or SDH) overhead bytes at the path level, simply switch over to the signal on the protection fiber.

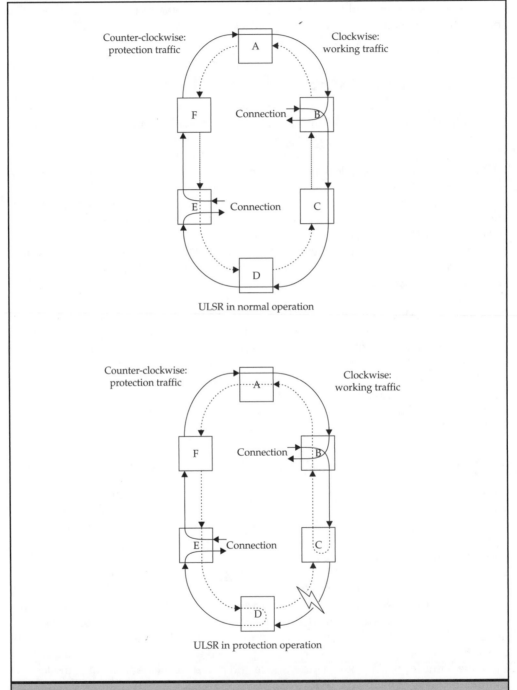

Counter-clockwise: protection traffic

Clockwise: working traffic

Connection

ULSR in normal operation

Counter-clockwise: protection traffic

Clockwise: working traffic

Connection

Connection

ULSR in protection operation

Figure 16-5. Examining unidirectional line-switched ring operation and protection

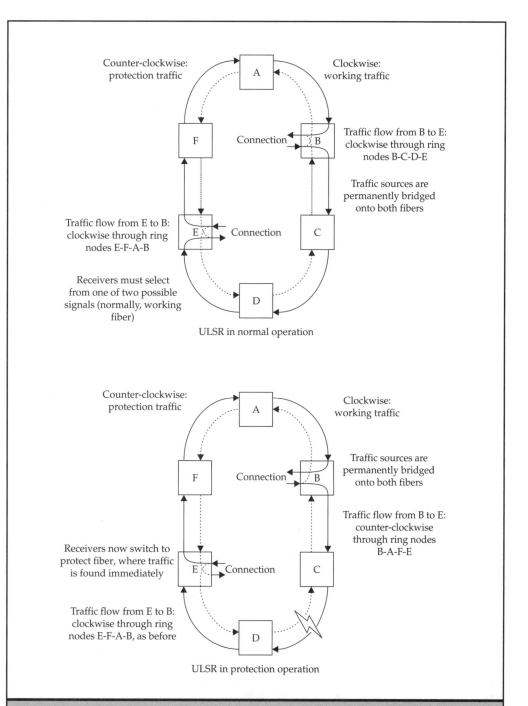

Figure 16-6. Examining unidirectional line-switched ring operation and protection

Sometimes the difference in signal quality between the working and protection pair is obvious to the receiver, especially when the working traffic vanishes altogether. Rings are allowed to exhibit "revertive behavior," however. This means that if a decent enough signal reappears on the working fiber, receivers can automatically revert to the signal on the working fiber. However, this behavior is usually optional, because as fiber repairs proceed, a fiber might appear to have a restored signal when it is not quite ready to be put back into service (in many cases, extensive testing is required on repaired fibers). Intermittent failures might cause "thrashing" between working and protection fibers. "Failover" is therefore often automatic from the working to the protection pair, but manual restore is required to return the ring node to the working pair once repairs are completed and testing is finished.

A related concept, applied mostly to line-switched rings, is known as "lockout". Lockout prevents a ring node in normal operation from switching from a working to a protection fiber (perhaps the protection fiber is being tested), or a ring node in protection operation from switching from a protection to a working fiber. Lockout messages are included in the K1/K2 APS bytes.

UPSRs protect without needing elaborate protection protocols. The ring nodes can perform their tasks locally, without worrying about the details of ring configuration. This made UPSRs very attractive for access networks with multi-vendor environments that could not rely on consistent APS protocol implementations. Consider a service provider wanting to deploy an access ring in a downtown area. If the service provider wants to allow some of the ring nodes to be customer premises equipment (CPE) owned and operated by the customer (simpler to deploy, but also beyond the control of the service provider to specify), UPSRs seem to be the way to go. This is especially true for connections leading to a central site.

As has been pointed out in this section, intermediate nodes are essentially transparent to UPSR protection switching. This local aspect of UPSRs leads directly to consideration of the "virtual ring." A virtual SONET ring consists of UPSR nodes linked by BLSR nodes. This might sound complicated, but it really isn't. A virtual ring is shown in Figure 16-7.

The central ring is a BLSR ring. No details on BLSRs have been discussed yet, but these rings will be investigated shortly. The important thing to realize is that line switching relies on the K1/K2 APS protocol to perform protection switching, so that BLSR ring nodes must generate, process, and exchange messages to coordinate the task of moving working traffic from one fiber to another. The UPSR nodes connect to the BLSR as shown, but form their own virtual ("logical") UPSR ring among those five UPSR nodes. As long as the BLSR distributes signals where the UPSR ring nodes expect them, the BLSR nodes just seem to be another (protected) span on the virtual UPSR ring. The local protection aspects of UPSR nodes make this type of ring possible.

To select the best signal at the receiver, UPSRs generally rely on four major path-level indicators. These are things such as the path-level AIS (AIS-P), path-level loss of pointer, path-level signal degrade, and excessive path-level BIP errors. Oddly, a protection-switched UPSR looks very much like a bidirectional ring when it comes to traffic flow. But there are differences, of course.

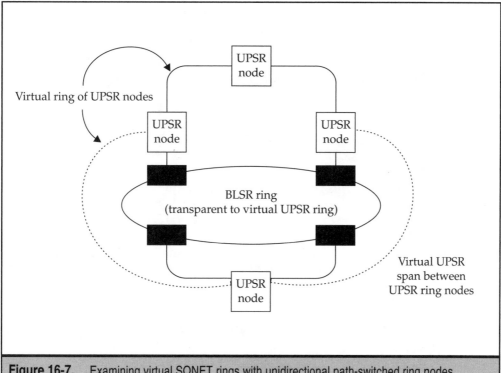

Figure 16-7.   Examining virtual SONET rings with unidirectional path-switched ring nodes and bidirectional line-switched ring nodes

BIDIRECTIONAL LINE-SWITCHED RINGS

SONET bidirectional line-switched rings can be built with two fibers and with four fibers. Two-fiber BLSRs look very much like any other ring that uses two fibers between each ring node. In fact, UPSRs, ULSRs, and two-fiber BLSRs are essentially identical. It is only the ring-node protection actions that distinguish the types. Only four-fiber BLSRs are fundamentally different from other ring types. Span switching capability is added to the usual ring switching behavior of all two-fiber rings. Like any other line-switched ring, BLSR rings can have up to 16 nodes, and 746 miles (1,200 km) of fiber in total on the ring perimeter.

Two-fiber BLSRs exhibit the same type of line-level ring switching behavior that ULSRs exhibit. In other words, fiber failures result in working traffic being shifted to protection channels. But, to appreciate the differences from unidirectional operation, those differences must be detailed.

Two-Fiber BLSR Protection

Two-fiber BLSRs provide protection against the failure of any individual fiber pair on the ring. Two-fiber BLSRs are most often used by local exchange carriers in major metropolitan areas. The bidirectionality gives a symmetrical delay for data users. These two-fiber rings are quite economical to deploy because only two fibers are used to close the ring.

However, a potential drawback with two-fiber BLSRs is that spare protection capacity must be reserved in all fibers to provide for the protection switching feature. This means that two-fiber BLSRs are not especially efficient in using installed bandwidth and must be "loaded" and configured carefully.

For example, in an OC-12 ring, the first 6 STS-1s are used for live traffic (the working fiber) and the last 6 STS-1s are used for the protection capacity. Under normal operating conditions, the SONET equipment will send the traffic on both outbound fibers in the usual bridging arrangement. However, in the vast majority of cases, both fibers are run in the same conduit. Therefore, both will fail at the same time.

When this happens, STS-1s 7 through 12 are used for the protection switched traffic. Figure 16-8 shows the general method by which two-fiber BLSRs provide protection switching. A lot is going on in the figure, and so a few words of explanation are in order.

Figure 16-8 shows a five-node OC-12 two-fiber BLSR. Naturally, the capacity of the fiber in each direction is 12 STS-1s. Again, the important point is that only 12 STS-1s support the *whole ring*. The figure makes this point by showing the third STS-1 on the ring (labeled "STS-1 #3") as configured (mapped) between node A and node C. In normal operation (shown in the upper portion of the figure), this working STS-1 would be mapped bidirectionally through node B (as is also shown in the figure). No user traffic could be mapped onto STS-1 #3 at node B, because this STS-1 is obviously in use, and user traffic must pass right through node B.

The first six STS-1s in each fiber are the working channels on the ring. Figure 16-8 shows the structure of the STS-12 frames as they travel around the ring, labeling each span with the working and protect STS-1s and the nodes they connect. Only the working STS-1s are protected. Other STS-1s can be used in the protect STS-1s; but, in case of a failure, such unprotected traffic is preempted. So STS-1 #8 (for example) could be mapped between node B and node C (or node D and node E) and work as long as the ring was intact. In fact, STS-1 #3 is available on the working STS-1s between node E and node D, regardless of whether operation is normal or not. All that is needed is careful configuration and mapping of the capacity on the ring as a whole.

It used to be claimed that two-fiber BLSRs limited capacity to half of what a two-fiber UPSR could furnish. But with careful loading and capacity management, two-fiber BLSRs can be very efficient. Today, software packages make this capacity planning much easier than in the early days of SONET rings, when the whole ring-mapping scheme was kept on a piece of paper and taped to the side of a node.

The lower portion of Figure 16-8 shows what would happen on the two-fiber BLSR if the pair of fibers between node A and node B were cut. Now the protection STS-1s, #7 through #12, carry the protected working STS-1s between node A and node B the long way around the ring, as shown and labeled in the figure. Note that any traffic mapped to STS-1 #3 between node D and node E is unaffected by the protection switch. However, any traffic mapped to STS-1 #8 between node D and node E would now be preempted (which is why unprotected traffic services on a SONET ring are frequently offered at a deep discount as compared with protected services).

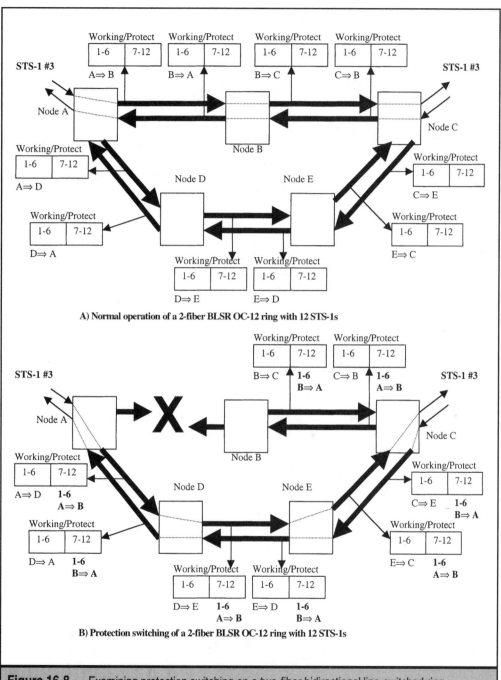

A) Normal operation of a 2-fiber BLSR OC-12 ring with 12 STS-1s

B) Protection switching of a 2-fiber BLSR OC-12 ring with 12 STS-1s

Figure 16-8. Examining protection switching on a two-fiber bidirectional line-switched ring

An apparent oddity is also illustrated in the lower portion of Figure 16-8. When the ring is in protection mode, it might seem (because STS-1 #3 is now taking the long way around the ring) that STS-1 #3 is now available between node B and node C! But configuring this STS-1 would destroy the mapping scheme when service was restored to normal operating conditions. It took a while to get rings to "squelch" (the official term) this oddity, but protocols in the SONET line overhead now prevent this from happening while protection mode is in effect.

It is sometimes claimed that two-fiber BLSRs impose a "bandwidth penalty" on the fibers, because protection bandwidth must be reserved on the ring. However, service connection can be routed around either side of the ring, so that the full OC-12 capacity is available at any ring node. This does not mean that loading such a ring is trivial, but no bandwidth is "lost" on two-fiber BLSRs as is sometimes claimed. The reuse pattern must be carefully planned, so that some services are mapped the "long way around" and other are mapped more directly. This makes two-fiber BLSRs ideal for regional or national distributed mesh connectivity, and not so well suited for local site–to–central office applications.

Figure 16-9 shows how bandwidth reuse is configured on a two-fiber BLSR. Other scenarios are possible; this is just an example.

Service from ring node A to node B is using STS-1 channel 1, but so is service from node B to C. The same applies to the service from node C to A. In fact, it could be claimed that the ring is now in a sense "oversubscribed," because the same STS-1 channel now carries the STS-3 amount of customer traffic.

Two-fiber BLSRs used in access networks with central office hubs have capacities equal to UPSRs at the same line rate. But a distributed mesh environment can "oversubscribe" the BLSR many times over, a nice feature, especially in best-case node-to–adjacent node connection scenarios.

Another nice feature of BLSRs (two-fiber and four-fiber alike) is the ability for the BLSR to carry unprotected "extra traffic." This extra traffic is removed from the protection channels whenever protection switching occurs. Naturally, the ability to sell extra traffic capacity offers a powerful financial incentive for BLSRs. Applications for extra traffic channels include diverse routing (it is hoped that everything does not fail at once), discounted services (a lower rate than protected traffic), and "degradable" data services (aggregated channels that exhibit lower data rates when channels fail).

As robust as two-fiber BLSRs are, the absolute maximum protection from fiber failures is offered by four-fiber BLSRs.

Four-Fiber BLSR Protection

Four-fiber BLSRs provide protection against the failure not only of any individual fiber pair between nodes on the ring (span switching), but also of *both* fiber pairs between any two nodes on the ring (the usual ring switching). Four-fiber BLSRs are most often used by inter-exchange carriers for regional or even national (and in at least one case, international) rings. The signal is propagated both ways around the ring. The destination picks the "best" signal to deliver to the user. This gives a symmetrical delay for data users. Four-fiber rings are quite expensive to deploy because four fibers are used to close the ring; however, four-fiber BLSRs can easily be loaded to capacity, making them very efficient once installed.

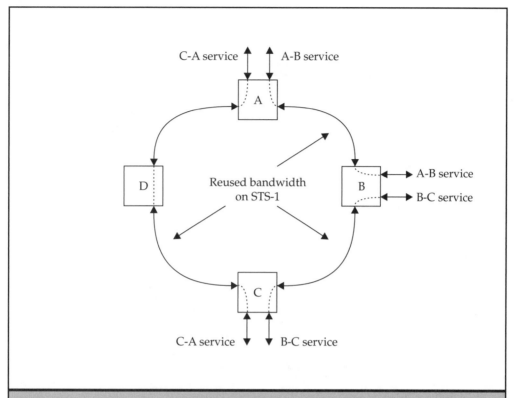

Figure 16-9. Illustrating bandwidth reuse on a two-fiber bidirectional line-switched ring

In normal operation, the signals are bridged and sent in both directions around the ring. If one pair were to fail, the signal would be sent *in the same direction* on the second pair of fibers (span switching). In the event that the second pair were also to fail between the same nodes, the signal now would be picked up on the protect pair of fibers on the "long way" around the ring (ring switching). Thus, four-fiber rings can continue to operate even in the face of multiple failures.

Four-fiber BLSRs can therefore span switch *and* ring switch. Typically, all that is needed is to swap ("span switch") the working pairs for the protection pairs between two SONET NEs. In other words, the traffic flow between ring nodes is not disrupted. But, if a ring switch is necessary, then all traffic between two ring nodes is disrupted, and the ring "wraps" to shunt traffic onto the protection fibers all the way around the ring. Figure 16-10 shows these two types of protection switching.

Figure 16-10 shows an STS-N that is provisioned on a SONET ring (perhaps OC-12 or OC-48) between two sites attached to ring node A and ring node F. The fibers labeled "working" and "protect" are two fibers in each case. In normal operation, the working fibers be-

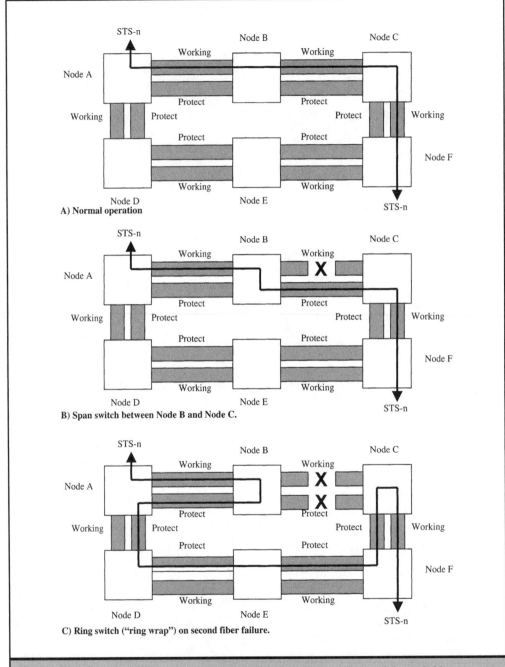

A) Normal operation

B) Span switch between Node B and Node C.

C) Ring switch ("ring wrap") on second fiber failure.

Figure 16-10.   Illustrating four-fiber bidirectional line-switched ring span and ring switching

tween nodes A and B, and nodes B and C are used to carry this traffic. Note that the protect fiber pairs might carry unprotected ("non-protected" in some documentation) traffic at the same time.

The middle portion of the figure shows how span switching works in the case of a fiber failure between nodes B and C. All working traffic between node B and node C is shunted onto the protect fiber pair. Any unprotected traffic mapped between the other nodes on the ring is unaffected by this outage.

The bottom of the figure shows the loss of both pairs of fibers between node B and node C. Typically, the fibers would be diversely routed between these nodes. However, because even 25 feet of facilities separation can qualify as "route diversity," road trenching and similar activities can easily wipe out both pairs of fibers. Now the ring wraps to provide ring switching and all of the protect pairs are used. All of the unprotected traffic is now preempted.

In many cases the OC-N traffic would be sent simultaneously on both the working and the protect fiber pairs. This would make it much easier and faster to wrap the ring when ring switching is necessary. The protection switching documentation has other scenarios for four-fiber BLSR failures, especially node failures. This section has outlined only the basics.

In all cases of ring wrap, whether on a two-fiber BLSR or a four-fiber BLSR, all the user sees is a very brief period of "garbage" traffic, followed by a jump in the end-to-end delay, reflecting the new flow of the traffic signal. SONET ring delay for user traffic is treated more fully later in this chapter.

RINGS OF RINGS

As SONET/SDH deployment progresses, it will become more common not to see isolated SONET/SDH rings, or bigger and bigger SONET/SDH rings (which just increase delay, as will be pointed out later), but "rings upon rings." That is, lower-speed SONET/SDH rings (access rings) will feed higher-speed SONET/SDH rings (backbone rings) in a virtually unlimited array of speeds and sizes.

In fact, it is quite common today in a SONET environment to have "low-speed" OC-3 rings feeding OC-48 rings. Both may even be four-fiber BLSRs. In some cases, the OC-48 rings feed an even higher-speed OC-192 ring spanning an entire regional service area or crossing national borders (into Canada, at least). Each ring is failure resistant on its own and has all of the SONET management advantages for service providers and customers alike. Figure 16-11 shows the structure of such an interconnected ring architecture.

Although the example shows interconnected four-fiber BLSRs, such interconnectivity is not limited to such rings, of course. In fact, it is possible to interconnect rings of any form, including two-fiber UPSRs to two- or four-fiber BLSRs. The procedure is not trivial, but can be done, and even provides the proper protection switching for traffic mapped across the mixed ring configurations.

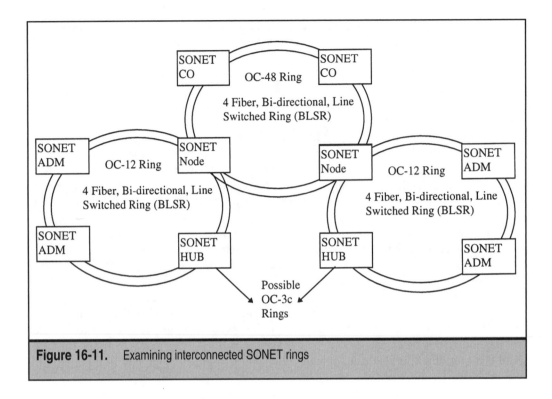

Figure 16-11. Examining interconnected SONET rings

RING AUTOMATIC PROTECTION SWITCHING

There is much more to SONET/SDH ring APS than just getting the OC-N or STM-n to detect a loss of signal (LOS) and to perform some signal reconfiguration. APS takes place at the line/multiplex section level in SONET/SDH anyway, and LOS is an electrical alarm. All of the electrical channels inside an optical fiber must therefore be able to decide if and when an APS should take place. This task is very complex, and the current ANSI document on the SONET APS issue, ANSI T1.105.01-2000, is more than 100 pages long. This section explores only some of the basics of SONET/SDH ring APS, emphasizing the role played by the K1/K2 bytes in the SONET/SDH overhead.

The K1/K2 bytes play a role in linear SONET/SDH APS as well, but only ring operation is detailed here. Figure 16-12 shows the structure and messages of the K1/K2 bytes as they are used in a ring configuration.

A lot is happening in what appears at first to be a simple figure. No full explanation of the entire APS message protocol can be attempted in a section of a chapter, but some of the overall goals and aims of the K1/K2 byte operation can be outlined. The full ANSI documentation contains many illustrations and should be consulted for details (which also apply to SDH, as mentioned earlier in this chapter). But the rest of this section applies only to SONET. SDH deserves a section of its own later on in this chapter.

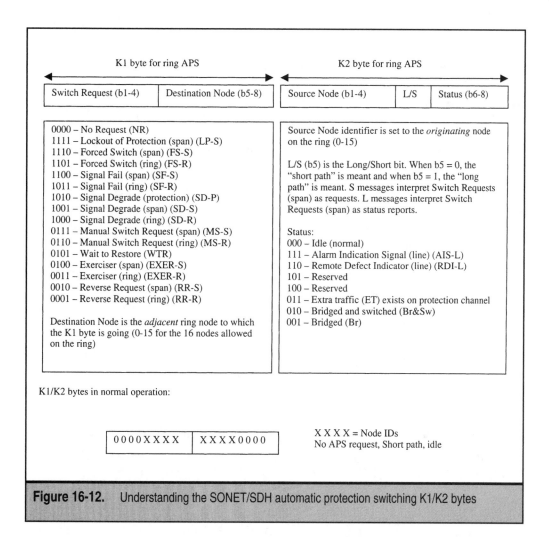

K1 byte for ring APS   K2 byte for ring APS

| Switch Request (b1-4) | Destination Node (b5-8) | | Source Node (b1-4) | L/S | Status (b6-8) |

0000 – No Request (NR)
1111 – Lockout of Protection (span) (LP-S)
1110 – Forced Switch (span) (FS-S)
1101 – Forced Switch (ring) (FS-R)
1100 – Signal Fail (span) (SF-S)
1011 – Signal Fail (ring) (SF-R)
1010 – Signal Degrade (protection) (SD-P)
1001 – Signal Degrade (span) (SD-S)
1000 – Signal Degrade (ring) (SD-R)
0111 – Manual Switch Request (span) (MS-S)
0110 – Manual Switch Request (ring) (MS-R)
0101 – Wait to Restore (WTR)
0100 – Exerciser (span) (EXER-S)
0011 – Exerciser (ring) (EXER-R)
0010 – Reverse Request (span) (RR-S)
0001 – Reverse Request (ring) (RR-R)

Destination Node is the *adjacent* ring node to which
the K1 byte is going (0-15 for the 16 nodes allowed
on the ring)

Source Node identifier is set to the *originating* node
on the ring (0-15)

L/S (b5) is the Long/Short bit. When b5 = 0, the
"short path" is meant and when b5 = 1, the "long
path" is meant. S messages interpret Switch Requests
(span) as requests. L messages interpret Switch
Requests (span) as status reports.

Status:
000 – Idle (normal)
111 – Alarm Indication Signal (line) (AIS-L)
110 – Remote Defect Indicator (line) (RDI-L)
101 – Reserved
100 – Reserved
011 – Extra traffic (ET) exists on protection channel
010 – Bridged and switched (Br&Sw)
001 – Bridged (Br)

K1/K2 bytes in normal operation:

| 0 0 0 0 X X X X | X X X X 0 0 0 0 |

X X X X = Node IDs
No APS request, Short path, idle

Figure 16-12. Understanding the SONET/SDH automatic protection switching K1/K2 bytes

Perhaps the most important piece of the K1/K2 APS scheme is right up front. The first four bits of the K1 byte are the "switch request" bits. Sixteen messages are possible, and all are listed in Figure 16-12. In normal ring operation, no APS request is being made, and the bits are 0000. Switch requests exist for span and line switching operations, some initiated externally (for example, MS-R), and some signaled automatically (for example, SF-R) as required by the nodes on the ring. Briefly, these are the externally initiated commands carried in the K1 byte. The lockout (LP-S) disables normal span switching. A forced switch (FS-S, FS-R) is just what it sounds like: working traffic forced to protection paths. A manual switch (MS-S, MS-R) is a type of override that forces the same thing but is supposed to be used only in particular circumstances. The exercise commands (EXER-S, EXER-R) try out the protection switching circuitry and operations without harming live traffic.

The automatically initiated commands can be briefly described as well. Signal failures (SF-S, SF-R) are triggered by loss of frame (LOF) or loss of signal (LOS) at the node. There is also a signal fail for protection that is the same as the lockout message. Signal degrades (SD-P, SD-S, SD-R) are caused by situations such as elevated BIP errors. The wait to restore (WTR) is used to make sure that all nodes go back to the working channels at the same time. Reverse requests (RR-P, RR-S) are used as acknowledgments to the original switch request.

The last four bits of the K1 byte are the destination ring node identifier, not to be confused with the SONET NE address in the J0 trace byte of the section overhead. By convention, the nodes are not numbered in most documentation, but are given letter values from A to P. The relationship between node letter and ring node number is not fixed. The four bits limit SONET rings to 16 nodes or fewer, but some efforts have been made to increase this limit. The destination node is always the *adjacent* ring node for which the K1 message is intended. But, owing to the failure, this message might have to be sent the long way around the ring. So on a five-node (A-B-C-D-E) ring, if a fiber failure occurs between node B and node C, the destination node field in the K1 bytes showing up at node A will be for node C, which is adjacent to node B (but not to node A).

The first four bits of the K2 byte are the source ring node identifier. In the preceding example, the source node identifier for the K1 message to node C would be node B, the node that originated the message.

The fifth bit of the K2 byte can be confusing. This is the long/short path bit ("L/S bit"). When set to 0, it indicates a message sent on the short path, and when set to 1, it indicates a message sent on the long path. But the use of this bit has little to do with ring circumference (although every ring node contains a "ring map" of the location of all ring nodes). In practice, this fifth bit is used only in four-fiber BLSRs that can perform both span and ring switching (two-fiber rings can ring switch only). When span switching, a four-fiber BLSR looks much like 1:1 linear protection, and the L/S bit is set to a 0 (short path). Ring switching messages set this bit to a 1 (long path). Furthermore, when span switch requests are sent with the L/S bit set to 1 (long path), then all switch requests coded in the K1 byte are expected to be interpreted as status reports, not as switch commands. When this bit is set to 0 (short path), and the switch request refers to span operation, then the switch request coded into the K1 byte is interpreted as a command. So the fifth bit just helps the ring nodes figure out how to process span-related switch requests, which are of little interest except to the ring nodes at each end of the span.

The last three bits of the K2 byte are interesting as well. Eight status messages are possible, and six have been defined. Normal operation is APS "idle," which is coded as 000. Two other status messages are for the line-level alarm indication signal (AIS-L) and the line-level remote defect indicator (RDI-L), use of which was outlined in the previous chapter. There is also a message (ET) to let other nodes know that extra traffic exists on the protection channels. The ANSI specification actually has three categories for extra traffic on protection channels on a ring. There can be no extra traffic, preempted traffic, or non-preempted unprotected traffic (NUT) on the protection channels. The presence of NUT users complicates APS considerably, but this category prevents users from being thrown off the ring because of a failure. (On the other hand, NUT users are never protection switched during ring failures, which is why these users are NUTs.) All four-fiber BLSR ring nodes can have basic or enhanced NUT tables (or both) to enable the ring nodes to figure out where the allowable protection channels are located. (Presumably, ring nodes with both basic and enhanced NUTs have mixed NUTs.)

The final complication is the presence of the bridged (Br) and bridged and switched (Br&Sw) status messages. Bridging puts duplicate signals onto protection fibers or channels. Switching always takes place *with* bridging (never by itself). Some ring nodes might permanently bridge, but in some cases (1:*n*) bridging must be done before the receiver can switch to the protection fiber to find the copy of the working traffic. The simplest way to explain the difference between these two messages is to understand that, with some ring node or fiber failures, bridging must be done by the sender before the switching is done by the receiver, and when the problem is clearing, the switching is removed before the bridging. This is just another way that all ring nodes are kept informed about what they should be doing at any time during the failure.

The bottom of Figure 16-12 shows what the K1/K2 bytes for each STS would look like on a SONET ring during normal (no APS) operation. The first and last four bits of the two bytes are all 0s, indicating no APS request, short path, and APS idle. The middle eight bits form the destination and source ring node identifiers.

Even the simplest APS scenario for a ring fiber failure would be far beyond the scope of this chapter to illustrate in any detail. So, Figure 16-13 simply shows a simple five-node SONET ring in normal operation. By normal ring convention, node A is shown twice for clarity. The arrows between each node are accompanied by the values of the K1/K2 bytes during normal ring operation. Instead of reproducing a bunch of 0 bits, the figure uses the no request (NR), short path, and idle status notation. Letters are used instead of bits, again by convention and to reinforce the fact that the letters and bits have no official correspondence.

SONET/SDH LINEAR PROTECTION

This chapter has emphasized SONET/SDH ring protection architectures and techniques. This is the type of protection usually employed by service providers that are traditional telephone companies and voice carriers. Rings of ADMs might form UPSRs or BLSRs, but they are all forms of rings.

But SONET and SDH offer another type of line switching in the K1/K2 APS bytes. This linear protection switching closely resembles the same 1+1 or 1:*n* protection available in T-carrier and E-carrier systems. But, if circles of ADMs are naturally deployed as SONET/SDH rings, and the value of rings is unsurpassed when it comes to service availability, what is linear protection for?

Linear protection still has a role in two main areas in a SONET/SDH network. First of all, SONET/SDH TMs are not always collocated with the ADMs they attach to. In many cases, TMs are located at many scattered customer sites, and it makes no sense to extend the ring to each location. Yet the TM to ADM link is the weak link in many SONET/SDH

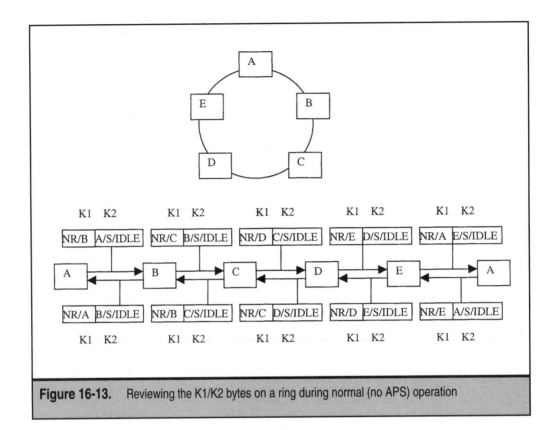

Figure 16-13. Reviewing the K1/K2 bytes on a ring during normal (no APS) operation

service offerings that make the claim for "ring protection." What good is a protected ring if the only link from TM to ADM ring node is down? Linear protection essentially extends SONET/SDH ring protection to the links connecting TMs to ADMs.

The second situation in a SONET/SDH network in which linear protection is helpful (or required) is when an IP router is used as the "TM" for packet-over-SONET/SDH (POS) applications. This is quickly becoming a more and more important application of SONET/SDH, especially as voice over IP (VoIP) becomes common among landline and wireless telephony service providers. Much the same point was made earlier in this book in Chapter 12 on SONET/SDH equipment providers. If a router carrying voice or video or data has a broken link to the rest of the backbone network, what good is that to anyone? Few routers (if any) are capable of SONET/SDH ADM operation, and they probably should not be expected to act this way, if routing is their most important task. Because of the increased role of routers in SONET/SDH networks, it is important to at least say a few words about SONET/SDH linear protection switching.

Routers are more likely to be involved in linear APS scenarios. Whether two routers are connected by a point-to-point SONET/SDH link, or a single router is acting as a TM connected to a ring of ADMs, the configuration is simple and easy to outline. SONET documentation defines a number of key concepts related to linear protection switching, including these:

▼ **1+1 Protection switching**   In this form of switching, each working fiber has a protection fiber with permanently bridged traffic. (Specifications often say "channel," but these examples use "fiber" terminology. The main points are still the same.) The receivers at each end can switch without communicating with the senders; and so an APS channel, although still nice to have, it not strictly needed in this architecture.

■ **1:*n* Protection switching**   This is a special form of general *m:n* protection switching. This just means that some number of protection fibers (*m*) is available for failures on some other number (*n*) of working fibers. With 1:*n*, there is one protection fiber for *n* working fibers between the NEs, and some form of APS protocol is necessary to coordinate the protection switching.

■ **1:1 protection switching**   This is another special form of *m:n* switching. The only real difference between 1+1 and 1:1 protection switching is that 1:1 protection requires the use of an APS protocol. In 1:1 architectures, the receiver *must* request the sender to bridge live traffic onto the protection fiber. In 1+1, this request is unnecessary.

■ **Head-end**   This node always bridges traffic onto both working and protection fibers.

▲ **Tail-end**   This node always initiates the APS request for the span.

A 1:*n* protection architecture is allowed to have high-priority and low-priority traffic. That is, high-priority working pairs of fibers can get preferential treatment over low-priority pairs when it comes to deciding which channel gets to use the one protection fiber available in the case of multiple fiber failures.

Linear protection switching may be unidirectional or bidirectional, but this has nothing to do with the same terms applied to rings, of course. Unidirectional switching in linear architectures shifts the working traffic from only broken fiber onto a protection fiber. The traffic on the return fiber is not changed. No endpoint coordination is required, of course. Bidirectional switching requires endpoint coordination, and this is exactly what the SONET/SDH K1/K2 APS bytes provide.

The role of head-end and tail-end could therefore easily be played by both devices at the ends of the span, given that the traffic runs in two directions. This is quite common, especially when one end of the link is an ADM. But routers are often tail-end devices only. This means that routers seldom bridge duplicate packets onto more than one fiber link, mainly because IP routing seeks a single next-hop interface for a packet, and that is that. Usually, routers are strictly tail-end devices. When fibers fail, they never bridge and generate APS requests to the ADMs that they attach to. Routers respond to APS requests received from head-end ADMs to switch to protection fibers when required.

Even when only one protection fiber is available, this link is always designated channel 15 by definition. So, if there are three working fibers and one protection fiber (1:3) between two routers connected by a point-to-point link, an APS request will always reference channel 15. In 1:*n* protection architectures, this channel can be used to carry extra traffic. Figure 16-14 shows the architecture.

How can a router link to a SONET/SDH ADM for linear protection purposes? Three distinct architectures are defined. The simplest way is just to terminate working and protection pairs on the same router interface card. However, it might make little sense to do so. If the interface card fails, both links are lost. Working and protection pairs can therefore be terminated on different interface cards. But now, if the router fails, both links are lost. The final configuration allows working and protection pairs to be terminated on different routers, usually as long as the routers themselves are directly connected. Figure 16-15 shows these various architectures.

In each case, the router sends only live packet traffic on the working link. The ADM can easily bridge, but not the router. There is just no easy way to get a copy of a packet to the protection pair interface, especially when that interface might be on another router altogether.

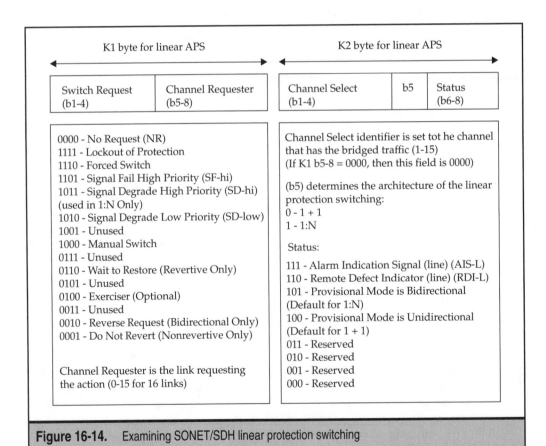

Figure 16-14.   Examining SONET/SDH linear protection switching

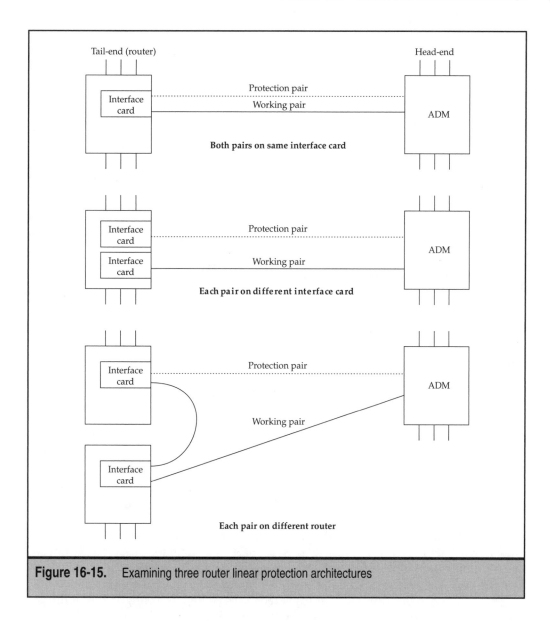

Figure 16-15. Examining three router linear protection architectures

Earlier in this chapter, Figure 16-12 listed SONET K1/K2 APS messages for ring architectures. Figure 16-16 lists the equivalent messages for linear protection switching. The SDH messages are essentially the same, except for the terminology. Appendix C gives the SDH APS messages.

Consider the K1 byte first. The first four bits (b1–b4) form the "switch request" action, and the last four bits (b5–b8) are the "channel requester" (the link requesting the action). This field can enumerate 16 links (0 to 15). Most of the messages are similar to ring-based

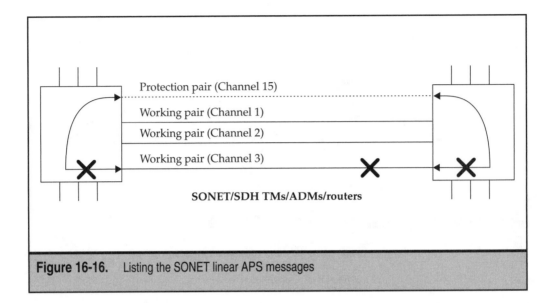

Protection pair (Channel 15)

Working pair (Channel 1)

Working pair (Channel 2)

Working pair (Channel 3)

SONET/SDH TMs/ADMs/routers

Figure 16-16. Listing the SONET linear APS messages

APS messages. No request is normal, and messages are available for lockout ("don't switch even if you want to"), forced switch ("switch because I said so"), and manual switch ("please try it").

These last three messages are issued as direct commands on the router. Lockout prevents protection switching, and if a link is already on the protection pair, it must revert to the working pair, even if "broken." A forced switch depends on initial conditions. If the traffic is on the working pair when this command is issued, the message forces a switch to the protection pair regardless of the status of that link. If the traffic is on the protection pair, a switch to the working pair is forced regardless of status. In both cases, a higher priority request (message type) can prevent any action from taking place at all. A manual switch will switch from working to protection pair, or protection to working, but only if the target pair is fault-free and unused.

High priorities apply to 1:*n* architectures only. The 1+1 protection architecture has low-priority links only; but, because there is one protection fiber for each working fiber in 1+1 APS, this makes little difference. The signal failure (SF) and signal degrade (SD) messages are the same as on rings. SF is triggered by an LOS, OOF, AIS-L, a settable BER threshold, or just "other" things. SD is triggered by exceeding the configured BER threshold. Finally, messages are available that affect revertive (automatic restore to working pair) behavior, and an optional exerciser.

The K2 byte has a channel select field (b1–b4). This identifies the channel that is carrying the bridged traffic in 1+1 operation. If the K1 channel requester field is 0, this field must be 0 as well (just another way of saying that channel 0 is special in linear protection). Bit 5 (b5) determines the two defined operational modes, 1+1 (0) and 1:*n* (1). The status bits (b6–b8) carry alarms (AIS-L and RDI-L), whether the architecture (provisional mode) is unidirectional (the default for 1+1) or bidirectional (the default for 1:*n*).

In 1+1 switching, the head-end always bridges traffic onto both fibers, and the tail-end might. The typical wait to restore (WTR) is five minutes, and this is usually settable, but only if revertive behavior to automatically restore traffic to the working pair is configured.

This section will close with a look at a simple 1+1 unidirectional failover between tail-end router and head-end ADM, as shown in Figure 16-17. The bit configurations for the K1/K2 bytes are shown for normal operation and during the switch, as is the interpretation of the K1 bytes (first line) and K2 bytes (second line).

Initially, there are no faults, and both ends are bridging duplicate channel 1 (0001) traffic onto channel 2 (0010). Channel 1 then degrades in the tail-to-head direction as a result of a high bit error rate (BER) that exceeds some pre-configured threshold on the fiber. As a result, the tail-end router sends a signal defect (SD) low-priority message (the only one defined for 1+1 protection) to the head-end, which basically sends the same thing back. Now both ends know where the traffic should be found: on the protection channel (channel 2 in this case). Because both ends bridge, K1/K2 APS messages are not strictly necessary, but are shown in this case for illustration.

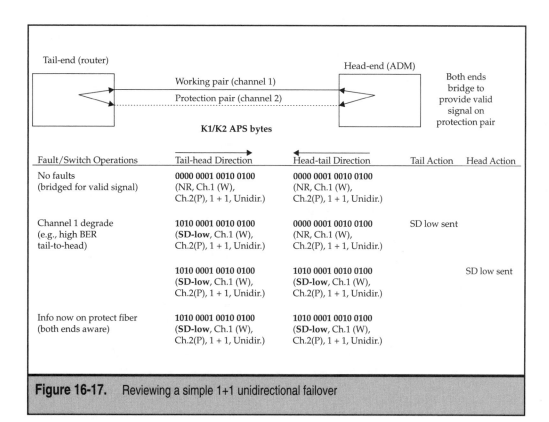

Fault/Switch Operations	Tail-head Direction	Head-tail Direction	Tail Action	Head Action
No faults (bridged for valid signal)	0000 0001 0010 0100 (NR, Ch.1 (W), Ch.2(P), 1 + 1, Unidir.)	0000 0001 0010 0100 (NR, Ch.1 (W), Ch.2(P), 1 + 1, Unidir.)		
Channel 1 degrade (e.g., high BER tail-to-head)	1010 0001 0010 0100 (**SD-low**, Ch.1 (W), Ch.2(P), 1 + 1, Unidir.)	0000 0001 0010 0100 (NR, Ch.1 (W), Ch.2(P), 1 + 1, Unidir.)	SD low sent	
	1010 0001 0010 0100 (**SD-low**, Ch.1 (W), Ch.2(P), 1 + 1, Unidir.)	1010 0001 0010 0100 (**SD-low**, Ch.1 (W), Ch.2(P), 1 + 1, Unidir.)		SD low sent
Info now on protect fiber (both ends aware)	1010 0001 0010 0100 (**SD-low**, Ch.1 (W), Ch.2(P), 1 + 1, Unidir.)	1010 0001 0010 0100 (**SD-low**, Ch.1 (W), Ch.2(P), 1 + 1, Unidir.)		

Figure 16-17. Reviewing a simple 1+1 unidirectional failover

SDH PROTECTION AND RINGS

SDH also has linear and ring protection (several major types each), with variations in the exact type of protection. Protection in SDH is generally one of three types:

1. **Line protection, known as multiplex section protection (MSP)**   MSP is just another name for having an alternative fiber route at the section (SONET line) level. Linear protection switching uses the K1/K2 APS bytes. This form of protection was discussed earlier in this chapter in a router context and need not be repeated here.

2. **Ring protection, mainly known as multiplex section shared protection rings (MS-SPRings)**   Older documents often reference these rings as MSPRings (or even MSPRINGs). Most SDH rings are two-fiber rings, and these are detailed in most specifications. Four-fiber rings identical to their SONET relatives are allowed as well, of course (although many SDH specifications reference them "for further study"). ETSI also defines a multiplex section dedicated protection ring (MS-DPRing); however, because MS-DPRings are used in special cases, they are not discussed further in this book. The focus of this discussion will be the two-fiber MS-SPRing architecture.

3. **Network connection protection, known as subnetwork connection protection (SNC-P)**   This form of protection establishes two fixed points in the network, forming the subnetwork. Across this subnetwork, one form of protection switching can be used. Across another subnetwork, another form can be used, and so on. In practice, this is just another form of diverse routing over multiple SDH sections and is not considered in any detail in this book.

Figure 16-18 shows the three forms of SDH protection.

In some documents, SDH MS-SPRings are designated with reference to their speeds. So, there are MS16 rings (a ring of ADMs running at the STM-16 rate of 2.4 Gbps), MS64 rings (a ring of ADMs running at the STM-16 rate or 10 Gbps), and even MS256 rings (a ring of ADMs running at the STM-16 rate of 40 Gbps).

Most of the differences between SDH rings and SONET rings concern terminology. This extends to more than just using the term "multiplex section alarm indication signal" (MS-AIS) for the SONET AIS line-level (AIS-L) alarm. For instance, instead of the term "channel" or "pair" for working and protection fibers, SDH uses the term "trail." SDH linear protection is "single-ended" instead of "unidirectional," and "dual-ended" instead of "bidirectional." There are other minor examples as well.

There is not much else to say about SDH rings that has not already been said about SONET rings. SDH rings tend to be less elaborate than SONET rings and are mostly the SDH equivalent of two-fiber bidirectional line-switched rings (2F BLSRs), although naturally variations occur, just as in SONET.

Appendix C gives the current ETSI SDH K1/K2 byte definitions.

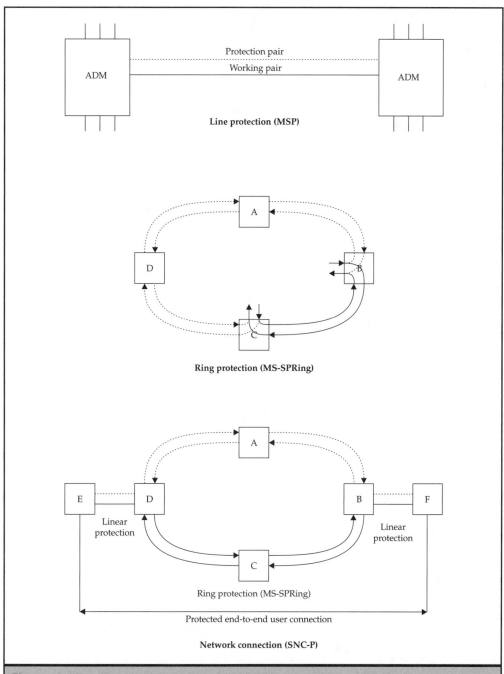

Figure 16-18.   Reviewing SDH protection capabilities

SONET/SDH RINGS AND DELAY

An interesting aspect of SONET/SDH rings concerns the delays experienced by users whose connections are delivered by these rings. Things are not as simple as in the T-carrier or E-carrier world. There, links were simple point-to-point spans, and a good idea of network delay could be worked out by the simple airline mileage between two endpoints and a few assumptions about equipment. All of these change with SONET/SDH rings. The topic is therefore worthy of some explanation.

First, it should be pointed out that delay in any network has three aspects: bandwidth, propagation delay, and nodal processing delay. Consider a point-to-point link operating at 1.5 Mbps. When a user is sending a 1,500-byte Ethernet frame($1,500 \times 8 = 12,000$ bits), the receiver generally cannot process it as it arrives off the network. It must first be complete. (The CRC error checking is at the end; memory needs to be allocated based on size; and so on.) Therefore, at 1.5 Mbps, the "bandwidth delay" to get a full frame off the network would be

12,000 bits/1,500,000 bps (close enough) = 0.008 seconds = 8 ms

Propagation delay is due to the finite speed of light. Most media (with the notable exception of wireless) do not propagate signals at the speed of light, but at the medium's nominal velocity of propagation (NVP), which is usually expressed as a fraction of the speed of light (c). Copper wire, for example, will typically have an NVP of 67 (sometimes written "0.67"), or two thirds of c. This is $0.67 \times 300,000$ km/s or 200,000 km/s. Strangely, fiber is about the same. The NVP for fiber is given by the formula

NVP = c/index of refraction of the fiber

Many SONET/SDH fibers have an index of refraction of 1.5, so that the NVP is still 200,000 km/s (300,000/1.5). This gives the fairly universal propagation delay of 5 μs/km (200,000/1 = 0.000005 second) for almost all media, including SONET/SDH fiber. (Notable exceptions are some coax cables with NVPs of about 75.)

Finally, the nodal processing delay is the delay added by going through SONET/SDH ADMs, DCSs, and other switching-office equipment. In the telecommunications world, the ITU says a signal must get in and out of a "switching node" in 450 μs or less (about 0.5 ms, given that 1,000 μs = 1 ms).

How does all of this help to calculate overall end-to-end delay? Well, when the endpoints are 100 km apart, with three switching nodes between them, the end-to-end delay that a user would experience is merely the sum of the three components:

8 ms + (100×0.005 ms) + (3×0.5 ms) = 8 + 0.5 + 1.5 = 10 ms

Note that the largest component by far is the bandwidth delay. This user application, whatever it is, is "bandwidth bound." That is, the limiting factor for the application is the network bandwidth, not the network delay. Applications limited by the network delay are "delay bound." Everything falls into one of those two categories. Either an application will not function properly because it lacks sufficient bandwidth, or it will not function properly because it lacks a sufficiently low delay.

What about SONET/SDH rings? Because the numbers are nearly universal, those numbers still apply. The problem that SONET/SDH rings pose for bandwidth/delay calculations has to do with these questions:

1. What is the actual cable mileage around the ring? (And how does it change on failure, when span or ring protection switching might be used?)
2. How many "switching nodes" are on the ring?

No SONET/SDH standard can answer these questions. Only the service provider or carrier can (and should). Sometimes a tariff or contract will specify network delay ("not to exceed 20 ms within the LATA ..."), but not always, especially with SONET rings. What can be done in this case? Either make assumptions (tricky) or push back ("tell me and I'll sign ..."). Beyond that, a customer or network designer needs to apply the general guidelines outlined earlier. Of course, most SONET/SDH rings will cut over to protection links in 50 ms (sometimes 60 ms is used for this figure, because standards allow 10 ms to detect the failure before protection switching takes place), but this is strictly a one-shot deal. The delay then stabilizes at the new (probably higher) value.

An example using SONET may be helpful. Consider that carrier A has three 100-km (cable length) SONET rings, interconnected, each with ten switching nodes per ring (not unwarranted assumptions, especially where two OC-3 rings feed an OC-12 ring). The end-to-end user network delay is then

$$3(0.005 \text{ ms/km} \times 100 \text{ km}) + 3(0.5 \text{ ms/node} \times 10 \text{ nodes}) = 1.5 \text{ ms} + 15 \text{ ms} = 16.5 \text{ ms}$$

The bandwidth delay for a full Ethernet frame (assuming the 1.5 Mbps DS-1 is now delivered on the SONET ring) is still 8 ms. Thus, the end-to-end user delay is now

$$8 \text{ ms} + 15.5 \text{ ms} = 23.5 \text{ ms}$$

Note that the application is now becoming more delay-bound, in the sense that the network delay is higher than the bandwidth delay. There will be a 50 ms (or 60 ms) outage "glitch" whenever any of the three rings "wrap" or "protect switch." The point to this exercise is this:

1. The route on a SONET ring is not point-to-point any more. Applications sensitive to network delay need to know more than just the airline mileage between source and destination. But only the carrier knows the exact ring route.
2. Owing to the nature of SONET rings as a mechanism for linking offices cost-effectively (the more offices, the more cost-effective), more "switching nodes" generally exist on the rings than on point-to-point links.
3. Some applications that are now bandwidth-bound may become delay-bound on SONET.

All of these points should be understood with regard to SONET (and SDH) rings and the services that they deliver to customers.

MORE RING ISSUES

The issue of SONET/SDH ring delay is only one of the issues that make services delivered on SONET/SDH rings fundamentally different from those delivered by T-carrier or E-carrier. The main point is that, although more service providers are building SONET/SDH rings, not all customers are necessarily buying SONET/SDH. Outside of high-end routers, almost no customer equipment can handle SONET/SDH speeds anyway. Most SONET/SDH links are therefore sold as virtual tributaries or containers delivering T-carrier/E-carrier services at lower speeds.

In today's competitive marketplace, network survivability is no longer an option—it is an entry-level requirement. At the same time, customers expect technology advances to result in lower costs. These expectations force many service providers to reduce their staffing levels. These two factors have forced many service providers to deploy SONET/SDH rings *and* digital cross-connect systems (DCSs) to interconnect their rings. But when service providers interconnect SONET/SDH rings, they can actually create environments where protection switching can mask problems on the network. Therefore, they must be very careful to ensure that interconnected rings do not affect survivability in user networks.

The problem is that some faults and alarms are not reported properly on interconnected rings. Dual-node interconnect with drop-and-continue wideband DCS equipment has become the standard to guarantee that inter-ring connectivity is secure in the event of a SONET/SDH NE failure. Drop-and-continue can be used between rings of different types. In many cases, as ring deployment grows, DCSs are necessary to groom inter-ring traffic at both the tributary/container and STS/STM levels. In other cases, just to mention a SONET example, M13 or VT multiplexers are used to multiplex DS-1 traffic from DS-1/VT1.5–based rings into other DS-3/STS-1s that are carried by OC-N interoffice rings.

The whole point of this architecture is to provide services that are at once survivable, economical, and manageable. In SONET, rings are typically interconnected at two points along the ring (known as the "primary" and "secondary" offices). Figure 16-19 shows this architecture connecting an OC-3 ring with customers to an OC-48 backbone ring.

However, when a SONET DCS or M13 is placed between rings to interconnect them, failures that occur in the switching office can actually be masked. All levels, from DS-3 to STS-1 to DS-1, signals are vulnerable to the same problem. Here's why: A simple failure of the ADM or intra-office wiring provides the DCS with a failed DS-3, STS-1, or DS-1. The SONET wideband digital cross-connect (WDCS) grooms at the VT level and inserts a VT/DS1 AIS inside a freshly created DS-3/STS-1.

Because the OC-48 ring monitors the STS-1 performance to select the better STS-1 signal, it is unaware of the failure to the DS-1 or other VT embedded in the STS-1. In the same fashion, when M13s are at the interconnection points between the rings, these devices will still multiplex a failed DS-1 into a good DS-3 and pass it to the OC-48 ring. Once again, the OC-48 is unaware of the failed DS-1 within the higher-speed signal.

An example of this situation can be illustrated using a path-switched OC-48 ring (but the same thing would happen with a line-switched ring). To select the good alternate path, a manual switch must be invoked at the OC-48 terminal. The failed DS-1 experiences a significant outage before this can occur.

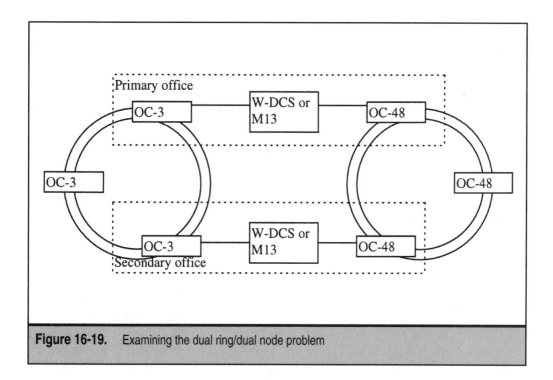

Figure 16-19.   Examining the dual ring/dual node problem

A failure at this DS-1 level in the primary office is multiplexed into a good DS-3 or STS-1 by the W-DCS or M13 multiplexer and is sent clockwise around the OC-48 ring. The counterclockwise direction for the signal originates from the secondary office and contains an STS-1 without any embedded DS-1 or VT failures, which is the whole point of doing rings in this fashion. The third ADM in the terminating OC-48 still sees two perfectly good STS-1s and does not protect switch, even though the still-selected STS-1 contains failed DS-1 or VTs. There is just no way to know.

What can be done? Vendors such as Alcatel have designed a new "fault escalation" feature into many of their SONET products. This feature provides visibility for embedded faults at the STS-1, DS-3, or VT levels. When a piece of equipment receives a "bad" input signal at the DS-1, DS-3, or STS-1 level, it escalates the failure to the output and provides a DS-3 or STS-1 loss of signal (LOS) and companion AIS to the OC-48 ring equipment. The escalated failure causes the OC-48 ring to perform a protection switch, restoring traffic.

Because fault escalation terminates all the embedded signals (good and bad alike), fault escalation should be performed at only one of the two W-DCS or M13 sites (in fact, the primary). If fault escalation is performed at both offices on the same DS-3 or STS-1, all of the embedded traffic would be lost. The likelihood of this "double fault" scenario occurring cannot be ignored, because each DS-3 carried on an STS-1 can receive DS-1s (as VT1.5s) from up to 28 different SONET NEs or other T-carrier facilities. For this to happen, the two faults must be within the same outbound DS-3/STS-1, but not necessarily the same inbound DS-1/VT.

To avoid this final double-fault situation, only the primary W-DCS is provisioned to perform the fault escalation procedure. Finally, the DS-3 or STS-1 switched rings should be provisioned to revert to the path containing the W-DCS that performs fault escalation (the primary one). Because such fault escalation is provisioned on a "per high-speed circuit" basis, a single-fault escalating network element can be both primary and secondary if it supports multiple rings. The same arguments can be extended to SDH lower-speed containers.

SONET/SDH RINGS AND TIMING

One more SONET/SDH ring topic should be mentioned briefly: network timing and synchronization. Most timing networks are built along T-carrier/E-carrier, linear, point-to-point architectures. Timing loops of all kinds are always bad and to be avoided. The inherent looped nature of SONET/SDH poses special challenges when timing signals are injected onto SONET/SDH rings.

Recall that five timing types are allowed in SONET/SDH equipment:

1. **External timing**   Direct stratum 1 (or stratum 3 in some cases) clock reference available.

2. **Line timing**   Derived clock from the scrambled SONET/SDH signal input.

3. **Through timing**   Also derived from input, but clocks the signal in a different outbound direction.

4. **Loop timing**   Same as line timing, with one major difference. Line timing can be used in an add/drop mux (ADM) used as CPE. Loop timing is used for straight CPE, and the bits must stop here. The ADM is configured for terminal mux (TM) mode only.

5. **Free running**   Internal clock only.

Consider only SONET for the moment. For SONET rings, there are two strategies. First, one of the NEs on a SONET ring may be free running; the others are through-timed. Second, external timing may be used. Either strategy avoids the creation of timing loops on the SONET rings.

THE SYNCHRONIZATION
STATUS MESSAGE AND SONET RINGS

The importance of timing on SONET (and SDH) rings is directly related to the need to add, drop, or cross-connect VTs. If all SONET/SDH links carried ATM cells or IP packets directly inside SPEs, timing would not be the crucial issue that it is. But as long as DS-1s/E-1s or DS-0s are in use on SONET/SDH links (and these DS-1s/E-1s or DS-0s might themselves be carrying ATM cells or IP packets!), timing will be important in SONET/SDH systems.

One of the bytes in the line overhead of SONET/SDH is the S1 byte. The S1 byte is used for carrying the synchronization status message (SSM) between SONET/SDH NEs.

Although the SSM is important on all SONET/SDH links, linear or ringed, the SSM is especially relevant to SONET/SDH rings and so is discussed at the close of this chapter.

When it comes to timing networks and clock distribution, the worst thing that can happen is the creation of "timing loops." A timing loop is like someone asking someone else the time, who then asks someone else, and so on, until the first person is asked! Slips causing jitter are bad enough. Jitter is annoying for voice, bad for data (resends needed), and deadly with compressed streams such as MPEG video. Timing loops are much worse because slips (jitter) build up, SONET/SDH payloads go wild with pointer adjustments, and *no alarms are generated*. This is the whole reason that the network synchronization hierarchy is arranged as a tree structure of various strata (layers). As long as networks consisted of linear, point-to-point links, it was easy enough to avoid the creation of timing loops. (Although they still did occur from time to time.) However, a SONET/SDH ring is *inherently* a timing loop just waiting to happen. That is the main reason that this chapter deals with the SSM issue, although the SSM is used in linear SONET/SDH systems as well.

It is impractical to expect a large mesh of SONET/SDH rings to draw timing from a single master clock. Most systems draw from several available sources, some of which might be direct stratum 1 sources such as GPS. The timing signal generator (TSG) available to each SONET/SDH NE will have a primary and a secondary source to choose from. If a primary clock source is lost, the secondary can be used.

So far so good. But the issue becomes more complex when SONET/SDH rings are considered. If only one clock source is injected into the ring, as discussed in the previous section, what happens when this reference is lost? And if multiple clock sources are available to a SONET/SDH ring node, which one should be used in preference to another? And what prevents timing from passing all the way around the SONET/SDH ring and ending up referencing *itself* as a clock source?

The SSM messages in the S1 byte answer all of these questions, especially with regard to SONET/SDH rings. When SSM is used, SONET/SDH rings can have multiple clock sources, ring nodes can choose the best stratum clock available, and timing loops are still avoided. As with APS, a full discussion of the SSM protocol is beyond the scope of this chapter. But a few basics of the S1 byte and how the SSM protocol is used on a SONET/SDH ring can still be covered.

The SSM information is carried in bits 5–8 of the S1 byte. Previously, this byte in the SONET/SDH overhead was required to be coded as 0000, so that one of the SSM codes can be 0000 for backward compatibility. There are 15 other possible combinations of bits, and 9 have been given values and abbreviations in SONET documentation. (SDH uses its own terminology, of course.) Each SSM is, in addition, assigned a "quality level" that helps SONET/SDH NEs figure out which timing source to use when a choice is available. Table 16-3 shows these SSMs.

Only a few more comments about Table 16-3 are needed. The RES quality level can be assigned by users (that is, service providers) to force timing signals arriving via a certain STS-N to take priority over others. Naturally, this must be done carefully. Most of the other levels follow logical preferences, but three points are noteworthy. First, the traceability unknown (STU) value, which is basically for backward compatibility, is given a fairly high priority, exceeded only if a stratum 1 clock is available on the STS-N. Second, the transit node clock (TNC) and stratum 3E clocks fall between stratum 3 and stratum 2 in quality.

Synchronization Status Message	S1 bits 5-6-7-8	Abbreviation	Quality Level
Stratum 1 traceable (source can be traced to stratum 1)	0001	PRS	1
Synchronized—traceability unknown (backward compatibility)	0000	STU	2
Stratum 2 traceable	0111	ST2	3
Transit node clock traceable (a nonstandard clock level)	0100	TNC	4
Stratum 3E traceable (another nonstandard clock level)	1101	ST3E	5
Stratum 3 traceable	1010	ST3	6
SONET minimum clock traceable	1100	SMC	7
Stratum 4 traceable (no value has been assigned for this)	N/A	ST4	8
DO NOT USE for synchronization	1111	DUS	9
Reserved for network synchronization use	1110	RES	User assigned

Table 16-3. The SONET synchronization status messages (SSMs)

These two clock levels are not even mentioned in many standards documents, but because they are used in the real world, it is only proper to include them at their correct quality level. Finally, although a placeholder has been established for stratum 4 clocks, no bit coding has been defined for this level. SONET is supposed to offer synchronous byte-interleaved multiplexing and fast, error-free cross-connections. This would be very hard to accomplish with stratum 4 clocks. Usually SONET rings and links are stratum 3 or better.

The most important part of the SSM set is the DO NOT USE (DUS) message (oddly, this message in all SONET documentation always has the "DO NOT USE" or "DON'T USE" in uppercase). This is the message that is most useful for preventing timing loops on SONET rings and is the main reason SSM is discussed at this point in the book.

When it comes to SDH, the SSM messages are different, as might be expected, but serve exactly the same purpose. Table 16-4 shows the SDH SSM messages and their bit configurations.

Several comparisons can be made between the SONET and SDH versions of the SSM messages tables. The SONET "synchronized—traceability unknown (backward compatibility)" is the same as the SDH "quality unknown (existing synchronization network)" message: both are 0000 and are intended for backward compatibility. Also, most of the SONET stratum message values are reserved in SDH, and SDH values are not used in SONET, a nice

Synchronization Status Messages	S1 bits 5-6-7-8
Quality unknown (existing synchronization network)	0000
Reserved	0001
G.811 (highest level clock)	0010
Reserved	0011
Synchronization supply unit A (SSU-A) (G.812 transit)	0100
Reserved	0101
Reserved	0110
Reserved	0111
Synchronization supply unit B (SSU-B) (G.812 local)	1000
Reserved	1001
Reserved	1010
G.813 Option I synchronous equipment clock (SEC)	1011
Reserved	1100
Reserved	1110
DO NOT USE for synchronization	1111

Table 16-4.   The SDH synchronization status messages (SSMs)

interoperability feature. Only one exception exists: the non-standard SONET TNC value (0100) is the same as the SDH synchronization supply unit A (SSU-A) value. (This is also seen as "G.812 transit" clock.) Because both serve the same function, this overlap is not a problem.

SSM has many applications in linear SONET/SDH systems as well, but this chapter closes with a look at SSM use on a SONET ring. The same messages and points could equally be made about SDH. The issue with timing and clock distribution on a SONET ring is that the ring forms a natural timing loop. If nothing is done on a SONET ring, a stratum 1 clock injected onto a SONET ring node will chase its tail right through all the ring nodes and end up back at the source. If the injected clock is lost, the source node might think "no problem, here's a clock traceable to stratum 1 coming in right over here." The end result would be disaster as the timing jitter reinforces itself around the ring.

It would also be better, on a 16-node SONET ring, to allow nodes closer to the primary reference source (PRS) clock to choose a signal that has passed through only 4 nodes around the ring, rather than 11 nodes. In other words, the short way around is always better than the long way around the ring when a choice of timing sources traceable to the same stratum level is available.

The trick is to configure two nodes on the SONET ring not to pass through the timing level available in the S1 byte, but to convert this to a DO NOT USE (DUS) SSM message at some point. This point should logically be halfway around the ring, but this is not a requirement. Figure 16-20 shows how the SSM protocol is used to prevent timing loops on a simple six-node SONET ring with an injected stratum 1 clock source.

As usual, the ring node with the injected timing (node A) passes the clock signal in both directions around the ring. Yet timing loops are prevented. As shown in the figure, the whole trick is to configure one of the ring nodes (node D in this case) *not* to pass the clock on to another node. And note that the nodes all change the returning clock SSM to

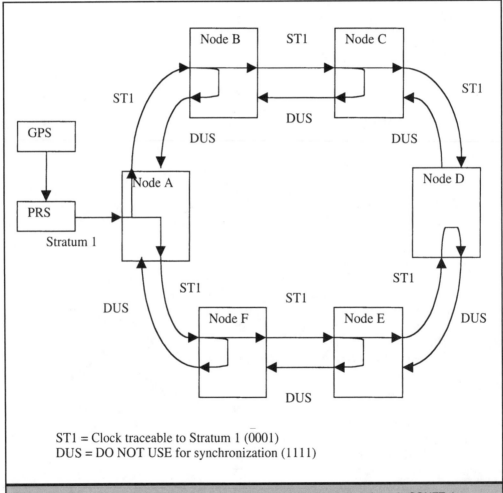

Figure 16-20. Understanding the synchronization status message on a six-node SONET ring

DO NOT USE (DUS). These methods prevent node F (for example) from deciding to take clock from node E, five nodes away from the PRS, instead of node A, the direct PRS node. Node D, halfway around the ring in this case, has a valid choice of timing through the same number of nodes.

Figure 16-20 shows the ring under normal operating circumstances. This example is quite simple, but the SSM protocol can be shown to work properly when two nodes on the ring have timing sources available, perhaps stratum 2 as a backup to stratum 1. The interplay between the S1 bytes can then be shown to properly detect the loss of stratum 1 clock, stabilize to stratum 2 clock holdover, and then return to stratum 1 when available.

PART V

SONET/SDH
Service Deployment

The previous parts of this book dealt with various aspects of SONET/SDH. Groups of chapters have discussed SONET/SDH standards, pieces of SONET/SDH equipment, SONET/SDH rings, and so forth. Little of the material, however, has given the SONET/SDH "big picture" in any detail. The term "big picture" simply means that no systematic attempt has been made to place SONET/SDH—and especially SONET—in its proper context as an access and trunking network delivering high bandwidth and highly reliable services to as many customers and users as possible.

This part of the book corrects that situation. SONET/SDH is still not everywhere—a fact that is even more true outside of the United States and Europe. SONET/SDH deployment is an ongoing process that will take many years yet to complete. This part of the book examines this deployment process in detail, showing many SONET/SDH network scenarios, with the intention of drawing together all of the concepts previously discussed. SONET/SDH networks are shown in a real-world, service-delivery environment.

Merely installing SONET/SDH fiber everywhere will not solve user problems. SONET/SDH, as has been pointed out in this work before, was originally part of a networking strategy where ATM provided switching and multiplexing, and B-ISDN defined the services users employed for their applications. However, in the last ten years, SONET/SDH has leapfrogged ATM (and especially B-ISDN) in the sense that customers and users more often look to SONET/SDH rather than to ATM or B-ISDN for the delivery of advanced network services.

Of course, it is always good to keep in mind that service providers with extensive SONET/SDH networks and rings in place may not be selling SONET/SDH directly to customers. But even a DS-1 or E-1 or DS-3 or E-3 delivered on a SONET or SDH ring offers many benefits to customers, mostly in the form of reduced pricing and increased reliability.

Also, the future of SONET/SDH is always of interest. For now, SONET/SDH seems to top out at the OC-768/STM-256 rate; but, sooner or later, equipment vendors might push digital links beyond this level. In fact, some techniques promise to make the future of SONET/SDH even brighter than it currently is.

This last major section of the book is divided into three chapters.

Chapter 17 deals with SONET/SDH service deployment. The chapter again emphasizes SONET, but only because the main geographic area explored is the United States. This chapter examines the typical SONET deployment pace in a given area. It also examines where SONET is encountered in most carrier networks in the United States today.

Chapter 18 deals with SONET/SDH services provided to customers. It explores more fully just what customers do with SONET/SDH links, such as connecting LANs through routers, and linking private or public ATM network switches. Other applications, some obvious, some not, are mentioned as well. SONET/SDH offerings from a sampling of service providers also are outlined.

Chapter 19 explores the future of SONET/SDH. It details some of the new optical networking techniques for extending SONET/SDH and looks at other aspects of SONET/SDH deployment in summary fashion.

CHAPTER 17

SONET/SDH Deployment

This chapter looks at SONET and SDH deployments in various network configurations. As before, the emphasis is on SONET, and comparisons to T-carrier networks are made at each step along the way, both for familiarity and to show how SONET improves on T-carrier scenarios. The same arguments can easily be extended to E-carrier environments.

The issue of SONET/SDH deployment is linked with the issue of fiber distribution. After all, SONET/SDH will not appear until fiber on which SONET/SDH can operate is in place. That is not to say that all fiber-optic cable is SONET/SDH fiber. Plenty of fiber still supports various kinds of DS-3 links and the proprietary higher levels from pre-SONET days. A more important consideration is the appearance of dense wavelength division multiplexing (DWDM) systems that require fibers fundamentally different than those for 1310 nm SONET/SDH operation (although some compatibility between fiber types exists). The majority of fiber miles are, however, intended for SONET/SDH signals.

Before this chapter proceeds to particular SONET/SDH deployment scenarios, now may be a good time to take a look at the constantly increasing pace of fiber installation in general and of SONET/SDH fiber in particular.

SONET/SDH, GIGABIT ETHERNET, DWDM, AND FIBER

Not too long ago, when fiber was installed by service providers, it was pretty much given that the fiber would be used for SONET, or SDH, or both. This is no longer true. As has been pointed out in an earlier chapter, DWDM and optical networking have changed the rules. A full exploration of the future of SONET/SDH in a world dominated by newer technologies and techniques such as Gigabit Ethernet (GBE) and DWDM will be given in the last chapter of this book. For now, it is enough to point out that, although fiber deployment rates still remain high, SONET and SDH are no longer the only game in town.

Some of the characteristics of running fiber intended for DWDM have been examined earlier and need not be repeated. GBE, which has been mentioned only briefly in earlier chapters, is basically an extension of 10/100Base-T LAN technology to run on fiber not only down the hall, but also over spans up to 13 miles in length. In fact, commercial GBE products exist to extend the reach of GBE to 75 miles without optical amplifiers, and up to 620 miles with them. The 10 Gigabit Ethernet waiting in the wings is as fast as an OC-192 or STM-64 link (although distance remains a real issue for 10 GBE).

The whole point is that, if GBE frames can be sent directly on fiber for 62 miles or 620 miles, what need is there for a WAN solution such as SONET/SDH when a LAN can do it all? The last chapter of this book has more to say on the implications of this whole issue. All that is needed here is an appreciation of the value of "dark fiber," as opposed to traditional service-provider offerings on "lit fiber" (such as SONET/SDH).

Fiber carries signals from transmitter to receiver. But who owns the transmitter and receiver? Who owns the fiber? In the past, the same entity almost universally owned both. Service providers who sold spare capacity on their SONET/SDH links and rings always controlled the transmitters and receivers (and any regenerators along the way, of course). The service provider "lit the fiber."

Today, a scramble to install and control dark fiber often occurs. Today, the fiber itself is seen to have intrinsic value, and not so much the transmitters and receivers. End-equipment ages, is upgraded, and changes capabilities quite often. And end equipment prices always seem to go down. Not so with fiber installation prices. The price of the fiber itself might tumble, but installation is still a labor-intensive activity. Although productivity might increase marginally, labor costs always escalate.

One large inter-exchange carrier estimates the cost of new fiber installation at about $70,000 per mile, or a little more than $13 per foot. Even when existing conduit or banks are used, costs still routinely exceed $30,000 per mile. No wonder the fiber "route miles" are what have value today, and not so much the signals that the fibers carry.

The pressure today is to possess as much fiber as possible, and not necessarily bandwidth. The same fiber could potentially carry much more bandwidth than in the past, especially with DWDM in the equation. The dark fiber has the value, not the SONET/SDH that could be put across it. Even stranger, because fiber is a cable, just having right-of-way to run fiber has tremendous value (based solely on the potential of, some day, running the fiber).

These points should always be kept in mind when reading this chapter. Plenty of SONET/SDH will be riding fiber for some time to come; but, whenever fiber is run today, SONET/SDH is not the only possible (or even probable) use.

The Function of SONET/SDH

Purely and simply, SONET/SDH represents a lot of bandwidth. What is all of this bandwidth needed for? Much of it will be used for the same types of applications in use today, together with many new services. Most of these new services are defined by B-ISDN specifications; but, in some cases, ATM specifications from the ATM Forum overlap somewhat with B-ISDN service definitions.

However, the use of SONET/SDH is in no way restricted to new services. Most existing services can still take good advantage of the enormous bandwidths SONET/SDH is capable of providing. Table 17-1 shows the results of a March 1996 survey from the Idea Factory.

Videoconferencing	64%
Graphic applications (CAD/CAM, visualization)	60%
Batch file transfer (FTP)	58%
Local database queries and access	57%
Remote database queries and access	39%
Transaction processing (SNA)	36%
Internal e-mail	33%
External e-mail	31%

Table 17-1.   Idea Factory survey of user applications needing more bandwidth

The study is becoming a little dated, but even as businesses and industries rise and fall, their needs tend to remain stable.

Not surprisingly, most user organizations claimed that they needed more bandwidth, especially on their WAN links, for visually intensive applications such as videoconferencing and graphics applications. Graphics applications include, for example, CAD/CAM and scientific visualization (for example, of molecules or seismic data).

User organizations also need more bandwidth for such mundane tasks as file transfer and database access. As distributed computing using client/server LAN applications becomes even more common, this need will only increase. This is often overlooked in the highly publicized concerns about bandwidth for the newer services.

Even such common and relatively low-speed applications as SNA transactions and e-mail are on the list. So much networked traffic means that more bandwidth is a requirement, not a luxury. After all, many of the LANs to which these users are attached now operate at 100 Mbps or better. It is difficult, if not impossible, to link these LANs with 64 Kbps leased lines. Web and e-commerce concerns, at least the ones that still exist, have only helped to extend the bandwidth pressure to residential users as well.

The next few sections will explore SONET use in the United States, but only because the deployment of SONET and fiber for SONET has been ongoing in the United States for more than ten years. Many areas of the world are just beginning this process (with SDH).

T-Carrier Trunking Network

Before showing how SONET is deployed in access, trunking, and regional networks, it might be a good idea to contrast SONET with T-carrier in these respects. After all, a recurring theme in this book has been to compare T-carrier methods with SONET modifications and improvements in all areas of network operations. Many of the ideas and scenarios in the following sections are based on Nortel Networks deployment plans and equipment categories. While acknowledging this debt, it should be kept in mind that other SONET vendors and users may differ somewhat in their major architectural features at the access, trunking, and regional networks.

The traditional T-carrier trunking network has been used in North America since the 1960s. In the T-carrier network, switching nodes in switching offices are linked by T-carrier (usually T-3s on fiber) in a service area. Most are local serving offices (central offices) with users attached to a switch. Others are various, more specialized, switching nodes. For instance, a "toll office" usually has no users (that is, telephones). Rather, these kinds of switching offices link other switches in a given geographic area. The wire center can also link all users in the area to one or several points of presence (POP) for an inter-exchange carrier (IXC).

Each T-carrier switching node could have many DSX-3 cross-connects or DS-3 drop-and-insert devices at each site. The fiber system is symmetrical in the sense that one terminal end is the mirror image of the other. Figure 17-1 shows a typical T-carrier trunking network.

The main point is that the channels between the offices are usually limited to some number of 64 Kbps channels, and all the other limitations of T-carrier apply. These limita-

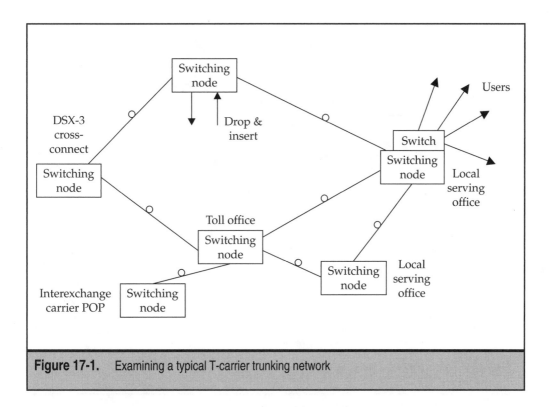

Figure 17-1.   Examining a typical T-carrier trunking network

tions include the need for back-to-back multiplexing equipment and complex drop-and-insert arrangements, as has already been detailed in earlier portions of this book.

T-Carrier Access Network

In a typical T-carrier network, customer access to the trunking network has its own distinctive features. The switching node, or local serving office, contains the endpoint of the trunking network fiber, which is normally a fiber-optic termination system (FOTS) device. This device is connected to the M13 multiplexer and fiber multiplexer terminals to other devices. The M13 multiplexer is connected, in turn, to the digital switch itself and to the channel banks for digitizing analog local loops leading to the users.

Access to the serving office involves a combination of methods. Telephony end-users with analog telephones are serviced by plain old telephone service (POTS) using unshielded twisted-pair (UTP) copper local loops. Alternatively, a T-1 could be run to a remote digital loop carrier (DLC) remote terminal (RT) device, and twisted-pair copper wire loops extend to all locations in a carrier service area (CSA). The DLC link could also be fiber, leading to a remote fiber terminal (RFT) arrangement, but the effect is the same.

UTP copper wire is also used to provide leased lines (that is, non-switched special services) for those users who desire them. Finally, newer fiber optic cable may extend directly to various customers' premises for providing DS-1 leased lines or other services. Figure 17-2 shows the architecture.

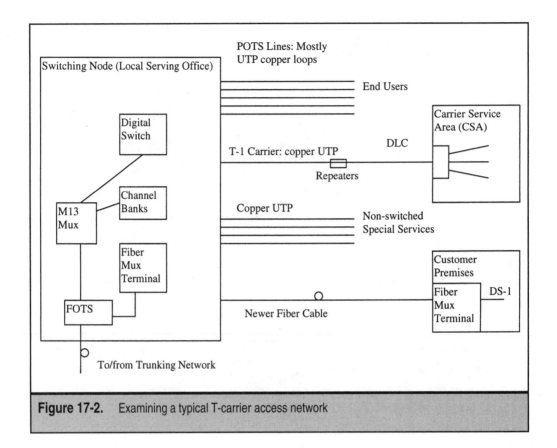

Figure 17-2. Examining a typical T-carrier access network

SONET Trunking Network

Having taken a brief look at how T-carrier trunking and access networks are put together, now is the time to see how SONET networks are deployed to accomplish the same ends. Comparing and contrasting T-carrier and SONET will now be much easier, making SONET advantages that much more concrete.

A SONET trunking network is much simpler, and yet much more robust, than a similar T-carrier trunking network. SONET switching nodes, containing either add/drop multiplexers (ADM) or SONET hubs (which combine SONET ADM capabilities with digital cross-connect system (DCS) capabilities for grooming and other purposes), are now configured as a series of interconnected SONET rings. The use of SONET rings in the trunking network assures higher availability and reliability for the trunks.

SONET OC-N links can also run directly to a customer's premises if the customer has SONET customer premises equipment (CPE) installed. This is admittedly unusual in customer environments that are more current. It is much more typical for the SONET rings to deliver T-1, fractional T-1, fractional T-3, and full T-3 speeds than any direct SONET service. But the rings make the T-carrier customer offerings that much more error-free and robust. Figure 17-3 shows a typical SONET trunking network.

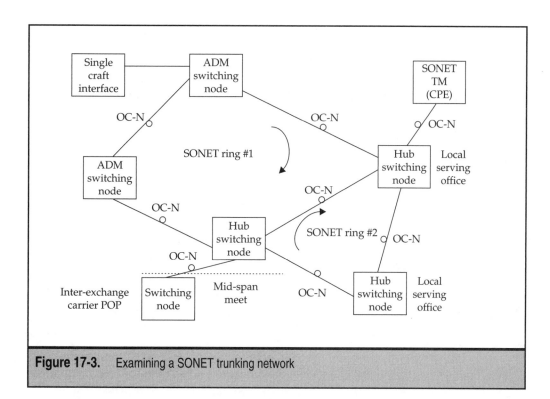

Figure 17-3.   Examining a SONET trunking network

SONET also offers the possibility of a mid-span meet for interfacing with other service providers. The SONET mid-span meet allows for one vendor's equipment to be at one end and another vendor's equipment to be at the other.

A single craft technician interface handles alarms and monitors network performance—another plus. On the whole, SONET is simpler and more efficient than the earlier T-carrier trunking network.

SONET Regional Network

An especially nice feature of SONET is its ability to extend the trunking network from centralized service areas to regional SONET networks that include suburban and even rural areas. Such regional area networks have no real counterpart in the T-carrier world, but are a structure and architecture that SONET can deploy easily and effectively.

The SONET rings and trunking network from the previous section are still in place, of course, and the links to the inter-exchange carriers are still SONET as well. However, it is now possible to extend the benefits of SONET to a wider (that is, regional) area by adding various OC-N links to SONET ADM and hub locations. Hubs combine SONET ADM equipment and DCS capabilities at the same location. Figure 17-4 shows this overall deployment for regional SONET networks.

These SONET ADMs and hubs are located away from high-density traffic areas and extend the benefits of SONET to locations that were previously unreachable for advanced

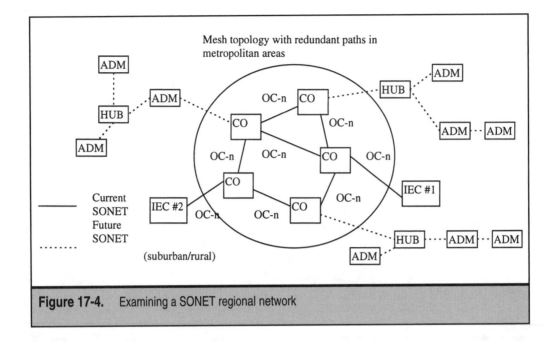

Figure 17-4.   Examining a SONET regional network

services. In urban areas, SONET rings provide redundant and robust paths. Suburban and rural areas can take advantage of the urban structure as well. SONET deployment at this level is an ongoing process—one that is based on the presence of SONET in an urban area before extending SONET services to less densely populated areas.

SONET Long-Haul Network

So far, the SONET networks mentioned have been concentrated in major metropolitan areas and, in spite of the promise of extending SONET to more rural areas and attaining regional coverage, these deployments mostly appeal to the LECs. But the flexibility and scalability of SONET also make it attractive to IXCs that often use SONET to create long-haul networks between major cities.

SONET is used in many long-haul networks built by IXCs or LECs entering the inter-LATA market. A long-haul SONET network deploys an OC-N (usually OC-48 or, on occasion, the even higher OC-192) high-bandwidth transport network over an existing point-to-point fiber network (usually a 565 Mbps or other-speed FOTS network with multiple DS-3s on a nonstandard fiber system). The main network equipment is a large (often very large) SONET ADM used in both terminal and repeater configurations. SONET broadband cross-connects are used for provisioning and managing the traffic on the long-haul network.

Figure 17-5 shows this overall structure. In the figure, SONET equipment is shown as square symbols. The non-standard DS-3 FOTS equipment is shown as circle symbols. The SONET construction can "overlay" the existing FOTS routes or consist of entirely new fiber construction intended to "close the rings." Naturally, such construction could occur

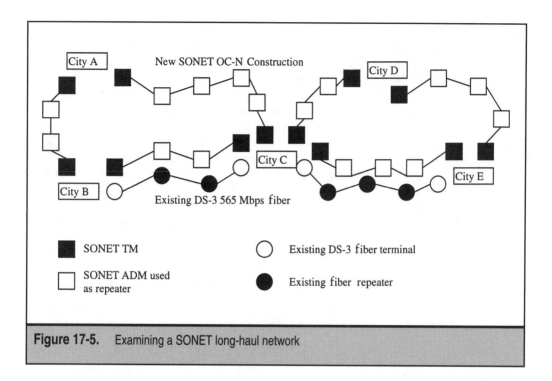

Figure 17-5. Examining a SONET long-haul network

much later, depending on the service provider's deployment schedule and budgeted capital expenditure.

These other routes between locations, including many not previously serviced by FOTS, could be added to the SONET network to close the rings. The older network may not have been extended to the other cities for a number of reasons, including cost, traffic patterns, and user densities.

In terms of robustness, the new SONET rings offer many advantages over simple point-to-point configurations.

SONET Access Network

Most SONET deployments by service providers, especially early ones, did not begin as trunking, regional, or long-haul networks, especially when SONET rings are considered. Most early SONET deployments occurred in downtown metropolitan areas, in rings from the very start. The intention was for the access network to take advantage of the benefits of SONET overall, and the benefits of rings in particular. The access network gathers traffic from a number of customer locations, each with distinctive services and requirements, and delivers that traffic to the service provider's switching nodes in the serving offices. Historically, the access network has been the source of most service interruptions.

SONET makes a lot of sense in the access network. This is especially true when new fiber is being installed or existing routes are becoming exhausted in terms of bandwidth. SONET provides a means not only for supporting new network capabilities in terms of protection and management, but also for supporting new services (for example, video). Figure 17-6 illustrates the three main scenarios for SONET access networks.

In a simple point-to-point configuration, SONET can be used (instead of a pair-gain link) to service CSA-containing DLC systems. The top of Figure 17-6 shows this situation. The SONET TM can support the DLC system, yet it has plenty of bandwidth to spare for supporting new services such as the video feed from a TV studio.

It is worth pointing out that technically sophisticated users (such as TV studios) can no longer be assumed to reside in business environments and commercial zones. Many communities have "high-tech" zoning that allows such businesses to operate near, or even within, otherwise mainly residential areas. In some areas of the United States, office parks and housing developments are interspersed so closely that the dividing line between them is often no more than a backyard fence. SONET offers a very good way of reaching both types of customers, even with their different servicing needs, in a cost-effective and scalable fashion.

The lower right portion of Figure 17-6 shows another type of access arrangement. In a hubbing or ADM application of this type, SONET can transport traffic from the switching node to a TM at a customer (or other) location having a SONET hub. The ADM could ser-

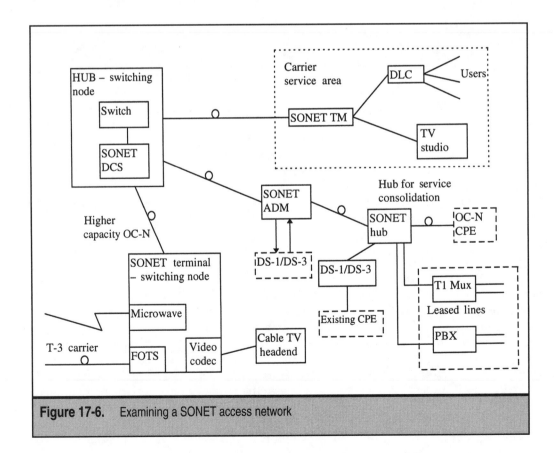

Figure 17-6.   Examining a SONET access network

vice existing DS-1s and DS-3s along the way, enabling them to take advantage of the low bit error rates, reliability, and management capabilities of SONET. Of course, the sale of new T-carrier services could continue as before, giving customers familiar technology and services, and also allowing them to purchase low-priced T-carrier end equipment.

The SONET hub offers a means of consolidating separate services in a cost-effective manner. The hub can support any customers with SONET OC-N CPE, although initially such customers may be few. Other DS-1s and DS-3s can also be supported with existing CPE, which is always attractive to customers. These customers do not even notice the arrival of SONET in an area. Finally, a customer's existing T-1 multiplexer or PBX can be supported using ordinary leased lines as before, but delivered on SONET. The ADMs could ultimately be connected in a closed OC-N loop to create a SONET ring.

Last, but by no means least, Figure 17-6 shows another type of SONET access in its lower left portion. Here, a higher capacity (perhaps an OC-48) SONET link can be used to connect to another SONET switching node, but with a different function. This node could access an existing microwave or cellular network, and existing FOTS facilities. More intriguingly, especially in the age of local service deregulation, this node could function as the entry point for a cable TV headend. Thus, SONET would provide the distribution network for cable TV video signals.

SONET ACCESS ARRANGEMENTS

The previous section presented a comprehensive array of access alternatives for SONET. This does not imply, however, that all types need to be used, or that some of them will ever become common. This may therefore be a good time to detail exactly which SONET access arrangements will be seen most often.

As can be seen from the previous section, SONET has many possible access arrangements. Nevertheless, SONET will mostly be used in a relatively routine manner in the access network, at least initially. SONET will be used either to service existing DLC RT to provide service for POTS users, or to provision DS-1s and DS-3s, all within a CSA. Figure 17-7 illustrates this situation.

Not all sites in a service area may be directly reachable from a given serving office. Even SONET fiber has distance limitations. In this case, SONET can be used in a straight ADM configuration, dropping and adding various DS-0, DS-1, or even DS-3 channels at hubbing sites along the way. These remote ADMs, in environment-hardened enclosures, may eventually loop back to the serving office to form a SONET ring.

A single SONET hub can branch out to several SONET TM arrangements, typically in office parks or to serve large housing tracts, again dropping DS-n channels along the way.

Private SONET/SDH Rings

It is not only for public network services that service providers are busy deploying SONET/SDH rings in access networks, trunking systems, and the like. Given the right circumstances, examples of totally private SONET/SDH infrastructures dedicated to one large customer can be found. The reason is that users, traditional applications (for exam-

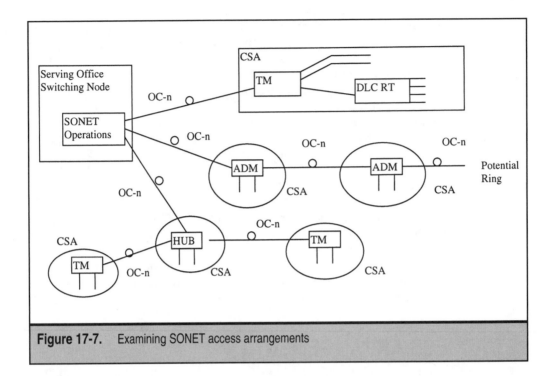

Figure 17-7.   Examining SONET access arrangements

ple, large file transfers), and new needs (for example, videoconferencing), are increasingly forcing many organizations to push to the limit the bandwidth delivered by their networks. Alternatives to existing private lines—for example, ATM and frame relay—are available. But it would be expensive and time-consuming for an organization to convert from a familiar private-line environment, especially when private-line solutions have been the rule rather than the exception, even outside of the United States.

ATM and frame-relay service offerings sometimes do not support the access bandwidth needed (for example, 45 Mbps) to improve application performance at an affordable price. ATM is still not always available on SONET/SDH fiber, and neither is frame relay. Conversely, SONET/SDH is, purely and simply, about more bandwidth. It offers a clear migration path to greater bandwidth. With SONET/SDH, the move from OC-3/STM-1 (155.52 Mbps) to OC-12/STM-4 (622.08 Mbps), and even on to OC-48/STM-16 (2.4 Gbps) is relatively painless and inexpensive. (Above those speeds, things can become very expensive, but so were all of the other speeds at some point in time.)

This ease of migration to greater speeds prompted the Miller Brewing Company of Milwaukee to deploy a private SONET ring infrastructure for their networking needs. Miller is the second-largest brewery in the United States and the third-largest in the world. Annually, it produces 45 million barrels of beer (amounting to about 19.5 billion 12-oz. glasses). The movement to a SONET ring from copper wire was prompted by a need for more bandwidth to accommodate additional users and new applications, and for reliable disaster prevention. Since 1995, the brewery has been using its own exclusive SONET OC-3 ring as the underlying transport backbone for its frame-relay switches.

Miller uses the SONET ring network at the Milwaukee campus for all voice and everyday data applications. The data applications include payroll, accounts payable and receivable, and applications that are more business-oriented, such as internal controls, production scheduling, and shipments.

Miller's SONET ring has all the usual benefits of SONET. Although the ring is dedicated to Miller, the company has access to multiple IXC offices for further disaster protection for switched services. Like many organizations, communication is vital to the company. Downtime is not necessarily fatal, but loss of data or lengthy outages can be enormously costly.

SONET and SDH are ultimately cheaper than other architectures, and this is especially true now that SONET and SDH have matured. Many users can recover the costs related to a SONET or SDH upgrade by way of total network savings in three to five years. Users with extremely heavy data traffic can easily compare the cost per megabit for SONET/SDH networks with copper costs. Running fiber is actually cheaper than running copper in most cases.

RUNNING SONET/SDH FIBER

At this point, it should be obvious that SONET/SDH deployment is intimately tied to the presence or absence of suitable fiber for the SONET/SDH signals. The good news is that plenty of fiber from FOTS systems can be found in the United States. The bad news is that this fiber is still used for FOTS traffic (and is often filled to capacity), and that older fiber may not be suitable for SONET. Fiber types and construction have evolved even more rapidly than achievable SONET/SDH speeds.

Thus, most new SONET/SDH deployments require new fiber deployments. It may be a good idea to close this chapter with a look at just where all of the existing fiber came from. After all, the aggressive SONET deployment plans of most carriers and service providers may lead some to believe that backhoes should be busily digging up every square inch of the United States sooner or later. Lately, however, a backlash has occurred regarding new fiber installations—and that backlash has mixed messages for SONET and SDH deployments. Many municipalities, especially in the United States, fed up with torn-up streets and construction detours, have declared a moratorium on new fiber installation permits. The downturn in the high-tech industry has contributed to this leveling-off effect.

What is the mixed message, then? First, if there is no new fiber to use, many service providers will turn to DWDM to increase fiber capacity where they need it—and DWDM is not SONET/SDH. So, SONET/SDH has had its day. But, second, DWDM does not run very well on fiber intended for SONET/SDH. If no new fiber can be run, service providers will have no choice but to light every fiber they have with SONET/SDH equipment. So, SONET/SDH will be more popular than ever.

Whatever happens will be interesting to watch. Just to show how fiber got where it is today, this chapter closes with a mainly historical look at the wild and crazy years of fiber installation: the 1990s. There are several ways of quickly installing the conduits and fiber cables needed for SONET/SDH services in a less disruptive fashion than trenching everything in sight. This section briefly introduces two methods (one for underground fiber runs and one for undersea fiber runs) and the organizations responsible for these advanced conduit and fiber installation techniques.

"Railblazers"

The Railblazers program was the product of SP Construction Services, a business unit of Qwest Communications of Denver. They took the art of installing SONET fiber to new heights. SP uses a "railplow" and more than one locomotive to lay fiber in conduits alongside railroad rights-of-way around the country, mostly in the Southwest. Railroads have found that using track rights-of-way for SONET fiber is a significant source of income. In some cases, it produces more profits for the company than actually running rolling stock down the line.

The Railblazer arrangement installs up to eight miles of multiple-strand fiber per day, in conduit, using a rail-mounted trenching plow, conduit and fiber supply cars, and two locomotives. In one smooth operation, the trench is dug, conduits and cable are laid from very large reels, and the conduit is buried. Galvanized pipes are used to span bridges where trenching is not viable. The conduit system is quite secure, being buried at a depth of four to five feet. Using railroad rights-of-way (usually with patrolled and secured access already in place) also protects the fiber from inadvertent contractor dig-ups.

The system can also be adjusted to dig parallel trenches without disturbing the first. Multiple fibers from several service providers, even competitors, can be handled in the same way.

The AT&T Fleet

AT&T had (and still has) a fleet of "fiber boats" that laid huge amounts of undersea international SONET/SDH fiber, mainly to "replace" the satellite system currently in use. SONET/SDH terrestrial links have much lower delays than satellites, as well as lower bit error rates.

The SONET/SDH undersea fiber, in spite of layers of protection, is still light enough that an installation crew of four can drag it onshore to termination points. A sonar arrangement alerts the captain to undersea mountains and the like, and adjusts the "slack" in the SONET/SDH fiber trailing behind the boat.

With recent fiber advances, no active and powered repeaters are now found underwater, which makes the whole process that much easier. The SONET/SDH fibers have a 13-year mean time to failure (MTTF) and a 25-year service lifetime. Breaks are handled by hauling up the broken ends (which are located by means of an optical time-domain reflectometer) and splicing them back together. (The difficulty is to *find* the cable under miles of seawater.)

A side benefit of this use of SONET/SDH is to make international leased lines actually affordable. Because SONET is the North American version of the official SDH international standard, it is relatively easy to create such links.

PLOWING CABLE

Despite AT&T ships and SP Qwest locomotives, it is only fitting to close this chapter with a tribute to those who installed all that fiber the old-fashioned way. They do it because it's their job, and someone has to do it.

Consider the case of northern Nevada. In 1930, 30-foot redwood poles snaked their way across the high plains, carrying toll calls from coast to coast. During World War II, buried analog toll cable was laid between bomb-hardened repeater buildings, but the aerial copper remained. It was not unusual for line maintenance personnel to drive 150 miles just to get to a 50-mile span of wire on poles servicing a few ranches.

The replacement of these wires with fiber has been mainly the work of one man. Art Brothers took over the maintenance of the copper wire because AT&T asked him to do it, and no one else wanted the job. One 107-mile stretch of I-80 has three exchanges with 100 customers. In 1992, 15 miles of fiber were "plowed in." In 1995, it was 11 miles. Another 52 miles got plowed in 1996, ending in the early snow.

Not just the elements make plowing tough. Another $1,000 per mile is needed for historical and cultural studies before trenching can begin. The remains of a Chinese work camp from the days when the railroad was built cannot be disturbed. And then there is the swarm of university students troweling and screening six feet down to see whether those slivers of material turned up by the fiber plow are of prehistoric value.

Art Brothers's fiber may never be used for SONET. But the fact that it exists and may be called upon if needed is a tribute to the ingenuity of outside-plant personnel everywhere. Thanks for all the bits.

CHAPTER 18

SONET/SDH Services

This chapter explores more fully what customers do with SONET/SDH links, whether delivered on SONET/SDH rings or not. The typical and somewhat obvious applications, such as asynchronous transfer mode (ATM) and LAN interconnectivity through routers, are considered in more detail, but from a more user-oriented approach than the pure technology basis used before. Less obvious applications, such as scientific visualization, medical imaging, and other graphical applications are mentioned. The most important point regarding these latter applications is the SONET/SDH support for entirely new services that were impossible before SONET/SDH bandwidths became available.

In earlier editions of this book, this chapter also provided details about the SONET offerings from various carriers throughout the United States. The first edition of this book contained a section some 12 pages long listing service providers—including local exchange carriers (LECs) and inter-exchange carriers (IXCs)—that offered SONET services for ATM, LAN interconnection, and the like. The point is that, not so long ago, SONET services were big news, and it was important that people knew that SONET was a working technology. Now, SONET offerings are greeted with a kind of shrug. Just as SONET equipment has ebbed from view on vendor web pages, so SONET services have receded into the depths of web sites run by service providers. One prominent SONET pioneer does not even list SONET as a separate service offering any longer. Searches on that site for "SONET" mention only its use internally. (You can still buy it from them, of course.)

Past editions investigated several service offerings in detail, not to advocate the offerings, but simply to give a flavor of what was available. In this edition, such service listings are no longer necessary, for two reasons. First, the collapse of telecommunications services offered by many "alternative" and non-traditional service providers (such as power companies) has contracted the available outlets for SONET/SDH services in many areas to the traditional telephone companies. It sounds simplistic, but if you want SONET or SDH, call the telephone company. This brings us to the second reason. Among the traditional telephone companies around the world, SONET and SDH services are readily available. Differences or restrictions in terms of service speeds and areas of coverage may exist, but almost anyone who wants SONET or SDH services in a major metropolitan area anywhere in the world can probably get it. (Someone once remarked that, if you need a list of sources for a technology, then that technology is probably not the best thing to use as a business strategy.) SONET and SDH have certainly moved into the technology mainstream.

SONET/SDH APPLICATIONS: ATM

ATM is the international standard for cell relay technology. In cell relay, information is transported through the ATM network in fixed-length cells. ATM is the multiplexing and switching mechanism for a B-ISDN network. SONET/SDH, in turn, is intended to be the trunking and access technology for ATM-based B-ISDN networks. Originally, B-ISDN standards were intended to contain the service definitions on which service providers would base service contracts for private network services and tariffs for public network services. However, detailed service definitions and functional descriptions never appeared to guide service providers in this area. Thus, B-ISDN has languished for a number of years.

ATM itself has had little success as an alternative desktop technology to shared-bandwidth LANs such as Ethernet. ATM has been much more successful as a backbone technology; but, in this respect, ATM is somewhat masked from users, and this use of ATM need not be considered further here. In fact, today, ATM is often used to link routers over an ATM-based core network. But newer technologies that seem better adapted for router connectivity seem destined to displace ATM from its last natural application.

ATM is therefore used in application environments other than a public B-ISDN. Until B-ISDN is more fully defined (if ever), this must be the case. Most ATM applications use the basic definitions of services derived from the ATM Forum, a vendor consortium faced with trying to fill the service gaps left by B-ISDN. Most of these definitions are essentially (but not exclusively) LAN-based data solutions.

Therefore, when it comes to linking ATM switches (or ATM-enabled routers) with SONET/SDH links and rings, the architectures are fairly consistent. For example, some LAN equipment vendors will use ATM in a LAN hub or router. The customer LAN itself may use ATM directly to the desktop, in which case each desktop device needs its own ATM interface board, usually operating at 155 Mbps or even 622 Mbps. (Note the alignment with SONET/SDH speeds.) More likely, the LAN itself is running some high-speed LAN technology other than ATM, such as Fast Ethernet at 100 Mbps, or even Gigabit Ethernet. ATM is used as a high-speed WAN technology because no high-speed LAN technology will link sites around the country. In either case, the problem now becomes one of effectively linking the ATM switches attached to the LANs at various locations. After all, it makes little sense to use ATM as a high-speed LAN or WAN technology when the traffic must be funneled through very low speed 56 or 64 Kbps links. SONET/SDH leased lines can be used, of course, but such huge amounts of bandwidth will be expensive when used exclusively for one customer—probably prohibitively so.

Figure 18-1 shows this use of SONET rings to interconnect ATM switches. (This example, and others that follow, use SONET, but the same results can be achieved with SDH.) Later in this chapter, an application using SDH (which could also be SONET) is presented.

In the figure, a service provider with a SONET ring has linked the customer premises ATM switches to a SONET ring and has thus provided the desired connectivity. The ring has more than enough bandwidth to satisfy the needs of many customers, especially given that the ring in the figure is an OC-12.

If the customer wishes to comply with existing B-ISDN and ATM specifications, the access link to the ring must be OC-3c or faster. This access is shown running from a SONET TM on the premises in each location. Technically, this TM would be SONET CPE. Thus, it could be the choice of the customer, because SONET standards allow for vendor interoperability. In many cases, this link would be a SONET link running directly to the nearest SONET ring node. In most major cities, that node would not be too far. Without some form of diversity, though, the node could become a major failure concern to the customer.

Note that other services besides ATM connectivity could be provided by the SONET TM at the customer site. These do not have to be ATM services, of course. Many traditional voice and data applications are possible, as are multimedia, videoconferencing, and others.

An alternative to this method of using SONET for ATM transport would be to supply simple ATM LAN connectivity through a DS-1 fractional T-1, fractional T-3, or full DS-3

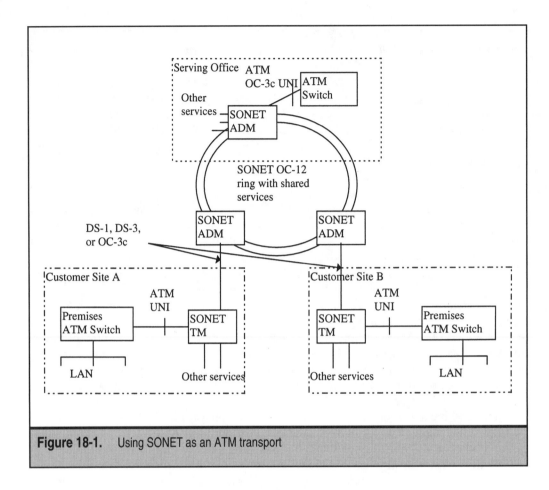

Figure 18-1. Using SONET as an ATM transport

link to the service provider's SONET terminal multiplexer (TM). This would not require SONET gear to be located at the customer's premises, and most users are quite comfortable with T-carrier. The service provider's TM would be located on a SONET ring.

Note also that, if the ring is connected to a serving office with an ATM switch of its own, then various ATM services could be offered to the ATM LAN users, along with the simple connectivity. For instance, the ATM switch could provide switched digital video (SDV) to any of the users on the ATM LANs equipped to receive it. (Most newer PCs would easily handle this.) This approach opens up all kinds of new revenue sources for the service providers, of course. However, the arrangement is not expected to become common for some time yet, as service providers wait to see exactly how common ATM equipment becomes with potential customers.

SONET/SDH APPLICATIONS: LAN INTERCONNECTION

The future of ATM is still an open issue. But not many suitable SONET/SDH applications are tied to ATM in any way. Most customer LANs are indifferent to ATM at best, and many position themselves as viable alternatives to ATM in any case. Gigabit Ethernet at 1 Gbps is a good example.

Another good application for the bandwidth that SONET/SDH provides is LAN interconnection. LANs today are no longer restricted to the older 4, 10, and 16 Mbps Ethernets and Token Rings. The Fast Ethernet of 100 Mbps has been around for quite a while. Gigabit Ethernet has arrived. (Some Apple computers come with a GBE port built in.) And 10 Gbps Ethernet is ready to appear. (It must be pointed out that Gigabit Ethernet frames can be put directly on fiber, and SONET/SDH is not required.) Even without matching these enormous LAN speeds, the delivery of LAN interconnection through routers acting as TMs on four-fiber, bidirectional, line-switched rings (BLSRs) has a lot to offer a customer.

Figure 18-2 shows the scenario for LAN interconnectivity using SONET. Note that, even in this scheme, the presence of ATM LANs and ATM-enabled routers is not ruled out. The major difference is that the service provider is furnishing only passive transport of a customer's ATM traffic, not processing or even switching it in any way. Any ATM functionality must be provided by the ATM CPE, which may actually be a plus from the customer's perspective.

When the DS-1 (or other-speed link, such as fractional T-3) is supported and delivered on the BLSR, few errors and service outages should be seen. Oddly, DS-2, never a common or popular T-carrier offering, is making somewhat of a comeback as a speed delivered on these SONET rings. At about 6 Mbps, DS-2 is a good match for existing Ethernet LANs operating at 10 Mbps. For customers needing higher speeds to link their LANs, whether with ATM LANs and routers, or just faster versions of Ethernet, SONET BLSR is really the only practical means of cost-effective delivery.

Note that the customer's existing TCP/IP routers still support traditional T-carrier interfaces. This cuts down on customer expense, because migrating to SONET OC-N interfaces is unnecessary, allowing the customer to retain the familiarity of working with T-carrier interfaces. Despite the added presence of the SONET ring, no customer network management software used for internal purposes should need to be changed.

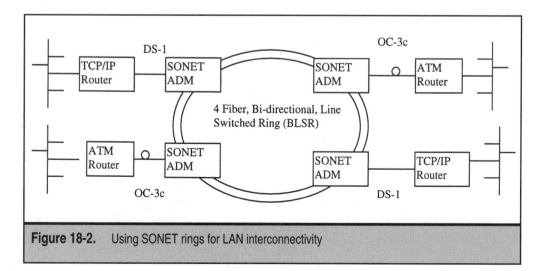

Figure 18-2. Using SONET rings for LAN interconnectivity

SONET/SDH APPLICATIONS: VIDEO AND GRAPHICS

Enough has been said about the bandwidth demands of video and graphics applications earlier in this work. They need not be repeated here. A brief summary of what users are doing to put such loads on LANs and WANs is worthwhile.

Users run many applications that can (and do) heavily load LANs and WANs. In the late 1980s, once PC video displays broke free of the severe limitations imposed by text-based DOS and UNIX interfaces, applications began to include sophisticated techniques for displaying eye-popping images and bitmapped graphics on the monitor screen. The availability of high-resolution, non-interlaced screens at a reasonable price fueled this movement.

Today, many insurance companies and financial institutions include visual data in the form of digital photographs or scanned documents as part of their applications. These may even be stored as a basic field type in a database. Even with compression, these images may be very large.

For example, a medical X-ray image may be represented by a 4,000×4,000-bit array. Each "gray scale" pixel is coded using eight bits. This 4K×4K×8-bit image yields about 128Mb or 16MB of data to transfer over a network. A typical 14.4 Kbps Internet link, with a transfer rate of 1,500 bytes/second would need more than 10,000 hours to send this image from one medical center to another. Over a high-speed network operating at 150 Mbps (such as a SONET OC-3c link), the transfer would take a second or so.

The publishing industry is another large user and generator of huge graphic files. These files represent the advertisements seen in any glossy magazine today. They are not produced at 600 and 1200 dots per inch (the resolution of many laser printers today). They are produced at 2,400 dots per inch. They do not use 64,000 colors, but 16 million or 4 billion colors. Even with compression, files that are hundreds of megabytes in size are not uncommon.

When the graphic or image changes over time—as in an animation or video sequence—the file size grows larger very rapidly. The 30 or so workers in a relatively modest rendering company (this graphic evolution process is known as "rendering") may overwhelm even a 10-Mbps Ethernet quite easily with images flowing frequently between clients and servers.

If two 10-Mbps LANs connected with bridges or routers are linked over a 64 Kbps point-to-point leased-line dedicated circuit, the bottleneck automatically created will be greater than 100:1 (rounding 64 Kbps to 100 Kbps, 10000 Kbps/100 Kbps = 100). Clearly, many of the files and graphics cannot be adequately used in a client–server environment with a 64 Kbps link. The medical image sent across this network from one hospital's LAN server to another hospital's LAN client will now take more than 11 hours (7 minutes×100 = 700 minutes = >11 hours)! The patient will, one hopes, wait.

Even at DS-1 speeds (a full 1.5 Mbps or so), the bottleneck is 6:1. The file transfer delay is cut to "only" 42 minutes or so, but it is still woefully inadequate for building enterprise-wide client–server networks that span the country. Even so, DS-1s have been widely embraced as a LAN interconnectivity option—to the extent that more than 3,000 corporations in America today have at least one DS-1. The search is therefore on for a more effective and efficient way to simultaneously implement LAN technologies and the WANs that connect them.

Lately, another class of application has demanded attention from the client–server community. This type of application demands not only high bandwidths, but also low network delays.

The new class of applications includes video (that is, animated graphics or full-motion frame sequences) as an integral part of the application. Usually called "multimedia" to reflect the fact that these applications consist of and deliver information that is not merely read, but also looked at and listened to, these new types of multimedia applications began to spring up in the early 1990s. Multimedia put strains on LAN stations as well, which more or less limited their deployment on hardware platforms based on 80286 chipsets with only 1MB or so of memory. Once 80386 chipsets became common, and memory sizes grew to the 2MB to 4MB range, multimedia applications became feasible. Today, of course, with Pentium and other chips running at between 1000 MHz and 2 GHz, multimedia is a given. WAN distribution is the problem.

If graphics and imaging applications produced large files than stretched PC hard drives and LANs to their limits, adding video to other applications just made matters worse. Video on a PC functions by presenting a series of still images, called frames, in rapid sequence on the PC monitor screen, usually in a Windows environment. Of course, this is exactly how TV and movies work as well. The smoothness of the motion depends on how fast the sequence changes. For PC video, this can vary from a jerky three frames per second to a smooth-looking and respectable 30 frames per second. The most common video file formats can easily take up to 10MB of hard disk space.

Adding video to applications stressed LANs as well. In fact, it proved so difficult to transfer video files over LANs that almost all videos come directly from a CD-ROM drive on the PC itself. The main difficulty was not even so much the bandwidth. Compared with large high-resolution graphics and image files, low-frame-rate videos placed relatively modest demands on the LANs. Instead, the problem proved to be the variable delay in delivering the video frames from a server to a particular client. For instance, when displaying video frames at a rate of 10 per second, the LAN must deliver frames consistently at that speed. The LAN cannot deliver 8 frames in one second, and then 12 in the next, even though the net result is still 10 frames per second. This concept of "isochronicity" (that is, events on the network happening with a consistent delay) is an important one, especially when applied to high-bandwidth networks.

When considering high-speed networks, the distinction between bandwidth and delay is an important one. Confusion comes when both concepts are spoken of in terms of "speed." "High bandwidth" is a measure of the amount of time between the arrival of the first bit at any point on a network and the arrival of the last bit at the same point. "Low delay" is a measure of the amount of time between the departure of the first bit from the source on a network and the arrival of this first bit at the receiver.

The point about the variability of delays on some networks, especially Ethernet-type LANs and router-based WANs, is worth emphasizing. Some applications, such as voice and video, cannot tolerate more than a few milliseconds of variation in delay across a network. Otherwise, the receiver will assume that something has gone wrong and take steps to react. This has historically been a problem in the decades-long struggle to "packetize voice" (that is, to send digitized voice samples over the same network used to transmit data). The problem is that the voice samples are variably delayed behind a large packet of data; thus, the receiver will not function correctly.

SONET, with its enormous bandwidths, stable delays (the links are essentially leased lines), and high reliability and availability, offered the best hope for productively running these types of applications across a WAN. (Other solutions using DWDM and Gigabit Ethernet exist today.)

Cinema of the Future

Excitement arises when SONET/SDH is asked to do things that were never possible before. For example, ATM switches can be combined with SONET/SDH to deliver major motion pictures to movie screens.

A nagging problem in the movie industry is that, if a theater has only one screen showing a suddenly popular film, or two screens showing a suddenly "dead" film, how can the situation be handled gracefully? For instance, instead of turning away paying customers, why not just "add" another showing in an otherwise almost empty theater? Why screen the same show in two theatres when one will do? This is impossible today, but soon may not be if the Motion Picture Association of America has its way.

Right now, movies are usually distributed in two or three cases containing the reels that run through the projectors. This has been the method of choice since the 1930s. However, if movies were digitized (most films contain a varying amount of digital "special effects" already), they could be "downloaded" directly from studios or pulled off video archives. Figure 18-3 shows the process.

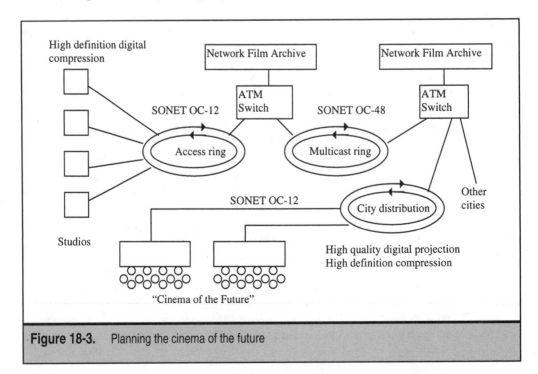

Figure 18-3. Planning the cinema of the future

Studios would produce not film, but bits. The bits would be stored in a high-definition digital format, and would be highly compressed until the feature was completed. Upon release, the studio would simply pipe the video to a SONET OC-12 access ring and through an ATM switch. The switch could archive the movie to allow for time zones and other factors. When shown, the movie would be multicast on a SONET OC-48 ring. Other ATM switches would distribute the signal onto various rings in cities across the country, where the theaters would be linked by multiple OC-12 access lines. The ATM multicasting switches could distribute the video streams to vast numbers of theaters simultaneously, potentially all across the country. This would involve the development of high-definition digital compression techniques and accompanying high-definition digital projection techniques, but the revenue potential is such that this is exactly what may happen soon. (SONET is no longer required to distribute cinema this way, but it might still play a role.)

And the future is getting closer than people think. Episode II of the *Star Wars* series of movies was shot entirely with digital cameras that produced no film, just bits. Nothing really new there: digital cameras have been used before. The big breakthrough is that, in some cases, the film was screened in a digital movie theater, on a digital screen that resembled nothing so much as a huge LCD monitor or digital widescreen television. The experience, according to all who had the good fortune to be invited to the 90 or so select places where this screening method was available, was very different from older analog projection methods—almost like the excitement of the early days of Cinemascope or IMAX. All that is really needed to extend this concept to the SONET/SDH application illustrated in the figure is a network.

Cable TV Distribution

In many places around the world, cable TV offers customers a seemingly endless list of hundreds of television channels, digital channels, movies on demand, and pay-per-view sports events. But even as these offerings diversify and expand, the cost to subscribers for each individual service, from basic cable packages to premium digital movie channels, continues to come down. To keep total revenues healthy, cable TV service providers all over the world are seeking the most cost-effective ways to distribute all of these signals to wider and wider serving areas. The more far-reaching a distribution network, the more cost-effective the entire network becomes.

A number of cable TV service providers have therefore turned to SONET/SDH combined with IP routers to construct infrastructures that can handle hundreds of video streams, distributing them anywhere in a service area. The example that follows uses SDH, but the same feat can be accomplished with SONET.

Consider the cable TV network shown in Figure 18-4. This network is based on an SDH ring running at STM-16, but an STM-64 ring could easily be used as well. Similar networks have already been constructed in Europe with 16-node MS-SPRings with ADMs and 70 or more IP routers acting as TMs attached to this ring.

The TM routers sit in the cable TV headends, which distribute the signals over the usual coaxial cable or fiber subscriber distribution network (or even both, as a hybrid fiber/coax distribution network). The router platforms can be used not only for video services, but also for cable modem Internet access and voice over IP. The routers use sparse-mode multicast to

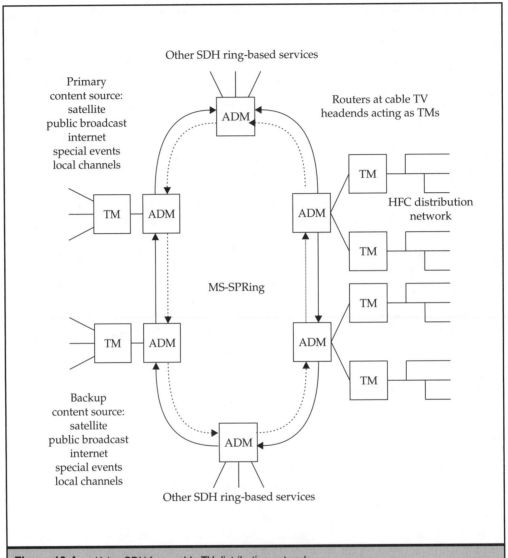

Figure 18-4. Using SDH for a cable TV distribution network

distribute their packets, so that channel "tuning" evolves into joining a specific multicast group. However, the main point here is not the IP router role, but the role of the SDH network, which has all of the bandwidth needed for present and future cable TV services. And the ring can always be upgraded to a higher speed if necessary.

People expect to be able to watch television any time they turn it on, day or night. The survivability features of the MS-SPRing therefore play an important role in this application. The ring can even have alternative sources for signals, a very nice failure recovery feature. Naturally, depending on speed, other SDH services can be offered on the same infrastructure, as is also shown in the figure.

Now, it might be claimed that Gigabit Ethernet (or some other form of optical networking) might be able to fulfill the same role as the SDH ring in this example. However, neither alternative has yet achieved the geographic penetration of SONET/SDH, and the equipment remains expensive. Also, most IP routers support SONET/SDH interfaces with ease, but struggle with early drafts of standards for DWDM optical interfaces. (This is not true for Gigabit Ethernet; but, in that case, distance is a concern.)

The role of SONET/SDH in a cellular telephony environment is much the same. In that case, cell towers and not cable TV headends are attached to the ring, but the end result is very similar architecturally.

SONET/SDH SERVICE ARCHITECTURES

All SONET/SDH service offerings are based on two distinct, but consistent, architectures. Lack of significant variation in SONET/SDH service architectures should come as no surprise. All SONET/SDH service architectures are based on the same array of SONET/SDH building blocks: ADMs, TMs, and related devices such as digital cross-connects (DCSs). Vendors of SONET/SDH equipment can be only so creative if their devices are to stay within the firm guidelines of specifications and standards. (One of the aspects of SONET/SDH currently under assault—and explored in the next chapter—is this apparent inflexibility of SONET/SDH.)

Two basic architectures for supporting SONET/SDH services are examined here. This section details the first major SONET/SDH service architecture, which has been used mostly by the traditional telecommunications service providers, the telephone companies. The next section investigates a more ambitious SONET/SDH service architecture, which integrates more types of access arrangements and even some higher-level services such as switching and routing into the basic SONET/SDH ring.

The first SONET/SDH service architecture can be called the "telephone-company model." This is only because it was the telephone companies, most notably the former pieces of the AT&T Bell System such as Bell Atlantic or Ameritech (the regional Bell operating companies, RBOCs), that invented and deployed this SONET architecture. (The same can be done with SDH.) Nothing prevents an ISP or even a major corporation from building and using this same SONET/SDH architecture.

Until recently, this service architecture was the traditional way that customers made use of SONET/SDH. So, it is also the "typical model." Figure 18-5 shows the overall architecture, again using just SONET for simplicity. This is a busy figure, and so a lengthy explanation is unavoidable.

The central feature is the SONET ring itself, which typically runs at OC-12 or OC-48, depending on the number of customer premises linked and the amount of traffic on the ring. Even OC-192 rings are appearing with this architecture. The ring itself could be a two-fiber UPSR, or a two- or four-fiber BLSR; this aspect does not matter to the overall architecture.

The service provider buildings on the ring, heavily outlined, are all telephone company central offices (COs), but COs are not a requirement. Any site suitable for an ADM

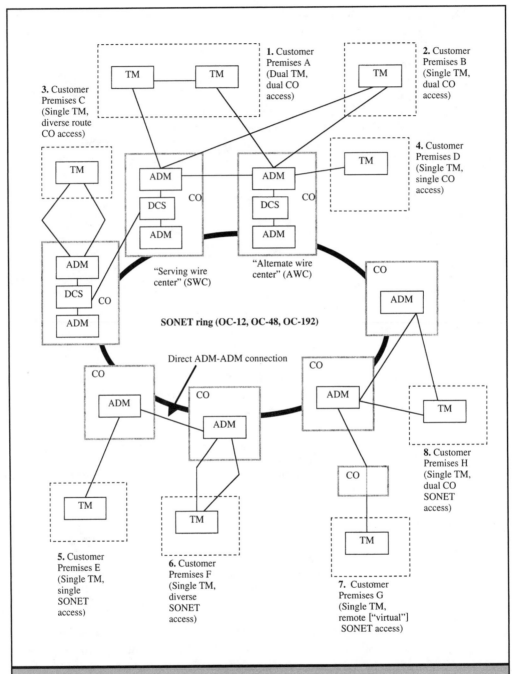

Figure 18-5. Understanding the service architecture of a typical SONET ring

or two will do; however, typically, SONET rings in this architecture link existing central offices. Minimally, the CO will have a SONET ADM, but some COs will have a DCS and more elaborate arrangements, also shown in the figure.

The customer premises are shown as dotted boxes. In this architecture, the customer premises equipment (CPE) would appear to be the SONET TM, but this need not be the case. It is more likely that the TM is owned, operated, configured, and fixed by the service provider. The TM is therefore network equipment in most cases, but nothing precludes major customers from having their own TM. The TM could be in a common equipment space (usually a basement) or in a special area to service an office complex or campus. In the figure, all of the links between offices and customer premises are SONET fiber links. Point-to-point links are shown as thin lines. Note that these links can run from ADM to ADM as well as from TM (customer site) to ADM.

Figure 18-5 shows eight distinct access arrangements to the SONET ring. The upper four are used when VT support is the main service provided, usually at DS-1 or DS-3. This is one good reason why the TM is usually service-provider equipment: customers retain their T-1 multiplexers and see no reason why they should also have to pay for a TM. The lower four access arrangements are primarily for ring access at native SONET speeds. Lower-speed VTs might be present on these SONET links, but all SONET frame content is just carried through the network untouched and delivered to another customer site. The lack of a DCS function is the distinguishing feature of these access arrangements.

Table 18-1 lists these eight customer access arrangements. The rest of this section discusses some of the major features of each arrangement. Usually, each arrangement is covered by a separate pricing plan. Any CO might actually support all eight scenarios. The use of many COs is just for convenience.

Scenario	Characteristics	Comment
1	Dual TM, dual CO access	Premises has two TMs linked together and linked separately to two COs (SWC, AWC). Hairpinning of VTs is possible.
2	Single TM, dual CO access	Premises has single TM, but two links to SWC and AWC. Hairpinning of VTs is possible.
3	Single TM, diverse route CO access	Premises has single TM and access to single CO. Diverse routing reduces link failure risk. Hairpinning of VTs is possible.

Table 18-1.   Eight access arrangements for SONET rings

Scenario	Characteristics	Comment
4	Single TM, single CO access	Premises has single TM and single link to CO. Hairpinning of VTs is possible.
5	Single TM, single CO SONET access	Premises has single TM, and access to ring ADM is direct. No VT cross-connection is possible.
6	Single TM, diverse route SONET access	Premises has single TM; diverse routing to CO reduces link failure risk. No VT cross-connection is possible.
7	Single TM, remote "virtual" SONET access	Premises has single TM, and ring ADM can be reached only through "tail-end" CO. No VT cross-connection is possible.
8	Single TM, dual CO SONET access	Premises has single TM, but links to two SONET COs. No VT cross-connection is possible.

Table 18-1.   Eight access arrangements for SONET rings *(continued)*

The first four scenarios consider customers whose primary use of SONET is for VT speeds—that is, mostly DS-1s, any DS-0s that ride inside the DS-1 channels, and perhaps even some DS-3s. The usage is channelized in the vast majority of cases, so that a key aspect of these scenarios is the ability to cross-connect at the VT level. This is the primary function of the DCS, of course. Now, the DCS might be a separate unit distinct from the SONET ADM, or the DCS could be a shelf or series of cards in the SONET ADM itself. No matter, it is the function that counts. The COs in each of the first four scenarios show *two* ADMs with a DCS in between. This is certainly a possible arrangement, given that a small ADM might gather traffic from a number of TMs, cross-connect and groom as needed, and then pass traffic on to a larger ADM with higher-speed SONET ports attached to the ring. Or the ADM could be one unit with the DCS function built in or separate. The precise equipment configurations are highly flexible.

The main differences between the first four SONET access scenarios are in the details of how the TMs on the customer premises are linked to the CO with the SONET ADM. A few words about each is all that is necessary.

1. There are two TMs on the customer premises, not only linked to each other, but also to the two physically separate COs. The TMs can therefore access the ring through two separate ring nodes. The ADMs might also be linked by a point-to-point fiber run. Also, the DCS might be linked directly to

another DCS in case of ring failure. This is the most bulletproof SONET access arrangement. The TMs on the customer premises are redundant, as are the point-to-point links to the COs. Because the COs are physically separate, route diversity is more or less automatic, except perhaps at the entrance facility of the building itself (even this is sometimes possible to diversify). If the ring fails, links mapped through DCS hairpinning (a cross-connection that turns right around and goes back the way it came) still works. If the main ring node fails, the other CO provides access. Usually, one of the COs is the serving wire center (SWC) for the customer and the other is the backup alternate wire center (AWC). The only truly weak link is that the SWC and AWC are usually a few miles apart, so that a major flood throughout the area (for example) can take out both the SWC and AWC.

2. Only one TM is located on the customer premises, but the route diversity and SWC/AWC structure is preserved. As long as the TM is reliable enough, this arrangement might be almost as good as the first. Hairpinning of VTs is also possible here.

3. Again, only one TM is located on the customer premises, and now only one CO and ADM are located at the other end. However, the SONET links between TM and ADM are two, and diversely routed. This arrangement can be much more inexpensive than running links to two COs, and many customers come to realize that if the outage condition is serious enough, even being hooked up to two COs in the same general area might not help. Hairpinning of VTs is possible here also.

4. The last scenario (essentially for VT support) is the simplest. One TM has one SONET fiber link to one CO with ADM. Cross-connecting and grooming still take place, but the weakest link here is the single-access fiber run, especially right at the premises entrance facility, where construction (stairways, ramps, etc.) frequently takes place. Hairpinning of VTs is also possible here.

The last four SONET service architecture scenarios emphasize native SONET speeds. No DCS is needed at the CO, because access is presumed to be at native OC-3 or OC-12 speeds. (OC-48 is also possible.) The OC-3 and OC-12 could be OC-3c or OC-12c, but the speed is what is important. Notice, too, that the ADMs might also groom and perform normal add/drop functions on the traffic inside an OC-3 and OC-12 on an STS-by-STS basis.

5. A single SONET TM is located on the customer premises and has a single link to the CO ADM. This is quite similar to scenario #4.

6. A single TM has two route-diverse links to the CO ADM, an arrangement similar to scenario #3.

7. This scenario is a little different from anything discussed already. The TM is located at a customer premises not ordinarily serviced by the CO with the SONET ring node ADM. The CO for the customer premises is not on the SONET ring. How can this customer receive SONET service? Answer: Simply route the

fiber through the intermediate CO to the SONET ring node CO. This is backhauling at its best, as has been done for ISDN and other services for many years. The arrangement is sometimes called "virtual SONET" or "remote SONET," but it is still basically backhauling. The risk here is that, if the CO in the middle is hit by an outage, even if TM, ADM, and ring are otherwise in fine condition, service might be disrupted.

8. A single TM is located on the customer premises, accessing two COs with SONET ADMs. This arrangement is similar to scenario #2.

The architecture described in this section has been challenged recently by a more integrated architecture. The new approach is interesting enough to deserve a section of its own.

SONET/SDH Integrated Architectures

One reason that the traditional architecture for SONET/SDH services emphasizes VTs (DS-1 and DS-3) or virtual container and native OC-N/STM-n private lines is that telephone companies in the United States have historically been restricted by regulation to offer only "basic transport" services. In this restricted environment (now changing worldwide), "enhanced services" such as routing IP packets, have been strictly off limits. While this climate has allowed ISPs to grow and flourish even in the shadow of the telecommunications giants, the trend today is toward a more permissive structure where service providers are allowed more latitude. So, the trend today with SONET/SDH service architectures is to create a ring structure that includes more switching and routing capability.

Service providers that favor the more integrated architecture can try to sell more than just private lines with lots of bandwidth or ring protection for circuits. These service providers try to integrate at least some routing of IP packets, or switching ATM cells, right into their network infrastructure. The services they sell are more inclusive than simple connectivity. Many ISPs have been actively pursuing this approach for years. In fairness to the telephone companies, many traditional telephone service providers, especially the United States RBOCs (former pieces of the giant AT&T network before 1984), have been limited by state and federal regulations with regard to the types of data services they can offer. As an example, BellSouth.net the ISP is not BellSouth the telephone company, and care must be taken to keep the two entities separate. Whether right or wrong, this situation is a reality, although things are slowly but surely changing.

Figure 18-6 shows the general structure of a more integrated SONET services architecture. While not quite as busy as the traditional SONET architecture, enough is going on in the figure to deserve some explanation.

The central feature is still the SONET ring, running at OC-3, OC-12, OC-48, or even OC-192. Traditional-model ADM and customer-premises TM might still be part of the overall picture. VT support and OC-N private lines, as well as all of the access arrangements detailed in the previous section might still be present as well.

But the important features of the integrated SONET services architecture are the presence, as SONET ring nodes, of devices labeled "integrated service platforms." These devices could also serve as SONET ADMs, but their real purpose is to add value to the basic bit transport service that the SONET ring provides. These devices are usually capable of

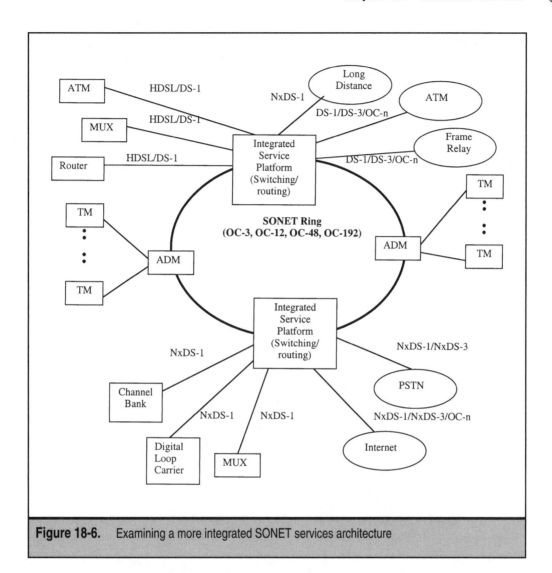

Figure 18-6.   Examining a more integrated SONET services architecture

switching any ATM cells *and* routing any IP packets that arrive in the payloads of the SONET frames. No separate routers or switches are needed: these functions are integrated into the fabric of the devices themselves. But the most important part of Figure 18-6 is what the integrated service platform attaches to.

Each integrated service platform in the figure attaches to a variety of customer premises equipment (shown as boxes). The platforms also have access to different types of networks (shown as circular clouds). That is the power of the integrated services platform—this bringing together of many types of CPE and many types of network services, all in the same package.

Consider CPE types first. Customers can link routers, T-1 multiplexers, and even more exotic CPE such as ATM switches to the integrated ring node. The links need not be fiber links; those shown are DS-1s or even high-bit-rate digital subscriber line (HDSL), a more cost-effective way of provisioning 1.5 Mbps T-1 rates on twisted copper pairs. (HDSL2 requires only one pair of copper.) Because only 5% of commercial office space in the United States has fiber access, this non-reliance on fiber is a nice feature. Some CPE arrangements normally require more than one DS-1 access link. These NxDS-1 arrangements, such as a digital voice channel bank, digital loop carrier, and larger T-1 multiplexers, are also shown. Naturally, variations exist in CPE access line speed requirements. Figure 18-6 is just a general diagram of connectivity, not a blueprint.

On the network service side, the integrated service platform can, by means of the SONET ring, link to a voice long-distance network, an ATM network separate from the ring, or even frame relay. The link speeds are appropriate to the service type: NxDS-1 (or DS-3 in some cases) for long distance, and DS-1 or DS-3 or OC-N for ATM and frame relay. Local voice services can be provided by direct links to the PSTN. Data services (for example, web access) are provided by the Internet. PSTN links would usually run at NxDS-1 or NxDS-3, depending on traffic load. Internet access would usually use NxDS-1, NxDS-3, or even OC-N, depending on the traffic load and the ISP at the other end of the link.

Admittedly, SONET rings could provide such connectivity in the traditional model. The attractive feature of the integrated services model is that the SONET ring nodes take a more active role in the services arena, not just in the connectivity arena. No additional routers, switches, or the like are needed.

It should be noted that there is nothing exclusive about either the "telephone company model" or the "integrated services model" of SONET service architecture. That is, many telephone companies are busily enhancing their legacy SONET architectures with integrated models, and many new service providers are seeking to capture new customers by offering private-line services.

Many of the same arguments can be extended and repeated to cover SDH. But the main points, except for some regulatory details, are unchanged.

This chapter has explored SONET/SDH service offerings and the architectures typically deployed to deliver them. But the SONET/SDH scene has been complicated by the arrival of more advanced optical networking techniques, trends introduced in an earlier chapter. It is time to consider the impact of DWDM and optical networking on SONET/SDH. Some observers feel that the rise of the IP protocol (attributable to the popularity of the Web and Internet), coupled with the appearance of Gigabit Ethernet and DWDM products, has left the future of SONET/SDH very much up in the air. This is the topic of the next chapter.

CHAPTER 19

SONET/SDH
Futures and Issues

This chapter takes a look at a few topics that are of definite interest, but that did not fit particularly well into the overall outline and plan of this work. They have been placed near the end of the book, not because they are an afterthought, but so that the reader can consider the overall picture and direction of SONET/SDH.

The foremost issue to be considered is the future of SONET/SDH with regard to current developments in optical networking. Two main areas are covered. First, the important developments of dense wavelength division multiplexing (DWDM) and optical networking are examined. This technology is not exclusive to SONET/SDH, but is generally applicable to all fiber-optic communications. However, this chapter investigates how SONET/SDH fits in with optical networks. Second, this chapter examines the position of SONET/SDH with regard to the IP protocol and the use of Gigabit Ethernet (GBE). IP is the protocol of the Internet and Web, and GBE is intended for fiber networks. Neither in any way requires SONET/SDH. What is the future of SONET/SDH in an IP/GBE world?

A second set of issues has to do with the testing of SONET/SDH links to make sure that they are within specifications. SONET/SDH standards with regard to OAM&P, protection switching, and, above all, SONET/SDH rings, would appear to make this a non-issue—but an issue it is. This is because SONET/SDH has been marketed and sold as a more advanced "private-line solution." Because purchaser organizations then see SONET/SDH as another way to build private networks, most want their network management personnel to monitor the SONET/SDH links, as they would any other leased line. Yet few customers have SONET/SDH equipment on their premises. (An exception might be routers with SONET/SDH interfaces, but these are not technically SONET/SDH devices in and of themselves.) What can be done with regard to SONET/SDH performance in this case?

This chapter tries bring a sense of closure to these SONET/SDH issues and to the future of SONET/SDH in general.

OPTICAL NETWORKING AND SONET/SDH

As has been pointed out previously, SONET/SDH is *not* an optical network. SONET and SDH are still firmly electrical networks with optical links. Any time anything worth mentioning happens in SONET/SDH, the electrical frame structures, and not the light waves themselves, are the heart of the issue.

By definition, optical networking has NEs and components that perform at least some of the essential networking tasks at an optical level. Optical network components currently include regenerators (really "optical amplifiers," REDFAs), optical ADMs (OADMs), and optical cross-connects (sometimes misleadingly called "optical switches"). REDFAs were discussed previously and will not be considered further. This section deals with OADMs and optical cross-connects. True optical switching has been a dream for many years, and optical switches in DWDM systems are almost, but not quite, ready to go.

This is not to say that DWDM is not an essential part of an optical network. DWDM allows for the packing of many wavelengths onto the same physical path—the optical equivalent of FDM. But if optical networking were DWDM alone, then optical networking would be as exciting as watching 100 TV channels whiz by on a coaxial cable. The excitement of optical networking is what happens before, during, and after the wavelengths come together on the fiber.

One of the problems when mentioning DWDM and optical networking in the same section is that the constantly evolving capabilities of DWDM (in terms of channels and aggregate speeds) make written statements almost instantly obsolete. Fortunately, this section is not so much concerned with feeds and speeds, but with how DWDM and optical NEs come together to make a true optical network. If readers need more detail, assume in the following descriptions that the DWDM is a 100-channel system (the current maximum is more than 1,000 channels), with each channel running at 10 Gbps, the OC-192/STM-16 rate (current serial maximum is 40 Gbps of OC-768/STM-256), for a total throughput of 1 Tb per second. Later in this chapter, a terabit router will be introduced as the perfect companion for this configuration.

After the operating principles of the OADM and optical cross-connect are outlined, an overall architecture for optical networking is introduced. The position of SONET/SDH in this architecture is of paramount interest for the purposes of this book. OADMs, in particular, make use of the in-fiber Bragg grating and optical coupler to selectively separate and merge wavelengths on a single strand of fiber. The details of these components have been examined in Chapter 3 and need not be repeated here.

The Optical ADM

The fundamental principle behind the optical add/drop multiplexer (OADM) is simple enough. A filter is used to isolate, or drop, the desired wavelength from the multiple wavelengths arriving on a fiber. Multiple wavelengths can be dropped, of course, or allowed to pass, depending on the physical configuration of the filters. Once a wavelength is dropped, another channel employing the same wavelength can be added, or inserted, onto the fiber as it leaves the OADM, as shown in Figure 19-1.

This simple OADM has only four input and output channels, each with four wavelengths (λs). Within the OADM, the absence of a dropped wavelength is indicated by a space in the wavelength profile on the fiber. No absolute connection exists between dropped or inserted wavelengths and the ports on the OADM. This is a configuration issue. Naturally, two channels using the same wavelength cannot share an output fiber. On their way through the OADM, the wavelengths might be amplified, equalized, or further processed. The important point is that all of this takes place in optical form, with no electrical/optical (E/O) conversion required.

Basically, all an OADM does is isolate wavelengths for some purpose. The ability to access a wavelength selectively is useful, but the real need is to rearrange wavelengths from fiber to fiber. Because the basic transport link is still the physical fiber itself, the OADM must have some means of rearranging the wavelengths from input fiber to output fiber. This is where the optical cross-connect ("optical switch") comes in.

Changing Fibers in the Middle of the Stream

Electricity can easily be made to follow paths through electronic components—a fact of nature that makes electrical cross-connecting a trivial exercise. Making electricity jump from wire to wire, or from path to path, is not a problem. Electricity always follows the

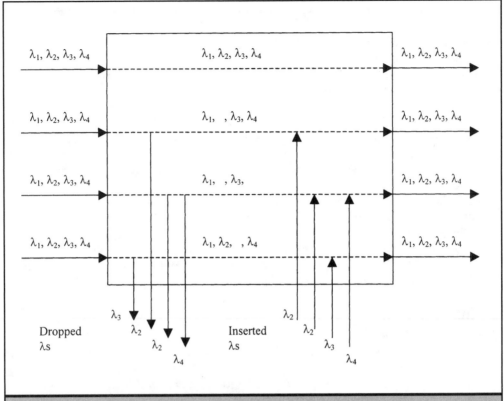

Figure 19-1. Examining an optical add/drop multiplexer (OADM)

easiest path when confronted with a choice. But the whole idea behind fiber-optics in general is to make light *stay* in the fiber. How can light waves be induced to change fibers without changing the light into electricity? This is what an optical cross-connect does.

An optical cross-connect can take four input fibers (for example), each carrying four wavelengths (not a lot for DWDM, but used for ease of illustration), and rearrange the 16 wavelengths onto four output fibers. Naturally, if all four fiber channels use the same wavelengths, then a risk exists that two input channels of the same wavelength will want to share the same output fiber. In this case, a simple "transponder" inside the optical cross-connect will shuffle one of the wavelengths to an available channel.

The optical cross-connect process is sometimes called "switching," but this is misleading. Switching (and routing, for that matter) takes place by examining some header or label inside the arriving traffic stream. Needless to say, cross-connecting does no such thing. It is a simple port-by-port rearrangement process, related much more to multiplexing than to switching and routing. The proof is that, if an output fiber can carry only four channels, then only four channels can be cross-connected onto the output fiber. In the example, 16 input channels are rearranged, and wavelengths are reassigned, but the result is always 16 output channels.

True switches can switch all arriving traffic onto one output port, if need be; and sometimes it seems as if this is always the case. Figure 19-2 shows the optical cross-connect principle applied to DWDM.

Note that a wavelength can arrive on one fiber and leave on another, as long as that is the way the cross-connection table has been configured. The wavelength used can also change (and often does). This "wavelength shift" requires just a simple transponder to detect one wavelength and transmit another. (Wavelengths can even shuffle within a fiber.)

The optical cross-connect performs exactly the same port-by-port reconfiguration functions as the electrical cross-connect. The traffic streams are not STS-1s (or whatever) inside an STS-12 (for example). The traffic streams cross-connected are wavelengths on the fibers. The trick is to do all this entirely in the optical domain. The key component is the "optical cross-connection element" or "fiber switching junction" (which is one reason the entire device is sometimes called a switch).

Ironically, plenty of electricity is involved in the optical cross-connection element—not so much in terms of electrical power, but in terms of role.

The trick, if there is one, is to fabricate the fibers (in some cases, these are actually optical channels etched onto the silicon substrate of a chip) in the optical cross-connection element so that the fiber cores actually come into contact with each other for a length. Usually, the length is no more than 3 mm to 4 mm, but a lot of variation is seen. Because light is optical, yet still within the realm of electromagnetic phenomena, the proper application of a modest voltage (or absence of voltage) across this junction area can make the light change (or prevent the light from changing) cores from one fiber to the other. Figure 19-3 illustrates the principle behind the optical cross-connection element.

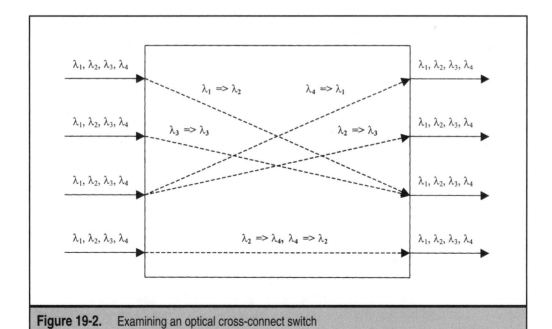

Figure 19-2. Examining an optical cross-connect switch

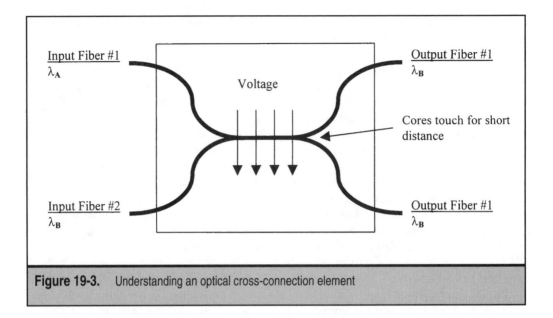

Figure 19-3. Understanding an optical cross-connection element

The wavelengths input into the optical cross-connect element are shown as λ_A and λ_B because optical cross-connects function at the same wavelengths. But the two input streams do a flip-flop onto the output fibers.

The basic optical cross-connect element can be combined into a simple optical cross-connect device, as shown in Figure 19-4. No DWDM is shown, just four wavelengths being shuffled around among input and output fibers according to how the device is configured. Only the path for λ_A is shown as a dashed line, but the others are routed in a similar fashion.

DWDM capabilities can be added to the basic optical cross-connecting, perhaps by the creative use of transponders to shift wavelengths in to and out of the optical cross-connect elements. This is a giant step closer to the optical "switch," because what is being "switched" is the multiple traffic streams on each input fiber. Care must be taken, of course, to prevent two wavelengths from trying to share an output port.

Whenever SONET/SDH is wanted as a wavelength in DWDM systems featuring OADMs and optical cross-connects, transponders are almost always used. This is because SONET/SDH signals all want to use the standard wavelength established for SONET/SDH, 1310 nm. But DWDM operates in the 1550 nm range. Transponders therefore not only shift the native SONET/SDH wavelength into the DWDM range, but also prevent collisions when two or more OC-12/STM-3 (for example) inputs are being carried.

Putting It All Together

The basic components of optical networks—DWDM, OADMs, transponders, and optical cross-connects (sometimes called ODCSs, although there is nothing really digital about them)—can be combined to create an optical node on a fiber ring. This fiber ring is no longer a

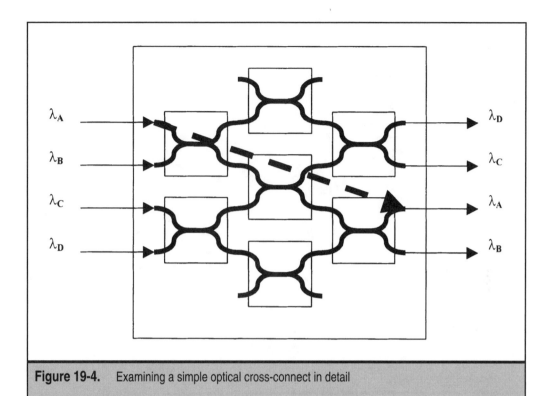

Figure 19-4.   Examining a simple optical cross-connect in detail

pure SONET or SDH ring, of course—although SONET/SDH ADMs and the like might still be present on the ring. The fiber is now carrying DWDM wavelengths and might even be new fiber installed just for DWDM (although if the spans are short enough, SONET/SDH single-mode fiber works just fine). Figure 19-5 shows this basic node structure.

The importance of Figure 19-5 is the role that SONET/SDH plays in the overall architecture. The ring architecture can still be a four-fiber BLSR or another ring type. Protection switching is still a key element. The nice things about the continued presence of SONET/SDH on the new DWDM ring are twofold. First, transponders can be used to allow one or more of the SONET/SDH ADMs to use the DWDM ring. In fact, ring capacity for SONET/SDH is greatly enhanced, because many wavelengths are available. Legacy SONET/SDH equipment can then share the ring, of course. Second, because only a single physical fiber ring needs protection, once the SONET/SDH line-level overhead makes the protect switching decision to use another span (or even ring-wrap), then all of the other DWDM channels go along for the ride.

What about all of the other wavelength channels? They need not be filled with SONET/SDH streams, although that is always possible. Perhaps other direct-to-fiber interfaces could be used, and this has always been a goal of DWDM and optical networking systems. Many frame structures other than SONET/SDH use fiber as a transport medium. For instance, the fiber distributed data interface (FDDI) is a 100 Mbps fiber ring architecture,

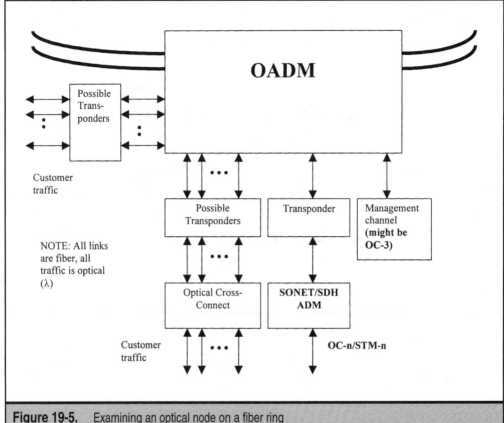

Figure 19-5. Examining an optical node on a fiber ring

but FDDI frames are not SONET/SDH frames. They fit very poorly into SONET STS-3c or SDH STM-1 payloads (about 55 Mbps are "wasted"), although virtual concatenation helps this situation. Fibre Channel is another architecture not aligned with SONET/SDH, and even 10-Mbps Ethernet has run on fiber for a number of years (10Base-F). None of these are SONET/SDH, or benefit from the presence of SONET/SDH directly. (SONET/SDH protection switching is really an indirect benefit, given that FDDI and some others have their own forms of protection switching.) The position of Gigabit Ethernet (GBE) will be considered in the next section. The point here is that, to allow direct-to-fiber traffic to most efficiently use a fiber ring, the requirement to use SONET/SDH framing should be dropped. If SONET/SDH nodes are present, just plug them into an OADM, assign a wavelength, and stand back.

Optical networking offers a way for service providers to sell *just a wavelength* to a customer for almost any form of traffic. The customer could use the wavelength for anything at all, SONET/SDH or not. The OADMs, transponders, optical cross-connects, and so on, take care of making sure that the traffic stream carried by the chosen wavelength finds its way through the ring nodes, across rings, between rings, and so forth, properly. The customer need not know (and need not care) what wavelengths are used for the traffic internally, although this knowledge might be provided routinely for troubleshooting and network management purposes.

This flexibility is the real promise of optical networking to customers and service providers. SONET/SDH runs at fixed speeds, but many things are not aligned with SONET/SDH speeds. SONET/SDH frames can carry many things, but in a very real sense both SONET and SDH are optimized for VT voice channels. Other things fit, but some only poorly, with much wasted bandwidth and many redundant overhead functions.

If all channels of a DWDM optical network become non-SONET/SDH, how then do the customer and service provider gain any benefit from SONET/SDH protection switching and network management functions? If no SONET/SDH is present on the optical network, then standard protection switching and network management must be added to each and every direct-to-fiber technology, right?

Probably not (although proposals along those lines have been made). The DWDM standards allow for a *network management channel* at 1310 nm, outside the normal DWDM traffic-bearing range. This channel coincides with the standard SONET/SDH wavelength, and use of SONET/SDH for this network management channel makes a lot of sense, but is not required. The presence of SONET/SDH on this channel protects and manages all of the other services on the single physical fiber.

This SONET/SDH management channel need not carry any live user traffic at all. This might seem quite wasteful, given that only the overhead is needed for protection and management purposes. However, some DWDM vendors have made creative use of this channel by placing forward error correcting (FEC) code information derived from blocks of bits on the other channels. The result is the added benefit of cutting down on the need for resends, which is the whole point of FECs in the first place. Some vendors even carry this FEC on in-band SONET/SDH channels, as discussed earlier in this book.

Summing Up

So, does SONET/SDH have a place in the optical networking world? If services are provisioned and sold by wavelength, who needs SONET/SDH? For now, the answers are "yes" and "anyone concerned with standard telecommunications network management," based on the role that legacy SONET/SDH equipment and services can play in the optical networking world. But this might not always be the case. One other consideration needs to be examined in the next section.

Many transport frame structures besides SONET/SDH exist—FDDI, Ethernet, and Fibre Channel are but a few. However, given the domination by the Internet and Web in the world of networking today, most frames exist for one purpose only: to carry IP packets between client (for example, web browser) and server (for example, web page). The IP packets themselves can carry voice, video, or data.

It is time to consider the impact on SONET/SDH of IP packets coupled with Gigabit Ethernet (GBE) direct-to-fiber interfaces.

SONET/SDH, IP, AND GIGABIT ETHERNET

The rise of the Internet and Web has all but eliminated concerns about OSI-RM layer 3 (network layer) protocols other than IP. IP packets have become the layer 3 protocol data unit (PDU) of choice. Once there were multiprotocol routers and multiprotocol networks and even "dual stack" clients and servers that understood more than one protocol. Today, the importance of the Web and the Internet, intranets for remote workers, and extranets for e-commerce (for consumers) and e-business (all inter-company networking) means that little room or need exists for anything but IP.

Other layer 3 protocols are often tolerated only if they look like IP. Tunneling (placing packets within packets) is a direct result of this need in many cases. Most routers handle just IP packets, period. The Internet is not a multiprotocol network; the Internet is a single-protocol network. IP only, please. Want to use the Internet and Web for sales, marketing, e-business and so on? Do you run IP? If so, welcome. If not ... well, nice of you to ask.

What has all of this to do with SONET/SDH? Both SONET and SDH are merely OSI-RM layer 1 (physical layer) transports, and as such are spectacularly unconcerned with the exact format of the information their frames carry. All that is needed is a standard way to package voice, video, ATM cells, or IP packets inside SONET/SDH frames, and that should be that.

But today that is not the whole story. The dominance of IP means that the very existence of the SONET/SDH frame structure, especially when DWDM and OADMs are factored in, has been called into question. Fiber links are more and more used to connect not SONET/SDH ADMs or ATM switches, but IP routers. Routers are the engines of the Internet in the same way that telephone company central offices were the engines of the analog voice network. The question today is not "Can IP routers be connected using SONET/SDH fiber links" but rather, "What is the *best* way in terms of cost and efficiency for fiber networks to be used to connect IP routers?" If it turns out to be more cost-effective and efficient to use DWDM fiber directly for IP than to use SONET/SDH, then so be it.

Why Not IP Direct-to-Fiber?

This section would be a lot shorter if there were a simple and easy way to put IP packets directly onto a fiber link. This direct-to-fiber interface for IP packet transmission would quickly replace every other form of WAN interface on a router. There would only be two remaining issues. First, how can the routers keep up with such high speeds (up to 40 Gbps per channel, and many channels with DWDM)? And, second, how could many such point-to-point links (paid for by speed and usually by the mile) ever be cost-justified? (More later on these related issues.)

The most widely used current definition of IP—IP version 4 (IPv4)—makes it difficult, but not impossible, to put IP packets directly onto a fiber link. This has not stopped people from trying some ingenious methods to achieve direct-to-fiber IPv4. It is also true that it is easier to place IPv6 packets onto fiber directly. However, this discussion will not consider direct IP-to-fiber methods, since most observers see that something else is needed between IP and the fiber, whether IPv6 is used or not.

Two problems arise with trying to put IP packets directly onto any physical transport, not just fiber. The same is true of wireless, coaxial cable, twisted-pair copper, and anything else. These problems are:

1. IP packet headers that contain no "start of packet" indicator.

2. IP standards that do not define what happens on the line *between* IP packets (true of both IPv4 and IPv6).

Both problems have equal impact and require equal consideration.

Packets (or any other PDU) need some way for a receiver to determine the beginning and end of the PDU. (The sender—originator of the PDU—is always assumed to know the length!). Because packets are, by definition, of variable length between some minimum and maximum, receivers must know, if nothing else, whether they have room in a buffer (memory area for communications) for the packet. The packet needs a way to tell receivers "this is the start of the packet" and "this is the end." This process is called "delineation" ("framing").

It might seem that simply adding start/end indicators to carry IP packet into a fiber would solve the whole issue. But this is really only half the problem. It is just as important to know what happens on the fiber *between* the IP packets. Packets are neither TDM time slots nor ATM cells. Time slots and ATM cells (once called "labeled time slots") are both the same length on a line. Once delineated (receiver locates head/tail boundary), time slots and ATM cells are easy to distinguish. Time slots use timing (naturally) to accomplish this; ATM uses a different method (whose details are of no concern here). The issue with time slots and ATM cells is whether the time slot of a cell has useful information, or is filled with an "idle pattern" that need not be processed.

IP packets, on the other hand, *always* contain information. Otherwise, there is no reason to compose and send one. So, packets are as bursty as the information they represent. Even if some packet streams carry 64 Kbps voice, there are ways to make this into a bursty IP packet stream. Links carrying IP packets (IPv4 or IPv6) will therefore sometimes be totally quiescent and sending no packets at all, if only for a short while.

Thus, the second issue boils down to the fact that *IP has no standard idle pattern.* No IP document spells out what the line is doing when not sending IP packets. Moreover, it is best to have the line doing *something* when not sending IP packets. Otherwise, long gaps between IP packet arrivals can be confused with outright line failures. This is not a good idea from the standpoint of network management. When idle patterns are used, line failures are instantly obvious.

And the issue runs deeper than just picking a bit pattern to fill the gaps between IP packets. Any idle pattern is obviously just a string of bits. What happens if this string of bits occurs *in the middle* of an IP packet? Would not a receiver think that a *new* IP packet had begun? So, the related issue of "transparency" always occurs when the idle pattern can appear within a packet.

Fortunately for networking, the related issues of packet length, idle pattern, and transparency have all been solved to everyone's satisfaction. They just have not been solved at the IP packet layer (layer 3 of the OSI-RM). These issues and functions are all properly concerns of the OSI-RM layer 2 (data-link layer). This is what the layer 2 frame is for.

All of the current IP documentation specifies placement of an IP packet into some form of OSI-RM layer 2 frame structure (even PPP) before transmission on a line. Frames have two main purposes related to this discussion. First, frames always have a length code. Second, frames always define an "interframe fill pattern" to be used on the line when no frames are being sent. (There are ways of making this interframe fill transparent as well, but this is not important here.) Frames also serve useful purposes for multiprotocol support and error control, but these aspects are not the focus in this section.

By itself, IP need not worry about the length and interpacket bit patterns. Everything works just fine if the IP packet is placed in a standard frame. Therefore, the rest of this section assumes that IP packets are placed into some form of frame before transmission. The question now is what kind of frame? And how many frame levels are needed?

How Many Layers Between IP and the Fiber?

How many layers are needed between IP and the physical fiber network? It would appear that only three layers are needed: the IP packet layer itself, the frame that contains the packet, and, finally, the physical fiber link itself. However, in most implementations of networks, there is more than just one framing layer between IP packet and fiber.

The time has come to examine specifics. Initially, international SONET/SDH standards did not recognize IP as a legitimate, standard protocol at all. The only way to carry IP packets inside a SONET/SDH frame was to carry IP packets inside either a VT (low speeds) or a stream of ATM cells (high speeds). Eventually, packet-over-SONET (POS) allowed IP packets to use a PPP frame to perform the same task at high speeds without requiring ATM cells.

A lot of attention has focused recently on Gigabit Ethernet (GBE), which is basically a series of GBE switches ("hubs," to some) connected by fiber links. The wavelength used by GBE is usually 850 nm, but 1300 nm can be used as well. DWDM transponders can easily shift these wavelengths onto DWDM systems with OADMs and optical cross-connects. No SONET/SDH required! Finally, there is the promise of direct-to-fiber IP interfaces (if standard "interframe fill" and transparency rules are established for IPv6).

Figure 19-6 shows these alternatives.

The real point when discussing SONET and SDH is that only the first two protocol stacks even have room for SONET/SDH. Once it is realized that the important parts of the stack are the IP top and fiber bottom, the only important consideration is how best to get from IP to fiber.

The third method uses the GBE frame to carry IP packets directly onto the fiber. This is fine and works very well, but there is one potential drawback: only one GBE stream can exist on the fiber. The trend today is toward DWDM and OADM optical networking. The fourth protocol stack therefore makes use of DWDM and OADM to allow many GBE streams onto the same physical fiber.

Finally, the GBE frame is nothing magical. It's a package for IP packets, just like SONET/SDH. If an easy way can be found to place delimited IP packets directly onto an optical network, then the ultimate stack might be the last. However, for the time being, IP in and of itself cannot replace all frame functions. So GBE will be needed for some time to come.

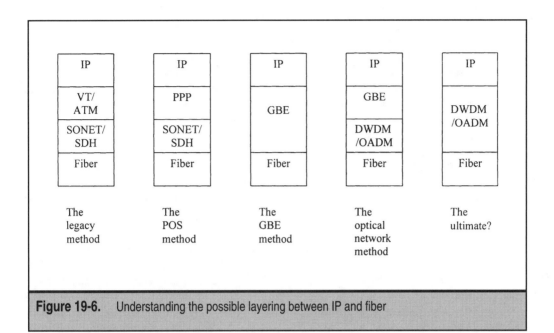

Figure 19-6.   Understanding the possible layering between IP and fiber

In fact, the GBE frame holds many attractions for IP packet transmission. Consider a normal 10 Mbps Ethernet frame the maximum 1,500 bytes in length. These 12,000 bits take $1/833^{rd}$ of second (or about 1.2 ms) to send. In 1.2 ms, a GBE link can send 1,200,000 bits (150,000 bytes).

Now, an IP packet can be about 64,000 bytes long, but it made no sense to even try to generate an IP packet that large. A packet that huge would have to span many 10 Mbps Ethernet frames. This packet fragmentation imposes a large performance penalty on the processor. Many IP protocol implementations tune the IP packet size to the 10 Mbps Ethernet frame size by default for this very reason. The largest IP packets routinely produced (by Sun's Network File System application) are about 10,000 bytes in length. Most IP stacks will fall on their knees if asked to produce or process an IP packet larger than 10,000 bytes or so. (Hackers routinely take advantage of this fact.)

But, because many files and audio and video clips are many megabytes in length, the larger the packet, the more efficient the transfer process in terms of overhead and effort. With a frame size in GBE on the order of 150,000 bytes, *multiple* very large IP packets can be sent in a single GBE frame over a fiber optic link, with no SONET or SDH required. Indeed, this practice is encouraged in GBE.

Once GBE is combined with DWDM and OADM, some amazing things can happen. A GBE switch (switching GBE frames) can send and receive multiple IP packet streams on separate wavelengths over the same fiber (up to 75 miles in many cases). This is sometimes called "photonic networking" or "IP over λ," or the like, but the key is that this application of optical networking requires no SONET/SDH at all.

Can the Routers Keep Up?

Exploring a world in which service providers sell wavelengths and GBE frames carry larger and more IP packets than ever before is all well and good, but all of these huge IP packets must come and go from an IP router. A GBE switch is not necessarily a router. How can IP routers possibly keep up with as many as 2,000 maximum-sized IP packets arriving and departing on a *single* GBE DWDM channel per second? Even ten channels used for this purpose would require the router to process and route 20,000 maximum-sized IP packets per second per link!

Fortunately, terabit (1,000 Gb) IP routers available today. These new routers are sometimes called "carrier-class" routers to distinguish between these fiber backbone nodes and the simple IP routers used as CPE for the enterprise. These routers are claimed to support link speeds as high as 20–60 Gbps, and extreme claims as high as 150 Gbps have been made. This is on a per-port basis, but all of these units support multiple links, of course. Most terabit router boxes are deployed in clusters and can be made to look like one unit operating at up to 19 Tbps in aggregate. The distance between units in a cluster can vary from "right here in this room" to about 10 miles (15 km) or even farther away, and so can be spaced much like ring nodes.

Ironically, most terabit routers still support SONET/SDH rates, from OC-3/STM-1 to OC-192/STM-16, and are not particularly aimed at GBE or DWDM. However, it is still impressive to see a router handle 15 or 16 OC-192/STM-16 ports running at 10 Gbps each. Table 19-1 shows a sampling of this new class of terabit routers.

Eventually, the terabit router, GBE switch, DWDM channels, and OADMs will come together to create a new form of optical networking. Whether this is called photonic networking or IP over λ or something else is not important. What is important is that there is little room for SONET/SDH in many of the new equations. SONET/SDH might remain useful as a network management channel for TMN-compliant OSSs, but the IP world uses other OAM&P techniques such as SNMP and Java.

Company	Product	Number in Cluster	Cluster Throughput	Max OC-3 Ports	Max. OC-12 Ports	Max. OC-48 Ports	Max. OC-192 Ports	Cluster Distance (miles)
Avici	TRS	14	1.4 Tbps	640	160	40	NA	NA
Lucent	NX64000	16	2.5 Tbps	128	64	64	16	1.25
Charlotte's Web	Aranea	32	5.1 Tbps	NA	128	64	NA	~3
Pluris	20000	128	19.2 Tbps	NA	NA	60	15	(feet)
Cisco	12016	16	2.39 Tbps	240	60	60	15	<0.1
Nortel	Verslar	96	4.8 Tbps	96	24	24	NA	~1

Table 19-1. Terabit router sampler

Of course, none of this may happen if the cost of optical networking components such as GBE direct-to-fiber interfaces remains higher than paying for the equivalent SONET/SDH capacity. Oddly, this is not the case. Newer fiber line cards such as GBE cost significantly *less* than their SONET/SDH counterparts—and offer higher speeds. How can this be?

The Cost Issue

Consider a customer or service provider given a choice between configuring a router/switch with a SONET OC-12 port (622 Mbps) or a GBE port (1,000 Mbps). Both will work, because either one can be carried on a separate wavelength over a DWDM system. In other words, the unit purchased and offered for sale is the λ, not the speed. In that case, the choice will more than likely boil down to the cost per bit and the cost per port (usually several ports will be desired, of course).

An OC-12/STM-4 interface for a SONET/SDH ADM can cost up to $12,000. A GBE interface can go for about $2,500. Once the overhead is stripped off the OC-12, the throughput available is about 600 Mbps, a little more than half that available on GBE. Why the difference?

To comply with specifications, the OC-12 board must be able to handle all SONET/SDH OAM&P, timing distribution, VTs, and so on, just in case someone might want to put channelized voice onto the SONET/SDH link. GBE suffers from no such requirement to support channelized voice. Want to do voice over GBE? Use voice over IP (VoIP), and forget about the need to cross-connect!

Strangely, this GBE-versus-SONET/SDH price differential closely mimics the early price differential between 10 Mbps Ethernet and 4 Mbps Token Ring LANs. Arguably, Token Ring was and is the better technology: no collisions occurred, delays were predictable, and so on. But Ethernet won the war. Ethernet boards cost about half (or less) what Token Ring boards cost, and the speed difference more than made up for the limitations of Ethernet. Why did not Token Ring vendors just lower prices to be competitive? They really could not, because extra circuitry was needed to perform token distribution, beaconing, and so on, as required for compliance with the Token Ring specification. Wherever chip prices fell for Token Ring, the same economies applied to the much simpler Ethernet chips, and so the differential held until Ethernet dominated the LAN marketplace. This same phenomenon might repeat itself with GBE and SONET/SDH.

This may be a good place to examine the issue of the relationship between OSI-RM frames, such as GBE and PPP frames, and transmission frames, such as SONET/SDH and DS-1 frames. Essentially, SONET/SDH frames exist because the type of traffic that both SONET and SDH are optimized for—namely 64 Kbps DS-0 voice—has no frame structure in and of itself. And the OSI-RM protocol concepts of the frame and packet apply to data networks, not voice networks. To get all of the benefits of frames in the voice world, such as error control and the like, it was necessary to invent a voice frame structure.

Such a structure is hardly a SONET/SDH innovation. The T-1 and E-1 frames are just voice-application frame structures to allow for voice access and cross-connections. Naturally, the issue is now whether such a frame structure for voice is really needed if VoIP becomes the rule, as appears to be the case.

The whole point is that, because GBE has a perfectly adequate frame structure for VoIP (and even video over IP, and anything else over IP), there is no longer any need to provide SONET/SDH framing on a fiber link. The potential absence of SONET/SDH might require the use of OAM&P methods other than TMN for the network, of course, but TMN had been under attack all along, for other reasons.

Sunset for SONET/SDH?

Will the rising sun of DWDM, optical networking, GBE, and the like, be mirrored by the sunset of SONET/SDH? All technologies give way to newer methods eventually. This section looks at the possible deployment of the missing piece of optical networking, the optical switch/IP router. But even in the photonic networking world, SONET/SDH might still have a place as a legacy, or even consumer, network method.

Optical cross-connects are sometimes called "optical switches" in the DWDM world, but that name is somewhat misleading. The shuffling of wavelengths and fiber links that occurs within optical cross-connects is done by configuration, not the true packet-by-packet or cell-by-cell routing and switching that is based on header content. But certain architectures actually route or switch packets through the device without ever changing the input photons to electricity (at least the payloads). Figure 19-7 shows one such architecture for a terabit router.

The trick of photonic switching and routing is to do it with mirrors—literally! The figure shows how. Note that the routed traffic enters and emerges as a stream of light, and yet individual packets present in the stream can be routed to different output ports. This fulfills

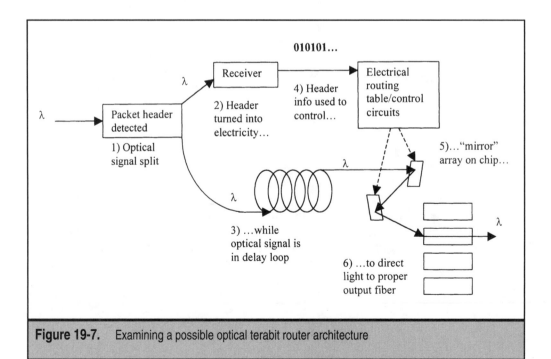

Figure 19-7.   Examining a possible optical terabit router architecture

the definition of photonic routing, but that is not the end of the story. True photonic switching and routing can occur only when a true optical computer and optical circuitry control the entire process. That development is still some way off. In the meantime, the best that can be done is to control the whole process with electrical circuits and table lookups.

Here is how the optical router works:

1. Light of a given wavelength enters an input port. The input fiber could have many wavelengths. If so, each is first isolated for routing purposes. When a packet header is detected in the input stream of photons, the optical signal is split through a simple process so that two copies of the header portion of the optical packet exist. The header can be detected by several means, including fixed-length data units (for example, cells) or a special optical pattern used as a delimiter.

2. The light representing the packet header is converted to electricity by a standard optical receiver. Only the header need be converted in this fashion. The process is quite efficient, because the header is presumably much smaller than the packet payload. This is sometimes called the "electrical wrapper" for the optical packet.

3. Meanwhile, the optical signal representing the packet is fed into a simple delay loop. Routing takes time, and this time is gained by delaying the optical packet until the path through the device is set up.

4. The electrical header information, now a string of 0's and 1's, is used to perform a normal routing table lookup. Routing table maintenance can be performed by existing routing protocols, either electrically or optically, given that most routing information flows between adjacent systems and is not routed itself.

5. Once the proper output port has been determined for the optical packet, an array of mirrors etched onto a silicon substrate is adjusted—electrically—so that the optical packet emerging from the delay loop is "aimed" at the proper output fiber.

6. Finally, the optical packet is directed to the proper output. Naturally, a stream of packets is conducted through the optical router very quickly, and capacities are in the terabit range.

Once the optical terabit router is added to the optical networking mix, many things change. Transport networks now carry IP by wavelength, and legacy traffic such as SONET/SDH by normal time division multiplexing (TDM). The basic access node is the DWDM/OADM device. The services provisioned and sold are either IP over λ (good for almost everything from VoIP to video to data) or legacy voice, frame relay, and ATM services (good for backward compatibility). GBE is easily added to the IP side of the ledger as a framed transport.

Once placed onto the DWDM fiber or fibers, the traffic is passed through an optical cross-connect node and onto a fiber ring (or mesh). This photonic core ring differs from SONET/SDH rings in the sense that it is the optical cross-connects that are the ring nodes, not the DWDM/OADMs themselves. This makes sense, because the basic transport unit of the photonic core is the wavelength, not the frame that rides the wavelength. Figure 19-8 shows this type of fiber network of the future.

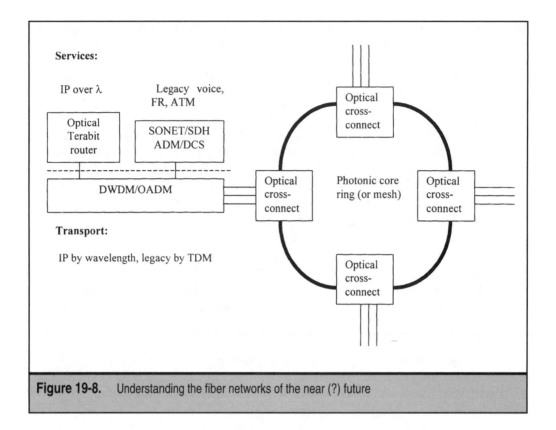

Figure 19-8. Understanding the fiber networks of the near (?) future

It is highly debatable how near or how far into the future this architecture might be placed. Some see it around every corner; others see such a bandwidth-rich infrastructure as total overkill, except in a handful of countries (the United States being one of the handful). One skeptic noted that plans to place a gigabit fiber link to the Fiji Islands was like running a gigabit link into a 30,000-person community in the United States—the economies are about the same size. Even with Internet demand what it is, it would be hard to cost-justify the link in the United States, let alone to the Fiji Islands. Not only that, the current moribund state of the entire telecommunications industry in the early 2000s has pushed the promise of this architecture even farther into the future than before.

So, SONET/SDH will provide enough bandwidth for present needs for some time to come. As for the future, many technologies make their way from service provider backbone to access network to consumer product. Modems have certainly made the trip, and DS-1 might follow. (Most plans for new residential networks feature speeds of 1.5 Mbps or better, at least downstream.) Why not SONET/SDH?

Stripping off the VT and timing support in SONET/SDH could reduce the price of an OC-12/STM-4 interface to about $3,300 for 622 Mbps. Perhaps the backbone SONET/SDH ADM once used as the ring node, and now replaced by the OADM, could make its way to the end of the access fiber link serving someone's house. This might be the best long-term hope for SONET/SDH, but competition will appear at every step along the way.

MISCELLANEOUS SONET/SDH ISSUES

The positioning of SONET/SDH as an advanced private-line solution has posed special challenges to the organizations that employ SONET/SDH in this fashion. SONET/SDH is optical, but most test equipment is electrical in nature. In the United States, for example, a customer may still buy a DS-3, but it is delivered over SONET fiber. How can this be tested without enormously expensive fiber-optic test equipment all over the place? And how are the optical test results to be squared with the end-to-end electrical observations?

When it comes to bit errors, SONET/SDH fiber all but eliminates them. How can technicians measure the bit error rates on links that have so few? The differences between SONET/SDH operating at a BER of 10^{-12} and 10^{-13} may seem trivial, but it may still be important for billing and service-level purposes. It is impractical just to wait for these rare bit errors to occur. They may be literally weeks apart. Yet some method of measurement must be developed.

What about jitter and wander? SONET/SDH is supposed to be synchronous, true, but these effects still exist. How should timing defects in a system designed to eliminate them be detected? This final section briefly explores some of these areas.

SONET/SDH Electrical Testing

The simple hierarchical structure of SONET/SDH signal levels has been a real benefit for maintenance and installation personnel: A technician needs simply to access the information embedded in the SONET/SDH overhead to find all the data needed to identify items such as payload types and to accurately detect error conditions. Access to SONET/SDH overhead has been one of the main challenges.

Many organizations with SONET/SDH links cannot afford to invest in equipment designed to access optical signals, particularly in the early stages of SONET/SDH use and deployment. Ironically, even though SONET/SDH rings are typically used with a traditional point-to-point structure, especially to the customer's premises, using a lot of existing test equipment would mean the loss of many real advantages of SONET/SDH, such as the ability to cross-connect an embedded lower-speed signal without demultiplexing the entire signal.

Fortunately for SONET/SDH users, almost all of the most critical overhead information needed to effectively troubleshoot and isolate problems on a SONET/SDH link circuit can be found and readily accessed at an electrical rather than an optical level. This allows approaches to SONET/SDH testing to be economical and practical alike.

Consider only SONET for a moment. In SONET, any STS-1 signal can be extracted directly from any other level of the hierarchy. The payload of each STS-1 signal may be composed of multiplexed virtual tributary (VT) signals containing lower-speed asynchronous signals, such as from T-carrier. Various VT sizes are defined to accommodate the various types of signals within the STS-1 format. This is what makes SONET so much more attractive than earlier asynchronous T-carrier (DS-1/DS-3) networks, where all "tributary channels" (usually DS-0s) had to be demultiplexed before any cross-connect or switching operation. With SONET, only the channels being acted upon need to be extracted. Other channels pass through the SONET network element undisturbed.

The most important part of the SONET STS-1 frame for testing purposes is the transport overhead bytes. These contain the most critical information regarding the performance of the SONET link on the section and line levels. The section overhead provides the diagnostic information needed to support the transportation of the user information in the synchronous payload envelope between repeaters on the link. The section overhead contains information about framing, error monitoring, data communications, and a local orderwire for point-to-point voice communication within the section.

The line overhead is also part of the transport overhead. These bytes are concerned with the movement of the synchronous payload envelope (SPE) with the STS frame as it moves between adjacent network elements and nodes. It also includes information for monitoring individual STS-1s in a higher level STS-N, line alarm indications, and remote defect indicators at the line level (RDI-L).

The SPE itself contains the path overhead. Path overhead is concerned with the end-to-end SONET circuit between termination points, such as add/drop multiplexers (ADMs) and channel service units. It includes its own remote alarm indicators as well as information on VT framing. All of these bytes allow for effective in-service testing of a SONET system, because the equipment passes failure indicators between these sections, lines, and paths.

These OAM&P functions have all been dealt with before and need not be repeated here in detail. The point is that the SONET/SDH overhead bytes provide a lot of information that allows a technician to reconstruct the sequence of alarms initiated by a failure and to trace the trouble back to the source. This information is not only useful for the maintenance technician troubleshooting the SONET/SDH link, but also for the installation technician performing routine adds, drops, and changes to the network.

The key to this whole process is to allow these technicians easy access to the SONET/SDH overhead bytes. This sort of access can be very difficult to do at the optical level. Here, individual elements and pieces of the signal cannot be easily extracted without disturbing the rest of the signal on the link. Special signal "splitters" can be used, but this involves expensive components and requires great care in installation to avoid disrupting the network. It also introduces the potential for annoying SONET/SDH signal degradation.

However, when the SONET/SDH optical signal is converted to an electrical environment, testing becomes much simpler. None of the original information is lost, because no intermediate signal-processing has occurred. Testing at the SONET/SDH electrical level can also be done with familiar methods from the T-carrier or E-carrier world at commonly deployed service monitoring points. Because the test equipment does not need to be designed to function with (often quite delicate) fiber optics, it can be much less costly and more hardened for frequent use.

A natural point for electrical testing on SONET/SDH is through the DCS or ADM at central office locations. Most of these DCSs will include access to the STS-1 or STM-1 signal level through an STX-1 or similar cross-connect point. Such a point can be inexpensively added to existing DS-3 bays using just cabling, connectors, and hardware. The use of the STX-1 point provides very simple access to SONET overhead byte information for fault location. All of the technicians can get as much information in the overhead bytes as they need for the problem.

Many SONET overhead byte functions also translate directly into messages or LED indicators that are common and familiar from the T-carrier world. Loss of signal or loss of pointer conditions (for example) are easily indicated with LEDs, and so are available on most transmission test sets. Where more detailed indications are needed, test sets with full displays of overhead bytes at all levels can be used.

Many traditional tests, such as loopbacks to verify simple circuit integrity, can be conducted from the STX-1 access port. The tests can be done at the DS-1 or DS-3 interface unit to verify the performance not just of the SONET link, but of the demultiplexed VTs that carry the DS-1 or DS-3 service. This is very useful in situations where an STS-1 frame is not being directly delivered to the customer, but is broken down into multiple DS-1s or a DS-3 before it is sent out to the customer site.

Electrical SONET testing puts the testing power where it is needed the most. Most SONET devices include built-in optical diagnostics that are always monitoring the status of the optical link. However, connection to the customer premises almost universally occurs at the electrical level. A technician equipped with both STS-1 and DS-1/DS-3 capabilities is better prepared to find and isolate a fault within any of the common digital speeds on the circuits actually delivered to the customer. The same applies to SDH.

Accessing optical circuits can be very complex; thus, such testing is not a routine activity. Everyday and routine add/drop activity does not need a testing program built around optical access. However, electrical testing is perfect for routine operations. It is similar to traditional T-carrier transmission testing, so that the learning curve is shorter and technicians can be more effective more quickly.

Electrical test sets for SONET/SDH, like many others, are often designed as hand-held packages with battery operation to simplify usage for field technicians. By making electrical SONET/SDH testers available to every technician, and by providing a shared optical tester for installation purposes, a service provider or even end-user can put together a sophisticated testing package. This also provides a significant cost savings compared with investing in strictly optical test equipment.

Bit Error Rates and SONET/SDH

The previous section should have convinced everyone that traditional electrical testing still has a role in SONET/SDH. This section explores this issue with respect to one important aspect of testing: bit error rates (BERs) and BER testing with BER testers (BERTs).

BERTs are a basic tool for testing not only SONET/SDH systems, but any transmission system, for signal quality in terms of bit errors. The BERT is the first tool used by a designer, and the last tool used by an installer to establish that the basic, physical transmission path is acceptable. A BERT consists of two parts, a transmitter and a receiver. Sometimes these parts are housed in separate devices for ease of use. The task of the BERT device is very simple: It feeds bit-by-bit test patterns across the transmission system and confirms that the transfer was error-free or, if not, what the BER was for that particular test.

At heart, a BERT is 0's and 1's. Generally, the BERT operates entirely on this level, with several different standard test patterns to choose from. This type of BERT is used to develop and test a new and as yet unspecified protocol. The tester can control the frequency and output ranges, and can adjust these values so as to "stress" the system. The process of stressing a system helps to determine the marginal conditions and boundaries of acceptable operation. With this type of test, data transmitted has no structure.

However, many BERTs are also designed to test link quality at the higher layers of a particular interface or electrical specification, such as T-1 or E-1. They can also use a higher-level protocol, such as ATM or ISDN. These testers will operate at the correct operating frequencies and tolerances for the standard chosen. Such BERTs are able to recognize frames or packets (or both) for the given protocol and to generate them. Typically, once a protocol standard is firmly established and common, the BERT device is replaced by a full protocol analyzer.

What has all this to do with SONET/SDH? Well, the most basic BERT measurement is the BER. This is just defined as the number of bit errors divided by the total number of bits sent during the test. The resulting number is very small, and positive—between 0 and 1. Usually, the BER is in the range 10^{-6} to 10^{-12}. The first number is typical of copper networks; the second is more typical of SONET/SDH networks.

BERT tests are very simple to perform. The technician connects the BERT to both ends of the transmission system or subsystem, and then the transmitter part of the BERT sends test signals across the system. At the other end, the receiver part listens to the incoming signal and determines the correct data level, voltage level, and timing used by the transmitter. The receiver can also compensate for things such as phase delay. The testing works by having both ends generate the test pattern locally using the same algorithm. Then the receiver synchronizes the transmitted pattern with the correct receive pattern.

Other measurements that a BERT can perform include a time interval within the test. SONET/SDH standards define measures such as errored seconds (ES). The ES is the number of seconds during which at least one bit was in error. Severely errored seconds (SES) have a BER of 10^{-3} or greater. A measure called consecutive SES is the measurement of SESs for which the previous two seconds were also SESs. Some BERTs will graph these measures for ease of interpretation.

Here is where the SONET/SDH comes in. Every time the BERT detects an error, the device logs it. However, how long should a BERT device run before a high level of confidence is achieved in regard to the BER? The fact that no errors are found does not mean that the link is perfect. The perfect link doesn't exist. Consider that with a SONET/SDH link engineered for a BER of 10^{-15} (not impossible today), an OC-24 system (an older, seldom seen SONET level used just for this example) operating at about 1 Gbps will encounter an error only every 11.5 days, on average.

A valid test certainly requires more than a single error to offer statistically relevant results. The key is that most BERT tests involve random sampling. A SONET/SDH span may encounter a burst of errors that will not occur again for another three months. If the sample period was not long enough, the test may indicate that the link is unacceptable. Conversely, a very bad link may appear to be good.

Statistically, testing until ten errors have accumulated offers only 68% confidence in the BER, and even 100 errors offers only 90% confidence. On very low-BER SONET/SDH links, imagine how long it would take to accumulate 100 bit errors, even at 1 Gbps. (At 1 Gbps with a BER of 10^{-15}, 100 errors will be logged in about 3.17 years.) Therefore, how can SONET/SDH links get accurate BER information?

SONET/SDH BIP checks can be used to determine BERs internally, but this is not the answer. Suppose the whole point is to verify that the BIP checks are working correctly? How could they be reliable? A better answer is to stress the system and increase the error rate. Error rates measured while a system is under stress can be extrapolated to lower error rates for the system when it is under no stress. By collecting error rates for the system under varying levels of stress, and comparing to normal testing results, a performance curve can be produced.

For example, under specific conditions of stress, a SONET/SDH link or ring may produce a BER of 10^{-5}, which can then be extrapolated to a BER of 10^{-10} for the link or ring without any stress. The goal of stressing the link is not to determine the absolute error rate, but rather the upper bound on the error rate. Stressing the system to increase the error rate can take significantly less time and still offer an accurate picture of link quality.

Any digital system has three sources of error: noise, jitter, and intersymbol interference. Noise is not a good option for stressing optical systems, because the systems are fairly impervious to noise in the first place. Jitter errors are caused by the active components within the system, such as the timing electronics and their associated errors. SONET/SDH is designed to minimize these effects as well. Intersymbol interference errors are caused when particular patterns (that is, "symbols") of bits interfere with one another, leading to "smearing" of signals. Fortunately, fiber systems are not invulnerable to this type of stressing. Intersymbol interference in SONET/SDH can be generated by cycling specific patterns through the design.

One major difficulty with using BERTs in SONET/SDH is getting the receiver side of the BERT hooked up to the other side of the link. Usually a person is needed at both ends to set up a test, although routers with SONET/SDH interfaces could conceivably run a BERT between them just through configuration and commands. One possible solution to the distance problem is to loop the signal back. The disadvantage here is that if an error were to occur, it would not be obvious whether the error occurred outbound or inbound. However, by looping the signal around, it is always possible to determine whether another person at the far end is necessary in the first place.

Another possible solution is to permanently integrate the BERT into the central office of the system and to remotely test lines. Most BERTs can be controlled by way of a modem from a PC. By using a digital access cross-connect, any embedded SONET/SDH circuit can be dropped into a test port. This method avoids the need for a person at each end of the line. Some BERTs can even be programmed to automatically test links one by one.

SONET/SDH Timing Errors

The path from the primary reference source (PRS) clocks to the network elements in a SONET/SDH network may be many. For example, in SONET, there are the PRS clocks themselves, building integrated timing supply (BITS) clocks, and derived DS-1 clocks. Consider only SONET for now. The American National Standards Institute (ANSI) and Bellcore both specify limits on the wander (the official term for what most people call "jitter") of each clock. The standards specify items such as time deviation (TDEV) and maximum time interval error (TIE) masks, and frequency offset and drift rate limits.

This section briefly describes typical synchronization distribution systems, explains various wander measurement techniques, and discusses a little about how the specified limits apply to a SONET system. Like any other network component, synchronization needs proper maintenance. SONET synchronization makes it possible to multiplex from an STS-1 at 51.84 Mbps up to an STS-48 operating at 2488.32 Mbps and back in one device. However, the synchronization system itself can cause problems when it is not properly monitored and maintained.

Probably the most common problem with SONET synchronization is faulty provisioning of the synchronization configuration itself. Every once in a while, clocks will fail, and the failure may go unnoticed. As a SONET network of many links and rings grows, wander will accumulate. Sometimes, wander may grow until it affects the payloads themselves. Thus, several points in a synchronization system must be monitored for preventive maintenance and possible trouble.

When a SONET network element (NE) receives bits and frames that have excessive wander relative to the synchronization clock provided to that network element, the SONET pointer adjustments are intended to be used to make up for any phase displacement. The greater the wander in the timing, the more frequent the pointer adjustments. When the times comes to extract the payload from the frame, these repeated pointer adjustments cause jitter in the payload. Excessive pointer adjustments cause excessive jitter and produce SES.

Not surprisingly, the applications most sensitive to delay variations (such as voice and video) are the most sensitive payload traffic. If a corporate PBX were to use a DS-1 payload as a timing reference, pointer adjustments could cause the local clock to lose lock on the payload bits. Therefore, service providers routinely warn customers not to use a DS-1 carried over SONET for timing purposes. Sometimes, however, the customer has no viable alternative, and the problems persist. In the case of video, pointer-induced jitter on the DS-3 payload used to carry the video can cause the colors to shift.

SONET rings can be a concern. Usually, a SONET ring will have two PRSs on opposite sides of a ring. Normally, the timing information from one PRS is carried clockwise on the ring to half of the SONET NEs. The timing information from the other PRS is carried clockwise to the other half of the devices. This use of two PRSs allows the timing of all the SONET NE devices to be traceable to a PRS when the ring experiences one "cut." Even when no cuts occur, one PRS can act as a backup in case the other PRS fails.

The SONET network elements on the ring are usually ADMs. In each ADM, a SONET minimum clock (SMC) can be provisioned to receive timing information from either the line signal on the ring or from an "external" DS-1 reference. In some nodes on the rings, a BITS clock takes care of the timing from the incoming DS-1, filtering the phase and providing some "holdover" capability. This means that the BITS clock "remembers" the timing frequency should the derived DS-1 clock fail. Because the SMCs provide some holdover capability themselves, it is not always necessary to have a BITS clock at every node. This is what SONET calls the "line timing" mode. A BITS clock also could be used to receive timing from a DS-1 signal not associated with the SONET ring.

Suppose a SONET ring passes traffic to another ring. Each has the same timing arrangement just described; thus, each is an "island" of timing. Whenever such "islands" of SONET pass payloads, one to another, payload jitter owing to pointer adjustments is bound to accumulate. As the number of interconnected SONET "islands" grows, jitter could become large (and ultimately excessive) if wander and the associated pointer adjustments were not limited in some way. Bellcore SONET documents specify how SONET equipment handles wander, and ANSI standards have limits for wander that is found at interface points.

Naturally, standards exist to ensure the integrity of SONET synchronization. Test equipment is also available to test conformance to the standards. Excessive pointer activity is an indication of excessive wander. Acquiring data about the synchronization of a SONET link or ring system at installation is very useful. If timing problems or concerns arise later, a known baseline is available. Measurement and consideration of jitter and wander will become increasingly important as links are made between SONET "islands" (especially rings) and as the complexity of the SONET network grows.

Much of the foregoing applies equally well to SDH, with just a few minor architectural and terminology differences.

APPENDIX A

Acronym and Abbreviation List

AAL5	ATM adaptation layer type 5
ACSE	Association control service element
A/D	Analog-to-digital (conversion)
ADM	Add/drop multiplexer
AIS	Alarm indication signal
AIS-L	Line-level alarm indication signal
AIS-P	Path-level alarm indication signal
AIS-V	Virtual tributary–level alarm indication signal
AIU	Access interface unit
AM	Administration module
ANSI	American National Standards Institute
APD	Avalanche photodiode (or photodetector)
API	Access point identifier
APS	Automatic protection switch
ATIS	Alliance for Telecommunications Industry Solutions
ATM	Asynchronous transfer mode
AU	Administrative unit
AU-n	Administrative unit level n (n = 3 or 4)
AU4-nc	n AU4 signals concatenated
AUG	Administrative unit group
AWC	Alternate wire center
BCH	Bose–Chaudhuri–Hocquenghem
BCH-3	Triple error-correcting BCH code
B-DCS	Broadband digital cross-connect system
BER	Bit error rate (or ratio)
BIM	Byte interleaved multiplexer
BIP	Bit interleaved parity
BIP-n	Bit interleaved parity with n bits
B-ISDN	Broadband ISDN
BITS	Building integrated timing supply
BLSR	Bidirectional line switched ring
BRI	Basic rate interface
C-n	Container level n (n = 11, 12, 2, 3, or 4)

CAS	Channel associated signaling
CBR	Constant bit rate
CC	Composite clock
CCC	Clear channel capability
CCITT	International Telegraph and Telephone Consultative Committee
CEPT-*n*	Conference of European Posts and Telecommunications level *n*
CEPT-1	See E1
CEPT-2	See E2
CEPT-3	See E3
CEPT-4	See E4
CLLI	Common language location identifier
CLMP	Connectionless network layer protocol
CMISE	Common management information service element
CO	Central office
CORBA	Common object request broker architecture
COT	Central office terminal
CPE	Customer premises equipment
CRC	Cyclic redundancy check
CU	Channel unit
CV	Coding violation
CV-P	Path-level coding violation
CV-S	Section-level coding violation
CV-V	Virtual tributary–level coding violation
DACS	Digital access cross-connect system
DBR	Distributed Bragg reflector
DCB	Digital channel bank
DCC	Data communication channel
DCOM	Distributed common object mode
DCS	Digital cross-connect system
DDS	Digital data services
DFB	Distributed feedback laser
DIU	Digital interface unit
DQDB	Distributed queue dual bus

DS-*n*	Digital signal level *n*
DS-0	Digital signal level 0 (64 Kbps)
DS-1	Digital signal level 1 (1.544 Mbps)
DS-1c	Digital signal level 1c (3.152 Mbps)
DS-3	Digital signal level 3 (44.736 Mbps)
DS-3c	Digital signal level 3c (91.053 Mbps)
DS-SMF	Dispersion shifted single-mode fiber
DSX-*m*	Digital signal cross-connect point for DS-*m* signals
DWDM	Dense wavelength division multiplexing
E1 (E-1)	ITU-T digital signal level 1 (2.048 Mbps)
E2 (E-2)	ITU-T digital signal level 2 (8.448 Mbps)
E3 (E-3)	ITU-T digital signal level 3 (34.368 Mbps)
E4 (E-4)	ITU-T digital signal level 4 (139.264 Mbps)
ECC	Embedded communication channel
EDFA	Erbium-doped fiber amplifier
ELED	Edge light-emitting diode
E/O	Electrical (signal) to optical (signal) conversion
EOC	Embedded operations channel
ERDI	Enhanced remote defect indicator
ES	Errored seconds
ESF	Extended superframe
ETSI	European Telecommunications Standards Institute
FBG	Fiber Bragg grating
FCS	Frame check sequence
FDDI	Fiber distributed data interface
FDM	Frequency division multiplexing
FEBE	Far end block error (now REI)
FEC	Forward error correction
FERF	Far end receive failure (now RDI)
FOTS	Fiber-optic transmission system
FSI	FEC (forward error correction) status indication
FWHM	Full width half maximum
FWM	Four wave mixing

GBE	Gigabit Ethernet
GFP	Generic framing procedure
GPS	Global positioning system
GRIN	Gradient index (fiber)
HCDS	High-capacity digital services
HDLC	High-level data-link control
HDTV	High-definition TV
HEC	Header error control
HO	Higher order
HOPOH	Higher order path overhead
HOVC	Higher order virtual container
HP	Higher order path
HVC	High-order virtual container (VC3/VC4)
ID	Identification
IDLC	Integrated digital loop carrier
IEC	Incoming error count
IETF	Internet Engineering Task Force
IM	Inverse multiplexer
IOS	Intra-office signal
IP	Internet protocol
ISDN	Integrated service digital network
ISF	Incoming signal failure
ISID	Idle signal identification
ISM	Integrated (intelligent) synchronous multiplexer
ISO	International standards organization
ISP	Internet service provider
ITU	International Telecommunication Union
ITU-R	ITU Radio Standardization Sector
ITU-T	ITU Telecommunications Standardization Sector (formerly CCITT)
IXC	Inter-exchange carrier
LAN	Local area network
LAPD	Link access procedure—D channel
LAPS	Link access procedure—SONET/SDH

LCAS	Link capacity adjustment scheme
LCD	Loss of cell delineation
LCD-P	Path-level loss of cell delineation
LCM	Least common multiplier
LCV	Line coding violations
LEAF	Large effective area fiber
LED	Light-emitting diode (as a light source)
LES	Line errored seconds
LLC	Logical link control
LO	Lower (or low) order
LOF	Loss of frame
LOH	Line overhead
LOM	Loss of multiframe
LOP	Loss of pointer
LOPOH	Lower (or low) order path overhead
LOP-P	Path-level loss of pointer
LOP-V	Virtual tributary–level loss of pointer
LOS	Loss of signal
LOVC	Lower (or low) order virtual container
LP	Lower (or low) order path
LSB	Least significant bit
LSES	Line severe errored seconds
LTE	Line termination equipment (See also MSTE)
LVC	Low-order virtual container (VC-11/VC-12/VC-2)
MAF	Management application function
MAN	Metropolitan area network
MCF	Message communications function
MD	Mediation device
MF	Mediation function
MFAS	Multiframe alignment signal
MO	Managed object
MS	Multiplex section
MS-AIS	Multiplex section AIS

MSB	Most significant bit
MSF-AIS	Multiplex section FEC alarm indication signal
MS-FERF	Multiplex section FERF
MSOH	Multiplex section overhead
MSPRing	Multiplex section-shared protection ring (older)
MS-RDI	Multiplex section remote defect indication
MS-REI	Multiplex section remote error indication
MS-SPRing	Multiplex section-shared protection ring (newer)
MSTE	Multiplex section terminating equipment (also LTE)
NDF	New data flag
NDF-P	Path-level new data flag
NDF-V	Virtual tributary–level new data flag
NE	Network element (or equipment)
NEF	Network element function
NLSS	Non-locally switched special
NNI	Network node interface
NOB	Network operator byte
NPC	Network parameter control
NPI	Null pointer indication
NRM	Network resource management
NUT	Non-preemptive unprotected traffic
NxDS-1	N-level DS-1 multiplexing (fractional T-1)
NZ-DSF	Nonzero dispersion-shifted fiber
OA&M	Operations, administration, and maintenance
OAM&P	Operations, administration, maintenance, and provisioning
OC-N	Optical carrier for Nth level
ODI	Outgoing defect indication
O/E	Optical (signal) to electrical (signal) conversion
OEI	Outgoing error indication
OFS	Optical fiber system; out-of-frame second
OH	Overhead
OLC	Optical loop carrier
OOF	Out-of-frame

ORL	Optical return loss
OS	Operations system
OSF	Operations system function
OSI	Open systems interconnection
OSI-RM	Open systems interconnection reference model
OTDR	Optical time-domain reflectometer
PC	Personal computer; protection channel
PCM	Pulse code modulation
PDA	Personal digital assistant
PDH	Plesiochronous digital hierarchy
PDI	Payload defect indication
PDI-P	Path-level payload defect indication
PIN	Positive–intrinsic–negative photodiode
PJ	Pointer justification
PJC	Pointer justification count
PLM	Payload label mismatch
PLM-P	Path-level payload label mismatch
PLM-V	Virtual tributary–level payload label mismatch
PM	Performance monitoring
PMO	Present method of operations
POH	Path overhead
PON	Passive optical network
POS	Packets on SONET (SDH)
POTS	Plain old telephone service
ppm	Parts per million
PPP	Point-to-point protocol
PRC	Primary reference clock
PRS	Primary reference source (for clocks)
PS	Protection switching
PSC	Protection switching count
PSD	Protection switching duration
PTE	Path-terminating equipment ("element" in SDH)
PTR	Pointer

PUC	Path user channel
PVC	Permanent virtual circuit
QA	Q-adapter
QOS	Quality of service
RAI	Remote alarm indication
RDI	Remote defect (degrade) indication (formerly FERF)
RDI-L	Line-level remote defect indication
RDI-P	Path-level remote defect indication
RDI-V	Virtual tributary–level remote defect indication
REDFA	Rare-earth-doped fiber amplifier ("regenerator")
REI	Remote error indication (formerly FEBE)
REI-L	Line-level remote error indication
REI-P	Path-level remote error indication
REI-V	Virtual tributary–level remote error indication
RFC	Request for comment
RFI	Remote failure indication
RFI-V	Virtual tributary–level remote failure indication
RI	Refractive index
RM	Resource management
ROSE	Remote operations service element
RSM	Remote switch module
RSOH	Regenerator section overhead
RSTE	Regenerator section-terminating equipment (also STE)
RSVP	Resource reservation protocol
RT	Remote terminal
SAPI	Service access point identifier
SBS	Stimulated Brillouin scattering
SCV	Section coding violations
SD	Signal degrade
SDH	Synchronous digital hierarchy
SEC	Synchronous equipment clock
SECB	Severely errored cell block
SEFS	Severely errored frame seconds

SES	Section errored seconds
SF	Signal fail
SF	Superframe
SIF	SONET Interoperability Forum
SLC	Synchronous line carrier
SLED	Surface light-emitting diode
SLM	Single longitudinal mode; synchronous line multiplexer; signal label mismatch
SM	Switching module
SMDS	Switched multimegabit data service
SMF	Single-mode fiber
SMN	SONET (SDH) management network
SMS	SDH management subnetwork
SNAP	Subnetwork access protocol
SNC	Subnetwork connection
SNMP	Simple network management protocol
SOH	Section overhead
SONET	Synchronous optical network
SPE	Synchronous payload envelope
SRS	Stimulated Raman scattering
SSM	Synchronization status message
SSU	Synchronization supply unit
STE	Section-terminating equipment (See also RSTE)
STM	Synchronous transfer mode (See also ATM); synchronous transport module
STM-n	Synchronous transport module level n ($n \times 155.52$ Mbps: $n = 1, 4, 16,$ or 64)
STS	Synchronous transport signal
STS-N	Synchronous transport signal level N (N$\times$51.84 Mbps: N = 1, 3, 12, 48, or 192)
STS-Nc	Contiguous concatenation of N STS-1 SPEs
STS-N-Xv	Virtual concatenation of X STS-N SPEs
SVC	Switched virtual circuit
SWC	Serving wire center

T-1	A digital carrier system for DS-1 signals
TC	Tandem connection
TCM	Tandem connection monitoring
TCOH	Tandem connection overhead
TCP	Transmission control protocol
TC-RDI	Tandem connection remote defect indication
TC-REI	Tandem connection remote error indication
TCT	Tandem connection trace
TCTE	Tandem connection terminating equipment ("element" in SDH)
TDM	Time-division multiplexing
TEI	Terminal endpoint identifier
TEST-P	SPE supervisory unequipped signal
TIM	Trace identifier mismatch
TIM-L	Line-level trace identifier mismatch
TIM-P	Path-level trace identifier mismatch
TIM-V	Virtual tributary–level trace identifier mismatch
TMN	Telecommunications management network
TOH	Transport overhead (SOH + LOH)
TS	Time slot
TSG	Timing signal generator
TSID	Test signal identification
TTI	Trail trace identifier
TU	Tributary unit
TU-n	Tributary unit level n (n = 11, 12, 2, or 3)
TUG-n	Tributary unit group n (n = 2 or 3)
UAS	Unavailable second
UAT	Unavailable time
UBN	Unlimited bandwidth network
UM	User-to-network interface
UNEQ	Unequipped
UNEQ-P	Unequipped at the path level
UNEQ-V	Unequipped at the virtual tributary level
UNI	User–network interface

UPC	Usage parameter control
UPSR	Unidirectional path switched ring
VBR	Variable bit rate
VC	Virtual channel; voice channel; virtual container
VC-*n*	Virtual container level n (n = 11, 12, 2, 3, or 4)
VC-*n*-X	Concatenation of X virtual containers at level n
VC-*n*-Xc	Contiguous concatenation of X virtual containers at level n
VC-*n*-Xv	Virtual concatenation of X virtual containers at level n
VCC	Virtual channel connection
VF	Voice frequency signal
VT	Virtual tributary
VT*n*	Virtual tributary of size n (n = 1.5, 2, 3, or 6)
VT*n*-Xv	Virtual concatenation of X VT*n* SPEs
WAN	Wide area network
WDCS	Wideband digital cross-connect system
WDM	Wavelength division multiplexing
X.25	ITU-T Recommendation for packet-switched user–network interface

APPENDIX B

SONET Quick Reference

This appendix gathers many figures and tables that are scattered throughout the text. Having information about SONET frame structures, overhead, and so on, in one place may be helpful to some readers, and that is the intent of this appendix.

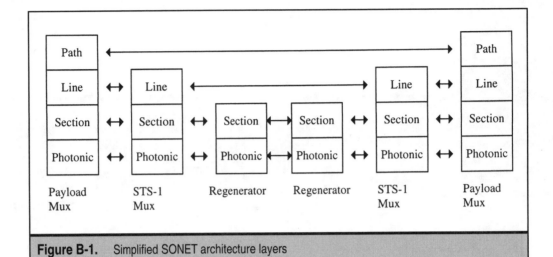

Figure B-1. Simplified SONET architecture layers

← 90 columns →

	Column 1	Column 2	Column 3	(Position varies)
Section Overhead	A1 Framing	A2 Framing	J0 Trace	J1 Trace
Section Overhead	B1 BIP-8	E1 Orderwire	F1 User Channel	B1 BIP-8
Section Overhead	D1 Data Com	D2 Data Com	D3 Data Com	C2 Signal Label
Line Overhead	H1 Pointer	H1 Pointer	H3 Pointer Action	G1 Path Status
Line Overhead	B2 BIP 8	K1 APS	K2 APS	F2 User Channel
Line Overhead	D4 Data Com	D5 Data Com	D6 Data Com	H4 Multiframe
Line Overhead	D7 Data Com	D8 Data Com	D9 Data Com	Z3 Growth
Line Overhead	D10 Data Com	D11 Data Com	D12 Data Com	Z4 Growth
Line Overhead	S1 Synch Status	M0 REI-L	E2 Orderwire	N1 Tandem Conn.

9 rows

Section and Line Overhead — Path Overhead

Figure B-2. SONET STS-1 overhead

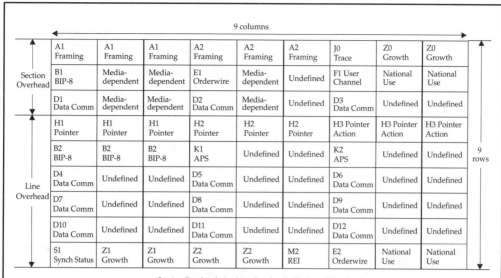

Figure B-3. STS-3 overhead

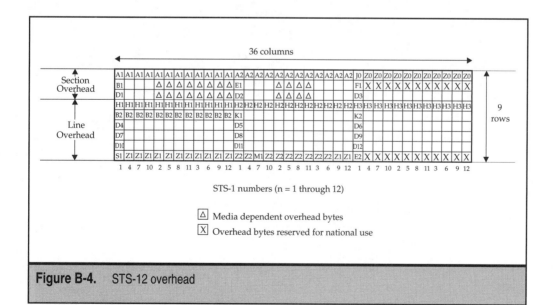

Figure B-4. STS-12 overhead

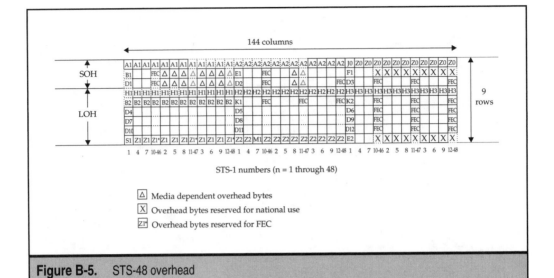

Figure B-5. STS-48 overhead

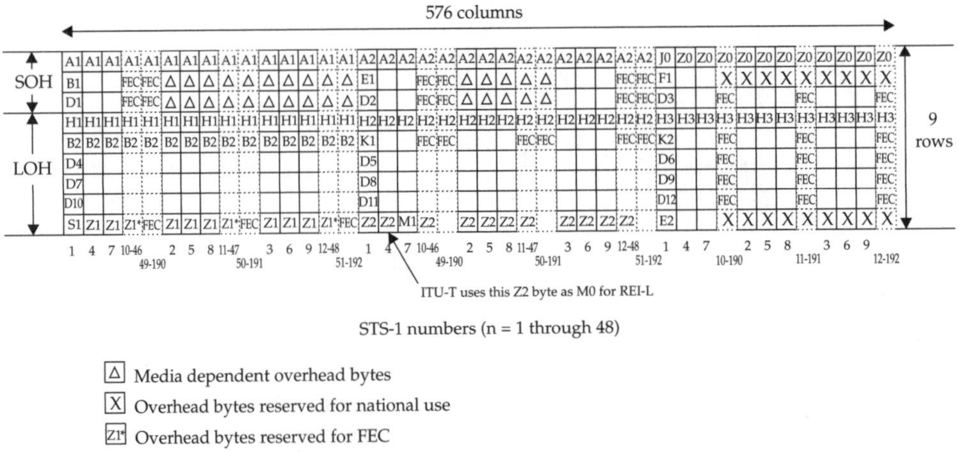

Figure B-6. STS-192 overhead

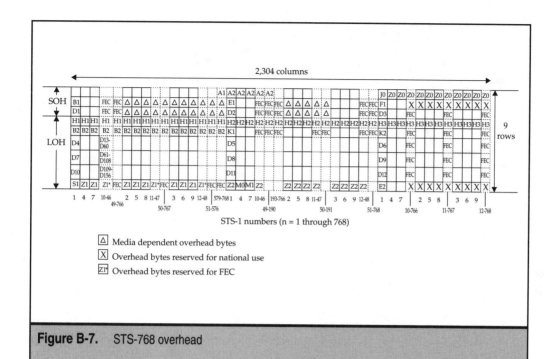

Figure B-7.   STS-768 overhead

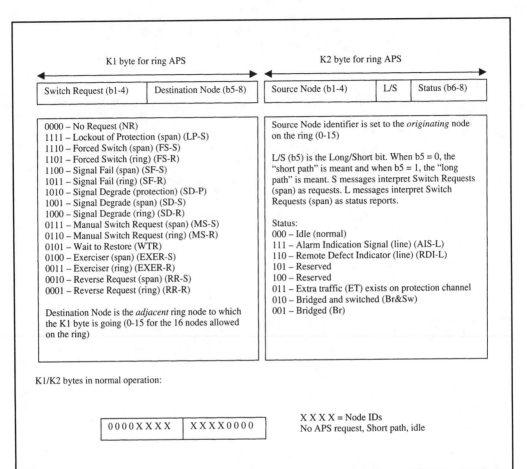

K1 byte for ring APS | K2 byte for ring APS

| Switch Request (b1-4) | Destination Node (b5-8) | | Source Node (b1-4) | L/S | Status (b6-8) |

0000 – No Request (NR)
1111 – Lockout of Protection (span) (LP-S)
1110 – Forced Switch (span) (FS-S)
1101 – Forced Switch (ring) (FS-R)
1100 – Signal Fail (span) (SF-S)
1011 – Signal Fail (ring) (SF-R)
1010 – Signal Degrade (protection) (SD-P)
1001 – Signal Degrade (span) (SD-S)
1000 – Signal Degrade (ring) (SD-R)
0111 – Manual Switch Request (span) (MS-S)
0110 – Manual Switch Request (ring) (MS-R)
0101 – Wait to Restore (WTR)
0100 – Exerciser (span) (EXER-S)
0011 – Exerciser (ring) (EXER-R)
0010 – Reverse Request (span) (RR-S)
0001 – Reverse Request (ring) (RR-R)

Destination Node is the *adjacent* ring node to which the K1 byte is going (0-15 for the 16 nodes allowed on the ring)

Source Node identifier is set to the *originating* node on the ring (0-15)

L/S (b5) is the Long/Short bit. When b5 = 0, the "short path" is meant and when b5 = 1, the "long path" is meant. S messages interpret Switch Requests (span) as requests. L messages interpret Switch Requests (span) as status reports.

Status:
000 – Idle (normal)
111 – Alarm Indication Signal (line) (AIS-L)
110 – Remote Defect Indicator (line) (RDI-L)
101 – Reserved
100 – Reserved
011 – Extra traffic (ET) exists on protection channel
010 – Bridged and switched (Br&Sw)
001 – Bridged (Br)

K1/K2 bytes in normal operation:

| 0 0 0 0 X X X X | X X X X 0 0 0 0 |

X X X X = Node IDs
No APS request, Short path, idle

Figure B-8. Understanding the SONET/SDH automatic protection switching K1/K2 bytes

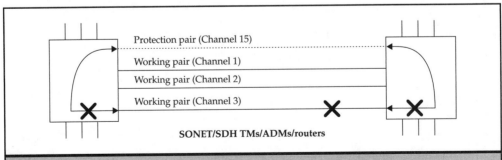

Protection pair (Channel 15)
Working pair (Channel 1)
Working pair (Channel 2)
Working pair (Channel 3)

SONET/SDH TMs/ADMs/routers

Figure B-9. Listing the SONET linear automatic protection switching messages

Synchronization Status Message	S1 bits 5-6-7-8	Abbreviation	Quality Level
Stratum 1 traceable (source can be traced to stratum 1)	0001	PRS	1
Synchronized—traceability unknown (backward compatibility)	0000	STU	2
Stratum 2 traceable	0111	ST2	3
Transit node clock traceable (a nonstandard clock level)	0100	TNC	4
Stratum 3E traceable (another nonstandard clock level)	1101	ST3E	5
Stratum 3 traceable	1010	ST3	6
SONET minimum clock traceable	1100	SMC	7
Stratum 4 traceable (no value has been assigned for this)	N/A	ST4	8
DO NOT USE for synchronization	1111	DUS	9
Reserved for network synchronization use	1110	RES	User assigned

Table B-1.   The SONET synchronization status messages (SSMs)

Byte #	Value (bit 1, 2, ..., 8)							
1	1	C_1	C_2	C_3	C_4	C_5	C_6	C_7
2	0	X	X	X	X	X	X	X
3	0	X	X	X	X	X	X	X
:	:	X	X	X	X	X	X	X
16	0	X	X	X	X	X	X	X

Alternate Method:
64 byte "frame"
Delimited by CR or LF CR
No checksum

(Receivers only check X values
if checksums do not match)

XXXXXXX = ASCII (T.50) character (15 in all)

C_1 - C_7 = CRC-7 on previous frame

Figure B-10.   The J0 16-byte message format

Binary value of s1s2 bits	Use
00	SONET (receivers ignore these bits)
01	Older ADMs might use this
10	AU-3 or AU-4 in SDH
11	Reserved

Table B-2.   Use of the s1s2 bits

Hex Code	Interpretation of VC-4-Xc/VC-4/VC-3 Content Mapping
00	Unequipped or supervisory unequipped: contains no payload on "open" connection
01	Equipped-nonspecific payload. Payload format determined by other means
02	Floating VT mode: mainly for "voice" tributaries
03	Locked VT-mode: for locked byte synch mode (no longer supported)
04	Asynchronous mapping for DS-3 (44.736 Mbps)
05	Mapping under development: an experimental category
12	Asynchronous E-4 (139.264 Mbps) mapping
13	ATM Cell mapping
14	Metropolitan Area Network (MAN) Dist. Queue Dual Bus (DQDB) mapping
15	Fiber Distributed Data Interface (FDDI) mapping
16	Mapping for HDLC over SONET: used for IP packets
17	Simple Data Link (SDL) for scrambler mapping (under study)
18	HDLC/LAPS (link access procedure – SDH) frame mapping (under study)
19	Simple Data Link (SDL) for scrambler mapping (under study)
1A	10 Gbps Ethernet frame mapping
1B	Flexible Topology Data Link mapping (called GFP in SDH)

Table B-3.   SONET C2 byte values

Hex Code	Interpretation of VC-4-Xc/VC-4/VC-3 Content Mapping
1C	Fibre Channel at 10 Gbps
CF	Reserved: former value used HDLC/PPP frame mapping (obsolete)
E1	STS-1 payload with 1 VT-x payload defect
D0-DF	Reserved for proprietary use
E2	STS-1 payload with 2 VT-x payload defects
:	:
FB	STS-1 payload with 27 VT-x payload defects
FC	STS-1 payload with 28 VT-x payload defects or a non-VT payload defect
FE	Test signal (also used, like 05, for experimental traffic)
FF	AIS (generated by source if there is no valid incoming signal)

Table B-3.   SONET C2 byte values *(continued)*

G1 b5–b7 (RDI-P)	Meaning	Event Triggers
000	No remote defect	No defect
001	No remote defect	No defect
010	Remote payload defect	LCD-P, PLM-P
011	No remote defect	No defect
100	Remote defect	AIS-P, LOP-P
101	Remote server defect	AIS-P, LOP-P
110	Remote connectivity defect	TIM-P, UNEQ-P
111	Remote defect	AIS-P, LOP-P

Table B-4.   The SONET G1 path status byte

K3 APS				Spare		Data Link	
1	2	3	4	5	6	7	8

Bits 1-2-3-4: APS channel for VC-4/VC-3 path
Bit 5-6:   Spare (receiver to ignore)
Bit 7-8:   Higher Order APS Data Link

Figure B-11.   The K3 byte

IEC				Data Link			
1	2	3	4	5	6	7	8

Bits 1-2-3-4: Incoming Error Counter (IEC)
Bits 5-6-7-8: TC Data Link

IEC Value	Interpretation
0000	0 BIP violation
0001	1 BIP violation
0010	2 BIP violations
0011	3 BIP violations
0100	4 BIP violations
0101	5 BIP violations
0110	6 BIP violations
0111	7 BIP violations
1000	8 BIP violations
1001	0 BIP violation
1010	0 BIP violation
1011	0 BIP violation
1100	0 BIP violation
1101	0 BIP violation
1110	0 BIP violation, Incoming AIS
1111	0 BIP violation

Figure B-12.   The N1 byte and the IEC

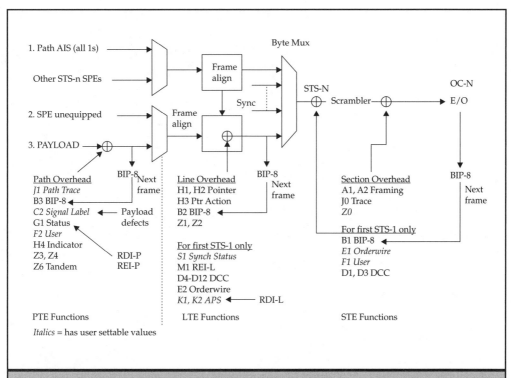

Figure B-13.   Building STS-1 overhead into a SONET frame

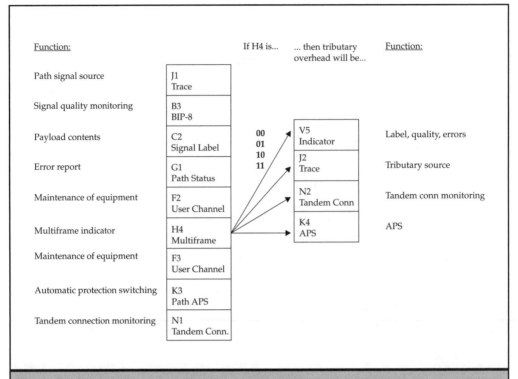

Figure B-14.   Tributary overhead and the H4 indicator bits

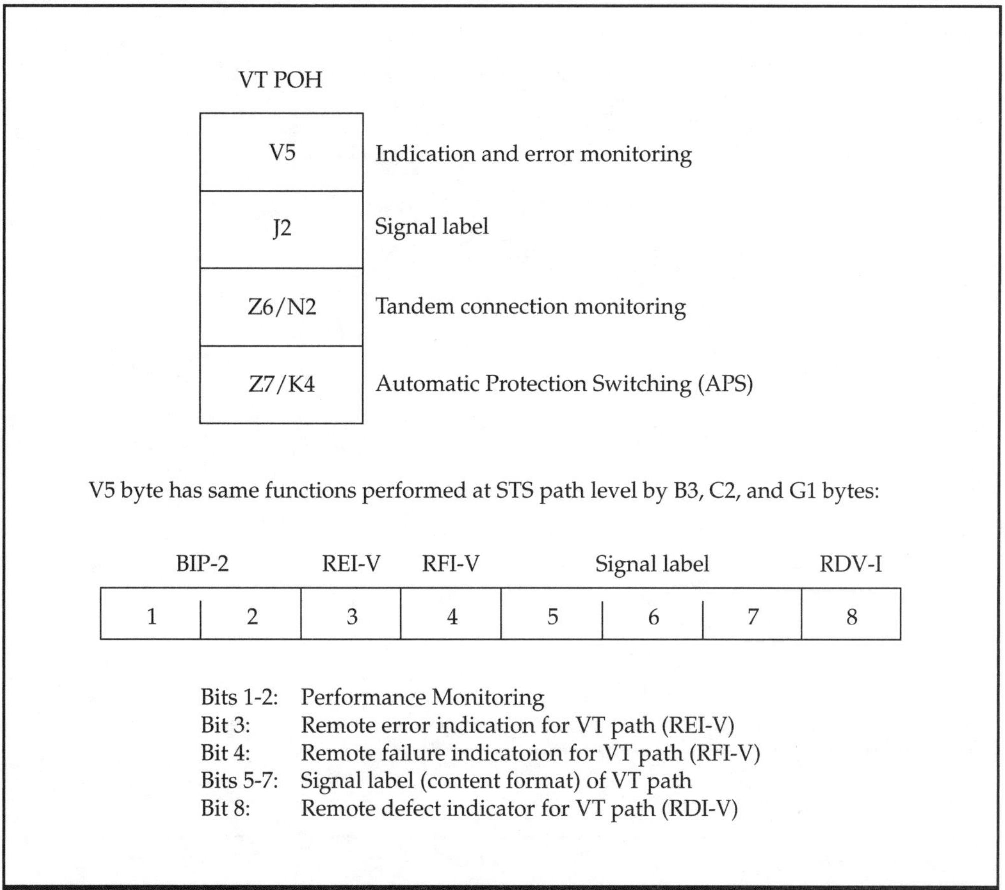

VT POH

V5	Indication and error monitoring
J2	Signal label
Z6/N2	Tandem connection monitoring
Z7/K4	Automatic Protection Switching (APS)

V5 byte has same functions performed at STS path level by B3, C2, and G1 bytes:

BIP-2		REI-V	RFI-V	Signal label			RDV-I
1	2	3	4	5	6	7	8

Bits 1-2:   Performance Monitoring
Bit 3:   Remote error indication for VT path (REI-V)
Bit 4:   Remote failure indicatoion for VT path (RFI-V)
Bits 5-7:   Signal label (content format) of VT path
Bit 8:   Remote defect indicator for VT path (RDI-V)

Figure B-15.   The V5 byte position and structure

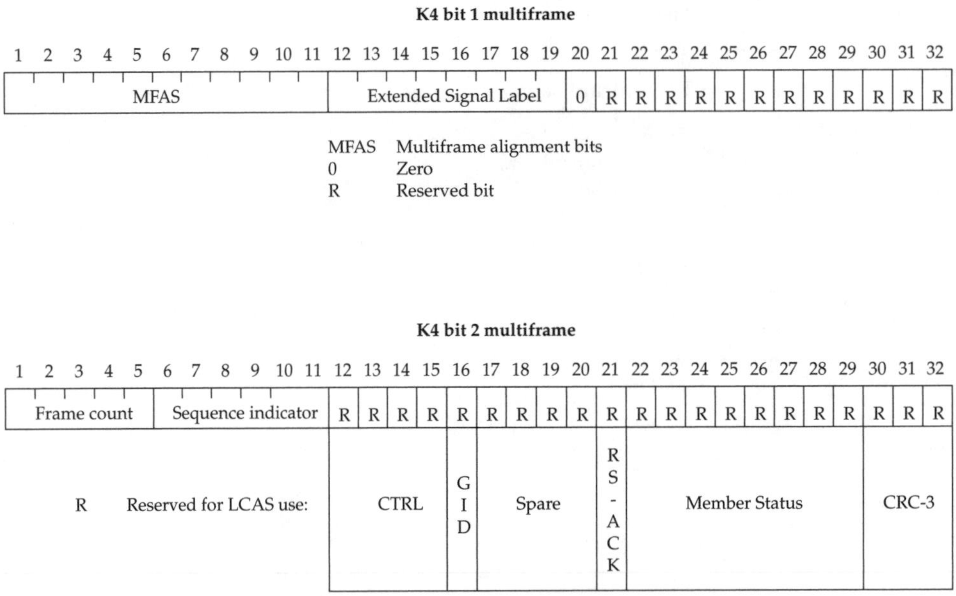

K4 bit 1 multiframe

1 2 3 4 5 6 7 8 9 10 11 12 13 14 15 16 17 18 19 20 21 22 23 24 25 26 27 28 29 30 31 32

| MFAS | Extended Signal Label | 0 | R | R | R | R | R | R | R | R | R | R | R | R |

MFAS Multiframe alignment bits
0   Zero
R   Reserved bit

K4 bit 2 multiframe

1 2 3 4 5 6 7 8 9 10 11 12 13 14 15 16 17 18 19 20 21 22 23 24 25 26 27 28 29 30 31 32

| Frame count | Sequence indicator | R | R | R | R | R | R | R | R | R | R | R | R | R | R | R | R | R | R | R | R | R |

R Reserved for LCAS use: | CTRL | GID | Spare | RS-ACK | Member Status | CRC-3 |

Figure B-16.   The K4(Z7) bits 1 and 2 multiframe

Hex Code	Interpretation of K4 bit 1 multiframe Extended Signal Label (bits 12-19)
00-07	Reserved so not to overlap with values 0-7 of V5 signal label
08	Mapping under development: content not defined in this table
09	ATM mapping
0A	HDLC/PPP mapping
0B	HDLC/LAPS
0C	Virtually concatenated test signal: for virtual concatenation content not defined here
0D	Flexible Topology Data Link mapping (under study)
FF	Reserved

Table B-5.   The interpretation of K4 bit 1 multiframe extended signal label (bits 12-19)

H4 byte								1st multi-frame number	2nd multi-frame number
Bit 1	Bit 2	Bit 3	Bit 4	Bit 5	Bit 6	Bit 7	Bit 8		
				1st multiframe indicator MFI1 (bits 1-4)					
Sequence indicator (SQ) MSBs (b1-4)				1	1	1	0	14	n-1 (ex.15)
Sequence indicator (SQ) LSBs (b5-8)				1	1	1	1	15	
2nd multiframe indicator MFI2 (b1-4)				0	0	0	0	0	n (ex.16)
2nd multiframe indicator MFI2 (b5-8)				0	0	0	1	1	
CTRL				0	0	1	0	2	
GID ("000x")				0	0	1	1	3	
Reserved ("0000")				0	1	0	0	4	
Reserved ("0000")				0	1	0	1	5	
CRC-8				0	1	1	0	6	
CRC-8				0	1	1	1	7	
Member status				1	0	0	0	8	
Member status				1	0	0	1	9	
RS-ACK				1	0	1	0	10	
Reserved ("0000")				1	0	1	1	11	
Reserved ("0000")				1	1	0	0	12	
Reserved ("0000")				1	1	0	1	13	
Sequence indicator (SQ) MSBs (b1-4)				1	1	1	0	14	
Sequence indicator (SQ) LSBs (b5-8)				1	1	1	1	15	
2nd multiframe indicator MFI2 (b1-4)				0	0	0	0	0	
2nd multiframe indicator MFI2 (b5-8)				0	0	0	1	1	n+1 (ex.17)
CTRL				0	0	1	0	2	
GID ("000x")				0	0	1	1	3	
Reserved ("0000")				0	1	0	0	4	
Reserved ("0000")				0	1	0	1	5	
CRC-8				0	1	1	0	6	
CRC-8				0	1	1	1	7	
Member status				1	0	0	0	8	

Figure B-17.   The H4 multiframe used with virtual concatenation

VTn-Xv	Carried in	X values	Capacity (Mbps)	Steps (Mbps)
VT1.5-Xv	STS-1	1 to 28	1.6 to 44.8	1.6
VT2-Xv	STS-1	1 to 21	2.176 to 45.969	2.176
VT3-Xv	STS-1	1 to 14	3.328 to 46.696	3.328
VT6-Xv	STS-1	1 to 7	6.784 to 47.448	6.784
VT1.5-Xv/VC-11-Xv	STS-3c	1 to 64	1.6 to 102.4	1.6
VT2/VC-12-Xv	STS-3c	1 to 63	2.176 to 137.088	2.176
VT3-Xv	STS-3c	1 to 42	3.328 to 139.776	3.328
VT6	STS-3c	1 to 21	6.784 to 142.464	6.784
VT1.5-Xv/VC-11-Xv	Unspec'd	1 to 64	1.6 to 102.4	1.6
VT2/VC-12-Xv	Unspec'd	1 to 64	2.176 to 139.264	2.176
VT3-Xv	Unspec'd	1 to 64	3.328 to 212.992	3.328
VT6	Unspec'd	1 to 64	6.784 to 434.176	6.784

Table B-6.   Capacities of virtually concatenated SONET VT-*n* payloads

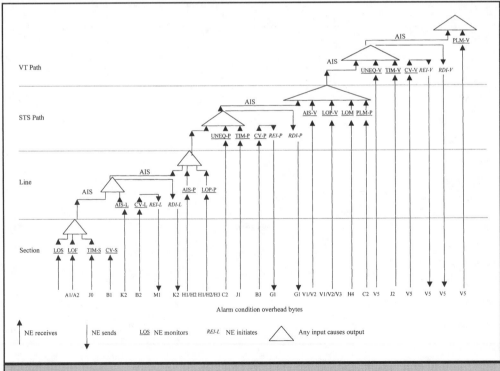

Figure B-18.   SONET alarm events and overhead

APPENDIX C

SDH Quick Reference

This appendix also gathers many figures and tables that are scattered throughout the text, but for SDH in this case. Having information about SDH frame structures, overhead, and so on, in one place may be helpful to the reader.

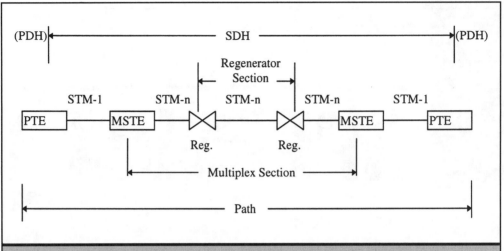

Figure C-1. Examining the SDH architecture layers

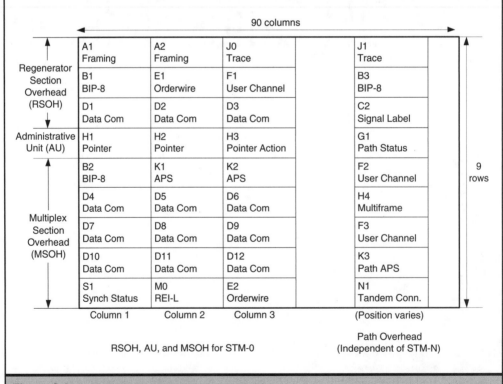

Figure C-2. The STM-0 overhead

9 columns

								9 rows
A1 Framing	A1 Framing	A1 Framing	A2 Framing	A2 Framing	A2 Framing	J0 Trace	National Use*	National Use*
B1 BIP-8	Media-dependent	Media-dependent	E1 Orderwire	Media-dependent		F1 User Channel	National Use	National Use
D1 Data Comm	Media-dependent	Media-dependent	D2 Data Comm	Media-dependent		D3 Data Comm		
Administrative Unit (AU) Pointer(s)								
B2 BIP-8	B2 BIP-8	B2 BIP-8	K1 APS			K2 APS		
D4 Data Comm			D5 Data Comm			D6 Data Comm		
D7 Data Comm			D8 Data Comm			D9 Data Comm		
D10 Data Comm			D11 Data Comm			D12 Data Comm		
S1 Synch Status	Z1 Growth	Z1 Growth	Z2 Growth	Z2 Growth	M1 REI	E2 Orderwire	National Use	National Use

RSOH (rows 1–3), MSOH (rows 5–9)

This is M2 byte in SONET

* First row of frame is unscrambled, so care is required

Figure C-3. STM-1 overhead

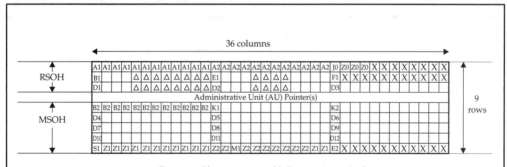

Figure C-4. STM-4 overhead

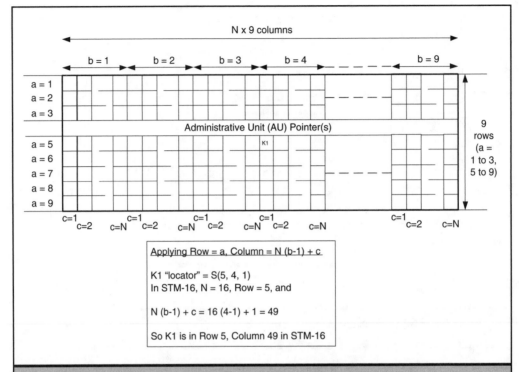

Figure C-5. The SDH a-b-c locator system

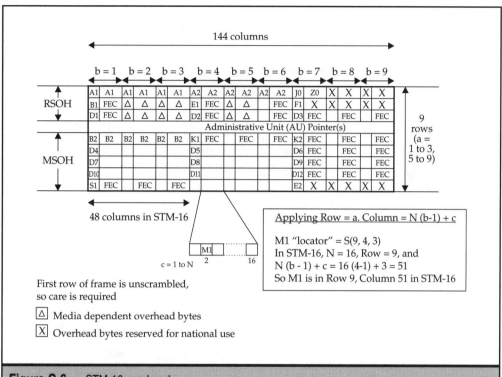

144 columns

b = 1   b = 2   b = 3   b = 4   b = 5   b = 6   b = 7   b = 8   b = 9

48 columns in STM-16

c = 1 to N

First row of frame is unscrambled,
so care is required

△ Media dependent overhead bytes

X Overhead bytes reserved for national use

Applying Row = a. Column = N (b-1) + c

M1 "locator" = S(9, 4, 3)
In STM-16, N = 16, Row = 9, and
N (b - 1) + c = 16 (4-1) + 3 = 51
So M1 is in Row 9, Column 51 in STM-16

Figure C-6.   STM-16 overhead

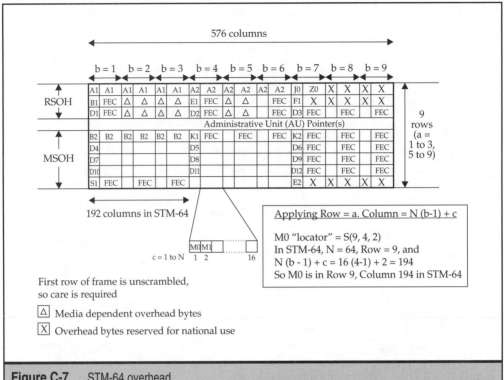

Figure C-7.   STM-64 overhead

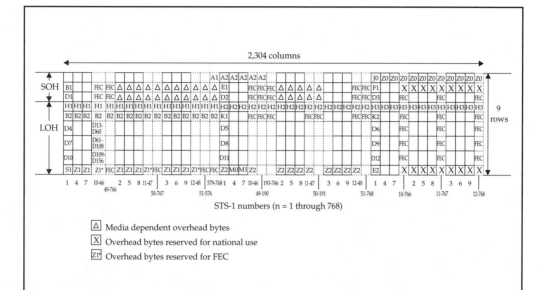

2,304 columns

STS-1 numbers (n = 1 through 768)

△ Media dependent overhead bytes
X Overhead bytes reserved for national use
Z1* Overhead bytes reserved for FEC

Figure C-8.   STM-768 overhead

Byte #	Value (bit 1, 2, ..., 8)							
1	1	C_1	C_2	C_3	C_4	C_5	C_6	C_7
2	0	X	X	X	X	X	X	X
3	0	X	X	X	X	X	X	X
:	:	X	X	X	X	X	X	X
16	0	X	X	X	X	X	X	X

Alternate Method:
64 byte "frame"
Delimited by CR or LF CR
No checksum

(Receivers only check X values
if checksums do not match)

XXXXXXX = ASCII (T.50) character (15 in all)

C_1 - C_7 = CRC-7 on previous frame

Figure C-9.   The J0 16-byte message format

Binary value of s1s2 bits	Use
00	SONET (receivers ignore these bits)
01	Older ADMs might use this
10	AU-3 or AU-4 in SDH
11	Reserved

Table C-1. Use of the s1s2 bits

M1 bits (bit 1 ignored by all but STM-16)	STM-0 interpretation	STM-1 interpretation	STM-4 interpretation	STM-16 interpretation
0000 0000	0 BIP violation	0 BIP violation	0 BIP violation	0 BIP violation
0000 0001	1 BIP violation	1 BIP violation	1 BIP violation	1 BIP violation
0000 0010	2 BIP violations	2 BIP violations	2 BIP violations	2 BIP violations
0000 0011	3 BIP violations	3 BIP violations	3 BIP violations	3 BIP violations
:	:	:	:	:
0000 1000	8 BIP violations	8 BIP violations	8 BIP violations	8 BIP violations
0000 1001	0 BIP violation	9 BIP violations	9 BIP violations	9 BIP violations
:	:	:	:	:
0001 1000	0 BIP violation	24 BIP violations	24 BIP violations	24 BIP violations
0000 1001	0 BIP violation	0 BIP violation	25 BIP violations	25 BIP violations
:	:	:	:	:

Table C-2. M1 for STM-0, STM-1, STM-4, and STM-16

M1 bits (bit 1 ignored by all but STM-16)	STM-0 interpretation	STM-1 interpretation	STM-4 interpretation	STM-16 interpretation
0110 0000	0 BIP violation	0 BIP violation	96 BIP violations	96 BIP violations
0110 0001	0 BIP violation	0 BIP violation	0 BIP violation	97 BIP violations
:	:	:	:	:
1111 1110	0 BIP violation	0 BIP violation	0 BIP violation	254 BIP violations
1111 1111	0 BIP violation	0 BIP violation	0 BIP violation	255 BIP violations

Table C-2. M1 for STM-0, STM-1, STM-4, and STM-16 *(continued)*

M0	M1	STM-64 interpretation	STM-256 interpretation
0000 0000	0000 0000	0 BIP violation	0 BIP violation
0000 0000	0000 0001	1 BIP violation	1 BIP violation
0000 0000	0000 0010	2 BIP violations	2 BIP violations
0000 0000	0000 0011	3 BIP violations	3 BIP violations
:	:	:	:
0000 0110	0000 0000	1536 BIP violations	1536 BIP violations
0000 0110	0000 0001	0 BIP violation	1537 BIP violations
:	:	:	:
0001 1000	0000 0000	0 BIP violation	6144 BIP violations
0000 0110	0000 0001	0 BIP violation	0 BIP violation
:	:	:	:
1111 1111	1111 1111	0 BIP violation	0 BIP violation

Table C-3. M0/M1 for STM-64 and STM-256

Hex Code	Interpretation of VC-4-*Xc*/VC-4/VC-3 Content Mapping
00	Unequipped or supervisory unequipped: contains no payload of "open" connection
01	Reserved: Formerly "Equipped-non-specific." New receivers ignore 01 from old NEs
02	TUG structure: mainly for "voice" tributaries
03	Locked TU-n: for locked byte synch mode (no longer supported)
04	Asynchronous mapping of E-3 (34.368 Mbps) or DS-3 (44.736 Mbps)
05	Mapping under development: an experimental category
12	Asynchronous E-4 (139.264 Mbps) mapping
13	ATM Cell mapping
14	Metropolitan Area Network (MAN) Dist. Queue Dual Bus (DQDB) mapping
15	Fiber Distributed Data Interface (FDDI) mapping
16	HDLC/PPP frame mapping: used for IP packets
17	Simple Data Link (SDL) for scrambler mapping (under study)
18	HDLC/LAPS (link access procedure – SDH) frame mapping (under study)
19	Simple Data Link (SDL) for scrambler mapping (under study)
1A	10 Gbps Ethernet frame mapping (under study)
1B	Generic Packet Framing (GFP) mapping
1C	Fibre Channel at 10 Gbps
CF	Reserved: former value used HDLC/PPP frame mapping (obsolete)
D0-DF	Reserved for proprietary use
E1-FC	Reserved for national use
FE	Test signal (also used, like 05, for experimental traffic)
FF	VC-AIS (generated by tandem connection source if there is no valid incoming signal)

Table C-4.   SDH C2 byte values

G1 REI				RDI	Reserved		Spare
1	2	3	4	5	6	7	8

Bits 1-2-3-4: Remote Error Indicator count from B3 (0-8 valid)
Bit 5:   Remote Defect Indicator (details of REI)
Bits 6-7:   Reserved
Bit 8:   Spare (receiver required to ignore)

Figure C-10.   The G1 path status byte

K3 APS				Spare		Data Link	
1	2	3	4	5	6	7	8

Bits 1-2-3-4: APS channel for VC-4/VC-3 path
Bit 5-6:   Spare (receiver to ignore)
Bit 7-8:   Higher Order APS Data Link

Figure C-11.   The K3 byte

	IEC			Data Link			
1	2	3	4	5	6	7	8

Bits 1-2-3-4: Incoming Error Counter (IEC)
Bits 5-6-7-8: TC Data Link

IEC Value	Interpretation
0000	0 BIP violation
0001	1 BIP violation
0010	2 BIP violations
0011	3 BIP violations
0100	4 BIP violations
0101	5 BIP violations
0110	6 BIP violations
0111	7 BIP violations
1000	8 BIP violations
1001	0 BIP violation
1010	0 BIP violation
1011	0 BIP violation
1100	0 BIP violation
1101	0 BIP violation
1110	0 BIP violation, Incoming AIS
1111	0 BIP violation

Figure C-12. The N1 byte and the IEC

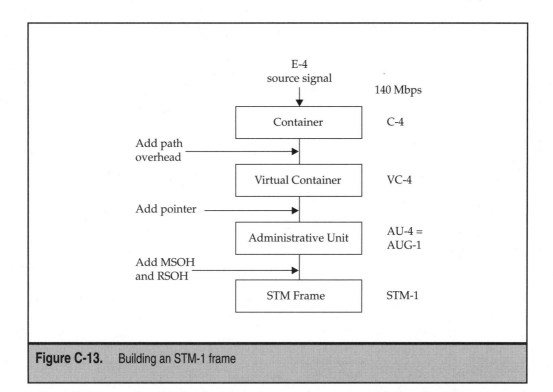

Figure C-13. Building an STM-1 frame

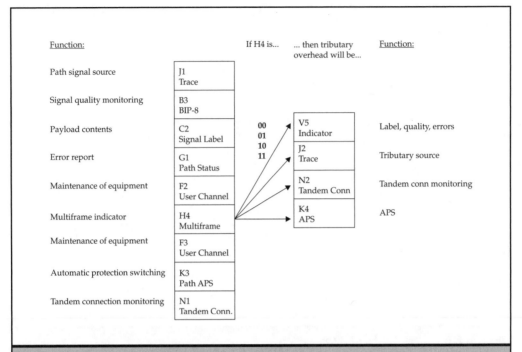

Figure C-14. Tributary overhead and the H4 indicator bits

b5-b6-b7	Interpretation of VC-2/VC-1 Content Meaning
0 0 0	Unequipped or supervisory-unequipped
0 0 1	Reserved: Formerly "Equipped-non-specific." New receivers ignore 001 from old NEs
0 1 0	Asynchronous mapping – see next chapter
0 1 1	Bit synchronous E-1 (2.048 Mbps) VC-12 mapping (no longer supported)
1 0 0	Byte synchronous mappings
1 0 1	Extended signal label: Used with K4 byte to provide mapping details at LO level
1 1 0	Test signal: for any non-virtual concatenated payload
1 1 1	VC-AIS (generated by tandem connection source if there is no valid incoming signal)

Table C-5. The SDH V5 signal label (b5-7)

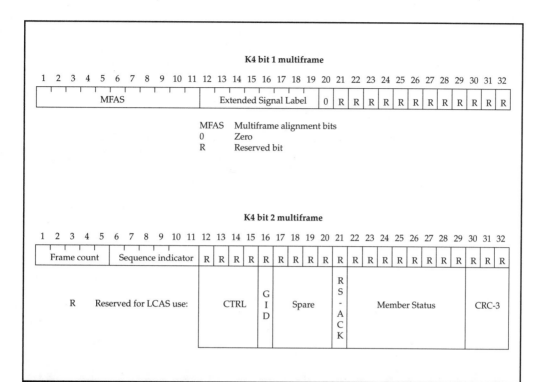

Figure C-15. The K4(Z7) bits 1 and 2 multiframe

Hex Code	Interpretation of K4 bit 1 multiframe Extended Signal Label (bits 12-19)
00-07	Reserved so not to overlap with values 0-7 of V5 signal label
08	Mapping under development: content not defined in this table
09	ATM mapping
0A	HDLC/PPP mapping
0B	HDLC/LAPS
0C	Virtually concatenated test signal: for virtual concatenation content not defined here
0D	Flexible Topology Data Link mapping (under study)
FF	Reserved

Table C-6. The interpretation of K4 bit 1 multiframe extended signal label (bits 12-19)

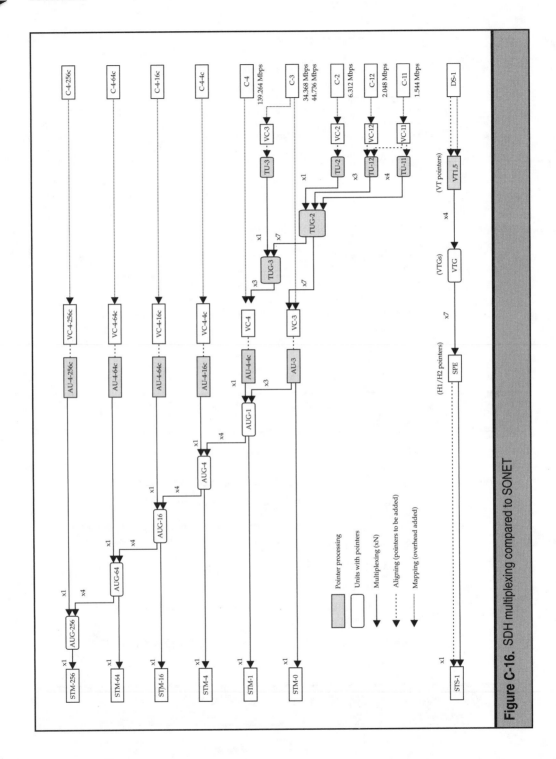

Figure C-16. SDH multiplexing compared to SONET

H4 byte								1st multi-frame number	2nd multi-frame number
Bit 1	Bit 2	Bit 3	Bit 4	Bit 5	Bit 6	Bit 7	Bit 8		
				1st multiframe indicator MFI1 (bits 1-4)					
Sequence indicator (SQ) MSBs (b1-4)				1	1	1	0	14	n-1 (ex.15)
Sequence indicator (SQ) LSBs (b5-8)				1	1	1	1	15	
2nd multiframe indicator MFI2 (b1-4)				0	0	0	0	0	n (ex.16)
2nd multiframe indicator MFI2 (b5-8)				0	0	0	1	1	
CTRL				0	0	1	0	2	
GID ("000x")				0	0	1	1	3	
Reserved ("0000")				0	1	0	0	4	
Reserved ("0000")				0	1	0	1	5	
CRC-8				0	1	1	0	6	
CRC-8				0	1	1	1	7	
Member status				1	0	0	0	8	
Member status				1	0	0	1	9	
RS-ACK				1	0	1	0	10	
Reserved ("0000")				1	0	1	1	11	
Reserved ("0000")				1	1	0	0	12	
Reserved ("0000")				1	1	0	1	13	
Sequence indicator (SQ) MSBs (b1-4)				1	1	1	0	14	
Sequence indicator (SQ) LSBs (b5-8)				1	1	1	1	15	
2nd multiframe indicator MFI2 (b1-4)				0	0	0	0	0	n+1 (ex.17)
2nd multiframe indicator MFI2 (b5-8)				0	0	0	1	1	
CTRL				0	0	1	0	2	
GID ("000x")				0	0	1	1	3	
Reserved ("0000")				0	1	0	0	4	
Reserved ("0000")				0	1	0	1	5	
CRC-8				0	1	1	0	6	
CRC-8				0	1	1	1	7	
Member status				1	0	0	0	8	

Figure C-17. The H4 multiframe used with virtual concatenation

VC-n-Xv	Carried in	X values	Capacity (Mbps)	Steps (Mbps)
VC-11-Xv	VC-3	1 to 28	1.6 to 44.8	1.6
VC-11-Xv	VC-4	1 to 64	1.6 to 102.4	1.6
VC-11-Xv	Unspec'd	1 to 64	1.6 to 102.4	1.6
VC-12-Xv	VC-3	1 to 21	2.176 to 45.696	2.176
VC-12-Xv	VC-4	1 to 63	2.176 to 137.088	2.176
VC-12-Xv	Unspec'd	1 to 64	2.176 to 139.264	2.176
VC-2-Xv	VC-3	1 to 7	6.784 to 47.448	6.784
VC-2-Xv	VC-4	1 to 21	6.784 to 142.464	6.784
VC-2-Xv	Unspec'd	1 to 64	6.784 to 434.176	6.784

Table C-7. Capacities of virtually concatenated SDH VC-*n* payloads

Synchronization Status Messages	S1 bits 5-6-7-8
Quality unknown (existing synchronization network)	0000
Reserved	0001
G.811 (highest level clock)	0010
Reserved	0011
Synchronization supply unit A (SSU-A) (G.812 transit)	0100
Reserved	0101
Reserved	0110
Reserved	0111
Synchronization supply unit B (SSU-B) (G.812 local)	1000
Reserved	1001
Reserved	1010
G.813 Option I synchronous equipment clock (SEC)	1011
Reserved	1100
Reserved	1110
DO NOT USE for synchronization	1111

Table C-8. The SDH synchronization status messages (SSMs)

Overhead Byte	Name	SONET Event	SDH Name
N/A	LOS	Loss of signal	LOS (same)
A1/A2	LOF	Loss of framing	LOF (same)
J0	TIM-S	Regenerator section trace identifier mismatch	RS-TIM (regenerator section trace identifier mismatch)
B1	CV-S	Coding violation—section	RS-BIP (regenerator section coding violation)
K2	AIS-L	Alarm indication signal—line	MS-AIS (multiplex section alarm indication signal)
B2	CV-L	Coding violation—line	MS-BIP (multiplex section coding violation)
M1	REI-L	Remote error indication—line	MS-REI (multiplex section remote error indication)
K2	RDI-L	Remote defect indication—line	MS-RDI (multiplex section remote defect indication)
H1/H2	AIS-P	Alarm indication signal—path	AU-AIS (administrative unit alarm indication signal)
H1/H2/H3	LOP-P	Loss of pointer—path	AU-LOP (administrative unit loss of pointer)
C2	UNEQ-P	Unequipped STS at path level	HP-UNEQ (higher path unequipped)
J1	TIM-P	Trace identifier mismatch at path level	HP-TIM (higher path trace identifier mismatch)
B3	CV-P	Coding violation—path	HP-BIP (higher path coding violation)
G1	REI-P	Remote error indication—path	HP-REI (higher path remote error indication)
G1	RDI-P	Remote defect indication—path	HP-RDI (higher path remote defect indication)
V1/V2	AIS-V	Alarm indication signal—virtual tributary	TU-AIS (tributary unit alarm indication signal)

Table C-9.   SDH alarm events and overhead

Overhead Byte	Name	SONET Event	SDH Name
V1/V2/V3	LOP-V	Loss of pointer—virtual tributary	TU-LOP (tributary unit loss of pointer)
H4	LOM	Loss of multiframe—virtual tributary	TU-LOM (tributary unit loss of multiframe)
C2	PLM-P	Path label mismatch—path	HP-PLM (higher path path label mismatch)
V5	UNEQ-V	Unequipped virtual tributaryat virtual tributarylevel	LP-UNEQ (lower path unequipped)
J2	TIM-V	Trace identifier mismatch at virtual tributarylevel	LP-TIM (lower path trace identifier mismatch)
V5	CV-V	Coding violation—virtual tributary	LP-B IP (lower path coding violation)
V5	REI-V	Remote error indication—virtual tributary	LP-REI (lower path remote error indication)
V5	RDI-V	Remote defect indication—virtual tributary	LP-RDI (lower path remote defect indication)
V5	AIS	Alarm indication signal—virtual tributary	TU-AIS (tributary unit alarm indication signal)

Table C-9. SDH alarm events and overhead *(continued)*

BIBLIOGRAPHY

TELCORDIA (BELLCORE) PUBLICATIONS

SONET started at Bellcore, and so that group deserves to lead off this section. However, most modern SONET research is more dependent on ANSI/ATIS documents. For that reason, the classic *GR-253-CORE, SONET Transport Systems: Common Generic Criteria* is not even listed here.

Family of Requirements

FR-476, *OTGR Section 6: Network Maintenance: Access and Testing* (Bellcore, 1998 edition; a subset of OTGR, FR-439).

FR-480, *OTGR Section 10: User System Interface* (Bellcore, 1998 edition; a subset of OTGR, FR-439).

Generic Requirements

GR-20-CORE, *Generic Requirements for Optical Fiber and Optical Fiber Cable,* Issue 2 (Bellcore, July 1998).

GR-63-CORE, *Network Equipment Building System (NEBS) Requirements: Physical Protection* (a module of LSSGR, FR-64, TSGR, FR-440, and NEBSFR, FR-2063), Issue 1 (Bellcore, October 1995).

GR-78-CORE, *Generic Requirements for the Physical Design and Manufacture of Telecommunications Products and Equipment* (a module of RQGR, FR-796 and NEBSFR, FR-2063), Issue 1 (Bellcore, September 1997).

GR-129-CORE, *Generic Requirements for Fiber Optic Branching Components,* Issue 2 (Bellcore, February 1998).

GR-149-CORE, *Generic Requirements on Security for OSI-Based Telecommunications Management Network Interfaces,* Issue 1 (Bellcore, September 1994).

GR-199-CORE, *OTGR Section 12.2: Operations Application Messages Memory—Administration Messages* (a module of OTGR, FR-439), Issue 2 (Bellcore, November 1996).

GR-303-CORE, *Integrated Digital Loop Carrier System Generic Requirements, Objectives, and Interface* (a module of TSGR, FR-440), Issue 2 (Bellcore, December 1998).

GR-326-CORE, *Generic Requirements for Single-Mode Optical Connectors and Jumper Assemblies,* Issue 2 (Bellcore, December 1996).

GR-409-CORE, *Generic Requirements for Premises Fiber Optic Cable,* Issue 1 (Bellcore, May 1994).

GR-418-CORE, *Generic Reliability Assurance Requirements for Fiber Optic Transport Systems* (a module of RQGR, FR-796), Issue 1 (Bellcore, December 1997).

GR-436-CORE, *Digital Network Synchronization Plan,* Issue 1 (Bellcore, June 1994), plus Revision 1 (June 1996).

GR-449-CORE, *Generic Requirements and Design Consideration for Fiber Distributing Frames,* Issue 1 (Bellcore, March 1995).

GR-454-CORE, *Generic Requirements for Supplier Provided Documentation* (a module of OTGR, FR-439, LSSGR, FR-64, and TSGR, FR-440), Issue 1 (Bellcore, December 1997).

GR-468-CORE, *Generic Reliability Assurance Requirements for Optoelectronic Devices Used in Telecommunications Equipment* (a module of RQGR, FR-796), Issue 1 (Bellcore, December 1998).

GR-472-CORE, *OTGR Section 2.1: Network Element Configuration Management* (a module of OTGR, FR-439), Issue 2 (Bellcore, November 1996), plus Revision 1 (July 1997).

GR-474-CORE, *OTGR Section 4: Network Maintenance: Alarm and Control for Network Elements* (a module of OTGR, FR-439), Issue 1 (Bellcore, December 1997).

GR-487-CORE, *Generic Requirements for Electronic Equipment Cabinets,* Issue 1 (Bellcore, June 1996).

GR-496-CORE, *SONET Add-Drop Multiplexer (SONET ADM) Generic Criteria* (a module of TSGR, FR-440), Issue 1 (Bellcore, December 1998).

GR-499-CORE, *Transport System Generic Requirements (TSGR): Common Requirements* (a module of TSGR, FR-440), Issue 2 (Bellcore, December 1998).

GR-765-CORE, *Generic Requirements for Single Fiber Single-Mode Optical Splices and Splicing Systems,* Issue 1 (Bellcore, September 1995).

GR-815-CORE, *Generic Requirements for Network Element/Network System (NEI NS) Security* (a module of LSSGR, FR-64, and OTGR, FR- 439), Issue 1 (Bellcore, November 1997).

GR-820-CORE, *OTGR Section 5.1: Generic Digital Transmission Surveillance* (a module of OTGR, FR-439), Issue 2 (Bellcore, December 1997).

GR-826-CORE, *OTGR Section 10.2: User Interface Generic Requirements for Supporting Network Element Operations* (a module of OTGR, FR-439), Issue 1 (Bellcore, June 1994).

GR-828-CORE, *OTGR Section 11.2: Generic Operations Interface—OSI Communications Architecture* (a module of OTGR, FR-439), Issue 1 (Bellcore, September 1994), plus Revision 2 (October 1996).

GR-831-CORE, *OTGR Section 12.1: Operations Application Messages—Language for Operations Application Messages* (a module of OTGR, FR-439), Issue 1 (Bellcore, November 1996).

GR-833-CORE, *OTGR Section 12.3: Network Maintenance: Network Element and Transport Surveillance Messages* (a module of OTGR, FR-439), Issue 2 (Bellcore, November 1996).

GR-834-CORE, *OTGR Section 12.4: Network Maintenance: Access and Testing Messages* (a module of OTGR, FR-439), Issue 2 (Bellcore, November 1996).

GR-836-CORE, *OTGR Section 15.2: Generic Operations Interfaces Using OSI Tools—Information Model Overview: Transport Configuration and Surveillance for Network Elements* (a module of OTGR, FR-439), Issue 2 (Bellcore, September 1996), plus revisions.

GR-836-IMD, *OTGR Section 15.2: Generic Operations Interfaces Using OSI Tools—Information Model Details: Transport Configuration and Surveillance for Network Elements* (a module of OTGR, FR-439), Issue 2 (Bellcore, September 1996).

GR-839-CORE, *Generic Requirements for Supplier-Provided Training* (a module of LSSGR, FR-64, TSGR, FR-440, and OTGR, FR-439), Issue 1 (Bellcore, July 1996).

GR-910-CORE, *Generic Requirements for Fiber Optic Attenuators*, Issue 2 (Bellcore, December 1998).

GR-1031-CORE, *OTGR Section 15.6: Operations Interfaces Using OSI Tools: Test Access Management*, Issue 2 (Bellcore, October 1997), plus Revision 1 (December 1998).

GR-1042-CORE, *Generic Requirements for Operations Interfaces Using OSI Tools Information Model Overview: Synchronous Optical Network (SONET) Transport Information Model*, Issue 3 (Bellcore, December 1998).

GR-1042-IMD, *Generic Requirements for Operations Interfaces Using OSI Tools Information Model Details: Synchronous Optical Network (SONET) Transport Information Model*, Issue 3 (Bellcore, December 1998).

GR-1093-CORE, *Generic State Requirements for Network Elements*, Issue 1 (Bellcore, October 1994), plus Revision 1 (December 1995).

GR-1230-CORE, *SONET Bidirectional Line Switched Ring Equipment Generic Criteria* (a module of TSGR, FR-440), Issue 4 (Bellcore, December 1998).

GR-1244-CORE, *Clocks for the Synchronized Network: Common Generic Criteria*, Issue 1 (Bellcore, June 1995).

GR-1250-CORE, *Generic Requirements for Synchronous Optical Network (SONET) File Transfer*, Issue 1 (Bellcore, June 1995).

GR-1253-CORE, *Generic Requirements for Operations Interfaces Using OSI Tools: Telecommunications Management Network Security Administration*, Issue 1 (Bellcore, June 1995).

GR-1309-CORE, *TSC/RTU and OTAU Generic Requirements for Remote Optical Fiber Testing* (a module of OTGR, FR-439), Issue 1 (Bellcore, June 1995).

GR-1332-CORE, *Generic Requirements for Data Communications Network Security*, Issue 2 (Bellcore, April 1996).

GR-1377-CORE, *SONET OC192 Transport System Generic Criteria* (a module of TSGR, FR-440), Issue 5 (Bellcore, December 1998).

GR-1400-CORE, *SONET Dual Fed Unidirectional Path Switched Ring (UPSR) Equipment Generic Criteria* (a module of TSGR, FR-440), Issue 2 (Bellcore, January 1999).

Technical References

TR-NWT-000057, *Functional Criteria for Digital Loop Carrier Systems* (a module of TSGR, FR-440), Issue 2 (Bellcore, January 1993).

TR-NWT-000078 (see GR78-CORE).

TR-NWT-000170, *Digital Cross-Connect System (DSC 1/0) Generic Criteria*, Issue 2 (Bellcore, January 1993).

TR-NWT-000357, *Generic Requirements for Assuring the Reliability of Components Used in Telecommunication Systems* (a module of RQGR, FR-796), Issue 2 (Bellcore, October 1993).

TR-NWT-000418 (see GR-418-CORE).

TR-NWT-000468 (see GR-468-CORE).

TR-NWT-000835, *OTGR Section 12.5: Network Element and Network System Security Administration Messages* (a module of OTGR, FR- 439), Issue 3 (Bellcore, January 1993).

TR-NWT-000917, *SONET Regenerator (SONET RGTR) Equipment Generic Criteria* (a module of TSGR, FR-440), Issue 1 (Bellcore, December 1990).

TR-NWT-000930, *Generic Requirements for Hybrid Microcircuits Used in Telecommunications Equipment* (a module of RQGR, FR-796), Issue 2 (Bellcore, September 1993).

TR-NWT-001112, *Broadband-ISDN User to Network Interface and Network Node Interface Physical Layer Generic Criteria*, Issue 1 (Bellcore, June 1993).

TR-OPT-000839 (see GR-839-CORE).

TR-TSY-000454 (see GR-454-CORE).

TR-TSY-000458, *Digital Signal Zero, "A" (DS-OA 64 kb/s) Systems Interconnection,* Issue 1 (Bellcore, December 1989).

TR-TSY-000782, *SONET Digital Switch Trunk Interface Criteria* (a module of LSSGR, FR-64, and TSGR, FR-440), Issue 2 (Bellcore, September 1989).

TR-TSY-000824, *OTGR Section 10.1: User System Interface—User System Access* (a module of OTGR, FR-439), Issue 2 (Bellcore, February 1988).

TR-TSY-000825, *OTGR Section 10.A: User System Interface—User System Language* (a module of OTGR, FR-439), Issue 2 (Bellcore, February 1988).

TR-TSY-000827, *OTGR Section 11.1: Generic Operations Interfaces-NonOSI Communications Architecture* (a module of OTGR, FR-439), Issue 1 (Bellcore, November 1988).

Technical Advisories/Framework Technical Advisories

TA-NPL-000286 (not available).

TA-NPL-000464, *Generic Requirements and Design Considerations for Optical Digital Signal CrossConnect Systems,* Issue 1 (Bellcore, September 1987).

TA-NWT-000487 (see GR-487-CORE).

TA-NWT-000983, *Reliability Assurance Practices for Optoelectronic Devices in Loop Applications,* Issue 2 (Bellcore, December 1993; replaced by GR-468-CORE).

FA-NWT-001345, *Framework Generic Requirements for Element Manager (EM) Applications for SONET Subnetworks,* Issue 1 (Bellcore, September 1992).

TA-NWT-001385 (not available).

TA-TSV-001 294, *Generic Requirements for Element Management Layer (EML) Functionality and Architecture,* Issue 1 (Bellcore, December 1992).

Special Reports

SR-104, *Bellcore Digest of Technical Information*, Volume 14, Issue 12 (Bellcore, December 1997).

SR-NWT-002224 (not available).

SR-TSV-002671, *EML Applications for Fault Management: Subnetwork Root Cause Alarm Analysis*, Issue 1 (Bellcore, June 1993).

SR-TSV-002672, *EML Applications for Fault Management: Intelligent Alarm Filtering for SONET*, Issue 1 (Bellcore, March 1994).

SR-TSV-002675, *EML Applications for Configuration Management: Resource Provisioning Selection and Assignment—Functional Description*, Issue 1 (Bellcore, December 1993).

SR-TSV-002678, *EML Applications for Configuration Management: Inventory Notification and Query—Functional Description*, Issue 1 (Bellcore, April 1994).

SR-NWT-002723, *Applicable TL1 Messages for SONET Network Elements*, Issue 1 (Bellcore, June 1993).

Note: All Telcordia (Bellcore) documents are subject to change, and this list reflects the most current information available. Readers are advised to check current status and availability of all documents. Documents can be ordered from Telcordia's online catalog or by contacting Telcordia.

To contact Telcordia:

> Telcordia Customer Service
> 8 Corporate Place, Room 3A-184
> Piscataway, NJ 088544156
> 1-800-521-CORE (2673) (U.S.A. and Canada)
> 1-732-699-5800 (All others)
> 1-732-336-2559 (Fax)

To order documents on line:

1. Enter the URL **telecominfo.telcordia.com**.
2. Click the Search button at the top of the page.
3. In the Keywords field, enter the document number (or keywords), and then click Submit Search.

OR

1. Enter the URL **telecominfo.telcordia.com**.
2. Click the Browse button located at the top of the page, then click the subject of interest.

EIA/TIA DOCUMENTS

Standards related to fiber-optic cable and lasers.

EIA/TIA-455-170, *Cutoff Wavelength of SingleMode Fiber by Transmitted Power.*

EIA/TIA-492, *Generic Specification for Optical Waveguide Fiber.*

EIA/TIA-559, *SingleMode Fiber Optic System Transmission Design.*

FOTP 127, *Spectral Characterization of Multimode Laser Diodes.*

OFSTP-2, *Effective Transmitter Output Power Coupled into SingleMode Fiber Optic Cable.*

OFSTP-3, *Fiber Optic Terminal Receiver Sensitivity and Maximum Receiver Input Power.*

OFSTP-10, *Measurement of Dispersion Power Penalty in SingleMode Systems.*

OFSTP-11, *Measurement of Single Reflection Power Penalty for Fiber Optic Terminal Equipment.*

These publications are available from

EIA/TIA Standards Sales Office
2001 Pennsylvania, NW
Washington, DC 20006
1-202-457-4963
URL: www.eia.org

AMERICAN NATIONAL STANDARDS INSTITUTE (ANSI) DOCUMENTS

These are today's main SONET documents, especially the T1.105 series.

ANSI T1.101-1999, *Synchronization Interface Standards for Digital Networks.*

ANSI T1.102-1993 (R 1999), *Digital Hierarchy—Electrical Interfaces.*

ANSI T1.102.01-1996 (R 2001), *Digital Hierarchy—VT1.5 Electrical Interface.*

ANSI T1.105-2001, *Synchronous Optical Network (SONET)—Basic Description including Multiplex Structure, Rates and Formats.*

ANSI T1.105.01-2000, *Synchronous Optical Network (SONET)—Automatic Protection Switching.*

ANSI T1.105.02-2001, *Telecommunications—Synchronous Optical Network (SONET)—Payload Mappings.*

ANSI T1.105.03-1994, *Synchronous Optical Network (SONET)—Jitter at Network Interfaces.*

ANSI T1.105.03a-1995, *Synchronous Optical Network (SONET)—Jitter at Network Interfaces—DS1 Supplement.*

ANSI T1.105.03b-1997, *Synchronous Optical Network (SONET)—Jitter at Network Interfaces—DS3 Supplement.*

ANSI T1.105.04-1995 (R 2001), *Synchronous Optical Network (SONET): Data Communication Channel Protocols and Architectures.*

ANSI T1.105.05-1994, *Synchronous Optical Network (SONET)—Tandem Connection Maintenance.*

ANSI T1.105.06-1996, *SONET Physical Layer Specifications.*

ANSI T1.105.07-1996 (R 2001), *Telecommunications—Synchronous Optical Network (SONET)—Sub STS-1 Interface Rates and Formats Specification.*

ANSI T1.105.07a-1997, *Telecommunications—Synchronous Optical Network (SONET)—Sub STS-1 Interface Rates and Formats Specification (inclusion of NxVT group interfaces).*

ANSI T1.105.08-2001, *Telecommunications—Synchronous Optical Network (SONET)—In-band Forward Error Correction Code.*

ANSI T1.105.09-1996, *Telecommunications—Synchronous Optical Network (SONET)—Network Element Timing and Synchronization.*

ANSI T1.107-2002, *Digital Hierarchy—Formats Specifications.*

ANSI T1.119-1994 (R 2001), *Telecommunications—Synchronous Optical Network (SONET)—Operations, Administration, Maintenance, and Provisioning (OAM&P) Communications.*

ANSI T1.119.01-1995 (R 2001), *Telecommunications—Synchronous Optical Network (SONET)—Operations, Administration, Maintenance, and Provisioning (OAM&P) Communications—Protection Switching Fragment.*

ANSI T1.119.02-1998, *Telecommunications—Synchronous Optical Network (SONET)—Operations, Administration, Maintenance, and Provisioning (OAM&P) Communications—Performance Management Fragment.*

ANSI TI.204-1994, *Operations, Administration, Maintenance, and Provisioning—Lower Layer Protocols for Telecommunication Management Network (TMN) Interfaces Between Operations Systems and Network Elements.*

ANSI TI.210-1993, *Operations, Administration, Maintenance, and Provisioning (OAM&P)—Principles of Functions, Architectures and Protocols for Telecommunications Management Network (TMN) Interfaces.*

ANSI TI.231-1997, *Digital Hierarchy—Layer 1 In-Service Digital Transmission Performance Monitoring.*

ANSI TI.245-1997, *Directory Service for Telecommunications Management Network (TMN) and Synchronous Optical Network (SONET).*

ANSI T1.416-1999, *Telecommunications—Network-to-Customer Installation Interfaces—Synchronous Optical Network (SONET) Physical Media Dependent Specification: Electrical.*

ANSI T1.416.01-1999, *Telecommunications—Network-to-Customer Installation Interfaces—Synchronous Optical Network (SONET) Physical Media Dependent Specification: Electrical.*

ANSI T1.416.02-1999, *Telecommunications—Network-to-Customer Installation Interfaces—Synchronous Optical Network (SONET) Physical Media Dependent Specification: Single Mode Fiber.*

ANSI T1.416.02a-2001, *Telecommunications—Network-to-Customer Installation Interfaces—Synchronous Optical Network (SONET) Physical Media Dependent Specification: Single Mode Fiber* (supplement).

ANSI T1.416.03-1999, *Telecommunications—Network-to-Customer Installation Interfaces—Synchronous Optical Network (SONET) Physical Media Dependent Specification: Electrical.*

ANSI T1.506A-1992, *Telecommunications—Network Performance— Specifications for Switched Exchange Network (Absolute Round-Trip Delay).*

ANSI T1.508-1992, *Telecommunications—Network Performance—Loss Plan for Evolving Digital Networks.*

ANSI TI.508A-1993, *Telecommunications—Network Performance—Loss Plan for Evolving Digital Networks.*

ANSI T1.514-1995, *Telecommunications—Network Performance Parameters and Objectives for Dedicated Digital Services—SONET Bit Rates.*

ANSI T1.646-1995, *Broadband ISDN—Physical Layer Specification for User–Network Interfaces Including DS1/ATM.*

ANSI Technical Report #6, *A Technical Report on Slave Stratum Clock Performance Measurement Guidelines.*

TI Technical Report #33, *A Technical Report on Synchronization Network Management Using Synchronization Status Messages.*

ANSI X3.216-1992, *Data Communications—Structure and Semantics of the Domain Specific Part (DSP) of the OSI Network Service Access Point (NSAP) Address.*

These publications are available from:

American National Standards Institute, Inc.
11 West 42nd Street
New York, NY 10036
URL: www.ansi.org

U.S. GOVERNMENT PUBLICATIONS

U.S. Department of Health, Education, and Welfare; Bureau of Radiological Health.
21 CFR 1040.10, *Performance Standard for Laser Products.*

This publication is available from:

Director, Division of Compliance
Bureau of Radiological Health
5600 Fishers Lane
Rockville, MD 20857
URL: www.access.gpo.gov/nara/cfr

ITU-T (AND CCITT) RECOMMENDATIONS

The main SDH document is Recommendation G.707, which —with its amendments and corrections —now incorporates information formerly found in separate recommendations, such as payload mapping details.

SDH Documents

G.703, *Physical/Electrical Characteristics of Hierarchical Digital Interfaces.*

G.707, *Network Node Interfaces for the Synchronous Digital Hierarchy (SDH)* (October 2000).

G.707.1a, *Network Node Interfaces for the Synchronous Digital Hierarchy (SDH)*, Amendment 1 (November 2001).

G.707.1c, *Network Node Interfaces for the Synchronous Digital Hierarchy (SDH)*, Corrigendum 1 (March 2001).

G.707.2c, *Network Node Interfaces for the Synchronous Digital Hierarchy (SDH)*, Corrigendum 2 (November 2001).

G.709, *Synchronous Multiplexing Structure* (now in G.707).

G.780, *Vocabulary of Terms for SDH Networks and Equipment.*

G.781, *Structure of Recommendations on Equipment for the Synchronous Digital Hierarchy (SDH).*

G.782, *Types and Characteristics of Synchronous Digital Hierarchy (SDH) Equipment.*

G.783, *Characteristics of Synchronous Digital Hierarchy (SDH) Equipment Functional Blocks.*

G.803, *Architecture of Transport Networks Based on the Synchronous Digital Hierarchy (SDH).*

G.810, *Definitions and Terminology for Synchronization Networks.*

G.811, *Timing Requirements at the Output of Primary Reference Clocks Suitable for Plesiochronous Operation of International Digital Links.*

G.813, *Timing Characteristics of SDH Equipment Slave Clocks (SEC).*

G.825, *The Control of Jitter and Wander in Digital Networks Based on the SDH.*

G.826, *Error Performance Parameters and Objectives for International, Constant Bit Rate Digital Paths at or Above the Primary Rate.*

G.832, *Transport of SDH Elements on PDH Networks.*

G.841, *Types and Characteristics of SDH Network Protection Architectures.*

G.842, *Interworking of SDH Network Protection Architectures.*

G.957, *Optical Interfaces for Equipments and Systems Relating to the Synchronous Digital Hierarchy.*

G.958, *Digital Line Systems Based on the Synchronous Digital Hierarchy for Use on Optical Fiber Cables.*

M.2101, *Performance Limit for Bringing into Service and Maintenance of International SDH Paths and Multiplex Sections.*

M.2110, *Bringing into Service International Paths Sections and Transmission Systems.*

M.2120, *Digital Path, Section and Transmission System Default Detection and Localization.*

M.3010, *Principles for a Telecommunications Management Network.*

M.3100, *Generic Network—Information Model.*

O.17s, *Jitter and Wander Measuring Equipment for Digital Systems Which are Based on the SDH.*

O.150, *General Requirements for Instrumentation for Performance Measurements on Digital Transmission Equipment.*

O.181, *Equipment to Assess Error Performance on STM-N SDH Interfaces.*

Q.811, *Lower Layer Protocol Profiles for the Q3 and X Interfaces.*

Q.921, *ISDN User–Network Interface—Data Link Layer Specification 0.*

X.121, *International Numbering Plan for Public Data Networks.*

X.226, *Presentation Protocol Specification for Open Systems Interconnection for CCITT Applications.*

TMN Documents

G.772, *Digital Protected Monitoring Points.*

G.773, *Protocol Suites for Q Interfaces for Management for Transmission Systems.*

G.774, *Synchronous Digital Hierarchy (SDH) Management Information Model for the Network Element View.*

G.774.01, *SDH Performance Monitoring for the Network Element View.*

G.774.02, *SDH Configuration of the Payload Structure for the Network Element View.*

G.774.03, *SDH Management of Multiplex Section Protection for the Network Element View.*

G.774.04, *SDH Management of Sub Network Connection Protection from the Network Element View* (Q3 interface network element view).

G.774.05, *SDH Management of Connection Supervision Functionality (HCS/LCS)* (Q3 interface network element view).

G.782, *Types and General Characteristics of SDH Equipment Functional Blocks.*

G.783, *Characteristics of SDH Equipment Functional Blocks* (replaces G.781, G.782, and G.783 version of January 1994).

G.784, *Synchronous Digital Hierarchy (SDH) Management.*

G.SHR-1, *SDH Self Healing Rings.*

G.SHR-2, *SDH Ring Interworking.*

G.77f, *Protocol Stack for F Interfaces for Transmission Equipment.*

G.77qiA, *Q Interface Adaptor for Transmission Equipment.*

G.803, *Architecture of Transport Networks Based on SDH.*

G.831, *Performance and Management Capabilities of Transport Network Based on the SDH.*

G.ATA, *Architecture of Transport Networks in the Access Application.*

G.ATMA, *Architecture of Transport Networks Based on ATM.*

G.ATME-1, *Types and General Characteristics of ATM Equipment.*

G.ATME-2, *Functional Characteristics of ATM Equipment.*

G.TNA, *Architecture of Transport Networks Based on SDH.*

G.atmm, *ATM Management* (Q3 interface network element view).

G.mtn1, *Management of Transmission Networks.*

These publications are available from:

> International Telecommunication Union General Secretariat—Sales Section
> Place des Nations, CH1211
> Geneva 20 (Switzerland)
> +41 22 730 5285
> URL: www.itu.int

ETSI DOCUMENTS.

ETSI interprets ITU-T SDH standards for application in the European telecommunications environment. The series contains almost 70 documents, and so this section leaves out some documents, mostly related to radio SDH links and video transmission.

ETS 300 147 ed. 1 (1992-03), Reference: DE/TM-03001, Source: TM3, Title: *Transmission and Multiplexing (TM): Synchronous Digital Hierarchy (SDH) Multiplexing Structure.*

ETS 300 147 ed. 2 (1995-01), Reference: RE/TM-03013, Source: TM3, Title: *Transmission and Multiplexing (TM): Synchronous Digital Hierarchy (SDH) Multiplexing Structure.*

ETS 300 147 ed. 3 (1997-04), Reference: RE/TM-03045, Source: TM3, Title: *Transmission and Multiplexing (TM): Synchronous Digital Hierarchy (SDH): Multiplexing Structure.*

ETS 300 147 V1.4.1 (2001-09), Reference: REN/TM-01063, Source: TM3, Title: *Transmission and Multiplexing (TM): Synchronous Digital Hierarchy (SDH): Multiplexing Structure.*

ETS 300 232/A1 ed. 1 (1996-03), Reference: RE/TM-01025, Source: TM1, Title: *Transmission and Multiplexing (TM): Optical Interfaces for Equipments and Systems Relating to the Synchronous Digital Hierarchy (SDH)* (ITU-T Recommendation G.957 [1995] modified).

ETS 300 272 ed. 1 (1994-03), Reference: DE/NA-053032, Source: NA5, Title: *Network Aspects (NA): Metropolitan Area Network (MAN), Physical Layer Convergence Procedure (PLCP) for 155.520 Mbit/s.* CCITT Recommendations G 707, G.708, and G.709, *SDH Based Systems Protocol Implementation Conformance Statement (PICS).*

ETS 300 276 ed. 1 (1994-03), Reference: DE/NA-053033, Source: NA, Title: *Network Aspects (NA): Metropolitan Area Network (MAN) Physical Layer Convergence Procedure (PLCP) for 622.080 Mbit/s.* CCITT Recommendations G.707, G.708, and G.709, *SDH Based Systems.*

ETS 300 277 ed. 1 (1994-03), Reference: DE/NA-053034, Source: NA5, Title: *Network Aspects (NA): Metropolitan Area Network (MAN) Physical Layer Convergence Procedure (PLCP) for 622.080 Mbit/s.* CCITT Recommendations G.707, G 708, and G.709, *SDH Based Systems Protocol Implementation Conformance Statement (PICS).*

ETSI ETS 300 300 ed. 1 (1995-02), Reference: DE/NA-052512, Source: NA5, Title: *Broadband Integrated Services Digital Network (BISDN): Synchronous Digital Hierarchy (SDH) Based User Network Access: Physical Layer Interfaces for BISDN Applications.*

ETSI ETS 300 300 ed. 2 (1997-04), Reference: RE/TM-03029, Source: TM3, Title: *Broadband Integrated Services Digital Network (BISDN): Synchronous Digital Hierarchy (SDH) Based User Network Access. Physical Layer User Network Interfaces (UNI) for 155 520 kbit/s and 622 080 kbit/s Asynchronous Transfer Mode (ATM) B ISDN Applications.*

ETS 300 304 ed. 1 (1994-11), Reference: DE/TM-022011, Source: TM2, Title: *Transmission and Multiplexing (TM): Synchronous Digital Hierarchy (SDH), Information Model for the Network Element (NE) View.*

ETS 300 304 ed. 2 (1997-02), Reference: RE/TM-02213, Source: TM2, Title: *Transmission and Multiplexing (TM): Synchronous Digital Hierarchy (SDH), SDH Information Model for the Network Element (NE) View.*

ETS 300 337 ed. 1 (1995-02), Reference: DE/TM-03007, Source: TM3, Title: *Transmission and Multiplexing (TM): Generic Frame Structures for the Transport of Various Signals (including Asynchronous Transfer Mode (ATM) Cells and Synchronous Digital Hierarchy (SDH) Elements) at the ITUT Recommendation G.702 Hierarchical Rates of 2 048 kbit/s, 34 368 kbit/s and 139 264 kbit/s.*

ETS 300 337 ed. 2 (1997-06), Reference: RE/TM-03056, Source: TM3, Title: *Transmission and Multiplexing (TM): Generic Frame Structures for the Transport of Various Signals (Including Asynchronous Transfer Mode (ATM) Cells and Synchronous Digital Hierarchy (SDH) Elements) at the ITUT Recommendation G.702 Hierarchical Rates of 2 048 kbit/s, 34 368 kbit/s and 139 264 kbit/s.*

ETSI ETS 300 417-2-1 ed. 1 (1997-04), Reference: DE/TM-01015-2-1, Source: TM1, Title: *Transmission and Multiplexing (TM): Generic Requirements of Transport Functionality of Equipment Part 2-1 Synchronous Digital Hierarchy (SDH) and Plesiochronous Digital Hierarchy (PDH) Physical Section Layer Functions.*

ETSI EN 300 417-2-1 V1.1.2 (1998-11), Reference: REN/TM-01015-2-1, Source: TM1, Title: *Transmission and Multiplexing (TM): Generic Requirements of Transport Functionality of Equipment; Part 2-1: Synchronous Digital Hierarchy (SDH) and Plesiochronous Digital Hierarchy (PDH) Physical Section Layer Functions.*

ETSI EN 300 417-2-1 V1.1.3 (1999-05), Reference: REN/TM-01015-2-1a, Source: TM1, Title: *Transmission and Multiplexing (TM): Generic Requirements of Transport Functionality of Equipment; Part 2-1: Synchronous Digital Hierarchy (SDH) and Plesiochronous Digital Hierarchy (PDH) Physical Section Layer Functions.*

ETSI EN 300 417-2-1 V1.2.1 (2001-10), Reference: REN/TM-01042-2-1, Source: TM1, Title: *Transmission and Multiplexing (TM): Generic Requirements of Transport Functionality of Equipment; Part 2-1: Synchronous Digital Hierarchy (SDH) and Plesiochronous Digital Hierarchy (PDH) Physical Section Layer Functions.*

ETSI ETS 300 417-2-2 ed. 1 (1997-11), Reference: DE/TM-01015-2-2, Source: TM1, Title: *Transmission and Multiplexing (TM): Generic Requirements of Transport Functionality of Equipment; Part 2-2: Synchronous Digital Hierarchy (SDH) and Plesiochronous Digital Hierarchy (PDH) Physical Section Layer Functions: Implementation Conformance Statement (ICS) Pro forma Specification.*

ETSI EN 300 417-2-2 V1.1.2 (1998-11), Reference: REN/TM-01015-2-2, Source: TM1, Title: *Transmission and Multiplexing (TM): Generic Requirements of Transport Functionality of Equipment; Part 2-2: Synchronous Digital Hierarchy (SDH) and Plesiochronous Digital Hierarchy (PDH) Physical Section Layer Functions: Implementation Conformance Statement (ICS) Pro forma Specification.*

ETSI EN 300 417-2-2 V1.1.3 (1999-02), Reference: REN/TM-01015-2-2a, Source: TM1, Title: *Transmission and Multiplexing (TM): Generic Requirements of Transport Functionality of Equipment: Part 2-2: Synchronous Digital Hierarchy (SDH) and Plesiochronous Digital Hierarchy (PDH) Physical Section Layer Functions: Implementation Conformance Statement (ICS) Pro forma Specification.*

ETSI EN 300 417-2-2 V1.1.4 (1999-05), Reference: REN/TM-01015-2-2b, Source: TM1, Title: *Transmission and Multiplexing (TM): Generic Requirements of Transport Functionality of Equipment: Part 2-2: Synchronous Digital Hierarchy (SDH) and Plesiochronous Digital Hierarchy (PDH) Physical Section Layer Functions Implementation Conformance Statement (ICS) Pro forma Specification.*

ETSI ETS 300 417-4-1 ed. 1 (1997-06), Reference: DE/TM-01015-4-1, Source: TM1, Title: *Transmission and Multiplexing (TM): Generic Requirements of Transport Functionality of Equipment; Part 4-1: Synchronous Digital Hierarchy (SDH) Path Layer Functions.*

ETSI EN 300 417-4-1 V1.1.2 (1998-11), Reference: REN/TM-01015-4-1 Source: TM1, Title: *Transmission and Multiplexing (TM): Generic Requirements of Transport Functionality of Equipment Part 4-1: Synchronous Digital Hierarchy (SDH) Path Layer Functions.*

ETSI EN 300 417-4-1 V1.1.3 (1999-05), Reference: REN/TM-01015-4-1a, Source: TM1, Title: *Transmission and Multiplexing (TM): Generic Requirements of Transport Functionality of Equipment: Part 4-1: Synchronous Digital Hierarchy (SDH) Path Layer Functions.*

ETSI EN 300 417-4-1 V1.2.1 (2001-10), Reference: REN/TM-01015-4-1a, Source: TM1, Title: *Transmission and Multiplexing (TM): Generic Requirements of Transport Functionality of Equipment: Part 4-1: Synchronous Digital Hierarchy (SDH) Path Layer Functions.*

ETSI EN 300 417-4-2 V1.1.1 (1999-06), Reference: DEN/TM-01015-4-2, Source: TM1, Title: *Transmission and Multiplexing (TM): Generic Requirements of Transport Functionality of Equipment; Part 4-2: Synchronous Digital Hierarchy (SDH) Path Layer Functions Implementation Conformance Statement (ICS) Pro forma Specification.*

ETSI EN 300 417-9-1 V1.1.1 (2001-09), Reference: DEN/TM-01015-9-1, Source: TM1, Title: *Transmission and Multiplexing (TM): Generic Requirements of Transport Functionality of Equipment; Part 9-1: Synchronous Digital Hierarchy (SDH) Concatenated Path Layer Functions; Sub-part 1: Requirements.*

ETSI EN 300 462-2-1 V1.2.1 (2002-01), Reference: REN/TM-01092, Source: TM1, Title: *Transmission and Multiplexing (TM): Generic Requirements for Synchronization Networks Part 2-1: Synchronization Network Architecture Based on SDH Networks.*

ETSI EN 300 462-4-1 V1.1.1 (1998-05), Reference: DEN/TM-03017-4-1, Source: TM3, Title: *Transmission and Multiplexing (TM): Generic Requirements for Synchronization Networks Part 4-1: Timing Characteristics of Slave Clocks Suitable for Synchronization Supply to Synchronous Digital Hierarchy (SDH) and Plesiochronous Digital Hierarchy (PDH) Equipment.*

ETSI EN 300 462-4-2 V1.1.1 (1999-12), Reference: DEN/TM-01057-4-2, Source: TM1, Title: *Transmission and Multiplexing (TM): Generic Requirements for Synchronization Networks; Part 4-2: Timing Characteristics of Slave Clocks Suitable for Synchronization Supply to Synchronous Digital Hierarchy (SDH) and Plesiochronous Digital Hierarchy (PDH) Equipment Implementation Conformance Statement (ICS) Pro forma Specification.*

ETSI ETS 300 462-5 ed. 1 (1996-09; withdrawn), Reference: DE/TM-03017-5, Source: TM3, Title: *Transmission and Multiplexing (TM): Generic Requirements for Synchronization Networks; Part 5: Timing Characteristics of Slave Clocks Suitable for Operation in Synchronous Digital Hierarchy (SDH) Equipment.*

ETSI EN 300 462-5-1 V1.1.2 (1998-05), Reference: REN/TM-0301-7-5-1, Source: TM3, Title: *Transmission and Multiplexing (TM): Generic Requirements for Synchronization Networks, Part 5-1: Timing Characteristics of Slave Clocks Suitable for Operation in Synchronous Digital Hierarchy (SDH) Equipment.*

ETSI EN 300 484 ed. 1 (1996-09), Reference: DE/TM-02220, Source: TM2, Title: *Transmission and Multiplexing (TM): Synchronous Digital Hierarchy (SDH) Network Information Model Connection Supervision Function (Higher Order Connection Supervision / Lower Order Connection Supervision [HCS/LCS]) for the Network Element (NE) View.*

ETSI 300 493 ed. 1 (1996-06), Reference: DE/TM-02216, Source: TM2, Title: *Transmission and Multiplexing (TM): Synchronous Digital Hierarchy (SDH) Information Model of the Sub Network Connection Protection (SNCP) for the Network Element (NE) View.*

ETSI ETS 300 635 ed. 1 (1996-10), Reference: DE/TM-04028, Source: TM4, Title: *Transmission and Multiplexing (TM); Synchronous Digital Hierarchy (SDH); Radio Specific Functional Blocks for Transmission of M x STM-N.*

ETSI EN 300 645 V1.2.1 (1998-10), Reference: REN/TMN-00010, Source: TMN, Title: *Telecommunication Network Management (TMN); Synchronous Digital Hierarchy (SDH); Radio Relay Equipment; Information Model for Use on Q Interfaces.*

ETSI ETS 300 746 ed. 1 (1997-02), Reference: DE/TM-03042, Source: TM3, Title: *Transmission and Multiplexing (TM); Synchronous Digital Hierarchy (SDH); Network Protection Schemes; Automatic Protection Switching (APS) Protocols and Operation.*

ETSI ETS 300 785 ed. 1 (1998-02), Reference: DE/TM-04029, Source: TM4, Title: *Transmission and Multiplexing (TM); Synchronous Digital Hierarchy (SDH); Radio Specific Functional Blocks for Transmission of M x sub-STM-N.*

ETSI I-ETS 300 801 ed. 1 (1999-01), Reference: DI/TMN-00047, Source: TMN, Title: *Telecommunication Network Management (TMN); Synchronous Digital Hierarchy (SDH) Network Information Model; Basic Trail and Sub-Network Connection (SNC) Configuration Management Ensemble.*

ETSI EN 301 164 V.1.1.1 (1999-05), Reference: DEN/TM-03072, Source: TM3, Title: *Transmission and Multiplexing (TM); Synchronous Digital Hierarchy (SDH); SDH Leased Lines; Connection Characteristics.*

ETSI EN 301 165 V.1.1.1 (1999-05), Reference: DEN/TM-03073, Source: TM3, Title: *Transmission and Multiplexing (TM); Synchronous Digital Hierarchy (SDH); SDH Leased Lines; Network and Terminal Interface Presentation.*

ETSI EN 301 165 V.1.1.2 (2000-06), Reference: REN/TM-01087, Source: TM1, Title: *Transmission and Multiplexing (TM); Synchronous Digital Hierarchy (SDH); SDH Leased Lines; Connection Characteristics.*

ETSI TS 101 009 V.1.1.1 (1997-11), Reference: DTS/TM-03025, Source: TM3, Title: *Transmission and Multiplexing (TM); Synchronous Digital Hierarchy (SDH); Network Protection Schemes; Types and Characteristics.*

ETSI TS 101 010 V.1.1.1 (1997-11), Reference: DTS/TM-03041, Source: TM3, Title: *Transmission and Multiplexing (TM); Synchronous Digital Hierarchy (SDH); Network Protection Schemes; Interworking: Rings and Other Schemes.*

ETSI ETR 114 ed. 1 (1993-11), Reference: DTR/TM-03006, Source: TM3, Title: *Transmission and Multiplexing (TM); Functional Architecture of Synchronous Digital Hierarchy (SDH) Transport Networks.*

These publications are available from:

> ETSI web site: www.etsi.org for download (ETSI On-Line, EOL)
> or purchase on paper or CD-ROM
> Address: ETSI Infocentre
> 06921 Sophia Antipolis
> CEDEX France
> Telephone: +33(0)4 92 42 22

ISO DOCUMENTS

ISO/IEC 74981:1994, *Information Technology—Open System Interconnection Reference Model: The Basic Model.*

ISO/IEC 8073:1988, *Information Processing Systems—Open Systems Interconnection—Connection Oriented Transport Protocol Specifications—Addendum 2: Class Four Operation over Connectionless Network Service.*

ISO/IEC 8073:1988/Addendum 2:1989, *Information Processing Systems—Open Systems Interconnection—Connection Oriented Transport Protocol Specifications.*

ISO/IEC 8073:1992, *Information Technology—Telecommunications and Information Exchange Between Systems—Open Systems Interconnection—Protocol for Providing the Connection-Mode Transport Service.*

ISO/IEC 8208/CCITT X.25, *Information Processing Systems—X.25 Packet Level Protocol for Data Terminal Equipment.*

ISO/IEC 8327-1:1988, *Information Processing Systems—Open Systems Interconnection—Connection Oriented Session Protocol Specification—Part 1: Protocol Specification.*

ISO/IEC DIS 8327-2, *Information Technology—Open Systems Interconnection—Basic Connection Oriented Session Protocol Specification.*

ISO 8348:1993, *Information Processing Systems—Data Communications—Network Service Definition.*

ISO/IEC 8473-1988, *Information Processing Systems—Data Communications Protocol for Providing the Connectionless-Mode Network Layer Service.*

ISO/IEC DIS 8473 1:1993, *Information Technology—Protocol for Providing the Connectionless-Mode Network Service.*

ISO 8648, *Internal Organization of the Network Layer.*

ISO/IEC 8650:1988, *Information Technology—Open Systems Interconnection Protocol Specification for the Association Control Service Element.*

ISO/IEC DIS 8650-2: *Information Technology—Open Systems Interconnection Protocol Specification for the Association Control Service Element.*

ISO 8802-2/ANSI/IEEE Std. 802.2:1989, *Information Processing Systems—Open Systems Interconnection—Local Area Networks—Part 2: Logical Link Control.*

ISO/IEC 8802-2, 1989/Amd 3, *Information Processing Systems—Local Area Networks—Part 2: Logical Link Control Amendment 3: Conformance Requirements.*

ISO 8802-3/ANSI/IEEE Std. 802.3-1990, *Information Processing Systems—Open Systems Interconnection—Local Area Networks—Part 3: Carrier Sense Multiple Access with Collision Detection (CSMA/CD) Access Method and Physical Layer Specifications.*

ISO/IEC 8823-1:1988, *Information Processing Systems—Open Systems Interconnection—Connection Oriented Presentation Protocol Specification—Part 1: Protocol Specification.*

ISO/IEC DIS 8823-2, *Information Technology—Open Systems Interconnection—Connection Oriented Presentation Protocol Specification.*

ISO/IEC 9542:1988, *Information Processing Systems—Telecommunications and Information Exchange between Systems—End System to Intel Mediate System Routing Exchange Protocol for Use in Conjunction with Protocol for Providing the Connectionless-Mode Network Service (ISO 8473).*

ISO/IEC TR 9577, *Information Processing Systems—Data Communications—Protocol Identification in the Network Layer.*

ISO/IEC 10040:1992, *Information Technology—Open Systems Interconnection—Systems Management Overview.*

ISO/IEC 10589:1992, *Information Technology—Telecommunications and Information Exchange Between Systems—Intermediate System to Intermediate System Intradomain Routing Information Exchange Protocol for Use in Conjunction with the Protocol for Providing the Connectionless-Mode Network Service (ISO 8473).*

ISO/IEC ISP 10607-1, *Information Technology—International Standardized Profiles AFTn—File Transfer, Access and Management.*

ISO/IEC ISP 10608-1:1992, *Information Technology—International Standardized Profile TAn—Connection-Mode Transport Service over Connectionless-Mode Network Service.*

ISO/IEC ISP 10608-2:1992, *Information Technology—International Standardized Profile TAn—Connection-Mode Transport Service over Connectionless-Mode Network Service.*

ISO DIS 10747:1992, *Information Technology—Telecommunications and Information Exchange Between Systems Protocol for Exchange of InterDomain Routing Information among Intermediate Systems to Support Forwarding of ISO/IEC 8473 PDUs.*

ISO/IEC ISP 111831, *Information Technology—International Standardized Profiles AOM1n OSI Management–Management Communications.*

ISO/IEC PDISP 111881, *Information Technology—International Standardized Profile—Common Upper Layer Requirements—Part 1: Basic Connection Oriented Requirements.*

ISO TR 10172:199 1, *Information Technology—Telecommunications and Information Exchange Between Systems—Network/Transport Protocol Interworking Specification.*

These publications are available from

> American National Standards Institute, Inc.
> 1430 Broadway
> New York, NY 10018
> URL: www.ansi.org

IEEE DOCUMENTS

IEEE 802.2, *Logical Link Control.*

IEEE 802.3, *Standard for Carrier Sense Multiple Access with Collision Detection (CSMA/CD) Access Method.*

IEEE P802.6_/D 14, *Distributed Queue Dual Bus (DQDB) Subnetwork of a Metropolitan Area Network (MAN)* (September 1990).

These publications can be obtained by calling:

> IEEE Standards Publications
> 1-800-678-IEEE, or
> 1-908-981-1393
> URL: www.ieee.org

SONET INTEROPERABILITY FORUM (SIF) DOCUMENTS

These documents are now obtained through ATIS.

SIF-002-1996, *Remote Login Implementation Requirements Specification,* Issue 1.

SIF-009-1997, *NE–NE Remote Login Implementation Requirements Specification.*

SIF-027-1999, *BLSR Interoperability Requirements.*

SIF-AR-9806-085, *BLSR Interoperability Requirements—Cross Connect, Squelch Table, and NUT (Non-preemptible, Unprotected Traffic).*

SIF-AR-9807-111, *Requirements for BLSR Map Generation Protocol.*

The SIF is an Alliance for Telecommunications Industry Solutions (ATIS)–sponsored committee. SIF-approved documents may be obtained directly from the ATIS WWW server at http://www.atis.org/atis/sif/sifdoc.htm.

For additional Information on SIF documents, contact

Alliance for Telecommunications Industry Solutions
Attn.: Lisa Colaianne
1200 G Street, NW
Suite 500
Washington, DC 20005
Tel: 1-202-434-8823

A Note on Obtaining Documents   Expect to pay (sometimes a lot) for print copies of these documents, and often even to download them. Most documents are available for download in Word or PDF formats. The good news is that SONET/SDH relies mostly on ANSI/ATIS specifications (for SONET) and ITU-T/ETSI specifications (for SDH). ETSI is the most open, offering free downloads of everything. The ITU-T allows individuals three free downloads per year (plan wisely) after registration, but addenda and the like to basic standards are usually not counted. ANSI expects payment for published specifications in any form, but allows free downloads of drafts and pre-publication (PP) documents (which are often exactly what is needed) if you know what they are called or where to find them. In many cases, however, figures and tables might be omitted, especially if these appeared in the previously published version. None of the research for this third edition relied on paid documentation.

Index

1:1 protection switching, 491–493, 519
1+1 protection switching, 519
1+1 unidirectional failover, 523
1:7 protection switching, 489–491
1:*n* protection switching, 490, 519
1s density, 92
10 Gigabit Ethernet (10 GBE), 540
1970s window (first window), 56, 57
1980s window (second window), 56, 57, 62
1990s window (third window), 56, 57, 59–60, 62, 65

 A

A1/A2 framing bytes, 196, 197, 435, 436
Abstract Syntax Notation version 1 (ASN.1), 140
access networks, 547–549
ActiveX, 473
ADC, 383
add/drop multiplexer (ADM)
    architecture, 371–372
    in feeder network, 362
    function in SONET, 153, 367, 370–372

    in SONET, 421, 422
    in SONET/SDH, 348, 349–350
    in T-carrier, 109
    and timing, 294
administrative unit (AU), in SDH, 145, 176, 202, 252, 319
administrative unit groups (AUG), in SDH, 252
administrative unit type 3 (AU-3), 254
administrative unit type 4 (AU-4), 254
agent, in network management, 459
alarm indication signal (AIS), 437, 440–441
alarms, in T-carrier networks, 92, 441, 443, 451
alarms, reporting in TL1, 484
alarm storms, 440
A-law, byte coding for E-carrier, 235, 254
Alcatel Network Systems, 383, 384, 529
    products, 385, 387–388
        sample product configuration, 398–399
Alliance for Telecommunications Industry Standards (ATIS), 126, 127, 129

C

D

▼ N

Q

R

▼ S

T

U

V

▼ W

X

INTERNATIONAL CONTACT INFORMATION

AUSTRALIA
McGraw-Hill Book Company Australia Pty. Ltd.
TEL +61-2-9417-9899
FAX +61-2-9417-5687
http://www.mcgraw-hill.com.au
books-it_sydney@mcgraw-hill.com

CANADA
McGraw-Hill Ryerson Ltd.
TEL +905-430-5000
FAX +905-430-5020
http://www.mcgrawhill.ca

GREECE, MIDDLE EAST,
NORTHERN AFRICA
McGraw-Hill Hellas
TEL +30-1-656-0990-3-4
FAX +30-1-654-5525

MEXICO (Also serving Latin America)
McGraw-Hill Interamericana Editores S.A. de C.V.
TEL +525-117-1583
FAX +525-117-1589
http://www.mcgraw-hill.com.mx
fernando_castellanos@mcgraw-hill.com

SINGAPORE (Serving Asia)
McGraw-Hill Book Company
TEL +65-863-1580
FAX +65-862-3354
http://www.mcgraw-hill.com.sg
mghasia@mcgraw-hill.com

SOUTH AFRICA
McGraw-Hill South Africa
TEL +27-11-622-7512
FAX +27-11-622-9045
robyn_swanepoel@mcgraw-hill.com

UNITED KINGDOM & EUROPE
(Excluding Southern Europe)
McGraw-Hill Education Europe
TEL +44-1-628-502500
FAX +44-1-628-770224
http://www.mcgraw-hill.co.uk
computing_neurope@mcgraw-hill.com

ALL OTHER INQUIRIES Contact:
Osborne/McGraw-Hill
TEL +1-510-549-6600
FAX +1-510-883-7600
http://www.osborne.com
omg_international@mcgraw-hill.com